Lectures on Quantum Mechanics

Second Edition

Lectures on Quantum Mechanics

Second Edition

Ashok Das
University of Rochester, USA

NEW JERSEY • LONDON • SINGAPORE • BEIJING • SHANGHAI • HONG KONG • TAIPEI • CHENNAI

Published by

World Scientific Publishing Co. Pte. Ltd.
5 Toh Tuck Link, Singapore 596224
USA office: 27 Warren Street, Suite 401-402, Hackensack, NJ 07601
UK office: 57 Shelton Street, Covent Garden, London WC2H 9HE

British Library Cataloguing-in-Publication Data
A catalogue record for this book is available from the British Library.

First edition was published by HBA in 2003.

LECTURES ON QUANTUM MECHANICS
Second Edition

ISBN-13 978-981-4374-38-5
ISBN-10 981-4374-38-5

Printed in India, bookbinding made in Singapore.

To
Mama, Papa, Litu
Burma, Jagu, Balu,
Ulfi and Gumlee

Contents

Preface

Over the years, I have taught a two semester graduate course as well as a similar two semester undergraduate course on quantum mechanics at the University of Rochester. The present book follows that material almost word-for-word. I have not attempted to polish the writing, and these lecture notes, therefore, reflect the informality of the class room. In fact, I even considered presenting the material in the original format, but lectures have a way of ending and starting in the middle of a topic, which is neither very appropriate nor expected for a book. Nonetheless, the subject is presented exactly in the order it was taught in class.

Some of the material is repeated in places, but this was deemed important for clarifying the lectures. The book is self-contained, in the sense that most of the steps in the development of the subject are derived in detail, and integrals are either evaluated or listed when needed. I believe that a motivated student can work through the notes independently and without difficulty. Throughout the book, I have followed the convention of representing three dimensional vectors by bold-faced symbols.

In preparing lectures for the course, I relied, at least partially, on the material contained in the following texts:

1. A. Das and A. C. Melissinos, "Quantum Mechanics: A Modern Introduction", Gordon and Breach, New York (1986).

2. L. I. Schiff, "Quantum Mechanics", McGraw-Hill, New York (1968).

3. R. Shankar, "Principles of Quantum Mechanics", Plenum, New York (1980).

Several of my colleagues at Rochester and at other universities, as well as many of my students, have influenced the development of these lectures. Most important were, of course, the excellent questions raised by students in class and during private discussions. I sincerely appreciate everyone's input.

The lecture notes were originally typed in LaTeX by Judy Mack, who deserves a lot of credit for her professionalism and sense of perfection. The present format of the book in LaTeX is largely due to the meticulous work of Dr. Alex Constandache, who succeeded in giving it a more "user friendly" appearance. Most of the figures were drawn using PSTricks, while a few were done using Gnuplot.

It is also a pleasure to thank the editors of the TRiPS series, as well as the publisher, for being so accommodating to all my requests in connection with the book.

Finally, I thank the members of my family, and in particular my younger sister Jhilli, for patient support and understanding during the completion of this work in Orissa, India.

Ashok Das
Rochester

Preface to the second edition

The modifications in this second edition of the book arose mainly from the requests by various readers. Several typos in the earlier version have been fixed and the presentation made clearer at some places. The figures now carry captions with references to them in the text. In addition to the numerous exercises that were already present in the text, I have now included a few selected problems at the end of every chapter in the present edition. Schrödinger equation with a periodic potential and Bloch functions are discussed in the chapter on symmetries (chapter **6**) as an example of finite translation symmetry in quantum mechanics.

Ashok Das
Rochester

CHAPTER 1

Review of classical mechanics

In this lecture, let us review some of the essential features of classical mechanics which we will use in the study of quantum mechanics.

1.1 Newton's equation

Let us consider a particle of mass m, moving in 1-dimension, which is subjected to a force F. Then, from Newton's law, we know that

$$ma = F, \tag{1.1}$$

where a denotes the acceleration of the particle. This is known as Newton's equation. If x denotes the coordinate of the particle, then, defining its potential energy as (we have chosen the reference point to be the origin for simplicity)

$$V(x) = -\int_{0}^{x} \mathrm{d}x'\ F(x'), \tag{1.2}$$

we can write Newton's equation, (1.1), also as

$$m\frac{\mathrm{d}^2x}{\mathrm{d}t^2} = F = -\frac{\mathrm{d}V}{\mathrm{d}x}. \tag{1.3}$$

This is a second order differential equation and can be solved uniquely provided we are given two initial conditions, namely, the position of the particle, x_0 at $t = 0$ as well as its initial velocity $\dot{x}_0$. In such a case, we can determine the trajectory of the particle, $x(t)$, uniquely.

1.2 Lagrangian approach

Another way of looking at the same problem is to define a scalar (scalar under Lorentz transformations) called the Lagrangian as

$$L = T - V \equiv L(x, \dot{x}), \tag{1.4}$$

where T and V represent, respectively, the kinetic and the potential energies of the particle. The Lagrangian may, in principle, also have explicit time dependence. However, we will not consider such systems in our discussions.

The integral of the Lagrangian along a trajectory defines an action associated with the Lagrangian for that particular trajectory, namely,

$$S[x] = \int_{t_i}^{t_f} \mathrm{d}t\, L(x, \dot{x}). \tag{1.5}$$

The square bracket is written to emphasize the fact that the action, S, is a function of a function. In mathematical language S is said to be a functional of x. As one can easily see, the value of the action depends on the path or the trajectory (see Fig. 1.1) along which the integration is carried out.

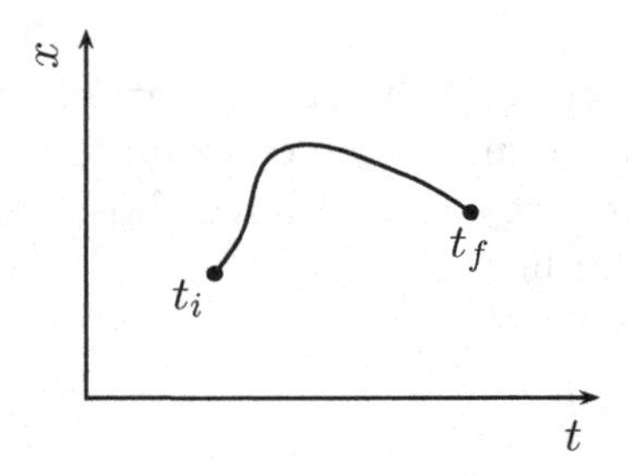

Figure 1.1: A trajectory between the initial time t_i and the final time t_f.

1.3 Principle of least action

The principle of least action says that the actual trajectory, which the particle follows, is such that the action associated with the Lagrangian along that trajectory is a minimum. In fact, what is strictly true is that the action is an extremum along the actual trajectory. In most familiar cases, however, it happens to be the minimum and hence the name. But there are situations where it can be a maximum as well.

Consequences. Suppose we have a function $f(x)$ which has a minimum (extremum) at x_0 as shown in Fig. 1.2. This clearly implies that the slope of the function at x_0 must be zero. What this means is that

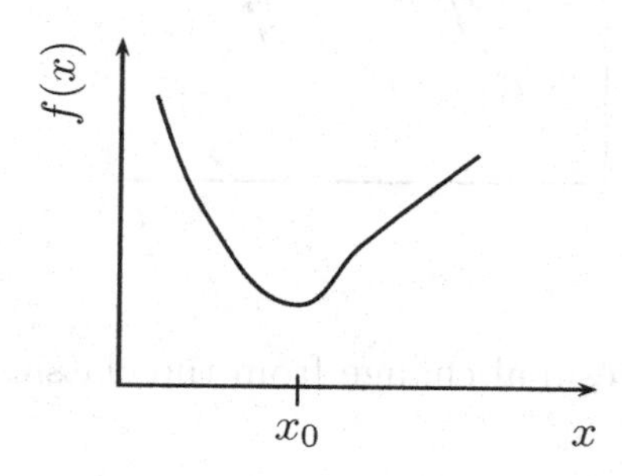

Figure 1.2: A function $f(x)$ with a minimum at x_0.

if we displace the function infinitesimally away from its minimum, we will obtain, from a Taylor expansion of the function,

$$f(x_0+\eta) = f(x_0) + \eta \left.\frac{\mathrm{d}f}{\mathrm{d}x}\right|_{x_0} + O(\eta^2),$$

$$\text{or,}\quad \delta f = f(x_0+\eta) - f(x_0) = 0, \tag{1.6}$$

to the lowest order in the displacement. Namely, the function is stationary at its minimum against infinitesimal displacements.

Let us now apply the same ideas to the case of the action. Let $x_{\rm cl}(t)$ be the actual trajectory of the particle, also known as the classical trajectory, which minimizes the action. Let $\eta(t)$ represent an infinitesimal displacement from the classical trajectory shown in Fig. 1.3. However, since in this case the end points of the trajectory are held fixed, the displacement has to satisfy the constraints

$$\eta(t_i) = \eta(t_f) = 0. \tag{1.7}$$

The infinitesimal change in the action is given by

$$\begin{aligned}\delta S &= S[x_{\rm cl}+\eta] - S[x_{\rm cl}] \\ &= \int_{t_i}^{t_f} \mathrm{d}t\, \left[L\left(x_{\rm cl}+\eta, \dot{x}_{\rm cl}+\dot{\eta}\right) - L\left(x_{\rm cl}, \dot{x}_{\rm cl}\right)\right] \\ &= \int_{t_i}^{t_f} \mathrm{d}t\, \left[\eta \left.\frac{\partial L}{\partial x}\right|_{x_{\rm cl}} + \dot{\eta} \left.\frac{\partial L}{\partial \dot{x}}\right|_{x_{\rm cl}} + \mathcal{O}(\eta^2)\right]\end{aligned}$$

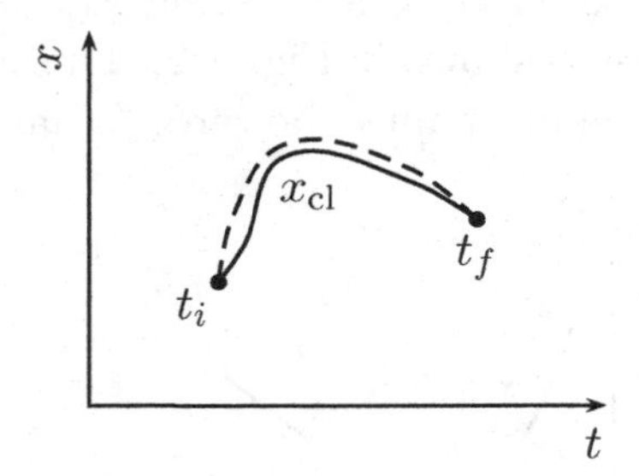

Figure 1.3: An infinitesimal change from the classical trajectory $x_{\rm cl}$.

$$= \int_{t_i}^{t_f} \mathrm{d}t \left[\eta \frac{\partial L}{\partial x} + \frac{\mathrm{d}}{\mathrm{d}t}\left(\eta \frac{\partial L}{\partial \dot{x}} \right) - \eta \frac{\mathrm{d}}{\mathrm{d}t} \frac{\partial L}{\partial \dot{x}} \right]_{x_{\rm cl}} + \mathcal{O}(\eta^2)$$

$$= \left(\eta \left. \frac{\partial L}{\partial \dot{x}} \right|_{x_{\rm cl}} \right) \Bigg|_{t_i}^{t_f} + \int_{t_i}^{t_f} \mathrm{d}t\, \eta \left[\frac{\partial L}{\partial x} - \frac{\mathrm{d}}{\mathrm{d}t} \frac{\partial L}{\partial \dot{x}} \right]_{x_{\rm cl}} + \mathcal{O}(\eta^2). \tag{1.8}$$

The first term, on the right hand side of (1.8) vanishes because of (1.7), namely,

$$\eta(t_i) = \eta(t_f) = 0.$$

Furthermore, we recognize from the previous example (see (1.6)) that the action functional must be stationary at its minimum. Thus, to the lowest order, we obtain

$$\delta S = 0 = \int_{t_i}^{t_f} \mathrm{d}t\, \eta \left[\frac{\partial L}{\partial x} - \frac{\mathrm{d}}{\mathrm{d}t} \frac{\partial L}{\partial \dot{x}} \right]_{x_{\rm cl}}. \tag{1.9}$$

Since $\eta(t)$ is an arbitrary function, the only way (1.9) can be satisfied is if

$$\left. \left(\frac{\partial L}{\partial x} - \frac{\mathrm{d}}{\mathrm{d}t} \frac{\partial L}{\partial \dot{x}} \right) \right|_{x_{\rm cl}} = 0. \tag{1.10}$$

In other words, the actual trajectory of the particle, $x_{\rm cl}$, must satisfy

$$\frac{\partial L}{\partial x} - \frac{\mathrm{d}}{\mathrm{d}t} \frac{\partial L}{\partial \dot{x}} = 0. \tag{1.11}$$

This is known as the Euler-Lagrange equation of motion for the theory described by the action (1.5) and $x_{\rm cl}$ is a solution of this equation.

So far, we have talked about particles in one dimension. For a particle in n dimensions, with coordinates x^i, $i = 1, 2, \ldots, n$, the Euler-Lagrange equations can be shown to have the form

$$\frac{\partial L}{\partial x^i} - \frac{\mathrm{d}}{\mathrm{d}t}\frac{\partial L}{\partial \dot{x}^i} = 0. \tag{1.12}$$

Connection with Newtonian mechanics. Let us consider the familiar example of the 1-dimensional particle in motion. Then, we have

$$T = \frac{1}{2}m\dot{x}^2, \qquad V = V(x),$$

so that

$$L = T - V = \frac{1}{2}m\dot{x}^2 - V(x), \tag{1.13}$$

which leads to

$$\frac{\partial L}{\partial x} = -\frac{\partial V}{\partial x}, \qquad \frac{\partial L}{\partial \dot{x}} = m\dot{x}. \tag{1.14}$$

The Euler-Lagrange (E-L) equation, (1.11), in this case, has the form,

$$-\frac{\partial V}{\partial x} - \frac{\mathrm{d}}{\mathrm{d}t}(m\dot{x}) = 0,$$

$$\text{or,}\quad m\frac{\mathrm{d}^2 x}{\mathrm{d}t^2} = -\frac{\partial V}{\partial x}. \tag{1.15}$$

Thus, we see that the Euler-Lagrange equation of motion actually gives rise to Newton's equation and is equivalent to it.

In discussions so far, we have assumed that $L = T - V$ and that the potential energy depends only on the position and not on the velocity. However, there are physical situations where the force does depend on the velocity. A familiar example is the force experienced by a charged particle moving in a magnetic field,

$$\mathbf{F} = \frac{q}{c}\mathbf{v} \times \mathbf{B}, \tag{1.16}$$

where q denotes the charge of the particle and c is the speed of light in vacuum. In such cases, to obtain the correct equations of motion, one has to introduce a velocity dependent generalized potential energy. For example, for the case of a charged particle with electromagnetic interactions, we have (we are assuming here that the particle experiences both an electric as well as a magnetic force)

$$L = T - U = \frac{1}{2}m\mathbf{v}\cdot\mathbf{v} - q\Phi + \frac{q}{c}\mathbf{v}\cdot\mathbf{A}, \tag{1.17}$$

where Φ represents the scalar potential while $\mathbf{A}$ is the vector potential and they are related to the electric and the magnetic fields through

$$\begin{aligned} \mathbf{E} &= -\boldsymbol{\nabla}\Phi - \frac{1}{c}\frac{\partial \mathbf{A}}{\partial t}, \\ \mathbf{B} &= \boldsymbol{\nabla}\times\mathbf{A}. \end{aligned} \tag{1.18}$$

Exercise. Work out the equation of motion for the particle, starting from the Lagrangian (1.17), namely, derive the Euler-Lagrange equations for such a particle.

It is clear, however, that the generalized potential energy U in (1.17) can not be interpreted as the potential energy of the particle, since the magnetic force does not do any work, being perpendicular to the velocity. Therefore, in general situations, it is improper to divide the Lagrangian into kinetic and potential energy terms. It is rather assumed that the Lagrangian, as a single entity, is a function of the position and the velocity, $L = L(x, \dot{x})$.

Advantages of the Lagrangian approach. We may ask at this point whether one gains anything by following this approach since, in the end, it seems to lead to the same Newton's equations of motion. The simple answer to this is that there are several nice features in the Lagrangian approach. First of all, the Lagrangian is a scalar and is, therefore, much easier to handle, in general, than vectors and tensors.

Second, the Lagrangian gives rise to equations of motion which have the same form independent of the coordinate system being used. We can easily convince ourselves of this by recognizing that nowhere, in the derivation of the equations from the principle of least action, did we utilize the fact that the coordinates are Cartesian. Thus, in terms of generalized coordinates q^i and $\dot{q}^i, i = 1, 2, \ldots n$, we can write $L \equiv L(q^i, \dot{q}^i)$ and the Euler-Lagrange equations take the forms

$$\frac{\mathrm{d}}{\mathrm{d}t}\left(\frac{\partial L}{\partial \dot{q}^i}\right) = \frac{\partial L}{\partial q^i}. \tag{1.19}$$

Here q^i can represent the coordinates of a particle in the Cartesian, or the polar or in any other coordinate system. This has to be contrasted with Newton's equations where the equations take very different forms in different coordinates.

We can try to bring these equations, (1.19), as close to Newton's equations as is possible by defining the generalized conjugate momentum

$$p_i = \frac{\partial L}{\partial \dot{q}^i}, \tag{1.20}$$

and the generalized force

$$F_i = \frac{\partial L}{\partial q^i}. \tag{1.21}$$

Then, equation (1.19) becomes

$$\frac{\mathrm{d}p_i}{\mathrm{d}t} = F_i. \tag{1.22}$$

Although the Euler-Lagrange equations take the same form in any coordinate system, we should always remember that, in general, p_i does not represent the momentum of the particle, and neither is F_i the force acting on it. In fact, if q^i (for a fixed i) corresponds to an angular variable θ, then the corresponding p_i would represent an angular momentum (component) of the particle and, similarly, F_i would denote the torque acting on it.

In the Lagrangian approach it is also easy to recognize quantities that are conserved. For example, if the Lagrangian is independent of a particular coordinate, then we say that the corresponding coordinate is a cyclic variable and the momentum conjugate to such a variable is conserved. This can be seen as follows. If q^i (for a fixed i) is cyclic, then, $\frac{\partial L}{\partial q^i} = 0$ and from the Euler-Lagrange equation, (1.19), we have (for the particular i)

$$\frac{\mathrm{d}}{\mathrm{d}t}\frac{\partial L}{\partial \dot{q}^i} = \frac{\mathrm{d}p_i}{\mathrm{d}t} = \frac{\partial L}{\partial q^i} = 0. \tag{1.23}$$

Hence p_i (for the particular i) is conserved. In Newtonian mechanics, if a Cartesian coordinate is cyclic the corresponding momentum is also conserved. However, the situation in the case of the Lagrangian is more general.

► **Example.** Let us now illustrate these with an example. Consider a particle moving in a plane (2 dimensions) and subjected to a force.

$$L = \frac{1}{2}m(\dot{x}^2 + \dot{y}^2) - V(x, y). \tag{1.24}$$

We also assume that the functional form of the potential is such that it depends only on the length of the vector. Thus,

$$V(x, y) = V(x^2 + y^2). \tag{1.25}$$

To determine the equations of motion, (1.19), in Cartesian coordinates, we need

$$\frac{\partial L}{\partial \dot{x}} = m\dot{x}, \qquad \frac{\partial L}{\partial \dot{y}} = m\dot{y},$$
$$\frac{\partial L}{\partial x} = -\frac{\partial V}{\partial x}, \qquad \frac{\partial L}{\partial y} = -\frac{\partial V}{\partial y}. \tag{1.26}$$

The Euler-Lagrange equations, in this case, take the forms

$$\frac{\mathrm{d}}{\mathrm{d}t}\left(\frac{\partial L}{\partial \dot{x}}\right) = \frac{\partial L}{\partial x} \Rightarrow m\frac{\mathrm{d}^2 x}{\mathrm{d}t^2} = -\frac{\partial V}{\partial x}, \tag{1.27}$$

$$\frac{\mathrm{d}}{\mathrm{d}t}\left(\frac{\partial L}{\partial \dot{y}}\right) = \frac{\partial L}{\partial y} \Rightarrow m\frac{\mathrm{d}^2 y}{\mathrm{d}t^2} = -\frac{\partial V}{\partial y}. \tag{1.28}$$

We recognize these to be the Newton's equations in the Cartesian coordinates and, in this formulation, conserved quantities of the system are not that obvious. Let us next derive the equations for the same system in polar coordinates within the Lagrangian formalism. In polar coordinates, the form of the Lagrangian, (1.24), can be determined by noting that the Cartesian and the polar coordinates are related by

$$x = r\cos\theta \Rightarrow \dot{x} = \dot{r}\cos\theta - r\dot{\theta}\sin\theta,$$
$$y = r\sin\theta \Rightarrow \dot{y} = \dot{r}\sin\theta + r\dot{\theta}\cos\theta, \tag{1.29}$$

so that

$$x^2 + y^2 = r^2,$$
$$\dot{x}^2 + \dot{y}^2 = \dot{r}^2(\cos^2\theta + \sin^2\theta) + r^2\dot{\theta}^2(\cos^2\theta + \sin^2\theta)$$
$$= \dot{r}^2 + r^2\dot{\theta}^2. \tag{1.30}$$

Therefore, the Lagrangian (1.24) takes the form

$$L = \frac{1}{2}m(\dot{x}^2 + \dot{y}^2) - V(x^2 + y^2)$$
$$= \frac{1}{2}m(\dot{r}^2 + r^2\dot{\theta}^2) - V(r^2). \tag{1.31}$$

It follows now from (1.31) that

$$\frac{\partial L}{\partial r} = -\frac{\partial V}{\partial r} + mr\dot{\theta}^2,$$
$$\frac{\partial L}{\partial \theta} = 0. \tag{1.32}$$

Clearly, θ is a cyclic coordinate and, therefore, the corresponding conjugate momentum must be conserved. The conjugate momenta have the forms

$$p_r = \frac{\partial L}{\partial \dot{r}} = m\dot{r},$$
$$p_\theta = \frac{\partial L}{\partial \dot{\theta}} = mr^2\dot{\theta}, \tag{1.33}$$

and we note that p_θ represents the angular momentum of the particle. In this case, the equations of motion, (1.19), take the forms

$$\frac{\mathrm{d}}{\mathrm{d}t}\frac{\partial L}{\partial \dot{\theta}} = \frac{\partial L}{\partial \theta} \Rightarrow \frac{\mathrm{d}p_\theta}{\mathrm{d}t} = \frac{\mathrm{d}}{\mathrm{d}t}(mr^2\dot{\theta}) = 0, \tag{1.34}$$

which shows that angular momentum is conserved and the dynamical equation follows from

$$\frac{\mathrm{d}}{\mathrm{d}t}\frac{\partial L}{\partial \dot{r}} = \frac{\partial L}{\partial r} \Rightarrow \frac{\mathrm{d}p_r}{\mathrm{d}t} = m\ddot{r} = mr\dot{\theta}^2 - \frac{\partial V}{\partial r}. \tag{1.35}$$

Here, the first term on the right hand side represents the centrifugal force while the second term gives a dynamical radial force. ◀

Exercise. Transform Newton's equations, (1.27)-(1.28), from Cartesian coordinates to polar coordinates.

1.4 Hamiltonian formalism

The Lagrangian is a function of the coordinates and the velocities, which are considered to be independent variables

$$L \equiv L(q^i, \dot{q}^i). \tag{1.36}$$

The conjugate momenta are defined in (1.20) as

$$p_i = \frac{\partial L}{\partial \dot{q}^i}, \tag{1.37}$$

and the dynamical equations are given by the Euler-Lagrange equations (1.19),

$$\frac{\mathrm{d}}{\mathrm{d}t}\frac{\partial L}{\partial \dot{q}^i} = \frac{\mathrm{d}p_i}{\mathrm{d}t} = \frac{\partial L}{\partial q^i}. \tag{1.38}$$

Given these, we can define another fundamental quantity associated with the system, called the Hamiltonian, which is a function of the coordinates and the momenta as

$$H \equiv H(q^i, p_i) = \sum_i p_i\dot{q}^i - L(q^i, \dot{q}^i). \tag{1.39}$$

Here q^i and p_i are treated as independent variables and $\dot{q}^i$'s become derived functions of q's and p's. Such a transformation is known as a Legendre transformation . It follows from (1.39) that

$$\begin{aligned}\frac{\partial H}{\partial p_j} &= \dot{q}^j + \sum_i p_i \frac{\partial \dot{q}^i}{\partial p_j} - \sum_i \frac{\partial L}{\partial \dot{q}^i}\frac{\partial \dot{q}^i}{\partial p_j}\\ &= \dot{q}^j + \sum_i p_i \frac{\partial \dot{q}^i}{\partial p_j} - \sum p_i \frac{\partial \dot{q}^i}{\partial p_j} = \dot{q}^j, \end{aligned} \tag{1.40}$$

$$\begin{aligned}\frac{\partial H}{\partial q^j} &= \sum_i p_i \frac{\partial \dot{q}^i}{\partial q^j} - \frac{\partial L}{\partial q^j} - \sum_i \frac{\partial L}{\partial \dot{q}^i}\frac{\partial \dot{q}^i}{\partial q^j}\\ &= \sum_i p_i \frac{\partial \dot{q}^i}{\partial q^j} - \frac{\partial L}{\partial q^j} - \sum_i p_i \frac{\partial \dot{q}^i}{\partial q^j} = -\frac{\partial L}{\partial q^j}. \end{aligned} \tag{1.41}$$

Using the Euler-Lagrange equations (see (1.19)), (1.41) becomes

$$\frac{\partial H}{\partial q^j} = -\frac{\partial L}{\partial q^j} = -\dot{p}_j. \tag{1.42}$$

Thus, from (1.40) and (1.42), we see that the equations of motion, in the Hamiltonian formalism, take the forms

$$\frac{\partial H}{\partial p_i} = \dot{q}^i, \qquad \frac{\partial H}{\partial q^i} = -\dot{p}_i. \tag{1.43}$$

which are known as the Hamiltonian equations of motion. It is clear that the n second order Euler-Lagrange equations in the Lagrangian formalism have become $2n$ first order equations in the Hamiltonian formalism. Given the initial values $q^i(0)$ and $p_i(0)$, one can determine the solutions of the Hamiltonian equations uniquely.

Interpretation of the Hamiltonian. In the absence of nonconservative forces, the Lagrangian can be written in Cartesian coordinates, in the form

$$L = T - V = \frac{1}{2}m(\dot{x}^i)^2 - V(x^i). \tag{1.44}$$

Furthermore, the conjugate momenta coincide with the actual momenta in this coordinate system, namely,

$$p_i = \frac{\partial L}{\partial \dot{x}^i} = m\dot{x}^i. \tag{1.45}$$

(The apparent mismatch of the indices is due to the fact that we are ignoring the metric, which is the trivial Kronecker delta in Euclidean space.) In this case, therefore, we see that

$$
\begin{aligned}
H &= \sum_i p_i \dot{x}^i - L = \sum_i \left(\frac{1}{2} m (\dot{x}^i)^2 + V(x^i) \right) \\
&= \sum_i \left(\frac{p_i^2}{2m} + V(x^i) \right) = T + V = E.
\end{aligned} \tag{1.46}
$$

Thus, the Hamiltonian corresponds to the total energy of the system in the absence of nonconservative forces. Even though we showed it in the Cartesian coordinates, this result holds in general. For example, let us assume that, in a general coordinate system, the kinetic energy has the form

$$
T = \frac{1}{2} \sum_{i,j} T_{ij}(q) \dot{q}^i \dot{q}^j, \qquad T_{ij} = T_{ji}. \tag{1.47}
$$

Then, from the definition of the Lagrangian for such a system (without any nonconservative forces)

$$
L = T - V(q) = \frac{1}{2} \sum_{i,j} T_{ij}(q) \dot{q}^i \dot{q}^j - V(q), \tag{1.48}
$$

it follows that

$$
p_i = \frac{\partial L}{\partial \dot{q}^i} = \sum_j T_{ij}\, \dot{q}^j, \tag{1.49}
$$

so that

$$
\sum_i p_i \dot{q}^i = \sum_{i,j} T_{ij} \dot{q}^j \dot{q}^i = 2T, \tag{1.50}
$$

and we obtain

$$
H = \sum_i p_i \dot{q}^i - L = 2T - (T - V) = T + V. \tag{1.51}
$$

This shows that the Hamiltonian for such a system can be identified with the total energy of the system.

Advantages of the Hamiltonian formalism. There are several advantages in using the Hamiltonian description of a dynamical system. First, the Hamiltonian equations of motion are first order equations and are, therefore, sometimes easier to handle.

Furthermore, the equations are symmetric in q^i and p_i. This is of considerable help when cyclic coordinates are present. We know from the Lagrangian description that when a cyclic coordinate is present, the corresponding conjugate momentum is a constant of motion. This continues to hold in the Hamiltonian formalism as well, since (if the Lagrangian is independent of a coordinate, the Hamiltonian also is)

$$\frac{\partial H}{\partial q^i} = -\dot{p}_i = 0. \tag{1.52}$$

However, the difference is that even in the presence of a cyclic coordinate q^i, the Lagrangian is a function of the corresponding velocity $\dot{q}^i$. And therefore, we still have to solve n-equations. On the other hand, in the Hamiltonian formalism if q^i is cyclic, p_i is a constant and, therefore,

$$H \equiv H\left(q^1, \ldots, q^{i-1}, q^{i+1} \ldots q^n, p_1, \ldots p_{i-1}, \alpha, p_{i+1}, \ldots p_n\right). \tag{1.53}$$

Consequently, the number of equations we have to solve is reduced. Furthermore, it is easier to recognize other conserved quantities in the Hamiltonian formalism.

► **Example.** Let us consider an arbitrary phase space variable $\omega = \omega(q,p)$ which does not depend on time explicitly. Then, its time evolution can be determined simply as

$$\begin{aligned}\frac{\mathrm{d}\omega}{\mathrm{d}t} &= \sum_i \left(\frac{\partial \omega}{\partial q^i}\dot{q}^i + \frac{\partial \omega}{\partial p_i}\dot{p}_i\right)\\ &= \sum_i \left(\frac{\partial \omega}{\partial q^i}\frac{\partial H}{\partial p_i} - \frac{\partial \omega}{\partial p_i}\frac{\partial H}{\partial q^i}\right)\\ &\equiv \{\omega, H\},\end{aligned} \tag{1.54}$$

where we have used Hamilton's equations and the curly bracket denotes the Poisson bracket of two variables. Explicitly, the Poisson bracket of two phase space variables $a(q,p)$ and $b(q,p)$ is defined to be

$$\{a, b\} = \sum_i \left(\frac{\partial a}{\partial q^i}\frac{\partial b}{\partial p_i} - \frac{\partial a}{\partial p_i}\frac{\partial b}{\partial q^i}\right). \tag{1.55}$$

Clearly,

$$\frac{\mathrm{d}\omega}{\mathrm{d}t} = 0, \quad \text{if} \quad \{\omega, H\} \equiv 0. \tag{1.56}$$

Namely, a quantity is conserved if it has a vanishing Poisson bracket with the Hamiltonian. Let us note that, since $\{H, H\} = 0$, this shows trivially that the Hamiltonian or the total energy of the system is a constant in time. ◄

From the definition of the Poisson bracket between any two variables on the phase space in (1.55), it is easy to check that they satisfy the following relations

$$\begin{aligned}
&\{\eta, \rho\} = -\{\rho, \eta\}, \\
&\{\eta, c\} = 0, \quad \text{where } c \text{ is a constant}, \\
&\{\eta_1 + \eta_2, \rho\} = \{\eta_1, \rho\} + \{\eta_2, \rho\}, \\
&\{\eta_1\eta_2, \rho\} = \eta_1 \{\eta_2, \rho\} + \{\eta_1, \rho\} \eta_2. \qquad (1.57)
\end{aligned}$$

Furthermore, it follows from (1.55) that the Poisson bracket between the coordinates and the momenta satisfy the simple relations

$$\begin{aligned}
&\{q^i, q^j\} = \{p_i, p_j\} = 0, \\
&\{q^i, p_j\} = \delta^i_j = -\{p_j, q^i\}. \qquad (1.58)
\end{aligned}$$

These are known as the canonical Poisson bracket relations and given the canonical Poisson bracket relations, the Poisson bracket between any two phase space variables can be trivially calculated using the identities given in (1.57).

So far, we have talked about the dynamics of particles. Particles, besides having a definite mass, are characterized by a definite momentum and energy (p, E). Namely, particles travel in well defined trajectories. When particles collide, they scatter in such a way as to conserve the total momentum and the total energy of the system. There is, of course, another kind of classical motion that we are aware of, namely, the wave motion. It is the propagation of a disturbance. Familiar examples of wave motion consist of ripples on the surface of water, sound waves, electromagnetic waves and so on. All such motions are governed by one equation, namely, the wave equation, which has the generic form

$$\begin{aligned}
&\frac{\partial^2 \psi}{\partial x^2} - \frac{1}{v^2}\frac{\partial^2 \psi}{\partial t^2} = 0, \quad \text{in one dimension}, \\
&\boldsymbol{\nabla}^2 \psi - \frac{1}{v^2}\frac{\partial^2 \psi}{\partial t^2} = 0, \quad \text{in higher dimensions}. \qquad (1.59)
\end{aligned}$$

Here ψ represents the disturbance, which can be the displacement (height) of water from its normal surface in the case of ripples, or the electric and the magnetic fields in the case of light waves. Furthermore, v represents the speed of propagation of the waves. Clearly, a solution to the wave equation (1.59) is a plane wave of the form (in one dimension)

$$\psi(x, t) = A\, e^{-i\omega t + ikx}, \qquad (1.60)$$

with

$$k^2 = \frac{\omega^2}{v^2}, \tag{1.61}$$

where

$$k = \text{wave number} = \frac{2\pi}{\lambda},$$
$$\omega = \text{angular frequency} = 2\pi\nu. \tag{1.62}$$

It follows from (1.61) and (1.62) that

$$\lambda\nu = v. \tag{1.63}$$

Thus, with each wave is also associated a pair of quantities (λ, ν) or (k, ω). The constant A in (1.60) is called the amplitude of the wave and $I = |\psi|^2 = |A|^2$ measures the intensity of the wave. As we can see from the form of the solution in (1.60), contrary to particle motion which is localized, the wave phenomenon is highly nonlocal in nature, i.e., at any given time the disturbance is spread over all space.

Thus, we see that most classical or macroscopic phenomena can be explained by either of these two descriptions of the system. Whereas planetary motion can be explained by particle mechanics, interference and diffraction of light are understood as phenomena in wave mechanics.

Around the turn of the twentieth century, however, this clear division of particle and wave mechanics ran into conflicts when applied to microscopic systems. Phenomena, such as blackbody radiation and photoelectric effect, needed the interpretation of electromagnetic radiation as consisting of particles called photons with quantized energy and momentum. All this led to a re-examination of the principles of classical mechanics, when applied to microscopic systems. But before talking about these in detail, let us get acquainted with the mathematical tools that we need.

1.5 Selected problems

1. Consider the Lagrangian for a particle of mass m interacting with static electromagnetic fields (not explicitly dependent on time) given by (see also (1.17))

$$L = \frac{1}{2}m\dot{\mathbf{x}}^2 - q\phi + \frac{q}{c}\,\dot{\mathbf{x}}\cdot\mathbf{A}, \tag{1.64}$$

where $\mathbf{x}$ represents the three dimensional coordinate vector and $\phi, \mathbf{A}$ denote the scalar and the vector potentials respectively, depending only on the coordinates (and not on velocities) and q represents the charge carried by the particle. In the static case (when $\phi = \phi(\mathbf{x}), \mathbf{A} = \mathbf{A}(\mathbf{x})$), the usual definitions (see (1.18))

$$\mathbf{E} = -\boldsymbol{\nabla}\phi - \frac{1}{c}\frac{\partial \mathbf{A}}{\partial t},$$
$$\mathbf{B} = (\boldsymbol{\nabla} \times \mathbf{A}), \tag{1.65}$$

reduce to

$$\mathbf{E} = -\boldsymbol{\nabla}\phi,$$
$$\mathbf{B} = (\boldsymbol{\nabla} \times \mathbf{A}). \tag{1.66}$$

(*a*) Show that the Euler-Lagrange equations following from the Lagrangian (1.64) give rise to Newton's equations for a particle with charge q interacting with an electric as well as a magnetic field (c is the speed of light).

(*b*) What is the Hamiltonian for this system?

2. If η, ρ represent two arbitrary phase space variables depending on (x^i, p_i), show explicitly from the definition of the Poisson bracket in (1.55) that the following Poisson bracket relations in (1.57) hold,

$$\{\eta, \rho\} = -\{\rho, \eta\},$$
$$\{\eta_1 + \eta_2, \rho\} = \{\eta_1, \rho\} + \{\eta_2, \rho\},$$
$$\{\eta_1\eta_2, \rho\} = \eta_1\{\eta_2, \rho\} + \{\eta_1, \rho\}\eta_2. \tag{1.67}$$

3. Prove the Jacobi identity for Poisson brackets, namely, if η, ρ and ζ denote three classical dynamical variables depending on the phase space variables (x^i, p_i), show explicitly from the definition in (1.55) that

$$\{\eta, \{\rho, \zeta\}\} + \{\rho, \{\zeta, \eta\}\} + \{\zeta, \{\eta, \rho\}\} = 0. \tag{1.68}$$

4. If a is a complex, classical dynamical variable (function of x^i, p_i) and a^* is its complex conjugate, and if

$$\{a, a^*\} = i, \tag{1.69}$$

calculate

$$\{a, aa^*\}, \quad \{a^*, aa^*\}, \quad \{a, a^*a\}, \quad \{a^*, a^*a\}, \tag{1.70}$$

using (1.69) as well as the properties of Poisson brackets in (1.67) (or (1.57)).

5. Calculate the Poisson brackets

$$\{j_x, j_y\}, \qquad \{j_y, j_z\}, \qquad \{j_z, j_x\}, \tag{1.71}$$

using the basic canonical Poisson bracket relations (1.58) (as well as properties of Poisson brackets). Here the classical angular momentum variables are defined as

$$j_x = yp_z - zp_y, \quad j_y = zp_x - xp_z, \quad j_z = xp_y - yp_x. \tag{1.72}$$

CHAPTER 2

Review of essential mathematics

In the following few lectures, we will recapitulate some of the mathematical concepts that we will need for a detailed understanding of quantum mechanics.

2.1 Linear vector spaces

Definition. *A set of quantities $\{V_i\}$, with a definite rule for addition and multiplication by scalars, is called a set of vectors if they satisfy*

1. $V_i + V_j = V_j + V_i$, (commutative law of addition),
2. $V_i + (V_j + V_k) = (V_i + V_j) + V_k$, (associative law of addition),
3. $\alpha(V_i + V_j) = \alpha V_i + \alpha V_j$,
4. $(\alpha + \beta)V_i = \alpha V_i + \beta V_i$, (distributive law),
5. $(\alpha\beta)V_i = \alpha(\beta V_i)$, (associative law of multiplication).

We are yet to enumerate the rules for addition and multiplication. But, let us go ahead and define a vector space.

Definition. *If $\mathbf{V}$ represents the set of vectors $\{V_i\}$ (namely, $V_i \in \mathbf{V}$) such that*

1. *$\alpha V_i + \beta V_j \in \mathbf{V}$, where α, β are constants,*
2. *there exists a unique null vector $\emptyset \in \mathbf{V}$ such that*

$$V_i + \emptyset = V_i = \emptyset + V_i,$$

3. *corresponding to every vector $V_i \in \mathbf{V}$, there exists a unique inverse $(-V_i) \in \mathbf{V}$ such that*

$$V_i + (-V_i) = \emptyset,$$

4. $0 \cdot V_i = \emptyset$,

5. $1 \cdot V_i = V_i$,

then, $\mathbf{V}$ *is called a linear vector space.*

The set over which the parameters α, β are defined is called the field, $F(\alpha)$, over which the linear vector space, $\mathbf{V}$, is defined. (Normally, one considers α, β to belong to real or complex numbers, although other possibilities are allowed.) For example, the familiar vectors of 3-dimensional Euclidean space represent a real, linear vector space. In that case, addition is simply defined by the familiar vector addition and multiplication by a real number corresponds to scaling the vector by that number as shown in Fig. 2.1. The null vector, in this case,

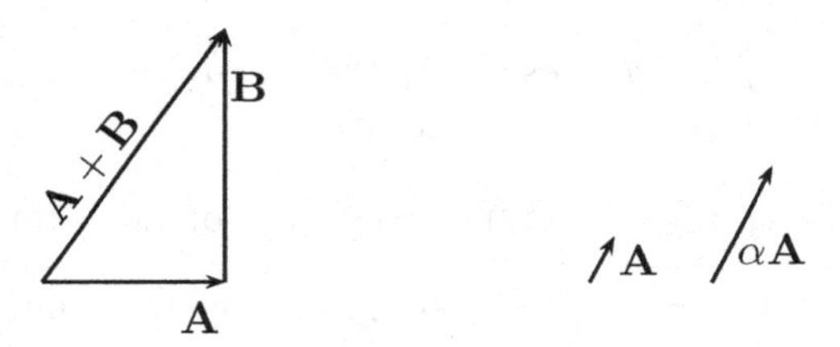

Figure 2.1: Addition of two vectors and multiplication of a vector by a scalar in 3-dimensional Euclidean space.

is a vector of zero magnitude and the inverse of a given vector is the vector with the arrow (direction) reversed.

Definition. *A set of vectors* $(V_1, \ldots, V_n)$ *in a vector space* $\mathbf{V}$ *is said to be linearly independent if the only solution to the relation*

$$\sum_{i=1}^{n} \alpha_i V_i = 0, \tag{2.1}$$

is that all the α_i*'s vanish, namely,*

$$\alpha_i = 0, \qquad i = 1, 2, \ldots, n. \tag{2.2}$$

Definition. *A vector space* $\mathbf{V}$ *is said to be* n*-dimensional and is denoted by* $\mathbf{V}^n$*, if the maximum number of linearly independent vectors that can be found in that space is* n*.*

Theorem. *An arbitrary nontrivial vector V in $\mathbf{V}^n$ can be expressed uniquely as a linear combination of n linearly independent vectors in $\mathbf{V}^n$.*

Proof. Let $(V_1, \ldots, V_n)$ be a set of n linearly independent vectors in $\mathbf{V}^n$. Then, a relation of the form

$$\alpha V + \sum_{i=1}^{n} \alpha_i' V_i = 0, \tag{2.3}$$

1. cannot imply $\alpha = \alpha_i' = 0$ for all i, since then that would imply that there are $n + 1$ linearly independent vectors which is impossible because we are dealing with an n-dimensional space, $\mathbf{V}^n$. Therefore, some of the parameters have to be nonzero.

2. cannot imply $\alpha = 0$ because then, with some of the other parameters not vanishing,

$$\sum_{i=1}^{n} \alpha_i' V_i = 0, \tag{2.4}$$

would imply that not all the V_i's are linearly independent, which is not true by assumption.

Therefore, at the most, some of the α_i' would be zero. This implies that we can write (2.3) also as

$$V = \sum_{i=1}^{n} \alpha_i V_i, \tag{2.5}$$

where

$$\alpha_i = -\frac{\alpha_i'}{\alpha}. \tag{2.6}$$

This shows that we can express an arbitrary vector in $\mathbf{V}^n$ as a linear combination of n linearly independent vectors in that space.

To prove the uniqueness of this expression, let us assume that there exists another expansion of the same vector in terms of the same linearly independent vectors as

$$V = \sum_{i=1}^{n} \beta_i V_i. \tag{2.7}$$

Then, subtracting the two expressions in (2.5) and (2.7), we obtain

$$\emptyset = \sum_{i=1}^{n} (\alpha_i - \beta_i) V_i. \tag{2.8}$$

However, since V_i's are linearly independent, using (2.1) and (2.2), we obtain

$$\alpha_i - \beta_i = 0, \quad \text{or,} \quad \alpha_i = \beta_i. \tag{2.9}$$

This proves that any arbitrary nontrivial vector in $\mathbf{V}^n$ can be expressed uniquely as a linear combination of n linearly independent vectors. ■

Definition. *Any set of n linearly independent vectors, $(V_1, \ldots, V_n)$, is said to form a basis in $\mathbf{V}^n$. The coefficients of expansion of any vector V in a given basis are said to be the components of V in that basis.*

For example, in Cartesian coordinates, a 3-dimensional vector is represented as $x = (x_1, x_2, x_3)$, where x_1, x_2, x_3 are the components of the vector in the Cartesian basis. It is worth noting here that once we have a basis, the rules for addition of vectors and multiplication by a scalar become simple. For example, suppose that in a basis, $\{V_i\}$, we can expand two vectors $V, \widetilde{V}$ as

$$V = \sum_{i=1}^{n} \alpha_i V_i, \qquad \widetilde{V} = \sum_{i=1}^{n} \beta_i V_i. \tag{2.10}$$

Then, it follows that

$$V + \widetilde{V} = \sum_{i=1}^{n} (\alpha_i + \beta_i) V_i, \quad cV = \sum_{i=1}^{n} \alpha_i' V_i, \quad \alpha_i' = c\alpha_i. \tag{2.11}$$

Namely, adding two vectors leads to a new vector whose components are the arithmetic sum of the components of the two vectors in the same basis. Similarly, multiplying a given vector by a number yields a new vector whose components correspond to the product of the components of the original vector by the given number.

2.2 Inner product

Definition. *An inner product is a procedure for assigning a number to two vectors in a vector space and is commonly denoted by (V_i, V_j). It satisfies the following properties:*

1. $(V_i, V_i) \geq 0, \quad (0 \text{ only if } V_i = \emptyset)$,

2. $(V_i, V_j) = (V_j, V_i)^*$,

3. $(V_i, \alpha V_j + \beta V_k) = \alpha (V_i, V_j) + \beta (V_i, V_k)$.

It follows from the properties (2) and (3) above that

$$\begin{aligned}
(\alpha V_j + \beta V_k, V_i) &= (V_i, \alpha V_j + \beta V_k)^* \\
&= \alpha^* (V_i, V_j)^* + \beta^* (V_i, V_k)^* \\
&= \alpha^* (V_j, V_i) + \beta^* (V_k, V_i) .
\end{aligned} \tag{2.12}$$

Definition. *A vector space, which admits an inner product, is called an inner product space.*

Definition. *The norm of a vector V, in an inner product space, is defined to be*

$$|V| = (V, V)^{\frac{1}{2}}. \tag{2.13}$$

A vector is said to be a unit vector or normalized, if its norm is unity.

Definition. *Two vectors are said to be orthogonal if their inner product vanishes. Namely, if*

$$(V_i, V_j) = 0, \tag{2.14}$$

then, V_i and V_j are said to be orthogonal.

Definition. *A set of vectors $(e_1, e_2, \ldots, e_n)$ in $\mathbf{V}^n$ is said to be orthonormal if*

$$(e_i, e_j) = \delta_{ij}, \quad i, j = 1, 2, \cdots, n. \tag{2.15}$$

Such a set consists of n linearly independent vectors and defines an orthonormal basis in $\mathbf{V}^n$.

Let $\{e_i\}$ denote an orthonormal basis in $\mathbf{V}^n$. Then, two arbitrary vectors V and W can be expanded in this basis as

$$V = \sum_{i=1}^{n} v_i e_i, \quad W = \sum_{j=1}^{n} \omega_j e_j. \tag{2.16}$$

In this case, it follows from (2.2), (2.12) and (2.15) that

$$
\begin{aligned}
(W,V) &= \Big(\sum_{i=1}^{n}\omega_i e_i, \sum_{j=1}^{n} v_j e_j\Big) = \sum_{i,j=1}^{n} (\omega_i e_i, v_j e_j)\\
&= \sum_{i,j=1}^{n} \omega_i^* v_j (e_i, e_j)\\
&= \sum_{i,j=1}^{n} \omega_i^* v_j \delta_{ij} = \sum_{i=1}^{n} \omega_i^* v_i. \qquad (2.17)
\end{aligned}
$$

In other words, in an orthonormal basis, the inner product of two vectors is completely determined, as in (2.17), by the components of the two vectors in that basis.

Let us note here parenthetically that the above discussion may give the naive impression that it is always possible to define an inner product. The catch, however, lies in the fact that to define an orthonormal basis, we must know how to define the norm and, therefore, the inner product.

2.3 Dirac notation

We recall (see (2.5) and (2.15)) that an arbitrary vector in $\mathbf{V}^n$, with an inner product, can be expressed uniquely in terms of an orthonormal basis $\{e_i\}$ as

$$
V = \sum_{i=1}^{n} v_i e_i. \qquad (2.18)
$$

A vector V can, then, be represented simply by its components in this basis as $V = (v_1, v_2, v_3, \ldots, v_n)$ - an ordered n-tuple. The familiar example in 3-dimensions is that of a vector written as $x = (x_1, x_2, x_3)$, where it is understood that the basis is Cartesian. We can collect the n-tuple into a column matrix and then the correspondence, in a given basis, becomes

$$
V = \begin{pmatrix} v_1 \\ v_2 \\ \vdots \\ v_n \end{pmatrix}. \qquad (2.19)
$$

Note here that the basis, in this case, is chosen such that

$$e_i = \begin{pmatrix} 0 \\ 0 \\ \vdots \\ 1 \\ 0 \\ 0 \end{pmatrix}. \tag{2.20}$$

where the non-zero entry is in the i-th row and

$$V = \sum_{i=1}^{n} v_i e_i,$$

as discussed in (2.18).

With such a representation of a vector, it is clear now that the addition of vectors and multiplication of a vector by a scalar obey matrix formulae. For example,

$$V + W = \begin{pmatrix} v_1 \\ v_2 \\ \vdots \\ v_n \end{pmatrix} + \begin{pmatrix} \omega_1 \\ \omega_2 \\ \vdots \\ \omega_n \end{pmatrix} = \begin{pmatrix} v_1 + \omega_1 \\ v_2 + \omega_2 \\ \vdots \\ v_n + \omega_n \end{pmatrix} = \begin{pmatrix} z_1 \\ z_2 \\ \vdots \\ z_n \end{pmatrix} = Z,$$

$$\alpha V = \begin{pmatrix} \alpha v_1 \\ \alpha v_2 \\ \vdots \\ \alpha v_n \end{pmatrix} = V'. \tag{2.21}$$

This should be compared with (2.11). It is worth emphasizing here that matrices satisfy all the properties of vectors and can, therefore, provide a representation for them.

A column representation of a vector is called a ket vector and is denoted by the correspondence

$$\text{ket } V \equiv |V\rangle = \begin{pmatrix} v_1 \\ v_2 \\ \vdots \\ v_n \end{pmatrix}. \tag{2.22}$$

However, given a column vector, we can also take its Hermitian conjugate to obtain a row vector as

$$\begin{pmatrix} v_1 \\ v_2 \\ \vdots \\ v_n \end{pmatrix}^\dagger = \begin{pmatrix} v_1 \\ v_2 \\ \vdots \\ v_n \end{pmatrix}^{*\,T} = \begin{pmatrix} v_1^* & v_2^* & v_3^* & \dots & v_n^* \end{pmatrix}. \tag{2.23}$$

Obviously, this can also provide a representation for V. It is called a bra vector and is denoted by

$$\begin{aligned} \text{bra } V &\equiv \langle V| = |V\rangle^\dagger = (\text{ket } V)^\dagger \\ &= \begin{pmatrix} v_1^* & v_2^* & v_3^* & \dots & v_n^* \end{pmatrix}. \end{aligned} \tag{2.24}$$

This operation is also known as taking the adjoint. Thus, corresponding to every ket vector, there exists a unique bra vector and *vice versa.* Let us now form the product of a bra vector with a ket vector, using the matrix laws of multiplication

$$\begin{aligned} \langle W|V\rangle &= \begin{pmatrix} \omega_1^* & \omega_2^* & \dots & \omega_n^* \end{pmatrix} \begin{pmatrix} v_1 \\ v_2 \\ \vdots \\ v_n \end{pmatrix} \\ &= \omega_1^* v_1 + \omega_2^* v_2 + \cdots + \omega_n^* v_n = \sum_{i=1}^{n} \omega_i^* v_i \\ &= (W, V). \end{aligned} \tag{2.25}$$

which we recognize from (2.17) to be the inner product of W with V.

Since we can express (see (2.18))

$$V = \sum_{i=1}^{n} v_i e_i,$$

where the components v_i are numbers and e_i's are the basis vectors, we can also define basis ket vectors as (the non-zero element is in the i-th row)

$$|e_i\rangle = \begin{pmatrix} 0 \\ \vdots \\ 1 \\ 0 \\ \vdots \\ 0 \end{pmatrix} \equiv |i\rangle, \tag{2.26}$$

and the basis bra vectors as

$$\langle e_i| = |e_i\rangle^\dagger = (0 \;\; 0 \;\; \ldots \;\; 1 \;\; 0 \;\; 0 \;\; \ldots 0) \equiv \langle i|. \tag{2.27}$$

With these, we can write

$$|V\rangle = \sum_{i=1}^{n} v_i |e_i\rangle, \qquad \langle V| = \sum_{i=1}^{n} v_i^* \langle e_i|, \tag{2.28}$$

which leads to

$$\begin{aligned}\langle W|V\rangle &= \sum_{i,j=1}^{n} (\omega_i^* \langle e_i|)\,(v_j|e_j\rangle) \\ &= \sum_{i,j=1}^{n} \omega_i^* v_j \langle e_i|e_j\rangle \\ &= \sum_{i=1}^{n} \omega_i^* v_i,\end{aligned} \tag{2.29}$$

where we have used (2.25) in the last step. This, then, implies that

$$\langle e_i|e_j\rangle = \delta_{ij}, \tag{2.30}$$

which is the orthonormality relation for the basis vectors in the Dirac notation.

Let us note next that if

$$|V\rangle = \sum_{j=1}^{n} v_j |e_j\rangle,$$

then, using (2.30), we obtain

$$\langle e_i|V\rangle = \sum_{j=1}^{n} v_j \langle e_i|e_j\rangle = \sum_{j=1}^{n} v_j \delta_{ij} = v_i. \tag{2.31}$$

Namely, the components of a ket vector, in a given orthonormal basis, can be obtained by taking the inner product of the vector with the appropriate bra basis vectors.

2.4 Linear operators

An operator is a map of a vector into another vector. Thus, if $|V\rangle$ and $|V'\rangle$ are two ket vectors and if Ω is a map which takes $|V\rangle$ to $|V'\rangle$, then, we write

$$|V\rangle \xrightarrow{\Omega} |V'\rangle, \quad \text{or,} \quad \Omega|V\rangle = |\Omega V\rangle = |V'\rangle. \tag{2.32}$$

Alternatively, one says that the operator Ω acting on the vector $|V\rangle$ transforms it to the vector $|V'\rangle$. Operators can also act on bra vectors to produce other bra vectors. Thus, for example, we can have

$$\langle V|\Omega = \langle V''|. \tag{2.33}$$

However, an operator cannot act on a ket vector to give a bra vector and *vice versa.*

Linear operators are operators, which obey the following rules

$$\Omega(\alpha|V_i\rangle) = \alpha(\Omega|V_i\rangle), \quad (\alpha \text{ is a scalar}), \tag{2.34a}$$
$$\Omega(\alpha|V_i\rangle + \beta|V_j\rangle) = \alpha(\Omega|V_i\rangle) + \beta(\Omega|V_j\rangle). \tag{2.34b}$$

Similarly, acting on the bra vectors, linear operators satisfy

$$(\langle V_i|\alpha)\Omega = (\langle V_i|\Omega)\alpha, \tag{2.35a}$$
$$(\langle V_i|\alpha + \langle V_j|\beta)\Omega = (\langle V_i|\Omega)\alpha + (\langle V_j|\Omega)\beta. \tag{2.35b}$$

The simplest linear operator is the identity operator, $\mathbb{1}$, which leaves all vectors invariant. Namely,

$$\mathbb{1}|V\rangle = |V\rangle, \qquad \langle V|\mathbb{1} = \langle V|.$$

Clearly, since our ket and bra vectors are column and row matrices respectively, a matrix representation of operators would involve square matrices with n^2 elements, in general. A knowledge of the transformation properties of a given set of basis vectors, under the action of an operator, determines completely the matrix elements of the operator in that basis. For example, if

$$\Omega|e_j\rangle = |e'_j\rangle, \quad j = 1, 2, \cdots, n, \tag{2.36}$$

then,

$$\langle e_i|e'_j\rangle = \langle e_i|\Omega|e_j\rangle = \Omega_{ij}, \quad i, j = 1, 2, \cdots, n. \tag{2.37}$$

Therefore, if all the $|e'_j\rangle$'s are known, this implies that all the Ω_{ij}'s are also known, which are called the matrix elements of the operator Ω in the particular basis. Once the Ω_{ij}'s are known, the transformation property of any arbitrary vector can be easily worked out. For example, if

$$|V\rangle = \sum_{i=1}^{n} v_i|e_i\rangle, \qquad \Omega|V\rangle = |V'\rangle = \sum_{i=1}^{n} v'_i|e_i\rangle, \tag{2.38}$$

then, the transformed components (*i.e.*, the components of the transformed vector in the given basis) can be obtained as

$$\begin{aligned}
v_i' = \langle e_i|V'\rangle = \langle e_i|\Omega|V\rangle &= \langle e_i|\Omega\sum_{j=1}^{n} v_j|e_j\rangle \\
&= \sum_{j=1}^{n} v_j\langle e_i|\Omega|e_j\rangle = \sum_j v_j\Omega_{ij} \\
&= \sum_{j=1}^{n} \Omega_{ij}v_j. \qquad (2.39)
\end{aligned}$$

When two or more operators act on a vector, the order in which they act is important. For example, $\Omega'\Omega|V\rangle$ stands for the action of Ω on $|V\rangle$ followed by the action of the operator Ω'. In general,

$$\Omega'\Omega|V\rangle \neq \Omega\Omega'|V\rangle. \qquad (2.40)$$

This is clearly reflected in the fact that matrix multiplication is not commutative (as we have seen, operators can be represented by square matrices). The object

$$\Omega'\Omega - \Omega\Omega' = [\Omega', \Omega], \qquad (2.41)$$

is called the commutator of Ω' with Ω and is, in general, nonzero. When it vanishes, the operators are said to commute.

We can also define the inverse Ω^{-1} of an operator Ω such that the action of Ω on any arbitrary vector followed by the inverse (or *vice versa*) leaves the vector unchanged. Namely,

$$\Omega^{-1}\Omega|V\rangle = |V\rangle = \Omega\Omega^{-1}|V\rangle, \qquad (2.42)$$

which implies that

$$\Omega^{-1}\Omega = \Omega\Omega^{-1} \equiv \mathbb{1} \equiv \text{ identity operator}. \qquad (2.43)$$

► **Example (Identity operator).** We have encountered the identity operator earlier. Here, let us analyze some of its properties. We know that we can write (see (2.31))

$$|V\rangle = \sum v_i|e_i\rangle,$$

with

$$v_i = \langle e_i|V\rangle.$$

It follows now that

$$\begin{aligned}|V\rangle &= \sum_{i=1}^{n} v_i|e_i\rangle = \sum_{i=1}^{n} |e_i\rangle v_i \\ &= \sum_{i=1}^{n} |e_i\rangle\langle e_i|V\rangle = \left(\sum_{i=1}^{n} |e_i\rangle\langle e_i|\right)|V\rangle \\ &= \mathbb{1}|V\rangle,\end{aligned} \tag{2.44}$$

where we have identified

$$\sum_{i=1}^{n} |e_i\rangle\langle e_i| = \mathbb{1} \equiv \text{ identity operator.} \tag{2.45}$$

Relation (2.45) is also known as the completeness relation for the basis vectors and corresponds to the outer product of the basis vectors.

The matrix elements of the identity operator can now be easily obtained as

$$\begin{aligned}\mathbb{1}_{jk} &= \langle e_j|\mathbb{1}|e_k\rangle = \langle e_j|\left(\sum_{i=1}^{n} |e_i\rangle\langle e_i|\right)|e_k\rangle \\ &= \sum_{i=1}^{n}\langle e_j|e_i\rangle\langle e_i|e_k\rangle = \sum_{i=1}^{n}\delta_{ij}\delta_{ik} \\ &= \delta_{jk}.\end{aligned} \tag{2.46}$$

Namely, the identity operator, as a square matrix, has only unit diagonal elements, which is what we expect intuitively. ◄

► **Example (Projection operator).** Let us note, from (2.45) that we can write

$$\mathbb{1} = \sum_{i=1}^{n} |e_i\rangle\langle e_i| = \sum_{i=1}^{n} P_i, \tag{2.47}$$

where we have defined

$$P_i = |e_i\rangle\langle e_i| = \text{projection operator onto the } i\text{-th state.} \tag{2.48}$$

As we have seen earlier, for an arbitrary vector $|V\rangle$, we can write the expansion

$$|V\rangle = \sum_{j=1}^{n} v_j|e_j\rangle.$$

It follows from this that

$$\begin{aligned}P_i|V\rangle &= \sum_{j=1}^{n} v_j P_i|e_j\rangle = \sum_{j=1}^{n} v_j|e_i\rangle\langle e_i|e_j\rangle \\ &= \sum_{j=1}^{n} v_j|e_i\rangle\delta_{ij} = v_i|e_i\rangle.\end{aligned} \tag{2.49}$$

Thus, acting on an arbitrary vector $|V\rangle$, the projection operator P_i projects out the i-th component of the vector, which is why it is called the projection operator.

Furthermore, note that

$$\begin{aligned}
P_i P_j &= |e_i\rangle\langle e_i|e_j\rangle\langle e_j| \\
&= |e_i\rangle\delta_{ij}\langle e_j| \\
&= \delta_{ij}|e_j\rangle\langle e_j| = \delta_{ij}P_j. \qquad (2.50)
\end{aligned}$$

Physically, what this means is that since P_j projects out the j-th component of a vector, action of P_i following P_j would be zero unless both i and j coincide and when $i = j, P_i$ acts like the identity operator. Symbolically, one can write

$$P^2 = P. \qquad (2.51)$$

Operators with such properties are called idempotent operators. ◄

2.5 Adjoint of an operator

If an operator Ω acting on a ket $|V\rangle$ gives a new ket $|V'\rangle$, then, the adjoint of Ω is defined to be that operator which transforms the bra $\langle V|$ to $\langle V'|$. Since, by definition,

$$\Omega|V\rangle = |V'\rangle \equiv |\Omega V\rangle, \qquad (2.52)$$

it follows, using (2.24), that

$$\langle\Omega V| = \langle V'| = \big(|V'\rangle\big)^\dagger = (\Omega|V\rangle)^\dagger = \langle V|\Omega^\dagger, \qquad (2.53)$$

where $\Omega^\dagger$ is known as the adjoint of Ω and its matrix elements are obtained to be

$$\begin{aligned}
\Omega^\dagger_{ij} &= \langle e_i|\Omega^\dagger|e_j\rangle = \langle\Omega e_i|e_j\rangle = \langle e_j|\Omega e_i\rangle^* \\
&= \langle e_j|\Omega|e_i\rangle^* = \Omega^*_{ji}. \qquad (2.54)
\end{aligned}$$

We recognize this to be the Hermitian conjugate of the matrix elements Ω_{ij} of the original operator Ω.

Exercise. Show that the adjoint of a product of operators is the product of the adjoint operators in the reversed order, namely,

$$(\Omega_1\Omega_2\ldots\Omega_n)^\dagger = \Omega^\dagger_n\Omega^\dagger_{n-1}\ldots\Omega^\dagger_1. \qquad (2.55)$$

Definition. *An operator is Hermitian if it is self-adjoint, i.e.,* $\Omega = \Omega^\dagger$.

Definition. *An operator is anti-Hermitian if* $\Omega = -\Omega^\dagger$.

Definition. *An operator is said to be unitary if* $\Omega\Omega^\dagger = \Omega^\dagger\Omega = \mathbb{1} =$ *identity operator.*

This implies that the adjoint of a unitary operator is the inverse of the operator.

Exercise. Show that a unitary operator U can be written as

$$U = e^{iH}, \tag{2.56}$$

where H is a Hermitian operator.

Theorem. *Unitary operators preserve the inner product between vectors they act on.*

Proof. Let U denote a unitary operator such that

$$|V'\rangle = U|V\rangle, \qquad |W'\rangle = U|W\rangle. \tag{2.57}$$

It follows from (2.24) that

$$\langle W'| = \langle W|U^\dagger. \tag{2.58}$$

Furthermore, since U is a unitary operator, it follows that

$$\langle W'|V'\rangle = \langle W|U^\dagger U|V\rangle = \langle W|\mathbb{1}|V\rangle = \langle W|V\rangle, \tag{2.59}$$

which proves that unitary operators preserve the inner product between two vectors. ∎

2.6 Eigenvectors and eigenvalues

In general, an operator acting on a particular vector takes it to a new vector

$$\Omega|V\rangle = |V'\rangle. \tag{2.60}$$

However, if the effect of an operator acting on a particular vector is to simply multiply it by a constant (scalar), *i.e.*,

$$\Omega|V\rangle = \omega|V\rangle, \qquad (\omega \text{ is a scalar}) \tag{2.61}$$

then, we say that $|V\rangle$ is an eigenvector of the operator Ω with the eigenvalue ω. Clearly, for linear operators, if $|V\rangle$ is an eigenvector, so is $\alpha|V\rangle$ where α is a scalar (since linear operators act only on vectors and not on scalars, as is evident from (2.34)) and this arbitrariness can be used to normalize an eigenvector. Note that we can write

(2.61) also as

$$
\begin{aligned}
& (\Omega - \omega\mathbb{1})|V\rangle = 0, \\
\text{or,} \quad & \langle e_i|(\Omega - \omega\mathbb{1})|V\rangle = 0, \\
\text{or,} \quad & \langle e_i|(\Omega - \omega\mathbb{1})\sum_{j=1}^{n} v_j|e_j\rangle = 0, \\
\text{or,} \quad & \sum_{j=1}^{n} v_j\langle e_i|(\Omega - \omega\mathbb{1})|e_j\rangle = 0, \\
\text{or,} \quad & \sum_{j=1}^{n} (\Omega_{ij} - \omega\delta_{ij})\, v_j = 0.
\end{aligned} \tag{2.62}
$$

This is a set of linear homogeneous equations (in the unknown variables v_i) known as the characteristic equation. A nontrivial solution, in this case, exists if the determinant of the coefficient matrix vanishes, *i.e.*,

$$\det(\Omega_{ij} - \omega\delta_{ij}) = 0. \tag{2.63}$$

Clearly, if we are working in an n dimensional vector space, this is an n-th order polynomial equation in ω and, therefore, would possess n solutions for ω, which will correspond to all the eigenvalues of the operator Ω. These roots need not all be distinct or real. However, once the eigenvalues are obtained, the eigenvectors can be derived from the characteristic equation (2.62) in a simple manner.

► **Example (Non-degenerate system).** In $\mathbf{V}^3$, let us consider an operator Ω which has the matrix representation,

$$\Omega = \begin{pmatrix} 1 & 0 & 0 \\ 0 & 0 & -1 \\ 0 & 1 & 0 \end{pmatrix}. \tag{2.64}$$

The characteristic equation,

$$\sum_{j=1}^{3} (\Omega_{ij} - \omega\delta_{ij})\, v_j = 0, \tag{2.65}$$

will have a nontrivial solution provided the determinant of the coefficient matrix vanishes (see (2.63)),

$$\det\left(\Omega_{ij} - \omega\delta_{ij}\right) = 0,$$

$$\text{or,}\quad \det\begin{pmatrix} 1-\omega & 0 & 0 \\ 0 & -\omega & -1 \\ 0 & 1 & -\omega \end{pmatrix} = 0,$$

$$\text{or,}\quad (1-\omega)(\omega^2+1) = 0,$$

$$\text{or,}\quad (1-\omega)(\omega+i)(\omega-i) = 0. \tag{2.66}$$

This determines that $\omega = 1, i, -i$ are the three distinct but complex eigenvalues of Ω.

For $\omega = 1$, the linear equations (2.65) become

$$\sum_{j=1}^{3}\left(\Omega_{ij} - \omega\delta_{ij}\right) v_j = 0, \quad \text{or,} \quad \sum_{j=1}^{3}\left(\Omega_{ij} - \delta_{ij}\right) v_j = 0, \tag{2.67}$$

and explicitly lead to the three equations

$$\left.\begin{aligned} 0 &= 0, \\ -v_2 - v_3 &= 0, \\ v_2 - v_3 &= 0, \end{aligned}\right\} \quad \Rightarrow v_2 = v_3 = 0, \quad v_1 \text{ is arbitrary.} \tag{2.68}$$

Thus, the eigenvector corresponding to the eigenvalue $\omega = 1$ has the form

$$\begin{pmatrix} v_1 \\ 0 \\ 0 \end{pmatrix}. \tag{2.69}$$

We can make use of the arbitrariness of v_1 in (2.69) to define a normalized eigenvector

$$|\omega = 1\rangle = \begin{pmatrix} 1 \\ 0 \\ 0 \end{pmatrix}, \tag{2.70}$$

such that

$$\langle\omega = 1|\omega = 1\rangle = 1. \tag{2.71}$$

◀

Exercise. Similarly, show that the normalized eigenvectors for the eigenvalues $\omega = \pm i$, in the above example, are

$$|\omega = i\rangle = \frac{1}{\sqrt{2}}\begin{pmatrix} 0 \\ i \\ 1 \end{pmatrix}, \qquad |\omega = -i\rangle = \frac{1}{\sqrt{2}}\begin{pmatrix} 0 \\ -i \\ 1 \end{pmatrix}, \tag{2.72}$$

so that together the three eigenvectors define an orthonormal basis.

► **Example (Degenerate system).** In the previous example, all the eigenvalues of Ω were distinct, which is an example of a non-degenerate operator. However, when two or more eigenvalues of an operator coincide we talk of a degenerate system. Let us consider an operator Ω in $\mathbf{V}^3$, which has the matrix representation

$$\Omega = \begin{pmatrix} 1 & 0 & 1 \\ 0 & 2 & 0 \\ 1 & 0 & 1 \end{pmatrix}. \tag{2.73}$$

In this case, for a nontrivial solution of the characteristic equation,

$$\sum_j (\Omega_{ij} - \omega\delta_{ij})\, v_j = 0, \tag{2.74}$$

the vanishing of the determinant of the coefficient matrix gives

$$\begin{aligned}
& \det(\Omega_{ij} - \omega\delta_{ij}) = 0, \\
\text{or,}\quad & \det\begin{pmatrix} 1-\omega & 0 & 1 \\ 0 & 2-\omega & 0 \\ 1 & 0 & 1-\omega \end{pmatrix} = 0, \\
\text{or,}\quad & (1-\omega)\left((2-\omega)(1-\omega)\right) + 1\left(-(2-\omega)\right) = 0, \\
\text{or,}\quad & (2-\omega)((1-\omega)^2 - 1) = 0, \\
\text{or,}\quad & (2-\omega)(2-\omega)(-\omega) = 0, \\
\text{or,}\quad & \omega = 0, 2, 2.
\end{aligned} \tag{2.75}$$

Thus, in this case, we see that all the eigenvalues are real, although two of them are degenerate.

For $\omega = 0$, the characteristic equation, (2.74), becomes

$$\sum_j \Omega_{ij} v_j = 0, \tag{2.76}$$

and leads to

$$\left.\begin{aligned} v_1 + v_3 &= 0, \\ 2v_2 &= 0, \\ v_1 + v_3 &= 0, \end{aligned}\right\} \quad \Rightarrow v_2 = 0, \quad v_3 = -v_1, \tag{2.77}$$

so that we can write the normalized eigenvector as

$$|\omega = 0\rangle = \frac{1}{\sqrt{2}} \begin{pmatrix} 1 \\ 0 \\ -1 \end{pmatrix}. \tag{2.78}$$

For $\omega = 2$, the characteristic equation, (2.74), takes the form

$$\sum_j (\Omega_{ij} - 2\delta_{ij})\, v_j = 0, \tag{2.79}$$

and explicitly leads to

$$\left.\begin{aligned} -v_1 + v_3 &= 0, \\ 0 &= 0, \\ v_1 - v_3 &= 0, \end{aligned}\right\} \quad \Rightarrow v_3 = v_1, \quad v_2 \text{ arbitrary}, \tag{2.80}$$

so that any vector of the form

$$\begin{pmatrix} v_1 \\ v_2 \\ v_1 \end{pmatrix}, \tag{2.81}$$

is an eigenvector of Ω with the eigenvalue $\omega = 2$. However, whenever possible, we would like the eigenvectors of an operator to form an orthonormal basis. With this in mind, we can choose the arbitrary constants in (2.81) such that the eigenvectors of Ω are not only normalized, but are orthogonal to one another as well. Thus, for example, for the present case, we can choose

$$|\omega = 0\rangle = \frac{1}{\sqrt{2}} \begin{pmatrix} 1 \\ 0 \\ -1 \end{pmatrix}, \tag{2.82a}$$

$$|\omega = 2\rangle_1 = \frac{1}{\sqrt{2}} \begin{pmatrix} 1 \\ 0 \\ 1 \end{pmatrix}, \tag{2.82b}$$

$$|\omega = 2\rangle_2 = \begin{pmatrix} 0 \\ 1 \\ 0 \end{pmatrix}, \tag{2.82c}$$

which would provide an orthonormal basis. However, this is not necessarily the unique choice. In fact, we could have chosen, as eigenvectors (2.81)

$$|\omega = 2\rangle_1 = \frac{1}{\sqrt{3}} \begin{pmatrix} 1 \\ 1 \\ 1 \end{pmatrix}, \quad |\omega = 2\rangle_2 = \frac{1}{\sqrt{6}} \begin{pmatrix} 1 \\ -2 \\ 1 \end{pmatrix}, \tag{2.83}$$

which would also provide an orthonormal basis. In fact, a general normalized eigenvector corresponding to $\omega = 2$ has the form

$$\frac{1}{\sqrt{2|\frac{v_1}{v_2}|^2 + 1}} \begin{pmatrix} \frac{v_1}{v_2} \\ 1 \\ \frac{v_1}{v_2} \end{pmatrix}, \tag{2.84}$$

and all such vectors will be orthogonal to (2.78). Thus, we see that there would be an infinite set of possible eigenvectors corresponding to different values of $\frac{v_1}{v_2}$ when degeneracy of eigenvalues occurs and we can no longer label the eigenvectors uniquely by the eigenvalues alone. ◀

Theorem. *A Hermitian operator has real eigenvalues.*

Proof. Let Ω represent a Hermitian operator. Then, by definition,

$$\Omega = \Omega^\dagger. \tag{2.85}$$

If $|\omega\rangle$ represents an eigenvector of Ω with eigenvalue ω, then, it follows that

$$\Omega|\omega\rangle = \omega|\omega\rangle, \quad \text{or,} \quad \langle\omega|\Omega|\omega\rangle = \omega\langle\omega|\omega\rangle. \tag{2.86}$$

Taking the Hermitian conjugate of the first equation in (2.86), we obtain,

$$\langle\omega|\Omega^\dagger = \langle\omega|\Omega = \omega^*\langle\omega|, \quad \text{or,} \quad \langle\omega|\Omega|\omega\rangle = \omega^*\langle\omega|\omega\rangle. \tag{2.87}$$

Taking the difference of (2.86) and (2.87), we obtain

$$(\omega - \omega^*)\langle\omega|\omega\rangle = 0. \tag{2.88}$$

Since the norm of a vector is positive semi-definite (see (2.13) or (2.17)),

$$\langle\omega|\omega\rangle \geq 0, \quad \Rightarrow \omega = \omega^*. \tag{2.89}$$

That is, all the eigenvalues of a Hermitian operator are real. Note, however, that the converse is not necessarily true, namely, operators with all real eigenvalues are not necessarily Hermitian. ■

► **Example.** Let us consider the operator Ω in $\mathbf{V}^2$, with the matrix representation,

$$\Omega = \begin{pmatrix} 1 & a \\ 0 & 1 \end{pmatrix}. \tag{2.90}$$

The eigenvalues of this operator are determined to be

$$\det\begin{pmatrix} 1-\omega & a \\ 0 & 1-\omega \end{pmatrix} = (1-\omega)^2 = 0, \quad \Rightarrow \omega = 1, 1. \tag{2.91}$$

All the eigenvalues of the operator Ω are real (although degenerate) in this case. However, as is obvious from (2.90), the operator is not Hermitian. ◄

Theorem. *Eigenvectors of a Hermitian operator with distinct eigenvalues are orthogonal.*

Proof. Let Ω represent a Hermitian operator and let $|\omega_1\rangle, |\omega_2\rangle$ represent two of its eigenstates with distinct eigenvalues ω_1, ω_2 respectively (namely, $\omega_1 \neq \omega_2$). Then, we have

$$\Omega|\omega_1\rangle = \omega_1|\omega_1\rangle, \quad \text{or,} \quad \langle\omega_2|\Omega|\omega_1\rangle = \omega_1\langle\omega_2|\omega_1\rangle. \tag{2.92}$$

Similarly, taking the adjoint of the second eigenvalue equation,

$$\Omega|\omega_2\rangle = \omega_2|\omega_2\rangle, \tag{2.93}$$

we have

$$\langle\omega_2|\Omega = \omega_2\langle\omega_2|, \quad \text{or,} \quad \langle\omega_2|\Omega|\omega_1\rangle = \omega_2\langle\omega_2|\omega_1\rangle, \tag{2.94}$$

where we have used the fact that Ω is Hermitian and that Hermitian operators have real eigenvalues. Now, taking the difference of the two relations (2.92) and (2.94), we obtain

$$(\omega_1 - \omega_2)\langle\omega_2|\omega_1\rangle = 0. \tag{2.95}$$

Since $\omega_1 \neq \omega_2$ by assumption, this implies that $\langle\omega_2|\omega_1\rangle = 0$, namely, the two eigenvectors are orthogonal. Clearly, if the vectors are degenerate, they don't automatically have to be orthogonal, but, as we have already seen, we can always choose them to be orthogonal in such a case. It follows, therefore, that the eigenvectors of a Hermitian operator can be chosen to provide an orthonormal basis. ■

Exercise. All eigenvalues of a unitary operator have unit norm. All eigenvectors corresponding to distinct eigenvalues of a unitary operator are orthogonal to one another.

Theorem. *The operator which transforms an orthonormal set of basis vectors into another is unitary.*

Proof. Let $|e_i\rangle$ be a set of orthonormal basis vectors and let U be the operator which takes it to another set of orthonormal basis vectors, denoted by $|\omega_i\rangle$. Therefore, we have

$$|\omega_i\rangle = U|e_i\rangle, \qquad \langle\omega_j| = \langle e_j|U^\dagger, \quad i.j = 1, 2, \cdots, n. \tag{2.96}$$

It follows now that

$$\langle\omega_j|\omega_i\rangle = \langle e_j|U^\dagger U|e_i\rangle, \quad \text{or,} \quad \delta_{ij} = \langle e_j|U^\dagger U|e_i\rangle. \tag{2.97}$$

We know that $|e_i\rangle$ represents an orthonormal basis so that

$$\langle e_j|e_i\rangle = \delta_{ij}. \tag{2.98}$$

It follows, therefore, that $U^\dagger U = \mathbb{1}$ and U is unitary. ■

Theorem. *If Ω is a Hermitian matrix, then there exists a unitary matrix U such that $U^\dagger\Omega U$ is diagonal.*

Proof. Let U be the matrix which changes the standard, orthonormal set of basis vectors in (2.26), $|e_i\rangle$, to the orthonormal eigenbasis $|\omega_i\rangle$ of Ω. Therefore, we have

$$|\omega_i\rangle = U|e_i\rangle, \quad i = 1, 2, \cdots, n, \tag{2.99}$$

where

$$\Omega|\omega_i\rangle = \omega_i|\omega_i\rangle. \tag{2.100}$$

Clearly, U is unitary since it takes one orthonormal basis into another. Now

$$\langle\omega_j|\Omega|\omega_i\rangle = \omega_i\langle\omega_j|\omega_i\rangle = \omega_i\delta_{ij}. \tag{2.101}$$

On the other hand, using (2.99), we also have

$$\langle\omega_j|\Omega|\omega_i\rangle = \langle e_j|U^\dagger\Omega U|e_i\rangle. \tag{2.102}$$

Comparing (2.101) and (2.102), we conclude that

$$\langle e_j|U^\dagger\Omega U|e_i\rangle = \omega_i\delta_{ij}. \tag{2.103}$$

This shows that $U^\dagger\Omega U$ is diagonal with the diagonal elements given by the eigenvalues of Ω. We say that U diagonalizes Ω. ∎

Theorem. *If Ω and Λ are two commuting Hermitian matrices (operators), they can be simultaneously diagonalized.*

Proof. Let $|\omega_i\rangle$ represent the complete set of eigenstates of Ω corresponding to the eigenvalues ω_i with $i = 1, 2, \cdots, n$ so that

$$\Omega|\omega_i\rangle = \omega_i|\omega_i\rangle. \tag{2.104}$$

Since Ω and Λ commute,

$$[\Omega, \Lambda] = \Omega\Lambda - \Lambda\Omega = 0, \tag{2.105}$$

and it follows that

$$\begin{aligned}
&(\Omega\Lambda - \Lambda\Omega)|\omega_i\rangle = 0,\\
\text{or, } &\Omega\left(\Lambda|\omega_i\rangle\right) = \Lambda\left(\Omega|\omega_i\rangle\right) = \omega_i\left(\Lambda|\omega_i\rangle\right), \quad i = 1, 2, \cdots, n.
\end{aligned} \tag{2.106}$$

In other words, $\Lambda|\omega_i\rangle$ is also an eigenvector of Ω with the eigenvalue ω_i. This is possible only if

$$\Lambda|\omega_i\rangle = \lambda_i|\omega_i\rangle, \quad i = 1, 2, \cdots, n. \tag{2.107}$$

Thus, all the eigenstates of Ω are also eigenstates of Λ and the same unitary matrix which diagonalizes Ω would also diagonalize Λ. ∎

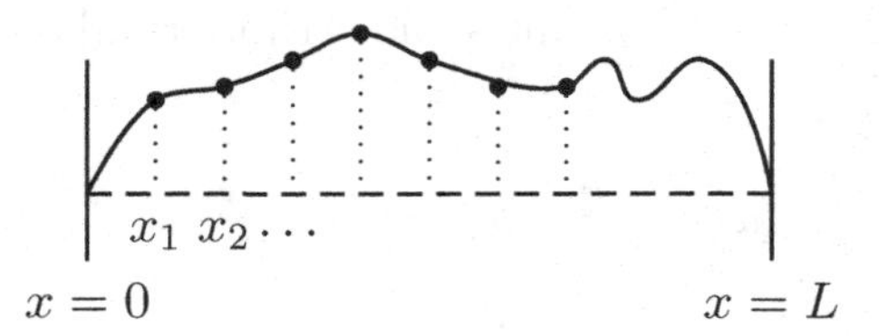

Figure 2.2: Displacement field of a string fixed at both ends $x = 0$ and $x = L$.

2.7 Infinite dimensional vector spaces

So far, we have talked about finite dimensional vector spaces. Let us now introduce the concept of an infinite dimensional vector space (which is important from the point of view of quantum mechanics) through the following example. Imagine a string fixed at two points $x = 0$ and $x = L$ as shown in Fig. 2.2. If we are talking about the displacement of the string from its equilibrium position, we can do so by dividing the interval into $n + 1$ equal parts and by describing the displacements at the n discrete (intermediate) points. Let us denote them by $f_n(x_i)$. Of course, this will not represent the true displacement $f(x)$ of the string, but as n is made larger and larger, it would come closer to the true description. (It is worth remarking here that any interval contains a non-countably infinite set of points. However, the only way we know how to do any practical calculation, such as integration etc, is by dividing the interval into subintervals of smaller and smaller lengths. This, in turn, treats the set as a countably infinite set of points, but works. Furthermore, the displacements of a string define a continuous function – they cannot be completely arbitrary at different points. This reduces the non-countably infinite set of displacements to a countably infinite set.)

We can think of the ordered n-tuple $(f_n(x_1), f_n(x_2), \ldots, f_n(x_n))$ as describing an n dimensional vector denoted by

$$|f_n\rangle = \begin{pmatrix} f_n(x_1) \\ f_n(x_2) \\ \vdots \\ f_n(x_n) \end{pmatrix}. \tag{2.108}$$

In terms of the basis vectors

$$|x_i\rangle = \begin{pmatrix} 0 \\ 0 \\ \vdots \\ 1 \\ 0 \\ 0 \\ 0 \end{pmatrix} \leftarrow i\text{-th place}, \tag{2.109}$$

we can write

$$|f_n\rangle = \sum_{i=1}^{n} f_n(x_i)|x_i\rangle, \tag{2.110}$$

where, as we have seen in (2.31)

$$f_n(x_i) = \langle x_i|f_n\rangle. \tag{2.111}$$

The basis vectors $|x_i\rangle$ obey the orthonormality and the completeness relations,

$$\langle x_i|x_j\rangle = \delta_{ij}, \quad \sum_i |x_i\rangle\langle x_i| = \mathbb{1}. \tag{2.112}$$

We can imagine dividing the interval into infinitesimal parts and in the limit of vanishing intervals, the position becomes a continuous variable and the displacements $f_\infty(x)$ would correspond to the true displacement of the string. This is now an infinite dimensional vector space. In this way, one can go from finite dimensions to infinite dimensions by letting $n \to \infty$. But, for this, certain modifications are necessary in some of the formulae which we discuss next.

2.8 Dirac delta function

Let us note that the inner product of two vectors in the n dimensional space of the form (2.108) is given by

$$\begin{aligned}\langle f_n|g_n\rangle &= \sum_{i,j=1}^{n} \langle x_j|f_n^*(x_j)g_n(x_i)|x_i\rangle \\ &= \sum_{i,j=1}^{n} f_n^*(x_j)g_n(x_i)\delta_{ij} \\ &= \sum_{i=1}^{n} f_n^*(x_i)g_n(x_i).\end{aligned} \tag{2.113}$$

In particular, the norm of a vector has the form

$$\langle f_n|f_n\rangle = \sum_{i=1}^{n} f_n^*(x_i)f_n(x_i) = \sum_{i=1}^{n} |f_n(x_i)|^2. \tag{2.114}$$

Clearly, this diverges as $n \to \infty$. One, therefore, needs a redefinition of the inner product such that a finite limit is obtained in (2.114) as $n \to \infty$. This is done by writing

$$\begin{aligned}\langle f_n|g_n\rangle &= \sum_{i=1}^{n} f_n^*(x_i)g_n(x_i)\frac{L}{n+1}\\ &\xrightarrow{n\to\infty} \int_0^L \mathrm{d}x\ f_\infty^*(x)g_\infty(x) = \int_0^L \mathrm{d}x\, f^*(x)g(x),\end{aligned} \tag{2.115}$$

where $\frac{L}{n+1}$ represents the length of each interval, which becomes smaller and smaller as n becomes larger.

Thus, for vectors defined within an interval $a \leq x \leq b$, the inner product takes the form

$$\begin{aligned}\lim_{n\to\infty} \langle f_n|g_n\rangle &\longrightarrow \lim_{n\to\infty} \int_a^b \mathrm{d}x\ f_n^*(x)g_n(x)\\ &= \int_a^b \mathrm{d}x\ f^*(x)g(x).\end{aligned} \tag{2.116}$$

The completeness relation, of course, still holds as in (2.112) (with the sum replaced by an integral)

$$\int_a^b \mathrm{d}x\, |x\rangle\langle x| = \mathbb{1}, \tag{2.117}$$

where $\mathbb{1}$ is the infinite dimensional identity matrix. Multiplying (2.117) on the left by $\langle x'|$ and by $|f\rangle$ on the right, we have

$$\int_a^b \mathrm{d}x\, \langle x'|x\rangle\langle x|f\rangle = \langle x'|f\rangle,$$

$$\text{or,}\quad \int_a^b \mathrm{d}x\, \langle x'|x\rangle f(x) = f(x'). \tag{2.118}$$

From the orthogonality relation of the basis vectors we know that

$$\langle x'|x\rangle = 0, \quad \text{if} \quad x \neq x'. \tag{2.119}$$

Therefore, we can limit the range of integration in (2.118) to an infinitesimal interval around x' to write

$$\lim_{\epsilon\to 0} \int_{x'-\epsilon}^{x'+\epsilon} \mathrm{d}x\, \langle x'|x\rangle f(x) = f(x'). \tag{2.120}$$

If $\langle x'|x\rangle$ is finite at $x = x'$ then the left hand side would vanish, since the range of integration is infinitesimally small. The only way this relation would make sense is if the inner product diverges at $x = x'$. Let us denote

$$\langle x'|x\rangle = \delta(x', x). \tag{2.121}$$

Thus, $\delta(x', x) = 0$ if $x \neq x'$ and it diverges when $x = x'$, but in such a way that the integral of $\delta(x', x)$ is unity, namely, since

$$\int \mathrm{d}x\, \delta(x', x) f(x) = f(x'), \tag{2.122}$$

which follows from (2.118), choosing $f(x) = 1$, we obtain

$$\int \mathrm{d}x\, \delta(x', x) = 1. \tag{2.123}$$

Furthermore, it only depends on the difference $x - x'$. The inner product in (2.121) is known as the Dirac delta function and is used to normalize continuous basis vectors as

$$\langle x|x'\rangle = \delta(x - x'). \tag{2.124}$$

2.9 Properties of the Dirac delta function

The Dirac delta function satisfies several interesting properties.

1. As we have already seen in (2.122), the defining relation gives

$$\int \mathrm{d}x\, \delta(x - x') f(x) = f(x'). \tag{2.125}$$

Namely, integrating with a delta function simply picks out the first term in the Taylor expansion of a function (around the point where the argument of the delta function vanishes).

2. It is an even function, which is easily seen as follows.

$$\begin{aligned}\delta(x'-x) &= \langle x'|x\rangle = \langle x|x'\rangle^* \\ &= \left(\delta(x-x')\right)^* = \delta(x-x'),\end{aligned} \tag{2.126}$$

where we have used the fact that the delta function is a real function.

3. Upon integration, the derivative of a Dirac delta function multiplied with a well behaved function leads to

$$\int \mathrm{d}x\, \delta'(x-x')f(x) = -f'(x'), \tag{2.127}$$

where prime denotes a derivative.

This can be seen as follows. Let us consider the integral in (2.127), which can be written as

$$\begin{aligned}&\int_a^b \mathrm{d}x \left[\frac{\mathrm{d}}{\mathrm{d}x}\,\delta(x-x')\right] f(x) \\ &= \int_a^b \mathrm{d}x \left[\frac{\mathrm{d}}{\mathrm{d}x}\left(\delta(x-x')f(x)\right) - \delta(x-x')\,\frac{\mathrm{d}f(x)}{\mathrm{d}x}\right] \\ &= \delta(x-x')f(x)\Big|_a^b - \int_a^b \mathrm{d}x\,\delta(x-x')\,\frac{\mathrm{d}f(x)}{\mathrm{d}x} \\ &= -\left.\frac{\mathrm{d}f(x)}{\mathrm{d}x}\right|_{x=x'} = -f'(x').\end{aligned} \tag{2.128}$$

It is clear that, since the delta function is even, the derivative of the delta function is an odd function,

$$\frac{\mathrm{d}}{\mathrm{d}x}\,\delta(x-x') = -\frac{\mathrm{d}}{\mathrm{d}x'}\,\delta(x-x'). \tag{2.129}$$

In general, we have

$$\int \mathrm{d}x\, \delta^{(n)}(x-x')f(x) = (-1)^n f^{(n)}(x'), \tag{2.130}$$

where the superscript (n) represents the number of derivatives acting on the delta function as well as on f.

General properties. Let us list below some of the properties of the Dirac delta function.

$$\delta(x) = \delta(-x). \tag{2.131a}$$

$$\delta'(x) = -\delta'(-x). \tag{2.131b}$$

$$x^n\delta(x) = 0, \quad n \geq 1. \tag{2.131c}$$

$$x\delta'(x) = -\delta(x). \tag{2.131d}$$

$$\delta(ax) = \frac{1}{|a|}\,\delta(x). \tag{2.131e}$$

$$\delta(x^2 - a^2) = \frac{1}{2|a|}[\delta(x-a) + \delta(x+a)]. \tag{2.131f}$$

$$f(x)\delta(x-a) = f(a)\delta(x-a). \tag{2.131g}$$

$$\int \mathrm{d}x\,\delta(x-b)\delta(a-x) = \delta(a-b). \tag{2.131h}$$

2.10 Representations of the Dirac delta function

The Dirac delta function is not a regular function. Rather, it is a generalized function, which can be thought of as the limit of a sequence of functions. In what follows, we will describe some of its representations that are used frequently.

Theorem.

$$\frac{1}{2\pi}\int_{-\infty}^{\infty} \mathrm{d}k\, e^{ikx} = \delta(x). \tag{2.132}$$

Proof. Let us assume that ϵ is infinitesimal and note that

$$\int_{-\epsilon}^{\epsilon} \mathrm{d}x\, f(x) \times \frac{1}{2\pi}\int_{-\infty}^{\infty} \mathrm{d}k\, e^{ikx}$$

$$= \int_{-\epsilon}^{\epsilon} \mathrm{d}x\; f(x) \times \lim_{g\to\infty} \frac{1}{2\pi}\int_{-g}^{g} \mathrm{d}k\, e^{ikx}$$

$$= \lim_{g\to\infty} \frac{1}{2\pi}\int_{-\epsilon}^{\epsilon} \mathrm{d}x\; f(x) \times 2\,\frac{\sin gx}{x}$$

$$= \lim_{g\to\infty} \frac{1}{\pi} \int_{-\epsilon}^{\epsilon} \mathrm{d}x\ f(x)\ \frac{\sin gx}{x}$$

$$= \lim_{g\to\infty} \frac{1}{\pi}\ \mathrm{Im} \int_{-\epsilon}^{\epsilon} \mathrm{d}x\ f(x)\ \frac{e^{igx}}{x}. \tag{2.133}$$

Let us define

$$x' = gx, \text{ then } \begin{cases} x = -\epsilon, & \Rightarrow x' = -g\epsilon, \\ x = \epsilon, & \Rightarrow x' = g\epsilon. \end{cases}$$

With this, the integral on the right hand side of (2.133) becomes

$$\lim_{g\to\infty} \frac{1}{\pi}\ \mathrm{Im} \int_{-g\epsilon}^{g\epsilon} \mathrm{d}x'\ f\left(\frac{x'}{g}\right) \frac{e^{ix'}}{x'}. \tag{2.134}$$

In the limit $g \to \infty$, we can use the method of residues. There is a pole at $x' = 0$, which yields the value of the integral to be (principal value has to be used)

$$\mathrm{Im}\ \left[\frac{1}{\pi} \times i\pi\ f(0)\right] = f(0). \tag{2.135}$$

Substituting this back into (2.133), we obtain

$$\lim_{\epsilon\to 0} \int_{-\epsilon}^{\epsilon} \mathrm{d}x\ f(x) \times \frac{1}{2\pi} \int_{-\infty}^{\infty} \mathrm{d}k\, e^{ikx} = f(0), \tag{2.136}$$

which shows that we can identify

$$\frac{1}{2\pi} \int_{-\infty}^{\infty} \mathrm{d}k\, e^{ikx} = \delta(x).$$

■

It is also clear from the above analysis that

Theorem.

$$\lim_{g\to\infty} \frac{1}{\pi} \frac{\sin gx}{x} = \delta(x). \tag{2.137}$$

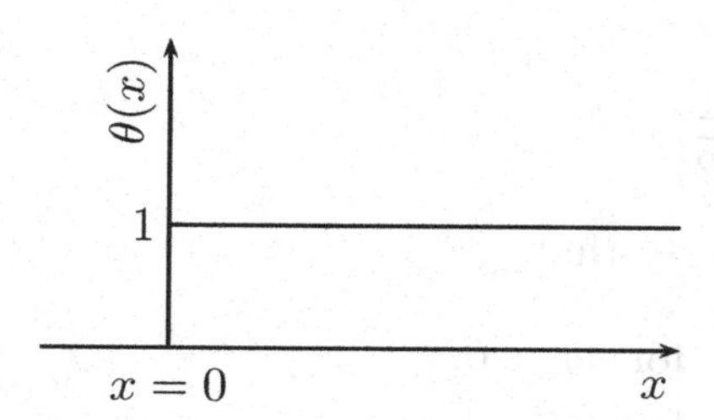

Let us consider the step function $\theta(x)$ defined as

$$\theta(x) = \begin{cases} 1 & \text{for } x > 0, \\ 0 & \text{for } x < 0. \end{cases} \tag{2.138}$$

Theorem. *The Dirac delta function is related to the step function as*

$$\frac{\mathrm{d}\theta(x)}{\mathrm{d}x} = \delta(x). \tag{2.139}$$

Proof. (All the test functions we use are assumed to be regular and vanish at infinity, namely, they satisfy $\lim_{|x|\to\infty} f(x) \to 0$.)

Let us consider the integral

$$\int_{-\epsilon}^{\epsilon} \mathrm{d}x \, \frac{\mathrm{d}\theta(x)}{\mathrm{d}x} \, f(x), \tag{2.140}$$

where ϵ is considered infinitesimal. Since the slope of the theta function vanishes away from the origin, we can easily extend the range of the integral and write

$$\begin{aligned} \int_{-\epsilon}^{\epsilon} \mathrm{d}x \, \frac{\mathrm{d}\theta(x)}{\mathrm{d}x} \, f(x) &= \int_{-\infty}^{\infty} \mathrm{d}x \, \frac{\mathrm{d}\theta(x)}{\mathrm{d}x} \, f(x) \\ &= -\int_{-\infty}^{\infty} \mathrm{d}x \, \theta(x) \, \frac{\mathrm{d}f(x)}{\mathrm{d}x} \\ &= -\int_{0}^{\infty} \mathrm{d}x \, \frac{\mathrm{d}f(x)}{\mathrm{d}x} = f(0). \end{aligned} \tag{2.141}$$

Sometimes one defines a step function slightly differently from (2.138) as

$$\epsilon(x) = \theta(x) - \frac{1}{2}, \tag{2.142}$$

which has the property that

$$\epsilon(x) = \begin{cases} \frac{1}{2}, & \text{for} \quad x > 0, \\ -\frac{1}{2}, & \text{for} \quad x < 0. \end{cases} \tag{2.143}$$

Such a step function is an odd function and is called the alternating step function. The delta function can also be defined as the derivative of the alternating step function, namely,

$$\frac{\mathrm{d}\epsilon(x)}{\mathrm{d}x} = \frac{\mathrm{d}\theta(x)}{\mathrm{d}x} = \delta(x). \tag{2.144}$$

■

Theorem. *The delta function can also be represented in terms of a Gaussian as*

$$\lim_{\alpha\to\infty} \sqrt{\frac{\alpha}{\pi}}\, e^{-\alpha x^2} = \delta(x). \tag{2.145}$$

Proof. First, let us note that,

$$\lim_{\alpha\to\infty} \int \mathrm{d}x\, \sqrt{\frac{\alpha}{\pi}}\, e^{-\alpha x^2} = \lim_{\alpha\to\infty} \sqrt{\frac{\alpha}{\pi}} \times \sqrt{\frac{\pi}{\alpha}} = 1. \tag{2.146}$$

Furthermore, using (2.146), let us note that

$$\begin{aligned}
&\int_{-\infty}^{\infty} \mathrm{d}x\, \sqrt{\frac{\alpha}{\pi}}\, e^{-\alpha x^2} f(x) - f(0) \\
&= \int_{-\infty}^{\infty} \mathrm{d}x\, \sqrt{\frac{\alpha}{\pi}}\, e^{-\alpha x^2} \left(f(x) - f(0)\right) \\
&\leq \int_{-\infty}^{\infty} \mathrm{d}x\, \sqrt{\frac{\alpha}{\pi}}\, e^{-\alpha x^2} \max\left\{\frac{\mathrm{d}f}{\mathrm{d}x}\right\} |x| \\
&= \max\left\{\frac{\mathrm{d}f}{\mathrm{d}x}\right\} \int_{-\infty}^{\infty} \mathrm{d}x\, \sqrt{\frac{\alpha}{\pi}}\, |x| e^{-\alpha x^2}
\end{aligned}$$

$$= \max\left\{\frac{\mathrm{d}f}{\mathrm{d}x}\right\}\sqrt{\frac{\alpha}{\pi}} \times 2\int_0^\infty \mathrm{d}x\, xe^{-\alpha x^2}$$

$$= \max\left\{\frac{\mathrm{d}f}{\mathrm{d}x}\right\}\sqrt{\frac{\alpha}{\pi}} \times \int_0^\infty \frac{\mathrm{d}z}{\alpha}\, e^{-z},$$

$$= \max\left\{\frac{\mathrm{d}f}{\mathrm{d}x}\right\}\frac{1}{\sqrt{\pi\alpha}}. \tag{2.147}$$

where $z = \alpha x^2$, $\mathrm{d}z = 2\alpha x \mathrm{d}x$. In the limit $\alpha \to \infty$, the right hand side of (2.147) vanishes leading to

$$\lim_{\alpha\to\infty}\int_{-\infty}^{\infty} \mathrm{d}x\, \sqrt{\frac{\alpha}{\pi}}\, e^{-\alpha x^2} f(x) - f(0)$$

$$= \lim_{\alpha\to\infty}\ \max\left\{\frac{\mathrm{d}f}{\mathrm{d}x}\right\}\frac{1}{\sqrt{\pi\alpha}} = 0. \tag{2.148}$$

Therefore, we obtain

$$\lim_{\alpha\to\infty}\int_{-\infty}^{\infty} \mathrm{d}x\, \sqrt{\frac{\alpha}{\pi}}\, e^{-\alpha x^2} f(x) = f(0), \tag{2.149}$$

so that we can identify

$$\lim_{\alpha\to\infty}\sqrt{\frac{\alpha}{\pi}} e^{-\alpha x^2} = \delta(x).$$

∎

Exercise. Show that

$$\int_{-\infty}^{\infty} \mathrm{d}x\, e^{-\alpha x^2} = \sqrt{\frac{\pi}{\alpha}}.$$

Theorem. *Another useful representation of the Dirac delta function is given by*

$$\lim_{\epsilon\to 0}\frac{1}{\pi}\frac{\epsilon}{x^2+\epsilon^2} = \delta(x). \tag{2.150}$$

Proof. First, let us note that we have

$$\lim_{\epsilon\to 0}\int_{-\infty}^{\infty} \mathrm{d}x\, \frac{1}{\pi}\frac{\epsilon}{x^2+\epsilon^2} = \lim_{\epsilon\to 0}\frac{\epsilon}{\pi} \times 2\pi i \times \frac{1}{2i\epsilon} = 1. \tag{2.151}$$

Next, let us note that

$$\lim_{\epsilon\to 0}\int_{-a}^{a} \mathrm{d}x\ \frac{1}{\pi}\ \frac{\epsilon}{x^2+\epsilon^2}\ f(x)$$

$$= \lim_{\epsilon\to 0}\ \frac{1}{\pi}\int_{-a}^{a}\frac{dx}{\epsilon}\ \frac{1}{\left(\frac{x^2}{\epsilon^2}+1\right)}\ f(x). \tag{2.152}$$

Let us redefine $\frac{x}{\epsilon}=x'$, which allows us to write the integral in (2.152) as

$$\lim_{\epsilon\to 0}\frac{1}{\pi}\int_{-\frac{a}{\epsilon}}^{\frac{a}{\epsilon}} \mathrm{d}x'\ \frac{1}{(x'^{\,2}+1)}\ f(x'\epsilon)$$

$$= \lim_{\epsilon\to 0}\ \frac{1}{\pi}\int_{-\infty}^{\infty} \mathrm{d}x'\ \frac{1}{x'^{\,2}+1}\ f(x'\epsilon). \tag{2.153}$$

The integral in (2.153) can now be evaluated using the method of residues. We recognize that the integrand has two poles at $x'=\pm i$, which yield the value of the integral to be

$$\lim_{\epsilon\to 0}\ \frac{1}{\pi}\times 2\pi i\times\frac{1}{2i}f(i\epsilon)=\lim_{\epsilon\to 0}\ f(i\epsilon)=f(0). \tag{2.154}$$

Substituting this back into (2.152), we can identify

$$\lim_{\epsilon\to 0}\ \frac{1}{\pi}\frac{\epsilon}{x^2+\epsilon^2}=\delta(x).$$

Let us also note that we can write

$$\frac{1}{\pi}\ \frac{\epsilon}{x^2+\epsilon^2}=\frac{1}{2\pi i}\left(\frac{1}{x-i\epsilon}-\frac{1}{x+i\epsilon}\right). \tag{2.155}$$

An alternate characterization is to note that

$$\lim_{\epsilon\to 0}\ \frac{1}{x\mp i\epsilon}=P\left(\frac{1}{x}\right)\pm i\pi\delta(x). \tag{2.156}$$

where P stands for the principal value. ■

2.11 Operators in infinite dimensions

We are now familiar with ket $|f\rangle$ and the basis vectors $|x\rangle$ in infinite dimensions. Let us next ask how linear operators act on this infinite dimensional space. As before, let us assume that an operator Ω takes a vector $|f\rangle$ to a new vector $|\tilde{f}\rangle$, namely,

$$\Omega|f\rangle = |\tilde{f}\rangle. \tag{2.157}$$

Since we can expand the vectors in the coordinate basis as

$$|f\rangle = \int \mathrm{d}x\ f(x)|x\rangle, \qquad |\tilde{f}\rangle = \int \mathrm{d}x\ \tilde{f}(x)|x\rangle, \tag{2.158}$$

we can also think of operators as taking functions $f(x)$ into $\tilde{f}(x)$

$$f(x) \xrightarrow{\Omega} \tilde{f}(x). \tag{2.159}$$

Let us denote by D the operator which takes $f(x)$ to $\frac{\mathrm{d}f(x)}{\mathrm{d}x}$, namely,

$$f(x) \xrightarrow{D} \tilde{f}(x) = \frac{\mathrm{d}f(x)}{\mathrm{d}x}. \tag{2.160}$$

Thus, we have

$$\begin{aligned}
&D|f\rangle = |\tilde{f}\rangle,\\
\text{or,}\quad &\langle x|D|f\rangle = \langle x|\tilde{f}\rangle = \tilde{f}(x) = \frac{\mathrm{d}f(x)}{\mathrm{d}x},\\
\text{or,}\quad &\int \mathrm{d}x'\langle x|D|x'\rangle\langle x'|f\rangle = \frac{\mathrm{d}f(x)}{\mathrm{d}x},\\
\text{or,}\quad &\int \mathrm{d}x'\langle x|D|x'\rangle f(x') = \frac{\mathrm{d}f(x)}{\mathrm{d}x}.
\end{aligned} \tag{2.161}$$

Recalling (2.127), we see that we can identify

$$\begin{aligned}
\langle x|D|x'\rangle &= D_{xx'} = \delta'(x-x')\\
&= \frac{\mathrm{d}}{\mathrm{d}x}\,\delta(x-x') = -\frac{\mathrm{d}}{\mathrm{d}x'}\delta(x-x').
\end{aligned} \tag{2.162}$$

This determines the representation of the operator in the $|x\rangle$ basis.

Let us next ask if D is Hermitian. In finite dimensional vector spaces, we know that D is Hermitian if $D = D^\dagger$. In the present case,

$$\begin{aligned}
(D^\dagger)_{xx'} &= D^*_{x'x} = \frac{\mathrm{d}}{\mathrm{d}x'}\delta(x'-x)\\
&= -\frac{\mathrm{d}}{\mathrm{d}x}\delta(x-x') = -D_{xx'}.
\end{aligned} \tag{2.163}$$

Thus, in fact, we see that the operator is naively anti-Hermitian. We can easily make it Hermitian by defining

$$K = -iD, \tag{2.164}$$

so that $K^\dagger = K$ and this would be naively Hermitian. But, we also know that for an operator to be Hermitian, it must satisfy

$$\begin{aligned}\langle g|K|f\rangle &= \langle g|Kf\rangle = \langle Kf|g\rangle^* \\ &= \langle f|K^\dagger|g\rangle^* = \langle f|K|g\rangle^*.\end{aligned} \tag{2.165}$$

Therefore, let us check whether this relation is satisfied as well. We see that the left hand side of (2.165) gives

$$\begin{aligned}\text{L.H.S.} &= \langle g|K|f\rangle \\ &= \int_a^b \mathrm{d}x \int_a^b \mathrm{d}x' \ \langle g|x\rangle\langle x|K|x'\rangle\langle x'|f\rangle \\ &= \int_a^b \mathrm{d}x \int_a^b \mathrm{d}x' \ g^*(x) \left(i\frac{\mathrm{d}}{\mathrm{d}x'}\delta(x-x')\right) f(x') \\ &= \int_a^b \mathrm{d}x \ g^*(x)(-i) \ \frac{\mathrm{d}f(x)}{\mathrm{d}x} \\ &= -i\int_a^b \mathrm{d}x \ g^*(x)\frac{\mathrm{d}f(x)}{\mathrm{d}x}.\end{aligned} \tag{2.166}$$

On the other hand, the right hand side of (2.165) gives

$$\begin{aligned}\text{R.H.S.} &= \langle f|K|g\rangle^* \\ &= \left[\int_a^b \mathrm{d}x \int_a^b \mathrm{d}x' \ \langle f|x\rangle\langle x|K|x'\rangle\langle x'|g\rangle\right]^* \\ &= \left[\int_a^b \mathrm{d}x \int_a^b \mathrm{d}x' \ f^*(x) \left(i\frac{\mathrm{d}}{\mathrm{d}x'} \ \delta(x-x')\right) g(x')\right]^* \\ &= \left[\int_a^b \mathrm{d}x \ f^*(x)(-i)\frac{\mathrm{d}g(x)}{\mathrm{d}x}\right]^*\end{aligned}$$

$$
\begin{aligned}
&= i\int_a^b \mathrm{d}x\,\frac{\mathrm{d}g^*(x)}{\mathrm{d}x}\;f(x)\\
&= i\int_a^b \mathrm{d}x\,\frac{\mathrm{d}}{\mathrm{d}x}\left(g^*(x)f(x)\right) - i\int_a^b \mathrm{d}x\; g^*(x)\frac{\mathrm{d}f(x)}{\mathrm{d}x}\\
&= ig^*(x)f(x)\Big|_a^b - i\int_a^b \mathrm{d}x\; g^*(x)\frac{\mathrm{d}f(x)}{\mathrm{d}x}.
\end{aligned}
\tag{2.167}
$$

Thus, comparing (2.166) and (2.167) we see that the operator K will satisfy (2.165), only if

$$
g^*(x)f(x)\Big|_a^b = 0. \tag{2.168}
$$

In this case, the operator K would be Hermitian. Thus, unlike in the finite dimensional case, in infinite dimensions, properties like Hermiticity depend on the space of functions on which the operators act. If the functions are like the displacements of a string which vanish at the end (the string is fixed at the ends), then, of course, (2.168) holds true. We can also think of periodic functions satisfying

$$
\begin{aligned}
&f(b) = f(a),\\
&g(b) = g(a),
\end{aligned}
\tag{2.169}
$$

for which (2.168) is also true and the operator K would be Hermitian. In quantum mechanics one works with functions defined on $-\infty \leq x \leq \infty$. Then, there are two kinds of functions that one deals with – those that vanish at infinity and others that are oscillatory. The first category, of course, does not create any problem with (2.168). But, for the second kind of functions, typically of the form e^{ikx}, it is not obvious whether

$$
e^{-ikx}\;e^{ik'x}\Big|_{-\infty}^{\infty} = 0. \tag{2.170}
$$

We note that we can write

$$
\begin{aligned}
e^{-i(k-k')x}\Big|_{-\infty}^{\infty} &= -i(k-k')\int_{-\infty}^{\infty} \mathrm{d}x\; e^{-i(k-k')x}\\
&= -2\pi i(k-k')\delta(k-k') = 0,
\end{aligned}
\tag{2.171}
$$

where we have used the definition of the delta function as well as the property of the delta function (see (2.131)) that

$$x\delta(x) = 0. \tag{2.172}$$

This shows that K is Hermitian in this space.

Let us now calculate the eigenvalues and the eigenfunctions of K. It would seem like a formidable task since K is an infinite dimensional matrix and, therefore, the characteristic equation would involve polynomials of infinite order. But, in practice it is not so bad. In fact, finding eigenvalues and eigenfunctions in infinite dimensions becomes equivalent to solving (partial) differential equations, which we can see in the following way. Let

$$\begin{aligned} & K|k\rangle = k|k\rangle, \\ \text{or,}\quad & \langle x|K|k\rangle = k\langle x|k\rangle, \\ \text{or,}\quad & \int \mathrm{d}x'\, \langle x|K|x'\rangle\langle x'|k\rangle = k\langle x|k\rangle. \end{aligned} \tag{2.173}$$

Defining

$$\langle x|k\rangle = \psi_k(x), \tag{2.174}$$

and using (2.162) (as well as the identification (2.164)), we obtain from (2.173)

$$\begin{aligned} & \int \mathrm{d}x' \left(i\frac{\mathrm{d}}{\mathrm{d}x'}\, \delta(x-x') \right) \psi_k(x') = k\psi_k(x), \\ \text{or,}\quad & -i\frac{\mathrm{d}\psi_k(x)}{\mathrm{d}x} = k\psi_k(x). \end{aligned} \tag{2.175}$$

The solution of (2.175) is clearly

$$\psi_k(x) = Ae^{ikx}. \tag{2.176}$$

Namely, any real number k is an eigenvalue of K with $\psi_k(x)$ defining the corresponding eigenfunction. Here, A is an arbitrary constant which we can choose to be $A = \frac{1}{\sqrt{2\pi}}$, yielding

$$\langle x|k\rangle = \psi_k(x) = \frac{1}{\sqrt{2\pi}}\, e^{ikx}, \tag{2.177}$$

so that the eigenvector $|k\rangle$ is normalized.

$$\begin{aligned}\langle k|k'\rangle &= \int \mathrm{d}x\ \langle k|x\rangle\langle x|k'\rangle \\ &= \int \mathrm{d}x\ \psi_k^*(x)\psi_{k'}(x) \\ &= \frac{1}{2\pi}\int \mathrm{d}x\ e^{-ikx}e^{ik'x} \\ &= \frac{1}{2\pi}\int \mathrm{d}x\ e^{-i(k-k')x} \\ &= \delta(k-k'). \end{aligned} \tag{2.178}$$

Let us note here that the eigenstates $|k\rangle$ define a complete basis, since K is a Hermitian operator.

Definition. *A Hilbert space is an infinite dimensional vector space such that every vector in this space can be normalized either to unity or to the Dirac delta function.*

2.12 Fourier transformation

Any vector $|f\rangle$ can be expanded in the $|x\rangle$ basis as well as in the $|k\rangle$ basis, which follows because both $|x\rangle$ and $|k\rangle$ define complete basis in the infinite dimensional space. Thus, we can write

$$|f\rangle = \int \mathrm{d}x\ f(x)|x\rangle, \qquad f(x) = \langle x|f\rangle. \tag{2.179}$$

Similarly, since $|k\rangle$ also defines a complete basis, we can write

$$|f\rangle = \int \mathrm{d}k\ g(k)|k\rangle, \tag{2.180}$$

where

$$\begin{aligned} g(k) &= \langle k|f\rangle \\ &= \int \mathrm{d}x\ \langle k|x\rangle\langle x|f\rangle \\ &= \int \mathrm{d}x\ \psi_k^*(x)f(x) = \frac{1}{\sqrt{2\pi}}\int \mathrm{d}x\ e^{-ikx}f(x). \end{aligned} \tag{2.181}$$

Similarly, we can show that

$$\begin{aligned} f(x) = \langle x|f\rangle &= \int \mathrm{d}k\ \langle x|k\rangle\langle k|f\rangle \\ &= \int \mathrm{d}k\ \psi_k(x)g(k) = \frac{1}{\sqrt{2\pi}}\int \mathrm{d}k\ e^{ikx}g(k). \end{aligned} \tag{2.182}$$

We realize that these are nothing other than Fourier transforms. This shows that the Fourier transformation takes us from one basis to another.

In the eigenbasis of K, the matrix elements of K are given by

$$\langle k'|K|k\rangle = k\langle k'|k\rangle = k\delta(k-k') = k'\delta(k-k'), \tag{2.183}$$

so that it is diagonal in this basis as we would expect. We can also ask what is the operator whose eigenfunctions form the basis $|x\rangle$. Let it be denoted by X. Then, by definition,

$$\begin{aligned} & X|x\rangle = x|x\rangle, \\ \text{or,}\quad & \langle x'|X|x\rangle = x\langle x'|x\rangle = x\delta(x-x') = x'\delta(x-x'). \end{aligned} \tag{2.184}$$

To find the action of this operator on an arbitrary vector, we note that

$$\begin{aligned} & X|f\rangle = |\tilde{f}\rangle, \\ \text{or,}\quad & \langle x|X|f\rangle = \langle x|\tilde{f}\rangle, \\ \text{or,}\quad & \int \mathrm{d}x'\ \langle x|X|x'\rangle\langle x'|f\rangle = \langle x|\tilde{f}\rangle, \\ \text{or,}\quad & \int \mathrm{d}x'\ x\delta(x-x')f(x') = \tilde{f}(x), \\ \text{or,}\quad & \tilde{f}(x) = xf(x). \end{aligned} \tag{2.185}$$

Thus, we see that the effect of X on a vector is to multiply its components in the basis $|x\rangle$ by x. We can ask what are the matrix elements of X in the $|k\rangle$ basis.

$$\begin{aligned} \langle k|X|k'\rangle &= \int \mathrm{d}x\mathrm{d}x'\ \langle k|x\rangle\langle x|X|x'\rangle\langle x'|k'\rangle \\ &= \int \mathrm{d}x\mathrm{d}x'\ \psi_k^*(x)x\delta(x-x')\psi_{k'}(x') \\ &= \frac{1}{2\pi}\int \mathrm{d}x\ xe^{-i(k-k')x} \end{aligned}$$

$$= \frac{1}{2\pi}\, i\frac{\mathrm{d}}{\mathrm{d}k}\int \mathrm{d}x\ e^{-i(k-k')x}$$

$$= i\frac{\mathrm{d}}{\mathrm{d}k}\ \delta(k-k'). \tag{2.186}$$

We see that, in the $|x\rangle$ basis, X acts as x and K as $-i\frac{\mathrm{d}}{\mathrm{d}x}$ when acting on functions, whereas in the $|k\rangle$ basis, K acts as k and X as $i\frac{\mathrm{d}}{\mathrm{d}k}$ on functions . Operators with such reciprocity are called conjugate operators . Clearly, conjugate operators do not commute, which can be seen as follows

$$\langle x|X|f\rangle = xf(x), \tag{2.187a}$$

$$\langle x|K|f\rangle = -i\frac{\mathrm{d}f(x)}{\mathrm{d}x}, \tag{2.187b}$$

$$\langle x|XK|f\rangle = -ix\frac{\mathrm{d}f(x)}{\mathrm{d}x}, \tag{2.187c}$$

$$\langle x|KX|f\rangle = -i\frac{\mathrm{d}}{\mathrm{d}x}(xf(x)). \tag{2.187d}$$

It follows now that

$$\langle x|(XK-KX)|f\rangle = -ix\frac{\mathrm{d}f(x)}{\mathrm{d}x} + if(x) + ix\frac{\mathrm{d}f(x)}{\mathrm{d}x}$$

$$= if(x) = i\langle x|f\rangle. \tag{2.188}$$

In other words, for any vector $|f\rangle$,

$$[X,K]|f\rangle = i|f\rangle, \quad \text{or,} \quad [X,K] = i\mathbb{1}. \tag{2.189}$$

Any two operators whose commutator is proportional to the identity operator are known as conjugate operators. As we will see, in quantum mechanics X corresponds to the position operator, while $P = \hbar K$ denotes the momentum operator.

► **Example.** As we have seen, Fourier transformation takes us from one basis to the conjugate basis,

$$g(k) = \frac{1}{\sqrt{2\pi}}\int \mathrm{d}x\ e^{-ikx}f(x). \tag{2.190}$$

Let us next consider a few examples of Fourier transformation. Let

$$f(x) = \delta(x), \tag{2.191}$$

then, from (2.190), we obtain

$$g(k) = \frac{1}{\sqrt{2\pi}}\int \mathrm{d}x\ e^{-ikx}\delta(x) = \frac{1}{\sqrt{2\pi}}. \tag{2.192}$$

Namely, the Fourier transformation of the Dirac delta function is a constant. For a Gaussian function,

$$f(x) = e^{-\frac{\alpha^2 x^2}{2}}, \tag{2.193}$$

the Fourier transformation leads to

$$\begin{aligned} g(k) &= \frac{1}{\sqrt{2\pi}} \int \mathrm{d}x \; e^{-ikx} \; e^{-\frac{\alpha^2 x^2}{2}} \\ &= \frac{1}{\sqrt{2\pi}} \int \mathrm{d}x \; e^{-\frac{\alpha^2 x^2}{2} - ikx + \frac{k^2}{2\alpha^2} - \frac{k^2}{2\alpha^2}} \\ &= \frac{1}{\sqrt{2\pi}} \int \mathrm{d}x \; e^{-\frac{1}{2}\left(\alpha x + \frac{ik}{\alpha}\right)^2 - \frac{k^2}{2\alpha^2}} \\ &= \frac{1}{\sqrt{2\pi}} \int \frac{\mathrm{d}x'}{\alpha} \; e^{-\frac{1}{2}x'^2} \; e^{-\frac{k^2}{2\alpha^2}} \\ &= \frac{1}{\sqrt{2\pi}} \times \frac{1}{\alpha} \sqrt{2\pi} \; e^{-\frac{k^2}{2\alpha^2}} = \frac{1}{\alpha} e^{-\frac{k^2}{2\alpha^2}}. \end{aligned} \tag{2.194}$$

Thus, we see that the Fourier transform of a Gaussian is a Gaussian, but with inverse width. ◀

Exercise. Show that

$$\frac{1}{\sqrt{2\pi}} \int_{-\infty}^{\infty} \mathrm{d}x \; e^{-\frac{1}{2}\left(\alpha x + \frac{ik}{\alpha}\right)^2} = \frac{1}{\alpha}. \tag{2.195}$$

2.13 Selected problems

1. If an operator Ω takes a vector $|\phi\rangle$ to another vector $|\tilde{\phi}\rangle$,

$$\Omega|\phi\rangle = |\tilde{\phi}\rangle,$$

and the vectors have the coordinate representations

$$|\phi\rangle = \int \mathrm{d}x \, \phi(x)|x\rangle, \quad |\tilde{\phi}\rangle = \int \mathrm{d}x \, \tilde{\phi}(x)|x\rangle,$$

one also says that

$$\phi(x) \xrightarrow{\Omega} \tilde{\phi}(x).$$

With this understanding, work out the following problems.

(*i*) Which of the following operators are linear?

$$
\begin{aligned}
&a) \quad \phi(x) \xrightarrow{\Omega} \phi(-x),\\
&b) \quad \phi(x) \xrightarrow{\Omega} \phi^2(x),\\
&c) \quad \phi(x) \xrightarrow{\Omega} \phi(x)+c, \qquad c \text{ is a constant},\\
&d) \quad \phi(x) \xrightarrow{\Omega} \phi(x+c), \qquad c \text{ is a constant},\\
&e) \quad \phi(x) \xrightarrow{\Omega} \phi(\frac{x}{2}),\\
&f) \quad \phi(x) \xrightarrow{\Omega} \int_{-\infty}^{\infty} \mathrm{d}x'\, K(x,x')\phi(x'), \quad \text{with}\\
&\qquad\qquad K(x,x') = K^*(x',x),\\
&g) \quad \phi(x) \xrightarrow{\Omega} \int_{-\infty}^{\infty} \mathrm{d}x'\, K(x,x')\phi(x'), \quad \text{with}\\
&\qquad\qquad K(x,x') = -K(x',x).
\end{aligned} \tag{2.196}
$$

(*ii*) Which of the following operators are Hermitian?

$$
\begin{aligned}
&a) \quad \phi(x) \xrightarrow{\Omega} \phi(x+c), \qquad c \text{ is a constant},\\
&b) \quad \phi(x) \xrightarrow{\Omega} \phi^*(x),\\
&c) \quad \phi(x) \xrightarrow{\Omega} \phi(-x),\\
&d) \quad \phi(x) \xrightarrow{\Omega} \int_{-\infty}^{\infty} \mathrm{d}x'\, K(x,x')\phi(x'), \quad \text{with}\\
&\qquad\qquad K(x,x') = -K(x',x) \text{ and real.}
\end{aligned} \tag{2.197}
$$

2. If A, B, C are Hermitian operators, determine which of the following combinations are Hermitian?

$$
\begin{aligned}
&a) \quad A+B,\\
&b) \quad \frac{1}{2i}[A,B] = \frac{1}{2i}(AB-BA),
\end{aligned}
$$

c) $(ABC - CBA)$,

d) $A^2 + B^2 + C^2$,

e) $A + iB$. (2.198)

3. Dtermine the constant B_α such that

$$\lim_{\alpha \to \infty} B_\alpha e^{-\alpha r} = \delta^3(\mathbf{r}). \tag{2.199}$$

Here r is the magnitude of the three dimensional vector $\mathbf{r}$ (namely, $r = |\mathbf{r}|$) and you should only check the normalization of the delta function.

4. Determine the value of the constant B such that, with

$$t_b(x) = \begin{cases} 0 & x^2 > b^2 \\ B|b - x| & x^2 < b^2 \end{cases}, \tag{2.200}$$

we have

$$\lim_{b \to 0} t_b(x) = \delta(x). \tag{2.201}$$

Once again, you are asked only to check the normalization.

5. a) Using the integral representation for $\frac{1}{|\mathbf{r}-\mathbf{r}'|}$, show that

$$\boldsymbol{\nabla}^2 \left(\frac{1}{|\mathbf{r} - \mathbf{r}'|} \right) = -4\pi \delta^3(\mathbf{r} - \mathbf{r}'). \tag{2.202}$$

b) Using Gauss' theorem, show that

$$\boldsymbol{\nabla}^2 \left(\frac{1}{r} \right) = -4\pi \delta^3(\mathbf{r}), \tag{2.203}$$

where r is the magnitude of $\mathbf{r}$.

6. If $g(k)$ represents the Fourier transform of $f(x)$ (in one dimension), determine the Fourier transforms of

$$a)\quad \frac{\mathrm{d}f(x)}{\mathrm{d}x}, \qquad b)\quad f(x+a), \quad c)\quad e^{i\mu x}f(x) \quad (\mu \text{ real}),$$
$$d)\quad f^*(x), \qquad e)\quad f(-x), \tag{2.204}$$

in terms of $g(k)$.

7. a) Calculate the Fourier transform of

$$f(x) = e^{-\mu|x|}, \qquad (\mu \text{ real, positive}). \tag{2.205}$$

Use this result to determine the Fourier transform of

$$\phi(x) = \frac{1}{\lambda^2 + x^2}, \tag{2.206}$$

where λ is a real and positive constant.

b) Calculate the Fourier transform of $\phi(x)$ in (2.206) by evaluating the integral using contour methods (residue theorem).

8. Determine the Fourier transform of

$$f(\mathbf{r}) = \frac{e^{-\mu r}}{r}, \tag{2.207}$$

where r is the magnitude of the three dimensional vector $\mathbf{r}$.

CHAPTER 3

Basics of quantum mechanics

In the next few lectures, we will introduce the basic concepts of quantum mechanics. However, let us first discuss the reasons for going beyond the classical description of physical systems, which we have discussed in the first chapter.

3.1 Inadequacies of classical mechanics

Classical mechanics works well when applied to macroscopic or large systems. However, around the turn of the twentieth century (1900-1920), it was observed that microscopic or small systems behaved very differently from the predictions of classical mechanics. We would, of course, discuss more quantitatively what we mean by microscopic systems. But, for the present, let us understand by a microscopic system, a system of atomic size or smaller and list below various difficulties that one runs into in applying the classical description to microscopic systems.

1. *Planetary model.* The planetary model of the atom, where electrons move in definite orbits around the nucleus, was in serious trouble. According to classical mechanics, a particle in such an orbit is being constantly accelerated. Furthermore, we also know that a classical charged particle, when accelerated, emits radiation. Therefore, an electron going around a nucleus would continuously emit radiation and become less and less energetic. This has the consequence that the radius of the orbit would constantly shrink in size, until the electron falls into the nucleus. Thus, according to classical mechanics, the planetary motion in atoms was unstable.

2. *Blackbody radiation.* The theoretical calculation of the blackbody radiation spectrum, which assumes that electromagnetic radiation is a wave and, therefore, can exchange energy in any

continuous amount, leads to a result which does not agree with the experimental measurement (curve). Planck, on the other hand, assumed that electromagnetic radiation of frequency ν can exchange energy only in units of $h\nu$, where the constant h is known as the Planck's constant,

$$h = 2\pi\hbar = 2\pi \times 1.054 \times 10^{-27} \text{ erg-sec}, \tag{3.1}$$

and his calculation led to a blackbody spectrum which agreed completely with the experimental measurement.

3. *Photo-electric effect.* Around the same time, it was also observed that it was possible to release electrons from a metal by irradiating the metal with electromagnetic waves or light. This was called the photo-electric effect. Furthermore, the interesting feature of these experiments was that it was not always possible to get electrons out of the metal. In fact, for any given metal, it was found that the light radiation which would free electrons had to have a frequency greater than a critical frequency, characteristic of the metal. With light of a lower frequency, one can make the radiation as intense as possible, but it would not lead to photo-electric effect (release of electrons). Einstein solved this puzzle and showed that this was consistent with Planck's hypothesis, namely, light with frequency ν can only exchange energy in the amount

$$E = h\nu. \tag{3.2}$$

For electrons to be released, therefore, we should have

$$E = h\nu = \text{BE} + \text{kinetic energy}, \tag{3.3}$$

where BE represents the binding energy for the metal under consideration. Writing BE $= h\nu_0$, therefore, we obtain from (3.3)

$$h(\nu - \nu_0) = \text{kinetic energy} \geq 0. \tag{3.4}$$

This implies that there cannot be any emission of electrons, unless $\nu \geq \nu_0$.

These two examples (blackbody radiation and photo-electric effect) clearly illustrate that although classically light is a wave, it can often behave like particles. This is further confirmed by the Compton effect.

4. *Compton effect.* If one considers the scattering of light by an electron (see Fig. 3.1), one finds that the experimental result can be explained only if a photon of frequency ν is considered to be a particle moving with energy E and momentum $p = |\mathbf{p}|$ given by

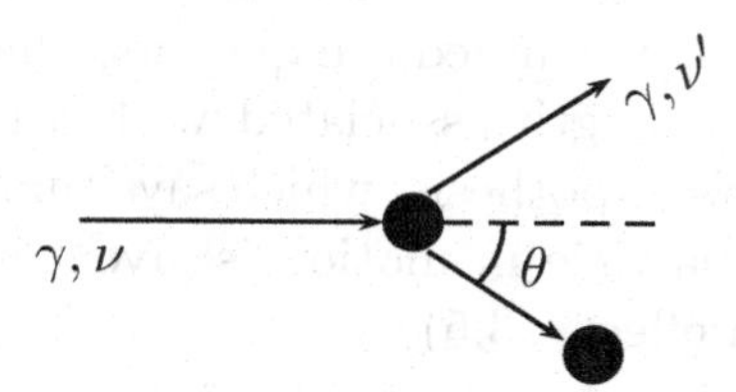

Figure 3.1: Scattering of light by an electron commonly known as the Compton effect. The solid blob represents the electron.

$$\begin{aligned} \text{energy}: &\quad E = h\nu, \\ \text{momentum}: &\quad p = \frac{E}{c} = \frac{h\nu}{c} = \frac{h}{\lambda}, \end{aligned} \tag{3.5}$$

where ν, λ denote respectively the frequency and the wavelength of the light wave. Furthermore, from special theory of relativity, we know that the energy and the momentum of any particle have to satisfy Einstein's relation,

$$E^2 = p^2c^2 + m^2c^4, \tag{3.6}$$

where m denotes the rest mass of the particle. In the present case, if we think of the photon as a particle, then, (3.5) implies that

$$m^2_{\text{photon}} = \frac{E^2}{c^4} - \frac{p^2}{c^2} = 0. \tag{3.7}$$

Namely, if we think of the photon as a particle, its rest mass must vanish.

From these discussions, it is clear that electromagnetic radiation does possess a dual behavior – sometimes it behaves like waves and sometimes as particles.

5. *Davisson-Germer experiment.* If one impinges a beam of electrons on a lattice of atoms (crystals), then, one observes a diffraction pattern. Diffraction, being a wave phenomenon, indicates that the electron, which is a particle, must sometimes behave like a wave. These observations lead to the general conclusion, that all objects must possess both wave and particle behavior. Of course, an immediate question that arises is what determines the wavelength associated with a particle. This is given by de Broglie's hypothesis, which says that the wavelength associated with a particle in motion is given by (compare also with the Compton effect, (3.5))

$$\lambda = \frac{h}{p}. \tag{3.8}$$

Furthermore, let us also note that experiments on atomic systems revealed that various measured quantities assumed only discrete (quantized) values, unlike the predictions of classical mechanics, where observable quantities take continuous values.

6. *Experiment with waves and particles.* Therefore, one believes at this point in the dual behavior of all materials – sometimes they behave as particles and sometimes they behave as waves. The main difference in the behavior of the two, at least classically, is that particles follow definite trajectories and hence do not show interference, whereas waves spread out and, therefore, interfere. Let us now consider the following experiment. Let us take a beam of particles moving towards a double slit arrangement. If one closes one of the slits, one obtains an intensity pattern. With the other slit closed, one also obtains a similar pattern. And when one opens both the slits simultaneously, then, one obtains a pattern which is the sum of the two patterns. This is a particle like behavior, namely, the intensities add up. So the distribution at any point, with both the slits open, is at least as big as with one of the slits open.

   Consider now the same experiment with the particle source replaced by a monochromatic light source. Reduce the intensity of the source to the extent that only one photon is emitted at a time. If one now performs the double slit experiment with one of the slits open, then, one obtains a distribution as in the previous case. However, when one repeats the experiment with both the slits open, then, one does not obtain the distribution to be the sum of the distributions when only one slit is open, as one would expect from particles. Rather, one obtains an interference pattern corresponding to waves. This is quite revealing. For it says that photons, even though particles, do not move in well defined trajectories. For if they did, then they would not exhibit interference.

The result of the experiment can be explained by assuming that with each particle is associated a wave function $\psi(x,t)$ such that $|\psi(x,t)|^2 \mathrm{d}x$ measures the probability of finding the particle between x and $x + \mathrm{d}x$. Since the particle is described by a wave function, it can interfere with itself and, as a consequence, one obtains an interference pattern, rather than just the sum of the intensities.

So, one of the first things we learn is that, unlike classical mechanics where the position and the momentum of particles are well determined quantities, in quantum mechanics, there is indeterminacy. Furthermore, a wave function is associated with a single particle rather than with a wave. Thus, we look for a description of microscopic dynamical systems which would accommodate such behavior and this is commonly known as quantum mechanics.

Microscopic systems. To determine the behavior of a system, one performs measurements which consist of a series of operations on the system. For example, the position of a particle is determined by radiating it with light or photons and then detecting the reflected light. The process of measurement, therefore, introduces a disturbance into the system. For example, the measurement of position would change the momentum of the system. If the system is such that the change or the disturbance is negligible, then, we say that it is a macroscopic system. On the other hand, if the disturbance due to the process of measurement is appreciable, then, we talk of a microscopic system.

Observables. Observables are results of measurements. As we have discussed, a measurement is some kind of an operation on the system. Therefore, the process of measurement can be thought of as an operator acting on a state of the system. The result of an operation is an eigenvalue of the operator corresponding to the specific measurement process and, since the results of measurements are real, the operators corresponding to measurements are assumed to be Hermitian (see (2.89)). However, we also know that operators do not commute and the identification of operators with the process of measurement would imply that the order of measurements in microscopic systems is crucial. This is, in fact, true. For example, suppose we determine the position of a system by radiating photons on it. This changes the momentum of the system. If we make a momentum measurement subsequently, we no longer obtain the true momentum. On the other hand, if we had measured the momentum first, we would have obtained a different value for it and it would have disturbed the position of the particle (momentum for a charged particle, for example, can be determined by applying a magnetic field which bends the trajectory) and hence a subsequent measurement of the position would have yielded a different value from the first measurement. This shows that the order of measurement is, in general, crucial in microscopic systems. Translated differently, if A and B are operators representing two measurements, then,

$$AB \neq BA. \tag{3.9}$$

Commutators. For quantum mechanics to be a good description of a physical system, it should be such that it reduces to classical mechanics when we are talking about macroscopic systems. Classically, of course, we know that the order of measurements and, therefore, the order of observable quantities do not matter. Therefore, let us see what we can deduce about the quantum commutators of operators from our knowledge of the classical Poisson brackets of observables. First of all, we note that commutators formally satisfy the same algebraic properties as the classical Poisson brackets (compare with (1.57)), namely,

$$\begin{aligned}
&[A, B] = AB - BA = -[B, A],\\
&[A, C] = 0, \qquad C = C\,\mathbb{1} = \text{constant},\\
&[A_1 + A_2, B] = [A_1, B] + [A_2, B],\\
&[A_1 A_2, B] = A_1 [A_2, B] + [A_1, B] A_2,\\
&[A, B_1 B_2] = B_1 [A, B_2] + [A, B_1] B_2.
\end{aligned} \tag{3.10}$$

Let us now consider the evaluation of the Poisson bracket

$$\{\eta_1 \eta_2, \rho_1 \rho_2\}. \tag{3.11}$$

Using (1.57), we can calculate (3.11) in two different ways. First, we have

$$\begin{aligned}\{\eta_1\eta_2, \rho_1\rho_2\} &= \eta_1\{\eta_2, \rho_1\rho_2\} + \{\eta_1, \rho_1\rho_2\}\eta_2 \\ &= \eta_1\rho_1\{\eta_2, \rho_2\} + \eta_1\{\eta_2, \rho_1\}\rho_2 \\ &\quad + \rho_1\{\eta_1, \rho_2\}\eta_2 + \{\eta_1, \rho_1\}\rho_2\eta_2.\end{aligned} \tag{3.12}$$

On the other hand, we can also evaluate (3.11) as

$$\begin{aligned}\{\eta_1\eta_2, \rho_1\rho_2\} &= \rho_1\{\eta_1\eta_2, \rho_2\} + \{\eta_1\eta_2, \rho_1\}\rho_2 \\ &= \rho_1\eta_1\{\eta_2, \rho_2\} + \rho_1\{\eta_1, \rho_2\}\eta_2 \\ &\quad + \eta_1\{\eta_2, \rho_1\}\rho_2 + \{\eta_1, \rho_1\}\eta_2\rho_2.\end{aligned} \tag{3.13}$$

Subtracting (3.13) from (3.12), we obtain

$$(\eta_1\rho_1 - \rho_1\eta_1)\{\eta_2, \rho_2\} = \{\eta_1, \rho_1\}(\eta_2\rho_2 - \rho_2\eta_2). \tag{3.14}$$

In classical mechanics, of course, the order of the quantities does not matter and, therefore, either way of doing gives the same answer (namely, the difference vanishes). However, when the order matters, we must have from (3.14)

$$\frac{\{\eta_1, \rho_1\}}{[\eta_1, \rho_1]} = \frac{\{\eta_2, \rho_2\}}{[\eta_2, \rho_2]}. \tag{3.15}$$

This relation must hold for any pair of observables and, therefore, the ratio must be equal to a universal constant. Furthermore, the constant must be imaginary since the ratio in (3.15) is anti-Hermitian. We also note (from the definition of the Poisson bracket in (1.55)) that the ratio has inverse dimensions of an action so that this constant must also have the dimensions of inverse action. It is experimentally determined to be $(i\hbar)^{-1}$ where $\hbar = 1.054 \times 10^{-27}$ erg-sec and is the Planck's constant defined in (3.1). Thus, we see that we can write

$$[\eta_1, \rho_1] = i\hbar\ \{\eta_1, \rho_1\}. \tag{3.16}$$

Quantum correspondence principle. This shows that the quantum commutator of two operators is $i\hbar$ times the value of their classical Poisson bracket. It is clear that, for macroscopic systems where effects of the order of $\hbar$ can be taken to be negligible, the commutator can be neglected and hence the order of quantities would not matter. Therefore, we see that the Planck's constant, $\hbar$, measures the non-classical nature of systems. More commonly, one says that one recovers classical mechanics in the limit $\hbar \to 0$.

3.2 Postulates of quantum mechanics

We have already seen that the laws of classical mechanics need some modification so that they can be applied to microscopic systems. And the modifications should be such that when we are considering a macroscopic system, we should get back the familiar predictions of classical mechanics. These modifications are implemented in the following way. Given a classical Hamiltonian system, one goes over to the quantum description through the following postulates. (We describe them for an one dimensional system for simplicity. The generalization to any higher dimension is straightforward.)

1. In classical mechanics a system at a fixed time is described by the coordinates $x(t)$ and the momenta $p(t)$.

   In quantum mechanics the state of a system at a fixed time is denoted by the infinite dimensional vector $|\psi(t)\rangle$ which belongs to a Hilbert space.

2. Every dynamical variable ω in classical mechanics is a function of the phase space variables x and p. Thus, $\omega = \omega(x, p)$.

   In quantum mechanics the observables x and p are replaced by the Hermitian operators X and P with the nontrivial commutation relation

   $$[X, P] = i\hbar = i\hbar\,\mathbb{1}. \tag{3.17}$$

   Furthermore, these operators have the following matrix elements in the eigenbasis of the operator X (see (2.162) and (2.184)),

   $$\begin{aligned} \langle x|X|x'\rangle &= x\delta(x - x'), \\ \langle x|P|x'\rangle &= -i\hbar\frac{\mathrm{d}}{\mathrm{d}x}\,\delta(x - x'). \end{aligned} \tag{3.18}$$

   Any operator Ω corresponding to the classical observable $\omega(x, p)$ is obtained as the same function of the operators X and P. Thus,

   $$\omega(x, p) \longrightarrow \Omega(X, P). \tag{3.19}$$

   However, we note that since in classical mechanics $xp = px$, there is an ambiguity of operator ordering in the definition of such a product of operators. The ambiguity is resolved by assuming that when dealing with products of two non-commuting operators, one symmetrizes them. Thus, the operator ordering is effected through

$$xp \longrightarrow \frac{1}{2}(XP + PX). \tag{3.20}$$

3. In classical mechanics if a system is in the state $x(t)$ and $p(t)$, then, the measurement of an observable $\omega(x, p)$ would yield a unique value and the system will be unaffected by the process of measurement.

   Quantum mechanics, on the other hand, gives probabilistic results. If a system is in a state $|\psi\rangle$, then a measurement corresponding to Ω yields one of the eigenvalues, ω_i, of Ω, with a probability

$$P(\omega_i) = \frac{|\langle\omega_i|\psi\rangle|^2}{\langle\psi|\psi\rangle}, \qquad \sum_i P(\omega_i) = 1. \tag{3.21}$$

   As a result of the measurement, the state of the system changes to the eigenstate $|\omega_i\rangle$ of the operator Ω.

4. In classical mechanics, the state variables change with time according to Hamilton's equations of motion (see (1.43))

$$\dot{x} = \frac{\partial H}{\partial p}, \qquad \dot{p} = -\frac{\partial H}{\partial x}. \tag{3.22}$$

   In quantum mechanics, the state vectors evolve with time according to the Schrödinger equation

$$i\hbar\frac{\mathrm{d}}{\mathrm{d}t}\,|\psi(t)\rangle = H|\psi(t)\rangle, \tag{3.23}$$

   where $H = H(X, P)$ is the Hamiltonian operator.

   In fact, this last postulate is not an independent postulate and can be argued to arise in the following manner. We know that $H(X, P)$ is the operator corresponding to the total energy of the system. Therefore, if $|\psi\rangle$ is an eigenstate with energy E then one can write

$$H|\psi\rangle = E|\psi\rangle, \tag{3.24}$$

just as, we have for momentum,

$$P|p\rangle = p|p\rangle. \tag{3.25}$$

In relativistic mechanics, however, we know that the energy and momentum form a four vector denoted by $P_\mu = (E, -\mathbf{p})$. They behave like different components of the same object. The relative negative sign is a consequence of the structure of the Lorentz group (or the structure of the four dimensional space-time). In the same way, we also know that space and time are similar in nature. We have seen in (2.187) that the momenta in the x basis (coordinate basis) correspond to operators of the form

$$P \to -i\hbar \, \frac{\mathrm{d}}{\mathrm{d}x}. \tag{3.26}$$

Based on arguments of Lorentz transformations, this, therefore, suggests that

$$H \to i\hbar \, \frac{\mathrm{d}}{\mathrm{d}t}. \tag{3.27}$$

Therefore, in the (x, t) basis, we expect

$$H|\psi(t)\rangle = i\hbar \, \frac{\mathrm{d}}{\mathrm{d}t} \, |\psi(t)\rangle, \tag{3.28}$$

which is the Schrödinger equation.

Implications of the postulates. In quantum mechanics, therefore, the physical system is described by a state vector belonging to a Hilbert space. By definition, a Hilbert space is an infinite dimensional vector space. So, it is natural to ask, how a classical two component system (specified by x, p) acquires infinite degrees of freedom in going to a microscopic system. The answer is easy to see if we go to a basis, say the x basis, where $\langle x|\psi(t)\rangle = \psi(x,t)$ is the coefficient of expansion of the state and this is also the wave function that we talked about in connection with the double slit experiment. We know that the particles in microscopic systems do not move in definite trajectories. Rather, they spread out and the spread can be infinite. $|\psi(x,t)|^2\mathrm{d}x$ measures not only the probability of finding a particle between x and $x + \mathrm{d}x$, but also how the probability changes with time. Thus, there is an infinite amount of information contained in the state and, consequently, it is infinite dimensional.

The probabilistic nature of quantum mechanics, of course, implies that two states $|\psi\rangle$ and $\alpha|\psi\rangle$ give the same probability for a particular measurement. Thus, corresponding to each physical state $|\psi\rangle$, there exists a set of states $\alpha|\psi\rangle$ for all possible complex values of α which define a ray in the Hilbert space. For a physical state, of course, we assume $\langle\psi|\psi\rangle = 1$ or equal to a Dirac delta function (namely, normalized states). This still allows for a ray of the form $e^{i\theta}|\psi\rangle$.

Since the state vectors define a Hilbert space, if $|\psi\rangle$ and $|\psi'\rangle$ define two states of the system, then, $\alpha|\psi\rangle + \beta|\psi'\rangle$ also defines a state because the Schrödinger equation, (3.23), is a linear equation. This is known as the principle of superposition. This is also quite common in classical mechanics. However, the implication of the principle of superposition here is quite different as we will see below. (Note that the principle of superposition is a consequence of the linear nature of the operators in quantum mechanics.)

If a state coincides with an eigenstate $|\omega_i\rangle$ of some operator Ω, then, the corresponding measurement would, for sure, yield the value ω_i. This follows immediately from the fact that

$$P(\omega_i) = \frac{|\langle\omega_i|\psi\rangle|^2}{\langle\psi|\psi\rangle} = \frac{|\langle\omega_i|\omega_i\rangle|^2}{\langle\omega_i|\omega_i\rangle} = 1. \tag{3.29}$$

Let us next consider the state formed by superposing two eigenstates $|\omega_1\rangle$ and $|\omega_2\rangle$ of the operator Ω. In other words, let

$$|\psi\rangle = \frac{\alpha|\omega_1\rangle + \beta|\omega_2\rangle}{\sqrt{|\alpha|^2 + |\beta|^2}}. \tag{3.30}$$

This state is normalized, if the eigenstates $|\omega_i\rangle$ are. If one makes a measurement corresponding to Ω in this state, then, from the definition in (3.21), we see that the measurement would yield a value ω_1 with probability $\frac{|\alpha|^2}{|\alpha|^2+|\beta|^2}$ and a value ω_2 with a probability $\frac{|\beta|^2}{|\alpha|^2+|\beta|^2}$. Thus, the measurements reveal that a superposed state sometimes behaves like it is in one of the eigenstates and sometimes in the other. This is quite different from the classical superposition principle. For example, if $f(x)$ and $g(x)$ correspond to two different configurations of our string example, then $\alpha f(x) + \beta g(x)$ also corresponds to a configuration of the string. However, measurements on this configuration are unique and distinct from those on $f(x)$ and $g(x)$.

If an operator is degenerate, say Ω is doubly degenerate with eigenstates $|\omega, 1\rangle$ and $|\omega, 2\rangle$, then the probability that a measurement would yield an eigenvalue ω is given by (compare with (3.21))

$$P(\omega) = \frac{1}{\langle\psi|\psi\rangle} \left[|\langle\omega, 1|\psi\rangle|^2 + |\langle\omega, 2|\psi\rangle|^2\right]. \tag{3.31}$$

Let us also note that there are other plausible explanations, such as the hidden variable theory, for the failures of classical mechanics. However, these do not help very much in practical calculations. Hence, we will not discuss them in these lectures.

3.3 Expectation value

Let us suppose that a physical system is in a quantum mechanical state $|\psi\rangle$ and that a measurement corresponding to the operator Ω is made. Then, clearly, from our earlier discussions, we conclude that we will obtain an eigenvalue ω_i with probability $P(\omega_i)$. Now suppose an infinite number of such experiments are performed. Then, one obtains a variety of values with different probabilities. We can ask what is the statistical mean of these measurements, which we can then, identify with the average value of the operator in the state, namely,

$$\langle\Omega\rangle = \sum_i P(\omega_i)\omega_i. \tag{3.32}$$

It now follows that (We are assuming that the state $|\psi\rangle$ is normalized and that the eigenvalues of Ω are discrete. The case of continuous eigenvalues can be handled by replacing the sums with integrals.)

$$\begin{aligned}
\langle\Omega\rangle &= \sum_i P(\omega_i)\omega_i = \sum_i |\langle\omega_i|\psi\rangle|^2\omega_i \\
&= \sum_i \langle\omega_i|\psi\rangle\langle\psi|\omega_i\rangle\omega_i = \sum_i \langle\psi|\omega_i\rangle\langle\omega_i|\psi\rangle\omega_i \\
&= \sum_i \langle\psi|\Omega|\omega_i\rangle\langle\omega_i|\psi\rangle = \langle\psi|\Omega\left(\sum_i |\omega_i\rangle\langle\omega_i|\right)|\psi\rangle \\
&= \langle\psi|\Omega|\psi\rangle,
\end{aligned} \tag{3.33}$$

where we have used the completeness relation of the eigenbasis of the operator Ω in the intermediate steps, namely,

$$\sum_i |\omega_i\rangle\langle\omega_i| = \mathbb{1}. \tag{3.34}$$

We note that $\langle\Omega\rangle$ is known as the expectation value of the operator Ω in the state $|\psi\rangle$ and gives the statistical mean of the measurements performed on the system in that state. It is clear from (3.33) that if the state is an eigenstate of the operator, then the expectation value would be the eigenvalue corresponding to that state.

We can similarly define the expectation value of the operator Ω^2 in the state $|\psi\rangle$.

$$\begin{aligned}\langle\Omega^2\rangle &= \langle\psi|\Omega^2|\psi\rangle = \langle\psi|\Omega^2\left(\sum_i |\omega_i\rangle\langle\omega_i|\right)|\psi\rangle \\ &= \sum_i \langle\psi|\Omega|\omega_i\rangle\langle\omega_i|\psi\rangle\omega_i = \sum_i \langle\psi|\omega_i\rangle\langle\omega_i|\psi\rangle\omega_i^2 \\ &= \sum_i P(\omega_i)\omega_i^2. \end{aligned} \tag{3.35}$$

Similarly, we can calculate the expectation value of any higher power of the operator and these would correspond to the familiar moments of a distribution in statistical mechanics.

3.4 Uncertainty principle

Let A and B be two non-commuting Hermitian operators with the commutator given by (we do not write explicitly the identity operator on the right hand side)

$$[A, B] = i\hbar. \tag{3.36}$$

As we have seen before, A and B, in such a case, are called conjugate operators. Let ΔA be the root mean square deviation for the operator A in a given quantum mechanical state, $|\psi\rangle$, so that

$$(\Delta A)^2 = \langle\psi|A^2|\psi\rangle - \langle\psi|A|\psi\rangle^2 = \langle A^2\rangle - \langle A\rangle^2. \tag{3.37}$$

Similarly, let ΔB be the root mean square deviation for B in the same quantum state so that

$$(\Delta B)^2 = \langle B^2\rangle - \langle B\rangle^2. \tag{3.38}$$

Then, Heisenberg's uncertainty relation says that

$$\Delta A\Delta B \geq \frac{\hbar}{2}. \tag{3.39}$$

To derive the uncertainty relation, (3.39), let us note, first of all, that we can write

$$(\Delta A)^2 = \langle A^2\rangle - \langle A\rangle^2 \tag{3.40}$$
$$= \langle (A - \langle A\rangle)^2\rangle, \tag{3.41}$$

and, similarly for $(\Delta B)^2$. Let us also define

$$\tilde{A} = A - \langle A\rangle, \qquad \tilde{B} = B - \langle B\rangle. \tag{3.42}$$

Then, clearly, since the expectation value is a constant,

$$[\tilde{A}, \tilde{B}] = [A, B] = i\hbar. \tag{3.43}$$

Furthermore, using (3.41) and (3.42), we can write

$$\begin{aligned}(\Delta A)^2(\Delta B)^2 &= \langle (A - \langle A\rangle)^2\rangle\langle (B - \langle B\rangle)^2\rangle \\ &= \langle \tilde{A}^2\rangle\langle \tilde{B}^2\rangle \\ &\geq |\langle \tilde{A}\tilde{B}\rangle|^2 \qquad\qquad \text{(Schwartz inequality)} \\ &= |\langle \frac{1}{2}(\tilde{A}\tilde{B} + \tilde{B}\tilde{A}) + \frac{1}{2}(\tilde{A}\tilde{B} - \tilde{B}\tilde{A})\rangle|^2 \\ &= |\langle \frac{1}{2}(\tilde{A}\tilde{B} + \tilde{B}\tilde{A}) + \frac{1}{2}[\tilde{A}, \tilde{B}]\rangle|^2 \\ &= |\langle \frac{1}{2}(\tilde{A}\tilde{B} + \tilde{B}\tilde{A}) + \frac{i\hbar}{2}\rangle|^2 \\ &= |\langle \frac{1}{2}\ (\tilde{A}\tilde{B} + \tilde{B}\tilde{A})\rangle|^2 + \frac{\hbar^2}{4}.\end{aligned} \tag{3.44}$$

Since the first term on the right hand side of (3.44) is positive semi-definite, this proves that

$$\Delta A \Delta B \geq \frac{\hbar}{2}. \tag{3.45}$$

This result tells us that for any pair of conjugate variables, there is a minimum of uncertainty associated with their measurements. Let us note that if $|\psi\rangle$ is an eigenstate of one of the operators, say A, then,

$$\Delta A = \langle\psi|A^2|\psi\rangle - \langle\psi|A|\psi\rangle^2 = 0. \tag{3.46}$$

This implies that we can measure the quantity A precisely or accurately in this state. The uncertainty principle, (3.45), then, says that the measurement of B in this state would be infinitely uncertain, since

$$\Delta B \geq \frac{\hbar}{2\Delta A} \to \infty. \tag{3.47}$$

► **Example.** Let us look for the form of the wave function for the state in which the uncertainty is the minimum. First of all, note that the Schwartz inequality becomes an equality if the vectors are parallel (recall that $|\mathbf{A}|^2|\mathbf{B}|^2 = |\mathbf{A}\cdot\mathbf{B}|^2$ if the two vectors, $\mathbf{A}$ and $\mathbf{B}$, are parallel). Thus, for minimum uncertainty, we must have

$$\tilde{A}|\psi\rangle = \lambda\tilde{B}|\psi\rangle, \qquad \langle\psi|\tilde{A} = \lambda^*\langle\psi|\tilde{B}. \tag{3.48}$$

This implies that

$$\begin{aligned}\frac{1}{2}\langle\psi|(\tilde{A}\tilde{B}+\tilde{B}\tilde{A})|\psi\rangle &= \frac{1}{2}\langle\psi|(\lambda^*\tilde{B}\tilde{B}+\lambda\tilde{B}\tilde{B})|\psi\rangle\\ &= \frac{1}{2}(\lambda+\lambda^*)\langle\psi|\tilde{B}\tilde{B}|\psi\rangle\\ &= \frac{1}{2}(\lambda+\lambda^*)\langle\tilde{B}\psi|\tilde{B}\psi\rangle.\end{aligned} \tag{3.49}$$

If this vanishes, then, we see from (3.44) that

$$(\Delta A)(\Delta B) = \frac{\hbar}{2}, \tag{3.50}$$

which is the minimum uncertainty. Since $\langle\tilde{B}\psi|\tilde{B}\psi\rangle > 0$, unless $|\tilde{B}\psi\rangle = 0$, for the vanishing of (3.49), we must have $\lambda+\lambda^* = 0$, which implies that λ is pure imaginary. Let $\lambda = -ic$ where c is real. Then we see from (3.48) that $\tilde{A}|\psi\rangle = \lambda\tilde{B}|\psi\rangle = -ic\tilde{B}|\psi\rangle$. Furthermore, by definition, (3.42),

$$\tilde{A} = A - \langle A\rangle, \quad \text{and} \quad \tilde{B} = B - \langle B\rangle.$$

Since A and B are conjugate variables, we can express them as differential operators, as we have seen earlier. As an example, let us consider

$$A = X, \quad \text{and} \quad B = P. \tag{3.51}$$

Then, in the coordinate basis, we have

$$\begin{aligned}&(x - \langle X\rangle)\psi(x) = -ic\left(-i\hbar\frac{\mathrm{d}}{\mathrm{d}x} - \langle P\rangle\right)\psi(x),\\ \text{or,}\quad &\frac{\mathrm{d}\psi(x)}{\mathrm{d}x} = \left(-\frac{1}{c\hbar}(x-\langle X\rangle) + \frac{i}{\hbar}\langle P\rangle\right)\psi(x),\\ \text{or,}\quad &\psi(x) = N\exp\left[-\frac{1}{2c\hbar}(x-\langle X\rangle)^2 + \frac{i}{\hbar}\langle P\rangle x\right].\end{aligned} \tag{3.52}$$

This is a Gaussian centered at $x = \langle X\rangle$ with a width Δx given by

$$\frac{1}{(\Delta x)^2} = \frac{2}{c\hbar}, \tag{3.53}$$

and the normalization N can be determined easily. ◄

Exercise. Determine N in the above example.

3.5 Ehrenfest theorem

Let us consider an operator Ω and its expectation value $\langle\Omega\rangle$ in a state $|\psi\rangle$. We can ask how the expectation value of Ω changes with time. We can, of course, write

$$\begin{aligned}\frac{\mathrm{d}}{\mathrm{d}t}\langle\Omega\rangle &= \frac{\mathrm{d}}{\mathrm{d}t}\langle\psi|\Omega|\psi\rangle \\ &= \langle\dot{\psi}|\Omega|\psi\rangle + \langle\psi|\frac{\partial\Omega}{\partial t}|\psi\rangle + \langle\psi|\Omega|\dot{\psi}\rangle. \end{aligned} \tag{3.54}$$

On the other hand, we know from the Schrödinger equation that

$$i\hbar\frac{\mathrm{d}}{\mathrm{d}t}|\psi(t)\rangle = i\hbar|\dot{\psi}\rangle = H|\psi\rangle, \qquad -i\hbar\,\langle\dot{\psi}| = \langle\psi|H. \tag{3.55}$$

Substituting this into (3.54), we have

$$\begin{aligned}\frac{\mathrm{d}}{\mathrm{d}t}\langle\Omega\rangle &= \langle\dot{\psi}|\Omega|\psi\rangle + \langle\psi|\frac{\partial\Omega}{\partial t}|\psi\rangle + \langle\psi|\Omega|\dot{\psi}\rangle \\ &= \frac{i}{\hbar}\langle\psi|H\Omega|\psi\rangle + \langle\psi|\frac{\partial\Omega}{\partial t}|\psi\rangle - \frac{i}{\hbar}\langle\psi|\Omega H|\psi\rangle \\ &= \langle\psi|\frac{\partial\Omega}{\partial t}|\psi\rangle + \frac{i}{\hbar}\langle\psi|[H,\Omega]|\psi\rangle. \end{aligned} \tag{3.56}$$

If the operator Ω has no explicit time dependence, then, the first term on the right hand side of (3.56) vanishes and we have

$$\frac{\mathrm{d}}{\mathrm{d}t}\langle\Omega\rangle = -\frac{i}{\hbar}\langle\psi|[\Omega,H]|\psi\rangle = \frac{1}{i\hbar}\langle\psi|[\Omega,H]|\psi\rangle. \tag{3.57}$$

This is known as the Ehrenfest theorem. Remembering the quantum correspondence principle (3.16), namely, $[A,B] = i\hbar\{A,B\}$ we note that this is quite analogous to the classical Hamiltonian equation (see (1.54))

$$\frac{\mathrm{d}\omega}{\mathrm{d}t} = \{\omega, H\}.$$

In other words, Ehrenfest's theorem says that expectation values of operators in quantum states obey classical equations.

► **Example.** As a direct application of the above discussions to particle motion in one dimension, where

$$H = \frac{P^2}{2m} + V(X), \tag{3.58}$$

we can calculate the time evolution of the expectation value of the coordinate operator using (3.57),

$$\begin{aligned}\frac{\mathrm{d}\langle X\rangle}{\mathrm{d}t} &= \frac{\mathrm{d}}{\mathrm{d}t}\langle\psi|X|\psi\rangle = \frac{1}{i\hbar}\langle\psi|[X,H]|\psi\rangle \\ &= \frac{1}{i\hbar}\langle\psi|\left[X,\ \frac{P^2}{2m}+V(X)\right]|\psi\rangle \\ &= \frac{1}{i\hbar}\langle\psi|\left[X,\frac{P^2}{2m}\right]|\psi\rangle \\ &= \frac{1}{i\hbar}\times\frac{1}{2m}\langle\psi|P[X,P]+[X,P]P|\psi\rangle \\ &= \frac{1}{i\hbar}\times\frac{1}{2m}\times 2i\hbar\langle\psi|P|\psi\rangle \\ &= \frac{1}{m}\langle\psi|P|\psi\rangle = \frac{\langle P\rangle}{m}. \end{aligned} \tag{3.59}$$

Let us recall from (1.43) that in classical mechanics we have $\dot{x} = \frac{\partial H}{\partial p} = \frac{p}{m}$, which can be compared with (3.59).

Furthermore, the time evolution of the expectation value of the momentum operator is obtained to be

$$\begin{aligned}\frac{\mathrm{d}\langle P\rangle}{\mathrm{d}t} &= \frac{\mathrm{d}}{\mathrm{d}t}\langle\psi|P|\psi\rangle = \frac{1}{i\hbar}\langle\psi|[P,H]|\psi\rangle \\ &= \frac{1}{i\hbar}\langle\psi|\left[P,\frac{P^2}{2m}+V(X)\right]|\psi\rangle \\ &= \frac{1}{i\hbar}\langle\psi|[P,V(X)]|\psi\rangle \\ &= \frac{1}{i\hbar}\langle\psi|(-i\hbar)\ \frac{\mathrm{d}V(X)}{\mathrm{d}X}\ |\psi\rangle \\ &= -\langle\psi|\frac{\mathrm{d}V(X)}{\mathrm{d}X}\ |\psi\rangle = -\langle\frac{\mathrm{d}V(X)}{\mathrm{d}X}\rangle. \end{aligned} \tag{3.60}$$

Once again, we note from (1.43) that in classical physics,

$$\dot{p} = -\frac{\partial H}{\partial x} = -\frac{\mathrm{d}V(x)}{\mathrm{d}x}, \tag{3.61}$$

which can be compared with (3.60).

Thus, we see that, for an operator Ω, which does not depend on time explicitly, the expectation value $\langle\Omega\rangle$ has the following evolution equation

$$\frac{\mathrm{d}\langle\Omega\rangle}{\mathrm{d}t} = -\frac{i}{\hbar}\langle[\Omega,H]\rangle = \frac{1}{i\hbar}\langle[\Omega,H]\rangle. \tag{3.62}$$

Let us consider a state $|\psi\rangle$ where the position measurement yields a value x with uncertainty Δx. And the momentum measurement yields a value p with uncertainty $\simeq \frac{\hbar}{2\Delta x}$. If the state is such that the uncertainties Δx and $\frac{\hbar}{2\Delta x}$ are negligible compared to the measured values x and p, then, we can replace

$$\left.\begin{aligned}\langle X\rangle &= x,\\ \langle P\rangle &= p,\end{aligned}\right\}\quad x \text{ and } p \text{ being classical quantities.} \tag{3.63}$$

For such a state, therefore, the fluctuation around the mean is negligible and we can write

$$\langle \Omega(X,P)\rangle = \Omega(\langle X\rangle, \langle P\rangle) = \Omega(x,p) = \omega(x,p). \tag{3.64}$$

Therefore, in such a case we can write the Ehrenfest equation as

$$\frac{\mathrm{d}\langle \Omega\rangle}{\mathrm{d}t} = \frac{\mathrm{d}\omega}{\mathrm{d}t} = \{\omega, H\}. \tag{3.65}$$

which is nothing other than Hamilton's equation, (1.54). Thus, we see that quantum mechanics, when applied to macroscopic systems (where uncertainties in measurements can be neglected), indeed reduces to classical Hamiltonian mechanics. ◀

3.6 Stationary state solutions

Let us consider a simple quantum mechanical system where the Hamiltonian does not depend on time explicitly, namely,

$$H \neq H(t). \tag{3.66}$$

The Schrödinger equation, (3.23), is given by

$$i\hbar \frac{\mathrm{d}}{\mathrm{d}t} |\psi(t)\rangle = H|\psi\rangle. \tag{3.67}$$

This is a first order differential equation. Therefore, given one initial condition, $|\psi(0)\rangle$, the solution can, in principle, be uniquely determined. Given an initial solution, one can also think of the solution at any finite time as resulting from the effect of an operator acting on the ket (vector) at $t = 0$. In other words, we can write

$$|\psi(t)\rangle = U(t)|\psi(0)\rangle, \tag{3.68}$$

where $U(t)$ is the operator which translates the time coordinate of the state vectors.

The main interest in quantum mechanics is to determine this operator $U(t)$. To find $U(t)$, let us work in the eigenbasis of the Hamiltonian H. Thus, let

$$H|E\rangle = E|E\rangle. \tag{3.69}$$

In this basis, we can expand the state vector as

$$|\psi(t)\rangle = \sum_E |E\rangle\langle E|\psi(t)\rangle = \sum_E a_E(t)|E\rangle, \tag{3.70}$$

where

$$a_E(t) = \langle E|\psi(t)\rangle. \tag{3.71}$$

Here, if the eigenvalues of the operator H are discrete, then one uses a summation. However, if the eigenvalues E take continuous values, then, one must replace the sum in (3.70) by an integral. We will assume the energy eigenvalues to be discrete for simplicity. In this basis, we can write

$$\begin{aligned}
& i\hbar \frac{\mathrm{d}}{\mathrm{d}t}|\psi(t)\rangle = H|\psi(t)\rangle, \\
\text{or,}\quad & i\hbar \sum_E \dot{a}_E(t)|E\rangle = \sum_E E a_E(t)|E\rangle,
\end{aligned} \tag{3.72}$$

where we have used the fact that the basis states $|E\rangle$ are time independent, since the Hamiltonian is. It follows from (3.72) that, for every energy mode, we can write

$$\begin{aligned}
& i\hbar\, \dot{a}_E(t) = E a_E(t), \\
\text{or,}\quad & \frac{\mathrm{d}a_E(t)}{\mathrm{d}t} = -\frac{i}{\hbar} E a_E, \\
\text{or,}\quad & a_E(t) = a_E(0) e^{-\frac{i}{\hbar}Et},
\end{aligned} \tag{3.73}$$

which allows us to express the state vector as

$$|\psi(t)\rangle = \sum_E a_E(t)|E\rangle = \sum_E a_E(0) e^{-\frac{i}{\hbar}Et}|E\rangle. \tag{3.74}$$

On the other hand, from the expansion of the state in the energy eigenbasis, (3.70), we also recognize that

$$|\psi(0)\rangle = \sum_E a_E(0)|E\rangle, \qquad a_E(0) = \langle E|\psi(0)\rangle. \tag{3.75}$$

Using this in (3.74), we see that we can write

$$\begin{aligned}
|\psi(t)\rangle &= \sum_E a_E(0) e^{-\frac{i}{\hbar}Et}|E\rangle \\
&= \sum_E e^{-\frac{i}{\hbar}Et}\langle E|\psi(0)\rangle|E\rangle \\
&= \sum_E e^{-\frac{i}{\hbar}Et}|E\rangle\langle E|\psi(0)\rangle \\
&= U(t)|\psi(0)\rangle,
\end{aligned} \tag{3.76}$$

where we have identified

$$U(t) = \sum_E e^{-\frac{i}{\hbar}Et}|E\rangle\langle E|. \tag{3.77}$$

This determines the operator $U(t)$, which leads to time evolution of physical states (and, therefore, is also known as the time evolution operator). If the eigenvalues of H are degenerate, then we can introduce a label α for the degeneracy and then, the time evolution operator can be determined in an analogous manner to be

$$U(t) = \sum_\alpha \sum_E e^{-\frac{i}{\hbar}Et}|E,\alpha\rangle\langle E,\alpha|. \tag{3.78}$$

Let us note here parenthetically that, since our main goal is to determine $U(t)$, this can be achieved if we know the energy eigenvalues and eigenstates of a quantum mechanical system.

Clearly, states of the form

$$|\psi_E(t)\rangle = |E(t)\rangle = e^{-\frac{i}{\hbar}Et}|E\rangle, \tag{3.79}$$

satisfy the Schrödinger equation. Such states, which are eigenstates of the Hamiltonian, are called stationary states because, in such states, the probability for measurement of any time independent operator Ω is independent of time. Namely,

$$\begin{aligned} P(\omega,t) &= |\langle\omega|\psi_E(t)\rangle|^2 = |\langle\omega|E(t)\rangle|^2 \\ &= |\langle\omega|E\rangle e^{-\frac{i}{\hbar}Et}|^2 = |\langle\omega|E\rangle|^2 \\ &= P(\omega,0), \end{aligned} \tag{3.80}$$

where $|\omega\rangle$ is assumed to be time independent since Ω is.

We can also write the time evolution operator in the form

$$\begin{aligned} U(t) &= \sum_E e^{-\frac{i}{\hbar}Et}|E\rangle\langle E| \\ &= \sum_E e^{-\frac{i}{\hbar}Ht}|E\rangle\langle E| = e^{-\frac{i}{\hbar}Ht}\sum_E |E\rangle\langle E| \\ &= e^{-\frac{i}{\hbar}Ht}. \end{aligned} \tag{3.81}$$

It is also clear that this expression is formally true even if the eigenvalues of H are degenerate. We say, formally, because the convergence of this series is hard to prove. Since H is Hermitian, it follows that $U(t)$ is a unitary operator. In other words,

$$U^{\dagger}(t)U(t) = \mathbb{1} = U(t)U^{\dagger}(t), \tag{3.82}$$

and, consequently,

$$\langle\psi(t)|\psi(t)\rangle = \langle\psi(0)|U^{\dagger}(t)U(t)|\psi(0)\rangle = \langle\psi(0)|\psi(0)\rangle. \tag{3.83}$$

If the Hamiltonian depends on time explicitly, namely, if

$$H = H(t), \tag{3.84}$$

then, the time evolution operator takes the following form

$$U(t) = \mathrm{T}\left(e^{-\frac{i}{\hbar}\int_0^t \mathrm{d}t'\, H(t')}\right), \tag{3.85}$$

where T is the time ordering operator which orders operators with larger times to the left.

It is clear that the operator U– both for time dependent and time independent Hamiltonians – depends on the initial time as well as the final time,

$$U = U(t_2, t_1), \tag{3.86}$$

so that

$$|\psi(t_2)\rangle = U(t_2, t_1)|\psi(t_1)\rangle. \tag{3.87}$$

Namely, it takes a state at time t_1 to a state at time t_2. Furthermore, the time evolution operator satisfies the following relations

$$\begin{aligned}
&U(t_1, t_1) = \mathbb{1},\\
&U(t_3, t_2)U(t_2, t_1) = U(t_3, t_1),\\
&U^{\dagger}(t_2, t_1) = U^{-1}(t_2, t_1) = U(t_1, t_2).
\end{aligned} \tag{3.88}$$

Let us now solve the same problem of a time independent Hamiltonian in the x-basis (coordinate basis). The Schrödinger equation leads to

$$\begin{aligned}
&i\hbar\,\frac{\mathrm{d}}{\mathrm{d}t}\,|\psi(t)\rangle = H|\psi(t)\rangle,\\
\text{or,}\quad &i\hbar\,\frac{\partial}{\partial t}\langle x|\psi(t)\rangle = \int \mathrm{d}x'\,\langle x|H|x'\rangle\langle x'|\psi(t)\rangle,\\
\text{or,}\quad &i\hbar\,\frac{\partial\psi(x,t)}{\partial t} = \left(-\frac{\hbar^2}{2m}\frac{\partial^2}{\partial x^2} + V(x)\right)\psi(x,t).
\end{aligned} \tag{3.89}$$

Since time and space derivatives are completely decoupled in (3.89), let us assume a separable solution of the form

$$\psi(x,t) = u(x)g(t). \tag{3.90}$$

Substituting this into (3.89), we obtain

$$\begin{aligned} i\hbar u(x)\,\frac{\mathrm{d}g(t)}{\mathrm{d}t} &= g(t)\left(-\frac{\hbar^2}{2m}\frac{\mathrm{d}^2}{\mathrm{d}x^2} + V(x)\right)u(x), \\ \text{or,}\quad i\hbar\,\frac{\dot{g}(t)}{g(t)} &= \frac{1}{u(x)}\left(-\frac{\hbar^2}{2m}\frac{\mathrm{d}^2}{\mathrm{d}x^2} + V(x)\right)u(x). \end{aligned} \tag{3.91}$$

Since the left hand side depends only on time and the right hand side depends only on spatial coordinates, for them to be equal for all times and space, both sides must be equal to a constant which we call E. Thus, the relation in (3.91) leads to two equations, the first of which has the form

$$\begin{aligned} i\hbar\,\frac{\dot{g}(t)}{g(t)} &= E, \\ \text{or,}\quad g(t) &= g(0)\,e^{-\frac{i}{\hbar}Et}. \end{aligned} \tag{3.92}$$

Furthermore, the second equation following from (3.91) gives

$$\left(-\frac{\hbar^2}{2m}\frac{\mathrm{d}^2}{\mathrm{d}x^2} + V(x)\right)u(x) = Eu(x). \tag{3.93}$$

Thus, we can write the wave function for the system as

$$\psi_E(x,t) = u(x)g(t) = u(x)e^{-\frac{i}{\hbar}Et}, \tag{3.94}$$

where the constant $g(0)$ has been absorbed into $u(x)$. Equation (3.93), in the coordinate basis, is also known as the time independent Schrödinger equation which determines E from the dynamics of the system.

Exercise. Convince yourself that E is the energy of the system. How is $u(x)$ related to the expansion in the energy basis?

3.7 Continuity equation

In electrodynamics, we know that the total charge of a system is a constant independent of time, namely,

$$Q = \text{ constant} \neq Q(t). \tag{3.95}$$

Such a relation is known as a global conservation law. Global conservation laws allow for the possibility of charges disappearing at some place and appearing suddenly at some other place within a given region. However, in electrodynamics, we also know of a continuity equation

$$\frac{\partial\rho(\mathbf{r},t)}{\partial t} = -\boldsymbol{\nabla}\cdot\mathbf{J}(\mathbf{r},t), \tag{3.96}$$

where ρ is the electric charge density and $\mathbf{J}$, the current density at some point. Integrating (3.96) over a small volume we have

$$\frac{\mathrm{d}}{\mathrm{d}t}\int_V \mathrm{d}^3r\,\rho(\mathbf{r},t) = -\int_V \mathrm{d}^3r\,\boldsymbol{\nabla}\cdot\mathbf{J} = -\int_{S_V}\mathrm{d}\mathbf{s}\cdot\mathbf{J}, \tag{3.97}$$

where S_V is the surface which bounds the volume V. This shows that any decrease in the charge, in an infinitesimal volume, must be accompanied by a flux of electric current out of the volume. In other words, charge has to be locally conserved. Such conservation laws prohibit simultaneous sudden destruction of charge at some point and creation of charge at some other point, which is allowed by global conservation laws.

In quantum mechanics, we also have globally conserved quantities. For example, we have already seen that (see (3.83))

$$\langle\psi(t)|\psi(t)\rangle = \langle\psi(0)|U^\dagger(t)U(t)|\psi(0)\rangle = \langle\psi(0)|\psi(0)\rangle. \tag{3.98}$$

If the states are normalized, this implies that the probability of finding a particle anywhere in space is unity at all times. Namely,

$$\begin{aligned}\langle\psi(0)|\psi(0)\rangle = 1 &= \langle\psi(t)|\psi(t)\rangle\\ &= \int \mathrm{d}x\,\langle\psi(t)|x\rangle\langle x|\psi(t)\rangle\\ &= \int \mathrm{d}x\,\psi^*(x,t)\psi(x,t)\\ &= \int \mathrm{d}x\,P(x,t).\end{aligned} \tag{3.99}$$

However, there are also local conservation laws in quantum mechanics. For example, let us consider the Schrödinger equation in three dimensions in the coordinate basis (see (3.89)),

$$i\hbar\,\frac{\partial\psi(\mathbf{r},t)}{\partial t} = H\psi(\mathbf{r},t) = \left(-\frac{\hbar^2}{2m}\,\boldsymbol{\nabla}^2 + V(\mathbf{r})\right)\psi(\mathbf{r},t). \tag{3.100}$$

Taking the complex conjugate of (3.100), we obtain

$$-i\hbar\,\frac{\partial\psi^*(\mathbf{r},t)}{\partial t} = (H\psi(\mathbf{r},t))^* = \left(-\frac{\hbar^2}{2m}\,\boldsymbol{\nabla}^2 + V(\mathbf{r})\right)\psi^*(\mathbf{r},t). \tag{3.101}$$

Multiplying (3.100) by $\psi^*(\mathbf{r},t)$ and (3.101) by $\psi(\mathbf{r},t)$ and subtracting one from the other, we have

$$\begin{aligned}
& i\hbar\left(\psi^*\frac{\partial\psi}{\partial t} + \psi\frac{\partial\psi^*}{\partial t}\right) = -\frac{\hbar^2}{2m}\left(\psi^*\boldsymbol{\nabla}^2\psi - \psi\boldsymbol{\nabla}^2\psi^*\right),\\
\text{or,}\quad & i\hbar\frac{\partial}{\partial t}(\psi^*\psi) = -\frac{\hbar^2}{2m}\,\boldsymbol{\nabla}\cdot\left(\psi^*\boldsymbol{\nabla}\psi - (\boldsymbol{\nabla}\psi^*)\psi\right),\\
\text{or,}\quad & \frac{\partial}{\partial t}(\psi^*\psi) = -\frac{\hbar}{2im}\,\boldsymbol{\nabla}\cdot\left(\psi^*\boldsymbol{\nabla}\psi - (\boldsymbol{\nabla}\psi^*)\psi\right).
\end{aligned} \tag{3.102}$$

We know that $\psi^*(\mathbf{r},t)\psi(\mathbf{r},t)\mathrm{d}^3r = P(\mathbf{r},t)\mathrm{d}^3r$ gives the probability of finding a particle between $\mathbf{r}$ and $\mathbf{r}+\mathrm{d}\mathbf{r}$ at a given time t. Therefore, $P(\mathbf{r},t)$ corresponds to the probability density. We can also define a probability current density associated with the particle as

$$\mathbf{S} = \frac{\hbar}{2im}\left(\psi^*(\mathbf{r},t)\boldsymbol{\nabla}\psi(\mathbf{r},t) - (\boldsymbol{\nabla}\psi^*(\mathbf{r},t))\psi(\mathbf{r},t)\right), \tag{3.103}$$

so that we have the continuity equation

$$\frac{\partial P(\mathbf{r},t)}{\partial t} = -\boldsymbol{\nabla}\cdot\mathbf{S}(\mathbf{r},t). \tag{3.104}$$

This again leads to a local conservation law, namely, locally the change in the probability density must be equal to the negative of the probability flux out of that volume. In other words, probability can not suddenly change at distinct points. The continuity equation, therefore, emphasizes that the solutions of the Schrödinger equation must be such that both $\psi(\mathbf{r},t)$ and $\boldsymbol{\nabla}\psi(\mathbf{r},t)$ should be continuous functions (for regular potentials). This can be seen in the following manner. Let us note that, for stationary states (see (3.93)), the Schrödinger equation is a second order differential equation, namely (consider one dimension for simplicity and note that the time dependent phase drops out in $P,\mathbf{S}$ for stationary states),

$$\frac{\mathrm{d}^2u(x)}{\mathrm{d}x^2} = \frac{2m}{\hbar^2}\left(V-E\right)u(x). \tag{3.105}$$

Integrating this over an infinitesimal interval around $x = \zeta$, we obtain

$$\lim_{\epsilon\to 0}\int_{\zeta-\epsilon}^{\zeta+\epsilon} \mathrm{d}x\, \frac{\mathrm{d}^2u(x)}{\mathrm{d}x^2} = \lim_{\epsilon\to 0}\int_{\zeta-\epsilon}^{\zeta+\epsilon} \mathrm{d}x\, \frac{2m}{\hbar^2}\,(V(x)-E)\,u(x),$$

$$\text{or,}\quad \lim_{\epsilon\to 0}\left.\frac{\mathrm{d}u(x)}{\mathrm{d}x}\right|_{\zeta-\epsilon}^{\zeta+\epsilon} = \lim_{\epsilon\to 0}\frac{2m}{\hbar^2}\int_{\zeta-\epsilon}^{\zeta+\epsilon} \mathrm{d}x\, V(x)u(x) \to 0. \tag{3.106}$$

which holds for continuous potentials.

3.8 Schrödinger picture and Heisenberg picture

We have seen that the time dependent Schrödinger equation has the form

$$i\hbar\,\frac{\mathrm{d}}{\mathrm{d}t}\,|\psi_S(t)\rangle = H|\psi_S(t)\rangle, \tag{3.107}$$

and implies that

$$|\psi_S(t)\rangle = e^{-\frac{i}{\hbar}Ht}|\psi_S(0)\rangle, \tag{3.108}$$

when the Hamiltonian is independent of time. Let us further assume that operators do not have any explicit time dependence. Then, the expectation value of operators change with time according to the Ehrenfest theorem (see (3.57)) as

$$\begin{aligned}\frac{\mathrm{d}}{\mathrm{d}t}\langle\Omega_S\rangle &= \frac{\mathrm{d}}{\mathrm{d}t}\langle\psi_S(t)|\Omega_S|\psi_S(t)\rangle\\ &= \frac{1}{i\hbar}\langle\psi_S(t)|[\Omega_S,H]|\psi_S(t)\rangle.\end{aligned} \tag{3.109}$$

Thus, even though the operators do not have time dependence, the expectation values of the operators, which are physically observable, change with time because the state vectors bring in time dependence. Such a description where time dependence is carried entirely by the state vectors is known as the Schrödinger picture (which is the reason for the subscript). On the other hand, we can think of a description where the state vectors do not have any time dependence. That is, they are constant in time. However, the operators are now functions of time. Such a description is known as the Heisenberg picture. The situation here is somewhat analogous to rotations which you are all familiar with. For example, while considering rotations of physical systems, one may consider the reference frame to be fixed and the system to be rotating. However, one can also view the same phenomenon as if the system were at rest and the reference frame was rotating in the opposite direction. Physics, of course, does not depend on what viewpoint one takes.

Thus, it is clear that if we want the state vectors to be independent of time, then, we can define them as

$$|\psi_H\rangle = |\psi_S(0)\rangle = e^{\frac{i}{\hbar}Ht}|\psi_S(t)\rangle. \tag{3.110}$$

Furthermore, using (3.108) and (3.110), we have

$$\begin{aligned}\langle\psi_S(t)|\Omega_S|\psi_S(t)\rangle &= \langle\psi_S(0)|e^{\frac{i}{\hbar}Ht}\Omega_S e^{-\frac{i}{\hbar}Ht}|\psi_S(0)\rangle \\ &= \langle\psi_H|\Omega_H|\psi_H\rangle,\end{aligned} \tag{3.111}$$

where we have defined

$$\Omega_H = e^{\frac{i}{\hbar}Ht}\Omega_S e^{-\frac{i}{\hbar}Ht}. \tag{3.112}$$

This defines the operators in the Heisenberg picture, so that the expectation values remain the same in the two descriptions. It is clear now that

$$\begin{aligned}\frac{\mathrm{d}}{\mathrm{d}t}\langle\psi_S(t)|\Omega_S|\psi_S(t)\rangle &= \frac{\mathrm{d}}{\mathrm{d}t}\langle\psi_H|\Omega_H|\psi_H\rangle = \langle\psi_H|\frac{\mathrm{d}\Omega_H}{\mathrm{d}t}|\psi_H\rangle \\ &= \langle\psi_H|\frac{\partial\Omega_H}{\partial t}|\psi_H\rangle - \frac{i}{\hbar}\langle\psi_H|\left[\Omega_H, H\right]|\psi_H\rangle,\end{aligned} \tag{3.113}$$

where we have identified

$$\frac{\partial\Omega_H}{\partial t} = e^{\frac{i}{\hbar}Ht}\,\frac{\partial\Omega_S}{\partial t}\,e^{-\frac{i}{\hbar}Ht}. \tag{3.114}$$

(Namely, we are allowing for some explicit time dependence of Ω_S.) Since, in the Heisenberg picture, the state vectors (bras and kets) do not depend on time, we can also write (3.113) as an operator equation

$$\frac{\mathrm{d}\Omega_H}{\mathrm{d}t} = \frac{\partial\Omega_H}{\partial t} - \frac{i}{\hbar}\,[\Omega_H, H] = \frac{\partial\Omega_H}{\partial t} + \frac{1}{i\hbar}\,[\Omega_H, H]. \tag{3.115}$$

This describes the time evolution for operators in the Heisenberg picture, which is quite analogous to the Hamiltonian description of classical mechanics. We can also derive (3.115) by direct differentiation of the defining relation for the operator Ω_H in (3.112).

Exercise. Show that the Hamiltonian remains the same in both the Heisenberg and the Schrödinger pictures.

3.9 Selected problems

1. Prove the Jacobi identity for non-commuting operators A, B, C, namely,

$$[A,[B,C]] + [B,[C,A]] + [C,[A,B]] = 0. \tag{3.116}$$

2. Prove the Cauchy-Schwarz inequality as well as the triangle inequality in a linear vector space with an inner product. Namely, if $|f\rangle$ and $|g\rangle$ denote two vectors in such a space, then show that

$$a) \quad |\langle f|g\rangle|^2 \leq \langle f|f\rangle\langle g|g\rangle, \ (\textit{Cauchy-Schwarz inequality}). \tag{3.117}$$

$$b) \quad || |f\rangle + |g\rangle || \leq || |f\rangle || + || |g\rangle ||, \ (\textit{Triangle inequality}),$$

where

$$|| |A\rangle ||^2 = \langle A|A\rangle. \tag{3.118}$$

3. Find a set of three 2×2 matrices satisfying

$$\begin{aligned}
[Q_x, Q_y] &= -iQ_z, \\
[Q_y, Q_z] &= -iQ_x, \\
[Q_z, Q_x] &= -iQ_y.
\end{aligned} \tag{3.119}$$

4. Consider a quantum mechanical operator represented by a 3×3 matrix

$$\Omega = \begin{pmatrix} 1 & 0 & \sqrt{3} \\ 0 & 2 & 0 \\ \sqrt{3} & 0 & 3 \end{pmatrix}. \tag{3.120}$$

a) If a quantum mechanical system is in a state (not normalized)

$$|\psi\rangle = \begin{pmatrix} 1 \\ 2i \\ -1 \end{pmatrix}, \qquad (3.121)$$

then, which of the eigenvalues of Ω is *most likely* to emerge in a measurement?

b) What would be the *average* value for the measurement of Ω in this state?

5. a) If Ω is a dynamical operator which does not depend on t explicitly, then, show that the expectation value of the operator in an energy eigenstate satisfies

$$\frac{\mathrm{d}\langle\Omega\rangle}{\mathrm{d}t} = 0. \qquad (3.122)$$

b) Using this result, show that in one dimension if a particle moves in a potential $V(x) = gx^n$, where g denotes a coupling constant, then, in an energy eigenstate

$$2\,\langle T\rangle = n\,\langle V\rangle. \qquad (3.123)$$

Here T, V denote respectively the kinetic and the potential energy operators and this relation is known as the quantum virial theorem.

CHAPTER 4

Simple applications of Schrödinger equation

In these lectures, we will apply the principles of quantum mechanics described in the last chapter to study some simple, one dimensional quantum mechanical systems. This will bring out the essential features of quantum mechanical systems and will also illustrate the differences of such systems from their corresponding classical counterparts.

4.1 Free particle

Let us consider a free particle of mass m moving in one dimension. The Hamiltonian, in this case, is simple and has the form

$$H = \frac{P^2}{2m}. \tag{4.1}$$

Therefore, the Schrödinger equation,

$$i\hbar\,\frac{\mathrm{d}|\psi(t)\rangle}{\mathrm{d}t} = H|\psi(t)\rangle, \tag{4.2}$$

in this case, takes the form (in the x-basis),

$$\begin{aligned} i\hbar\frac{\partial\psi(x,t)}{\partial t} &= -\frac{\hbar^2}{2m}\frac{\partial^2\psi(x,t)}{\partial x^2}, \\ \text{or,}\quad \frac{\partial\psi(x,t)}{\partial t} &= \frac{i\hbar}{2m}\frac{\partial^2\psi(x,t)}{\partial x^2}. \end{aligned} \tag{4.3}$$

This is like the heat equation except that the coefficient, on the right hand side, is imaginary. The solution of (4.3) can be written as

$$\psi(x,t) = \frac{N}{\sqrt{a^2 + \frac{i\hbar t}{m}}}\, e^{-\frac{x^2}{2(a^2+\frac{i\hbar t}{m})}}, \tag{4.4}$$

where "a" is a constant, which can be determined from the initial value of the wave function. The constant N is determined from the normalization of the probability (wave function). The probability density, in this case, is given by

$$\psi^*\psi = \frac{|N|^2}{a\sqrt{a^2 + \frac{\hbar^2 t^2}{m^2 a^2}}} e^{-\frac{x^2}{a^2 + \frac{\hbar^2 t^2}{m^2 a^2}}} . \tag{4.5}$$

Thus, we see that the wave function for a free particle, also called a wave packet, is such that the probability density is described by a Gaussian. The probability of finding the particle peaks around $x = 0$ and the Gaussian has a mean width of

$$\Delta = \frac{1}{\sqrt{2}} \left(a^2 + \frac{\hbar^2 t^2}{m^2 a^2} \right)^{\frac{1}{2}} . \tag{4.6}$$

Therefore, we see that, by choosing an appropriate "a", we can localize the particle fairly well initially (at time $t = 0$), but as time grows, the width of the Gaussian increases. This is known as the dispersion of the wave packet and is a general feature of most quantum mechanical systems.

Exercise. Derive the same result by working in the momentum basis and then transforming to the x-basis.

4.2 Infinite square well

Let us consider the one dimensional potential given by (see Fig. 4.1)

$$V(x) = \begin{cases} 0, & \text{for } x^2 \leq a^2, \\ \infty, & \text{for } x^2 \geq a^2. \end{cases} \tag{4.7}$$

To examine the motion of a particle in this potential, let us attempt to solve the system with

$$V(x) = \begin{cases} 0, & \text{for } x^2 \leq a^2, \text{ or, } -a \leq x \leq a, \\ V_0, & \text{for } x^2 \geq a^2, \text{ or, } x \geq a,\ x \leq -a, \end{cases} \tag{4.8}$$

and take the limit $V_0 \to \infty$. In the x-basis, the Hamiltonian has the form

$$H = -\frac{\hbar^2}{2m} \frac{\mathrm{d}^2}{\mathrm{d}x^2} + V(x), \tag{4.9}$$

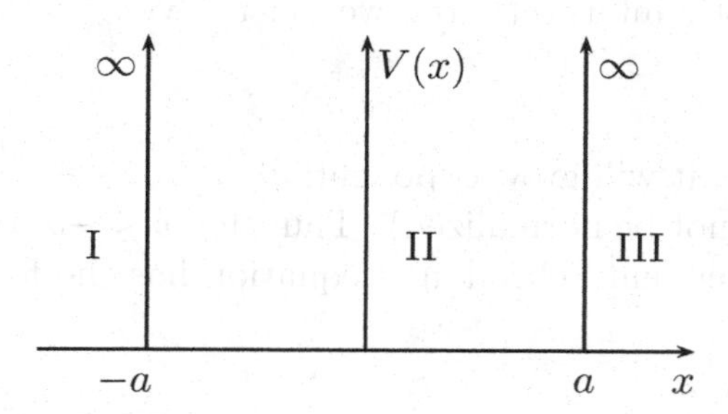

Figure 4.1: Particle moving in an infinite square well.

with $V(x)$ given in (4.8). The time independent Schrödinger equation, (3.93), in the present case, takes the form

$$\left(-\frac{\hbar^2}{2m}\frac{\mathrm{d}^2}{\mathrm{d}x^2}+V(x)\right)u(x)=Eu(x). \tag{4.10}$$

Since the potential in the problem is different in different regions, we have to solve the Schrödinger equation separately in the three regions.

Region I*:* $x \leq -a$. In this region, the potential is a constant and the Schrödinger equation has the form

$$-\frac{\hbar^2}{2m}\frac{\mathrm{d}^2u}{\mathrm{d}x^2}+V_0u=Eu,$$

$$\text{or,}\quad \frac{\mathrm{d}^2u}{\mathrm{d}x^2}=\frac{2m}{\hbar^2}(V_0-E)u, \tag{4.11}$$

where we are assuming $V_0 > E > 0$. We note here that, since inside the potential well the potential vanishes, E represents the kinetic energy of the particle in that region which has to be positive, leading to $E > 0$. Furthermore, we assume $V_0 > E$ because we are ultimately going to take the limit $V_0 \to \infty$. The solution of the equation (4.11), in region I, can be written in the form

$$u_{\mathrm{I}}(x)=Ae^{\kappa x}+Be^{-\kappa x}, \tag{4.12}$$

where A and B are constants and we have defined

$$\kappa=\sqrt{\frac{2m}{\hbar^2}(V_0-E)}. \tag{4.13}$$

It is clear from the solution (4.12) that if the wave function has to retain a probabilistic interpretation, we must have

$$B = 0, \tag{4.14}$$

because, otherwise, it will grow exponentially as $x \to -\infty$ and would not converge (can not be normalized). Thus, for $x \leq -a$, the solution for the time independent Schrödinger equation has the form

$$u_{\rm I}(x) = A e^{\kappa x}. \tag{4.15}$$

However, as $V_0 \to \infty$, $\kappa \to \infty$. Therefore, in this limit,

$$u_{\rm I}(x) = 0, \qquad x \leq -a. \tag{4.16}$$

Region III*:* $x \geq a$. As in region I, we can show that, in this region as well, the wave function vanishes,

$$u_{\rm III}(x) = 0, \qquad x \geq a. \tag{4.17}$$

Region II*:* $-a \leq x \leq a$. In this region, the potential vanishes so that the Schrödinger equation takes the simple form,

$$-\frac{\hbar^2}{2m}\frac{\mathrm{d}^2 u}{\mathrm{d}x^2} = Eu,$$

$$\text{or,} \quad \frac{\mathrm{d}^2 u}{\mathrm{d}x^2} = -\frac{2mE}{\hbar^2}\,u = -k^2 u, \qquad k^2 = \frac{2mE}{\hbar^2},$$

$$\text{or,} \quad u_{\rm II}(x) = C\sin kx + D\cos kx, \tag{4.18}$$

where C and D are arbitrary constants.

We know from our earlier discussions that the solutions of the Schrödinger equation have to be continuous everywhere and, in particular, at the boundaries. Therefore, matching the solutions at $x = \pm a$ we have

$$u_{\rm II}(a) = C\sin ka + D\cos ka = 0 = u_{\rm III}(a),$$

$$u_{\rm II}(-a) = -C\sin ka + D\cos ka = 0 = u_{\rm I}(-a). \tag{4.19}$$

These linear homogeneous equations (in the unknowns C and D) can have nontrivial solutions only if the determinant of the coefficient matrix vanishes, namely,

$$\sin ka\,\cos ka = 0. \tag{4.20}$$

There are two possible nontrivial solutions to the set of conditions in (4.20).

1. Even solution: $\cos ka = 0$.

   In this case, it follows from (4.19) that

   $$C = 0,$$

   and we must have (for $\cos ka = 0$ to hold)

   $$ka = \frac{(2n+1)\pi}{2}, \quad n \text{ integer}$$

   $$\text{or,} \quad k_n^2 = \frac{(2n+1)^2\pi^2}{4a^2},$$

   $$\text{or,} \quad E_n = \frac{\hbar^2 k_n^2}{2m} = \frac{(2n+1)^2\pi^2\hbar^2}{8ma^2}. \tag{4.21}$$

   In this case, the nontrivial solution in region II has the form

   $$u_n(x) = D_n \cos k_n x, \qquad -a \leq x \leq a, \tag{4.22}$$

   which is an even function of x.

2. Odd solution: $\sin ka = 0$.

   In this case, it follows from (4.19) that

   $$D = 0,$$

   and we must have (for $\sin ka = 0$ to hold)

   $$ka = n\pi, \quad n \text{ integer } \neq 0,$$

   $$\text{or,} \quad k_n^2 = \frac{n^2\pi^2}{a^2},$$

   $$\text{or,} \quad E_n = \frac{\hbar^2 k_n^2}{2m} = \frac{n^2\pi^2\hbar^2}{2ma^2}. \tag{4.23}$$

   The nontrivial solution in region II, in this case, takes the form

   $$u_n(x) = C_n \sin k_n x, \qquad -a \leq x \leq a, \tag{4.24}$$

   which is an odd function of x.

We note that corresponding to every value of E_n given in (4.21) or (4.23), we will have a physical solution and we can write a general solution of the Schrödinger equation as a linear superposition

$$\psi(x,t) = \sum_n A_n u_n(x) e^{-\frac{i}{\hbar}E_n t}, \tag{4.25}$$

where A_n's are constants.

One of the things that we notice immediately from this analysis is that, whereas classically particle motion is allowed for any $E > 0$, quantum mechanically particle motion is allowed only for discrete values of the energy, namely, energy for the system is quantized (see (4.21),(4.23)). We also see that for this system,

$$\lim_{|x|\to\infty} u(x) \to 0. \tag{4.26}$$

Such a system, where the wave function vanishes (beyond a certain range or asymptotically), is called a bound state solution and, for every bound state, we have quantization of energy. A very familiar example is the Hydrogen atom which we will study in detail later. In fact, in the present system,

$$u(x) = 0, \qquad \text{for } x^2 \geq a^2. \tag{4.27}$$

Therefore, this system is also referred to as a particle in a box of length $2a$. (The probability for finding the particle outside this region is zero.)

Exercise. Normalize the solutions obtained for the infinite square well potential. Calculate Δx for the ground state. Estimate the ground state energy from the uncertainty principle and compare it with the actual value.

Exercise. Plot the first few solutions and describe their qualitative features. In particular, show that the nth state has $(n-1)$ nodes inside the well.

4.3 Finite square well

Let us next consider the same potential as in the previous case, but with a finite value for the height of the potential (see Fig. 4.2), namely,

$$V(x) = \begin{cases} 0, & \text{for } -a \leq x \leq a, \\ V_0, & \text{for } x \geq a, \quad x \leq -a. \end{cases} \tag{4.28}$$

Let us consider the bound state motion of a particle in this potential for which,

$$V_0 > E > 0. \tag{4.29}$$

Once again, we have to solve the equation in different regions separately, since the potential is different in different regions.

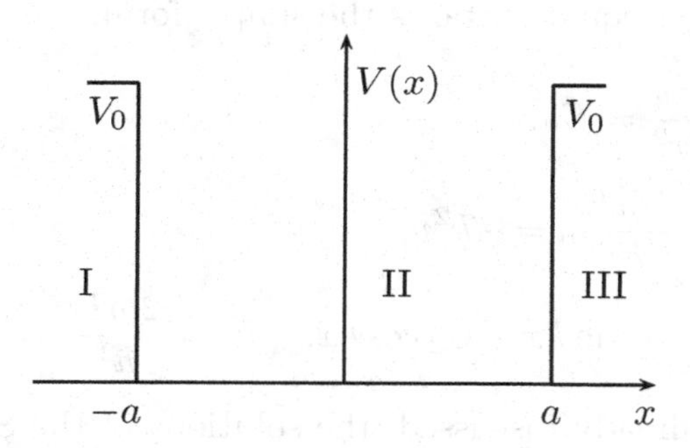

Figure 4.2: Particle moving in a square well of height V_0.

Region I*:* $x \leq -a$. In this region, the Schrödinger equation has the form,

$$-\frac{\hbar^2}{2m}\frac{\mathrm{d}^2u}{\mathrm{d}x^2} = -(V_0 - E)u,$$

$$\text{or,}\quad \frac{\mathrm{d}^2u}{\mathrm{d}x^2} = \frac{2m}{\hbar^2}(V_0 - E)u = \kappa^2 u,$$

$$\text{or,}\quad u_{\mathrm{I}}(x) = A_1 e^{\kappa x} + A_2 e^{-\kappa x}, \quad \kappa^2 = \frac{2m}{\hbar^2}(V_0 - E), \tag{4.30}$$

where A_1, A_2 are constants and the boundedness of the solution (as $x \to -\infty$) requires that $A_2 = 0$, so that we have

$$u_{\mathrm{I}}(x) = A_1 e^{\kappa x}, \qquad x \leq -a. \tag{4.31}$$

Region III*:* $x \geq a$. In this region, the Schrödinger equation has the same form as in (4.30) so that

$$\frac{\mathrm{d}^2u}{\mathrm{d}x^2} = \kappa^2 u,$$

$$\text{or,}\quad u_{\mathrm{III}}(x) = B_1 e^{\kappa x} + B_2 e^{-\kappa x}. \tag{4.32}$$

Since the solution has to be bounded as $x \to \infty$, we must have $B_1 = 0$, which gives

$$u_{\mathrm{III}}(x) = B_2 e^{-\kappa x}, \qquad x \geq a. \tag{4.33}$$

Region II*:* $-a \leq x \leq a$. In this region, the potential vanishes and the Schrödinger equation takes the simple form,

$$-\frac{\hbar^2}{2m}\frac{\mathrm{d}^2u}{\mathrm{d}x^2} = Eu,$$

$$\text{or,}\quad \frac{\mathrm{d}^2u}{\mathrm{d}x^2} = -\frac{2mE}{\hbar^2}u = -k^2u,$$

$$\text{or,}\quad u_{\mathrm{II}}(x) = C_1\sin kx + C_2\cos kx,\quad k^2 = \frac{2mE}{\hbar^2}. \tag{4.34}$$

As we have already discussed, the solutions of the Schrödinger equation and their derivatives must be continuous across the boundaries. Thus, matching solutions at $x = a$, we have

$$u_{\mathrm{III}}(a) = u_{\mathrm{II}}(a),$$

$$\text{or,}\quad B_2e^{-\kappa a} = C_1\sin ka + C_2\cos ka. \tag{4.35}$$

Similarly, matching the solutions at $x = -a$, we obtain

$$u_{\mathrm{I}}(-a) = u_{\mathrm{II}}(-a),$$

$$\text{or,}\quad A_1e^{-\kappa a} = -C_1\sin ka + C_2\cos ka. \tag{4.36}$$

Furthermore, matching the derivatives $\frac{\mathrm{d}u}{\mathrm{d}x}$ at $x = a$, we have

$$\left.\frac{\mathrm{d}u_{\mathrm{III}}(x)}{\mathrm{d}x}\right|_{x=a} = \left.\frac{\mathrm{d}u_{\mathrm{II}}(x)}{\mathrm{d}x}\right|_{x=a},$$

$$\text{or,}\quad -\kappa B_2e^{-\kappa a} = k(C_1\cos ka - C_2\sin ka). \tag{4.37}$$

Similarly, matching of the derivatives at $x = -a$ gives

$$\left.\frac{\mathrm{d}u_{\mathrm{I}}(x)}{\mathrm{d}x}\right|_{x=-a} = \left.\frac{\mathrm{d}u_{\mathrm{II}}(x)}{\mathrm{d}x}\right|_{x=-a},$$

$$\text{or,}\quad \kappa A_1e^{-\kappa a} = k(C_1\cos ka + C_2\sin ka). \tag{4.38}$$

Adding and subtracting (4.35) and (4.36), we obtain

$$2C_2\cos ka = (A_1 + B_2)e^{-\kappa a},$$

$$2C_1\sin ka = -(A_1 - B_2)e^{-\kappa a}. \tag{4.39}$$

Similarly, adding and subtracting (4.37) and (4.38) we have

$$2kC_1\cos ka = \kappa(A_1 - B_2)e^{-\kappa a},$$

$$2kC_2\sin ka = \kappa(A_1 + B_2)e^{-\kappa a}. \tag{4.40}$$

To simplify the analysis, let us next define

$$A_1 - B_2 = -D_1,$$
$$A_1 + B_2 = D_2. \tag{4.41}$$

Then, the two sets of equations in (4.39) and (4.40) can be combined and rewritten as

$$2C_1 \sin ka - D_1 e^{-\kappa a} = 0,$$
$$2kC_1 \cos ka + \kappa D_1 e^{-\kappa a} = 0, \tag{4.42}$$

and

$$2C_2 \cos ka - D_2 e^{-\kappa a} = 0,$$
$$2kC_2 \sin ka - \kappa D_2 e^{-\kappa a} = 0. \tag{4.43}$$

The set of homogeneous linear equations in (4.42) has nontrivial solutions (for C_1 and D_1) only if

$$\det \begin{vmatrix} 2\sin ka & -e^{-\kappa a} \\ 2k\cos ka & \kappa e^{-\kappa a} \end{vmatrix} = 0,$$

$$\text{or,}\quad 2\kappa \sin ka e^{-\kappa a} + 2k\cos ka e^{-\kappa a} = 0,$$

$$\text{or,}\quad k\cot ka = -\kappa. \tag{4.44}$$

The other set of homogeneous linear equations in (4.43) has nontrivial solutions (for C_2 and D_2) only if

$$\det \begin{vmatrix} 2\cos ka & -e^{-\kappa a} \\ 2k\sin ka & -\kappa e^{-\kappa a} \end{vmatrix} = 0,$$

$$\text{or,}\quad -2\kappa \cos ka e^{-\kappa a} + 2k\sin ka e^{-\kappa a} = 0,$$

$$\text{or,}\quad k\tan ka = \kappa. \tag{4.45}$$

It is clear that it is impossible to satisfy both (4.44) and (4.45) simultaneously and, therefore, we have two classes of solutions. In either of the relations (4.44) and (4.45), however, we see that energy is the only unknown quantity. Therefore, we see again that quantum mechanically motion is allowed only if the energy satisfies certain conditions. Let us analyze the two classes of solutions in detail

1. Even solution: $k\tan ka = \kappa$.

In this case, we have

$$C_1 = D_1 = 0.$$

Let us define the dimensionless variables, $\xi = ka$ and $\eta = \kappa a$. Then, it follows that (see the definitions in (4.30) and (4.34))

$$\begin{aligned}\eta^2 + \xi^2 &= a^2(\kappa^2 + k^2)\\ &= a^2\left(\frac{2m}{\hbar^2}(V_0 - E) + \frac{2m}{\hbar^2}E\right)\\ &= \frac{2mV_0a^2}{\hbar^2} = \text{constant},\end{aligned}\tag{4.46}$$

which is the equation of a circle.

Furthermore, relation (4.45) can be written in these variables as

$$\begin{aligned}&ka\tan ka = \kappa a,\\ \text{or,}\quad &\xi\tan\xi = \eta.\end{aligned}\tag{4.47}$$

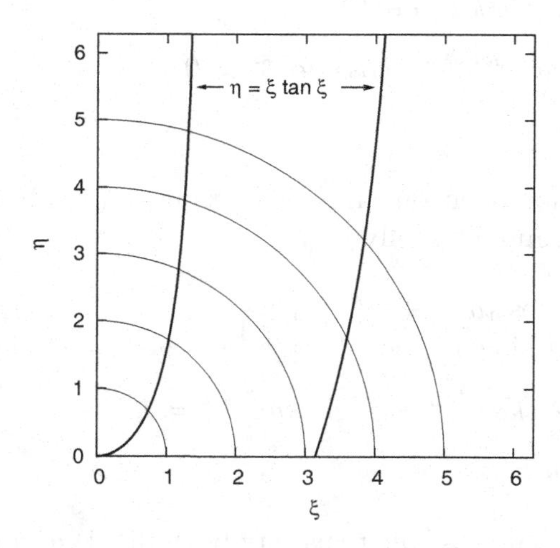

Figure 4.3: Existence of even solutions corresponds to intersection points of the two curves.

The simultaneous solutions of the two equations, (4.46) and (4.47), can be obtained by plotting the two functions and determining their points of intersection. We note that both ξ and

η take positive values and, therefore, plotting the two functions on the first quadrant only, we obtain the graph shown in Fig. 4.3. It is clear that they intersect once if $0 \leq \frac{2mV_0a^2}{\hbar^2} < \pi^2$. There are two intersections if $\pi^2 \leq \frac{2mV_0a^2}{\hbar^2} < (2\pi)^2$. And the number of solutions (intersections) keeps on increasing as the value of the parameter $\frac{2mV_0a^2}{\hbar^2}$ grows. For each such allowed value of the energy (determined from the point of intersection), we will have an even solution in the present case.

2. Odd solution: $k \cot ka = -\kappa$.

   In this case, it follows that

$$C_2 = D_2 = 0.$$

We note that (4.46) continues to hold, namely,

$$\eta^2 + \xi^2 = \frac{2mV_0a^2}{\hbar^2} = \text{constant},$$

and, in addition, we can write (4.44) in the dimensionless variables as

$$ka \cot ka = -\kappa a,$$

$$\text{or,} \quad \xi \cot \xi = -\eta. \tag{4.48}$$

Once again, the simultaneous solutions of (4.46) and (4.48) can be obtained graphically as before. Plotting these functions in the first quadrant as shown in Fig. 4.4, we see that if $0 \leq \frac{2mV_0a^2}{\hbar^2} < \left(\frac{\pi}{2}\right)^2$, there is no solution. For, $\left(\frac{\pi}{2}\right)^2 \leq \frac{2mV_0a^2}{\hbar^2} \leq \left(\frac{3\pi}{2}\right)^2$, there is one solution (intersection) and so on. Thus, we see that the existence of solutions depends on the parameters of the problem like the mass, V_0 as well as the range of the potential. A simultaneous solution, of course, determines the allowed energy for which the quantum mechanical motion is described by an odd wave function, in the present case.

It is interesting to note here that, for both the even and the odd solutions, $u(x)$ is nonzero outside the well so that there will be a nonzero probability for finding the particle there, which is different from what we would expect in classical mechanics. Let us also note here parenthetically that, if $V_0 \to \infty$, then, it is easy to see that the intersections will occur at $\xi = (n + \frac{1}{2})\pi$, $n\pi$, which is what we have seen in our earlier analysis (see (4.21) and (4.23) respectively).

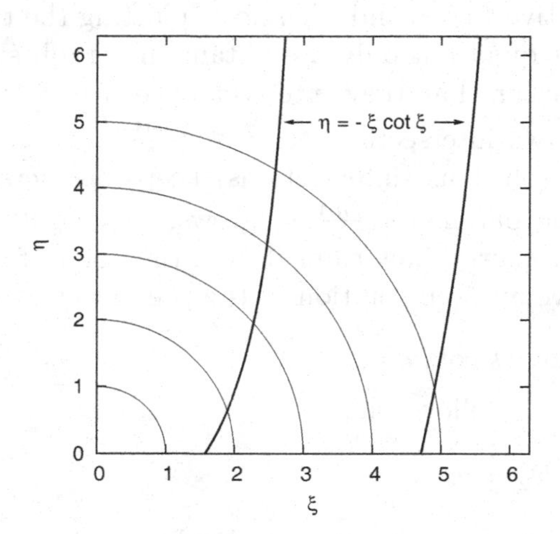

Figure 4.4: Existence of odd solutions corresponds to the intersection points of the two curves.

4.4 Parity

In the case of the infinite square well as well as the finite square well, we have classified solutions as even and odd. Let us examine the origin of this. The time independent Schrödinger equation has the general form

$$\left[-\frac{\hbar^2}{2m}\frac{\mathrm{d}^2}{\mathrm{d}x^2}+V(x)\right]u(x)=Eu(x). \tag{4.49}$$

Let us assume now that our potential is symmetric about $x=0$, namely, $V(-x)=V(x)$, which is, in fact, the case in the two examples which we have studied. In this case, upon changing $x\rightarrow -x$, the Schrödinger equation, (4.49), becomes

$$\left[-\frac{\hbar^2}{2m}\frac{\mathrm{d}^2}{\mathrm{d}x^2}+V(x)\right]u(-x)=Eu(-x). \tag{4.50}$$

This shows that if $u(x)$ is an eigenstate of the Hamiltonian, so is $u(-x)$ with the same energy eigenvalue. Therefore, they must be related by a multiplicative constant,

$$u(-x)=\epsilon u(x). \tag{4.51}$$

Furthermore, iterating this twice, we are led to

$$u(x)=\epsilon u(-x)=\epsilon^2 u(x). \tag{4.52}$$

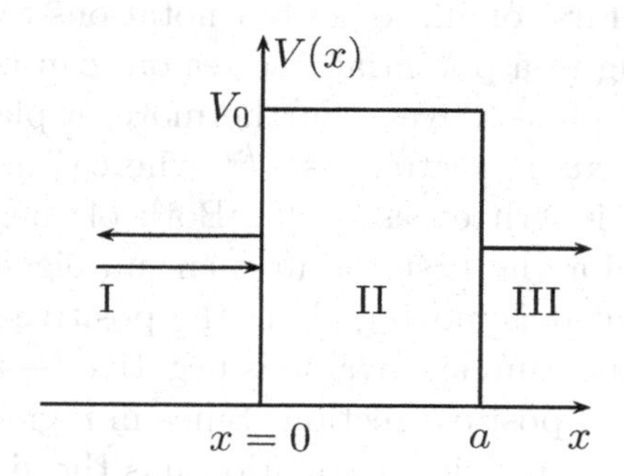

Figure 4.5: Particle incident on a barrier of height V_0 with energy $E < V_0$.

Therefore, this leads to

$$\epsilon^2 = 1, \quad \text{or,} \quad \epsilon = \pm 1. \tag{4.53}$$

We conclude, therefore, that all the eigenfunctions of a Hamiltonian with a symmetric potential are either even or odd under a reflection of the coordinate (changing $x \to -x$), and are said to have even or odd parity respectively. The transformation $x \to -x$ is known as space reflection or parity. We will study more about parity later when we discuss symmetries in quantum mechanical systems.

4.5 Penetration of a barrier

Let us next consider the physical example where a particle with energy $E > 0$ is incident on a potential barrier (see Fig. 4.5) given by

$$V(x) = \begin{cases} 0, & x \leq 0, \quad x \geq a, \\ V_0, & a \geq x \geq 0, \end{cases} \tag{4.54}$$

where V_0 represents the height of the barrier. If $E < V_0$, then classically we expect the particle to be reflected. But, as we will see, quantum mechanics allows for a transmitted wave in addition to the reflected beam. In other words, quantum mechanically a particle can actually tunnel through a barrier.

Since the potential is different in different regions, we have to solve the Schrödinger equation separately in each of these regions.

Regions I *and* III. First of all, to get our notations right, we note that a particle, traveling in a potential free region, can be represented by a superposition of plane waves. Furthermore, a plane wave moving in the positive x axis is written as e^{ikx} whereas one moving in the opposite direction is written as e^{-ikx}. Both of them have the same energy. However, for the first the momentum eigenvalue is positive ($\hbar k$) indicating that it is moving along the positive x-axis, while, for the latter, the momentum eigenvalue is negative ($-\hbar k$) implying that it is moving in the opposite direction. Since in regions I and III there is no potential, the Schrödinger equation has the simple form

$$-\frac{\hbar^2}{2m}\frac{\mathrm{d}^2u}{\mathrm{d}x^2} = Eu, \tag{4.55}$$

so that

$$\begin{aligned} u_{\mathrm{I}}(x) &= Ae^{ikx} + Be^{-ikx}, && x \leq 0, \\ u_{\mathrm{III}}(x) &= Ce^{ikx}, && x \geq a, \end{aligned} \tag{4.56}$$

where we have used our physical intuition that in region I there will be an incident as well as a reflected wave, while in region III there can at most be a transmitted wave. As before, we have also identified

$$k = \sqrt{\frac{2mE}{\hbar^2}}. \tag{4.57}$$

We note that A, B, C, in (4.56) are constants, but to obtain their physical meaning, let us write down the probability current densities in both these regions (see (3.103))

$$\begin{aligned} S_{\mathrm{I}} &= \frac{\hbar}{2im}\left(u_{\mathrm{I}}^*(x)\frac{\mathrm{d}u_{\mathrm{I}}(x)}{\mathrm{d}x} - u_{\mathrm{I}}(x)\frac{\mathrm{d}u_{\mathrm{I}}^*(x)}{\mathrm{d}x}\right) \\ &= \frac{\hbar}{2im}\Big[ik\left(A^*e^{-ikx} + B^*e^{ikx}\right)\left(Ae^{ikx} - Be^{-ikx}\right) \\ &\qquad +ik\left(Ae^{ikx} + Be^{-ikx}\right)\left(A^*e^{-ikx} - B^*e^{ikx}\right)\Big] \\ &= \frac{\hbar k}{2m}\Big[2(AA^* - BB^*) + (-A^*B + A^*B)e^{-2ikx} \\ &\qquad +(B^*A - B^*A)e^{2ikx}\Big] \\ &= \frac{\hbar k}{m}\left(|A|^2 - |B|^2\right) = v\left(|A|^2 - |B|^2\right). \end{aligned} \tag{4.58}$$

Similarly, in region III we obtain

$$\begin{aligned} S_{\rm III} &= \frac{\hbar}{2im}\left(u_{\rm III}^*(x)\,\frac{{\rm d}u_{\rm III}(x)}{{\rm d}x} - u_{\rm III}(x)\,\frac{{\rm d}u_{\rm III}^*(x)}{{\rm d}x}\right) \\ &= \frac{\hbar k}{m}\,CC^* = v|C|^2. \end{aligned} \tag{4.59}$$

Here, we have defined v as the velocity corresponding to the wave number k. Furthermore, the expressions for $S_{\rm I}$ and $S_{\rm III}$ suggest that they can be interpreted as the net probability flux to the right. This is also consistent with the physical picture of an incident, a reflected and a transmitted wave. In other words, A represents the amplitude for the incident wave, B the amplitude for the reflected wave and C the amplitude for the transmitted wave, so that we can also write

$$\begin{aligned} S_{\rm inc} &= \frac{\hbar k}{m}\,|A|^2, \\ S_{\rm refl} &= \frac{\hbar k}{m}|B|^2, \\ S_{\rm trans} &= \frac{\hbar k}{m}|C|^2. \end{aligned} \tag{4.60}$$

Region II. Here the potential is a positive constant and, correspondingly, the time independent Schrödinger equation takes the form

$$\begin{aligned} & -\frac{\hbar^2}{2m}\frac{{\rm d}^2u}{{\rm d}x^2} + V_0u = Eu, \\ \text{or,}\quad & \frac{{\rm d}^2u}{{\rm d}x^2} = \frac{2m}{\hbar^2}(V_0 - E)u = \kappa^2 u, \quad \kappa = \sqrt{\frac{2m}{\hbar^2}(V_0 - E)}, \\ \text{or,}\quad & u_{\rm II}(x) = D_1e^{-\kappa x} + D_2e^{\kappa x}, \qquad 0 \le x \le a. \end{aligned} \tag{4.61}$$

Since the region in which this solution is valid is of finite extension, both the exponentially growing and the damped solutions are allowed.

The solutions and their first derivatives have to be continuous. Therefore, matching the solutions and their derivatives at $x = 0$, we have

$$\begin{aligned} u_{\rm I}(0) = u_{\rm II}(0), &\quad\Rightarrow\quad A + B = D_1 + D_2, \\ u_{\rm I}'(0) = u_{\rm II}'(0), &\quad\Rightarrow\quad ik(A - B) = -\kappa(D_1 - D_2), \end{aligned} \tag{4.62}$$

where "prime" denotes a derivative (with respect to x). Multiplying the first relation in (4.62) by ik and taking the sum and the difference of the two relations, we obtain,

$$\begin{aligned} 2ikA &= (ik - \kappa)D_1 + (ik + \kappa)D_2, \\ 2ikB &= (ik + \kappa)D_1 + (ik - \kappa)D_2. \end{aligned} \tag{4.63}$$

Furthermore, matching the solutions and their derivatives at $x = a$ we have

$$\begin{aligned} u_{\rm III}(a) &= u_{\rm II}(a), \quad \Rightarrow \quad Ce^{ika} = D_1\, e^{-\kappa a} + D_2\, e^{\kappa a}, \\ u'_{\rm III}(a) &= u'_{\rm II}(a), \quad \Rightarrow \quad ikCe^{ika} = -\kappa\big(D_1\, e^{-\kappa a} - D_2\, e^{\kappa a}\big), \end{aligned} \tag{4.64}$$

so that, from the above two relations, we obtain a relation between D_1 and D_2 of the form

$$(ik + \kappa)D_1\, e^{-\kappa a} + (ik - \kappa)D_2\, e^{\kappa a} = 0, \tag{4.65}$$

which leads to

$$\frac{D_1}{D_2} = -\frac{(ik - \kappa)}{(ik + \kappa)} e^{2\kappa a}. \tag{4.66}$$

Furthermore, using (4.66) in (4.63), we have

$$\begin{aligned} \frac{B}{A} &= \frac{(ik + \kappa)D_1 + (ik - \kappa)D_2}{(ik - \kappa)D_1 + (ik + \kappa)D_2} \\ &= \frac{(ik + \kappa)\frac{D_1}{D_2} + (ik - \kappa)}{(ik - \kappa)\frac{D_1}{D_2} + (ik + \kappa)} \\ &= \frac{-(ik + \kappa)\frac{(ik-\kappa)}{ik+\kappa} e^{2\kappa a} + (ik - \kappa)}{-(ik - \kappa)\frac{(ik-\kappa)}{ik+\kappa} e^{2\kappa a} + (ik + \kappa)} \\ &= \frac{-(ik - \kappa)(e^{2\kappa a} - 1)}{-\frac{(ik-\kappa)^2}{ik+\kappa} e^{2\kappa a} + (ik + \kappa)} \\ &= \frac{(\kappa^2 + k^2)(e^{2\kappa a} - 1)}{-(ik - \kappa)^2 e^{2\kappa a} + (ik + \kappa)^2} \\ &= \frac{(\kappa^2 + k^2)e^{\kappa a}(e^{\kappa a} - e^{-\kappa a})}{(\kappa^2 - k^2)(1 - e^{2\kappa a}) + 2i\kappa k(1 + e^{2\kappa a})} \\ &= \frac{(\kappa^2 + k^2)(e^{\kappa a} - e^{-\kappa a})}{(\kappa^2 - k^2)(e^{-\kappa a} - e^{\kappa a}) + 2i\kappa k(e^{-\kappa a} + e^{\kappa a})} \\ &= \frac{(\kappa^2 + k^2)\sinh \kappa a}{-(\kappa^2 - k^2)\sinh \kappa a + 2i\kappa k \cosh \kappa a}. \end{aligned} \tag{4.67}$$

This leads to

$$\left|\frac{B}{A}\right|^2 = \frac{(\kappa^2+k^2)^2\sinh^2\kappa a}{(\kappa^2-k^2)^2\sinh^2\kappa a + 4\kappa^2k^2\cosh^2\kappa a}$$
$$= \frac{(\kappa^2+k^2)^2\sinh^2\kappa a}{(\kappa^2+k^2)^2\sinh^2\kappa a + 4\kappa^2k^2}, \tag{4.68}$$

where we have used $\cosh^2 x = 1 + \sinh^2 x$. Putting in the values for k^2 and κ^2 from (4.57) and (4.61) respectively, we have

$$\left|\frac{B}{A}\right|^2 = \frac{\left(\frac{2m}{\hbar^2}V_0\right)^2\sinh^2\kappa a}{\left(\frac{2m}{\hbar^2}V_0\right)^2\sinh^2\kappa a + \frac{16m^2}{\hbar^4}E(V_0-E)}$$
$$= \frac{V_0^2\sinh^2\kappa a}{V_0^2\sinh^2\kappa a + 4E(V_0-E)}. \tag{4.69}$$

Similarly, using (4.66) and the second equation in (4.64), we obtain

$$e^{ika}\frac{C}{A} = \frac{-2\kappa\,(D_1\,e^{-\kappa a} - D_2\,e^{\kappa a})}{(ik-\kappa)D_1 + (ik+\kappa)D_2}$$
$$= \frac{-2\kappa\left(e^{-\kappa a}\,\frac{D_1}{D_2} - e^{\kappa a}\right)}{(ik-\kappa)\,\frac{D_1}{D_2} + (ik+\kappa)}$$
$$= \frac{-2\kappa\left(-e^{-\kappa a}\,\frac{(ik-\kappa)}{ik+\kappa}e^{2\kappa a} - e^{\kappa a}\right)}{-(ik-\kappa)\,\frac{(ik-\kappa)}{ik+\kappa}e^{2\kappa a} + (ik+\kappa)}$$
$$= \frac{2\kappa e^{\kappa a}\left(\frac{ik-\kappa}{ik+\kappa}+1\right)}{\frac{1}{(ik+\kappa)}\times(-(ik-\kappa)^2e^{2\kappa a} + (ik+\kappa)^2)}$$
$$= \frac{2\kappa e^{\kappa a}(2ik)}{((\kappa^2-k^2)(1-e^{2\kappa a}) + 2i\kappa k(1+e^{2\kappa a}))}$$
$$= \frac{2i\kappa k}{(-(\kappa^2-k^2)\sinh\kappa a + 2i\kappa k\cosh\kappa a)}, \tag{4.70}$$

which leads to

$$\left|\frac{C}{A}\right|^2 = \frac{4\kappa^2k^2}{(\kappa^2-k^2)^2\sinh^2\kappa a + 4\kappa^2k^2\cosh^2\kappa a}$$
$$= \frac{4\kappa^2k^2}{(\kappa^2+k^2)^2\sinh^2\kappa a + 4\kappa^2k^2}$$

$$= \frac{4E(V_0 - E)}{V_0^2 \sinh^2 \kappa a + 4E(V_0 - E)}. \tag{4.71}$$

The quantities $\left|\frac{B}{A}\right|^2$ and $\left|\frac{C}{A}\right|^2$ are known respectively as the reflection and the transmission coefficients. (They, basically, measure the fractions of the incident flux that are reflected and transmitted respectively.) It is obvious from (4.69) and (4.71) that

$$\left|\frac{B}{A}\right|^2 + \left|\frac{C}{A}\right|^2 = 1, \tag{4.72}$$

as it should be, since we do not expect probability to disappear suddenly. The fact that the particle is able to get to the other side of the barrier, even when the energy of the particle is lower than the height of the barrier, is known as tunneling and is a quantum mechanical phenomenon. Looking at the transmission coefficient, (4.71), it is clear that for $E \ll V_0$, tunneling is highly suppressed. Nonetheless, tunneling effects are quite physical, particularly in processes such as radioactive decays.

4.6 Selected problems

1. Justify why the Schrödinger wave function, $u(x)$, and its derivative, $\frac{du(x)}{dx}$, have to be continuous across the boundary for the finite square well potential, even though the potential is discontinuous.

2. Solve the Schrödinger equation for a particle bound in an attractive delta potential in one dimension,

$$V(x) = -\gamma\,\delta(x), \qquad \gamma > 0. \tag{4.73}$$

   Calculate the average kinetic and potential energy for such a particle in this state (namely, calculate $\langle T\rangle$ as $\langle E\rangle - \langle V\rangle$).

3. Calculate $\langle T\rangle$ directly for the previous problem, namely, evaluate

$$\langle T\rangle = \int \mathrm{d}x\,\psi^*(x)\left(-\frac{\hbar^2}{2m}\frac{\mathrm{d}^2}{\mathrm{d}x^2}\right)\psi(x), \tag{4.74}$$

   and compare with the previous result. (You have to be careful, since the wave function may not be a smooth function.)

4. Show that, in one dimension, bound states cannot be degenerate.

5. A particle, moving in an infinite square well of width $2a$ (discussed in class), is in a state whose wavefunction at time $t = 0$ is given by

$$\psi(x, t = 0) = \frac{1}{\sqrt{2}} \left(u_0(x) + u_1(x)\right), \tag{4.75}$$

where $u_0(x)$ and $u_1(x)$ are the normalized ground state and the first excited state wave functions respectively. What is its average kinetic energy and average potential energy in this state as a function of time? Calculate Δx in this state.

6. A particle moves in a potential, in one dimension, of the form

$$V(x) = \begin{cases} \infty, & x^2 \geq a^2, \\ \gamma\delta(x), & x^2 \leq a^2, \end{cases} \tag{4.76}$$

where $\gamma > 0$. For sufficiently large γ, calculate the time required for the particle to tunnel from being in the ground state of the well extending from $x = -a$ to $x = 0$ to the ground state of the well extending from $x = 0$ to $x = a$.

7. The first excited state of a particle moving in a potential in one dimension is given by

$$\psi_1(x) = (\sinh \alpha x)\, \psi_0(x), \tag{4.77}$$

and is a state corresponding to the energy eigenvalue E_1. Here α is a constant and $\psi_0(x)$ is the ground state wavefunction corresponding to the energy eigenvalue E_0. Determine the potential (in terms of x, α, E_0, E_1) in which the particle moves.

8. A particle, moving in one dimension, has a ground state wave function (not normalized and do not normalize) given by

$$\psi_0(x) = e^{-\frac{\alpha^4 x^4}{4}}, \tag{4.78}$$

(where α is a real constant) belonging to the energy eigenvalue

$$E_0 = \frac{\hbar^2 \alpha^2}{m}. \tag{4.79}$$

Determine the potential in which the particle moves.

9. In one dimension a particle moves in a potential

$$V(x) = \begin{cases} \infty, & \text{for} \quad x \leq 0, \\ -\gamma\delta(x-a), & \text{for} \quad x > 0, \end{cases} \tag{4.80}$$

where the constants $\gamma, a > 0$. What is the *minimum* value of γa above which a bound state is possible?

CHAPTER 5

Harmonic oscillator

5.1 Harmonic oscillator in one dimension

In quantum mechanics, the harmonic oscillator plays an important role, much like in classical mechanics. Furthermore, this is also one of the few physical systems that can be exactly solved. Therefore, we will spend a few lectures in studying various features of this system. We will begin with the harmonic oscillator in one dimension before generalizing the results to higher dimensions.

There are two distinct ways of studying the harmonic oscillator. Let us first discuss the operator solution for this problem after which we will also solve the Schrödinger equation for this system explicitly.

5.2 Matrix formulation

We are familiar with the motion of a classical harmonic oscillator. In one dimension, the Hamiltonian for the system is given by

$$H = \frac{p^2}{2m} + \frac{1}{2} m\omega^2 x^2 = T + V. \tag{5.1}$$

Quantum mechanically, of course, P and X become operators. Let us now define two new operators from linear combinations of these as

$$\begin{aligned} a &= \sqrt{\frac{m\omega}{2\hbar}} \left(X + \frac{i}{m\omega} P \right), \\ a^\dagger &= \sqrt{\frac{m\omega}{2\hbar}} \left(X - \frac{i}{m\omega} P \right). \end{aligned} \tag{5.2}$$

These are dimensionless complex operators and, clearly,

$$
\begin{aligned}
a^\dagger a &= \frac{m\omega}{2\hbar}\left(X^2 + \frac{P^2}{m^2\omega^2} + \frac{i}{m\omega}(XP - PX)\right)\\
&= \frac{m\omega}{2\hbar}\left(X^2 + \frac{P^2}{m^2\omega^2} + \frac{i}{m\omega}\times i\hbar\right)\\
&= \frac{P^2}{2\hbar\omega m} + \frac{1}{2\hbar}\, m\omega X^2 - \frac{1}{2}\\
&= \frac{1}{\hbar\omega}\left(\frac{P^2}{2m} + \frac{1}{2}m\omega^2X^2\right) - \frac{1}{2}\\
&= \frac{1}{\hbar\omega}H - \frac{1}{2},
\end{aligned}
\tag{5.3}
$$

where we have used the basic commutation relations of quantum mechanical operators, (3.17), in the intermediate steps. Therefore, we recognize that we can write

$$
H = \hbar\omega\left(a^\dagger a + \frac{1}{2}\right). \tag{5.4}
$$

Furthermore, using the basic canonical commutation relations given in (3.17), we obtain

$$
\begin{aligned}
[a, a^\dagger] &= \frac{m\omega}{2\hbar}\left[X + \frac{i}{m\omega}P, X - \frac{i}{m\omega}P\right]\\
&= \frac{m\omega}{2\hbar}\left\{\left[X, -\frac{i}{m\omega}P\right] + \left[\frac{i}{m\omega}P, X\right]\right\}\\
&= \frac{m\omega}{2\hbar}\left(-\frac{i}{m\omega}\times i\hbar + \frac{i}{m\omega}\times(-i\hbar)\right)\\
&= \frac{1}{2} + \frac{1}{2} = 1.
\end{aligned}
\tag{5.5}
$$

It is, of course, trivially true that

$$
[a, a] = 0 = [a^\dagger, a^\dagger]. \tag{5.6}
$$

From the commutation relations in (5.5) and (5.6), we note that we have

$$
\begin{aligned}
\left[a, a^\dagger a\right] &= \left[a, a^\dagger\right] a = a,\\
\left[a^\dagger, a^\dagger a\right] &= a^\dagger\left[a^\dagger, a\right] = -a^\dagger\left[a, a^\dagger\right] = -a^\dagger.
\end{aligned}
\tag{5.7}
$$

Therefore, we can define a new operator

$$N = a^\dagger a. \tag{5.8}$$

such that the Hamiltonian can also be written as

$$H = \hbar\omega\left(N + \frac{1}{2}\right). \tag{5.9}$$

Using the relations in (5.7) as well as the definitions in (5.8) and (5.9), we obtain

$$\begin{aligned}
[a, N] &= a = \frac{1}{\hbar\omega}[a, H],\\
[a^\dagger, N] &= -a^\dagger = \frac{1}{\hbar\omega}[a^\dagger, H],\\
[H, N] &= 0.
\end{aligned} \tag{5.10}$$

These commutation relations imply that H and N can be simultaneously diagonalized (or that they have a common set of eigenvectors). Let us denote these simultaneous eigenvectors by $|n\rangle$, which we assume to be normalized, such that

$$N|n\rangle = n|n\rangle, \tag{5.11}$$

where n denotes the eigenvalue of N. Clearly, then,

$$H|n\rangle = \hbar\omega\left(N + \frac{1}{2}\right)|n\rangle = \hbar\omega\left(n + \frac{1}{2}\right)|n\rangle = E_n|n\rangle. \tag{5.12}$$

Thus, we see that the energy associated with a state $|n\rangle$ is $\hbar\omega\left(n + \frac{1}{2}\right)$, namely,

$$E_n = \hbar\omega\left(n + \frac{1}{2}\right). \tag{5.13}$$

Let us now consider the effect of the following commutator on the state $|n\rangle$,

$$\begin{aligned}
&[a, H]|n\rangle = \hbar\omega a|n\rangle,\\
\text{or,}\quad &(aH - Ha)|n\rangle = \hbar\omega a|n\rangle,\\
\text{or,}\quad &E_n a|n\rangle - Ha|n\rangle = \hbar\omega a|n\rangle,\\
\text{or,}\quad &Ha|n\rangle = (E_n - \hbar\omega)a|n\rangle,\\
\text{or,}\quad &H(a|n\rangle) = (E_n - \hbar\omega)(a|n\rangle).
\end{aligned} \tag{5.14}$$

This shows that the state $a|n\rangle$ is also an eigenstate of the Hamiltonian if $|n\rangle$ is, but the energy associated with this state is $(E_n - \hbar\omega)$. Thus, we see that the effect of the operator a on an energy eigenstate is to lower its energy by a unit of $\hbar\omega$. Therefore, a is called the lowering operator.

Let us note from the value of E_n in (5.13) that

$$\begin{aligned} E_n - \hbar\omega &= \hbar\omega\left(n + \frac{1}{2}\right) - \hbar\omega \\ &= \hbar\omega\left((n-1) + \frac{1}{2}\right) \\ &= E_{n-1}, \end{aligned} \tag{5.15}$$

so that the state $a|n\rangle$ must be proportional to the energy eigenstate $|n-1\rangle$. Thus, let us write

$$a|n\rangle = C_n|n-1\rangle, \quad \Rightarrow \quad \langle n|a^\dagger = C_n^*\langle n-1|. \tag{5.16}$$

The norm of this state leads to

$$\begin{aligned} &\langle n|a^\dagger a|n\rangle = C_n^* C_n \langle n-1|n-1\rangle, \\ \text{or,}\quad &\langle n|N|n\rangle = |C_n|^2, \\ \text{or,}\quad &n\langle n|n\rangle = |C_n|^2, \\ \text{or,}\quad &|C_n|^2 = n. \end{aligned} \tag{5.17}$$

We can choose C_n to be real, in which case, we have

$$C_n = C_n^* = \sqrt{n}, \tag{5.18}$$

and we can write

$$\begin{aligned} &a|n\rangle = \sqrt{n}|n-1\rangle, \\ \text{or,}\quad &|n-1\rangle = \frac{a}{\sqrt{n}}|n\rangle. \end{aligned} \tag{5.19}$$

This shows that, given the state $|n\rangle$, we can obtain a state with a lower energy by applying the lowering operator with some suitable normalization constant.

Similarly, we can consider the effect of the following commutator on an energy eigenstate,

$$\begin{aligned}
& [a^\dagger, H]|n\rangle = -\hbar\omega a^\dagger|n\rangle, \\
\text{or,}\quad & (a^\dagger H - Ha^\dagger)|n\rangle = -\hbar\omega a^\dagger|n\rangle, \\
\text{or,}\quad & E_n a^\dagger|n\rangle - Ha^\dagger|n\rangle = -\hbar\omega a^\dagger|n\rangle, \\
\text{or,}\quad & H(a^\dagger|n\rangle) = (E_n + \hbar\omega)(a^\dagger|n\rangle).
\end{aligned} \tag{5.20}$$

This shows that $(a^\dagger|n\rangle)$ is also an eigenstate of the Hamiltonian with energy $(E_n + \hbar\omega)$. Thus, the operator $a^\dagger$, acting on an energy eigenstate, raises its energy by a unit of $\hbar\omega$. Correspondingly, $a^\dagger$ is known as the raising operator.

Furthermore, since the state $|n+1\rangle$ corresponds to the energy eigenstate with eigenvalue $E_n + \hbar\omega = E_{n+1}$, we must have

$$a^\dagger|n\rangle = C'_n|n+1\rangle, \quad \Rightarrow \quad \langle n|a = C'^*_n\langle n+1|. \tag{5.21}$$

Thus, the norm of this state leads to

$$\begin{aligned}
& \langle n|aa^\dagger|n\rangle = C'^*_n C'_n\langle n+1|n+1\rangle, \\
\text{or,}\quad & \langle\left(n|[a, a^\dagger] + a^\dagger a\right)|n\rangle = |C'_n|^2, \\
\text{or,}\quad & |C'_n|^2 = \langle n|(N+1)|n\rangle = (n+1), \\
\text{or,}\quad & C'_n = C'^*_n = \sqrt{n+1},
\end{aligned} \tag{5.22}$$

so that we can write

$$\begin{aligned}
& a^\dagger|n\rangle = \sqrt{n+1}|n+1\rangle, \\
\text{or,}\quad & |n+1\rangle = \frac{a^\dagger}{\sqrt{n+1}}|n\rangle.
\end{aligned} \tag{5.23}$$

This again shows that, given an eigenstate of energy, we can obtain a state with a higher energy by operating with $a^\dagger$ with a suitable normalization constant.

Thus, we have shown that, given an energy eigenstate, we can obtain an infinite sequence of states by repeated application of a's and $a^\dagger$'s. These states would, in general, have energy $E_n - \hbar\omega\ell$ or $E_n + \hbar\omega\ell$. Thus, it would appear that energy can take a series of values between $-\infty$ and $+\infty$ with a spacing of $\hbar\omega$ between the adjacent levels. Let us now consider a general energy eigenstate $|n\rangle$. Then,

$$\begin{aligned}
E_n &= \langle n|H|n\rangle \\
&= \langle n|\hbar\omega\left(N+\frac{1}{2}\right)|n\rangle \\
&= \hbar\omega\langle n|a^\dagger a+\frac{1}{2}|n\rangle \\
&= \hbar\omega\langle n|a^\dagger a|n\rangle+\frac{\hbar\omega}{2}\langle n|n\rangle \\
&= \hbar\omega\langle an|an\rangle+\frac{\hbar\omega}{2}\langle n|n\rangle>0, \qquad \omega>0.
\end{aligned} \tag{5.24}$$

Namely, the right hand side is the sum of two squares (or positive quantities since each term involves the norm of a state) and, therefore, has to be positive definite. This tells us that the energy can only be positive definite and that the infinite sequence of states must terminate.

Let us assume that the ground state or the state with the lowest energy is denoted by $|n_{\rm min}\rangle$. Then, clearly, what we mean by this is that we can not obtain a state with a lower energy by applying the lowering operator on this state. In other words, the operator a acting on $|n_{\rm min}\rangle$ should give us a null vector for this to be true,

$$a|n_{\rm min}\rangle=0, \quad \Rightarrow \quad \langle n_{\rm min}|a^\dagger=0. \tag{5.25}$$

Furthermore, from (5.25), we obtain

$$\begin{aligned}
& a^\dagger a|n_{\rm min}\rangle=0, \\
\text{or,} \quad & N|n_{\rm min}\rangle=0, \\
\text{or,} \quad & n_{\rm min}|n_{\rm min}\rangle=0, \quad \Rightarrow \quad n_{\rm min}=0.
\end{aligned} \tag{5.26}$$

Thus, we can denote the ground state as $|0\rangle$ which satisfies

$$H|0\rangle=\hbar\omega\left(N+\frac{1}{2}\right)|0\rangle=\frac{\hbar\omega}{2}|0\rangle. \tag{5.27}$$

In other words, what we have learnt so far is that the eigenvectors of the Hamiltonian can be labeled by the eigenvalues of the operator N, which take values

$$n=0,1,2,\ldots,\infty. \tag{5.28}$$

The states are denoted by $|n\rangle$ and the energy eigenvalue associated with such a state is

$$E_n = \hbar\omega\left(n + \frac{1}{2}\right), \qquad n = 0, 1, 2, \cdots, \tag{5.29}$$

so that the ground state has the energy

$$E_0 = \frac{\hbar\omega}{2}. \tag{5.30}$$

The ground state energy of a system is also commonly known as the zero point energy associated with the system.

We see that, in this system, energy is quantized (contrast this with the classical oscillator which can have any continuous energy), and that the quantum of energy is $\hbar\omega$. The eigenstate $|n\rangle$ corresponds to a state with n quanta of energy. Therefore, the operator N is also known as the number operator, since it counts the number of quanta of energy present in a state. The lowering operator, acting on an eigenstate of energy, takes us to another state with a lower energy, namely, $a|n\rangle \sim |n-1\rangle$. Therefore, the operator a acting on a state reduces the number of quanta by one and, correspondingly, it is also known as the annihilation operator or the destruction operator. Similarly, the raising operator, acting on an eigenstate of energy, takes us to a state with a higher energy, namely, $a^\dagger|n\rangle \sim |n+1\rangle$. Hence $a^\dagger$ raises the number of quanta by one and, consequently, it is also known as the creation operator.

Let us now try to find out how various operators look in this energy eigenbasis. We know that

$$a|n\rangle = \sqrt{n}|n-1\rangle, \tag{5.31}$$

so that the matrix elements of the annihilation operator can be written as

$$\begin{aligned}\langle n'|a|n\rangle &= \sqrt{n}\langle n'|n-1\rangle\\ &= \sqrt{n}\delta_{n',n-1}, \quad n, n' = 0, 1, 2, \cdots.\end{aligned} \tag{5.32}$$

Similarly, since

$$a^\dagger|n\rangle = \sqrt{n+1}|n+1\rangle, \tag{5.33}$$

it follows that

$$\langle n'|a^\dagger|n\rangle = \sqrt{n+1}\langle n'|n+1\rangle = \sqrt{n+1}\delta_{n',n+1}. \tag{5.34}$$

Writing out explicitly in the energy basis, these operators, therefore, have the matrix representation of the following off-diagonal forms,

$$a = \begin{pmatrix} 0 & \sqrt{1} & 0 & 0 & 0 & 0 & 0 \\ & 0 & \sqrt{2} & 0 & 0 & 0 & 0 \\ & & 0 & \sqrt{3} & 0 & 0 & 0 \\ & & & 0 & \sqrt{4} & 0 & 0 \\ & & & & 0 & \sqrt{5} & 0 \\ & & & & & 0 & \ddots \end{pmatrix},$$

$$a^{\dagger} = \begin{pmatrix} 0 & & & & \\ \sqrt{1} & 0 & & & \\ 0 & \sqrt{2} & 0 & & \\ 0 & 0 & \sqrt{3} & 0 & \\ 0 & 0 & 0 & \sqrt{4} & 0 \\ 0 & 0 & 0 & 0 & \ddots \end{pmatrix}. \tag{5.35}$$

Furthermore, since by definition (see (5.2)),

$$X = \sqrt{\frac{\hbar}{2m\omega}}(a + a^{\dagger}),$$
$$P = -i\sqrt{\frac{\hbar m\omega}{2}}(a - a^{\dagger}), \tag{5.36}$$

we obtain the matrix elements of the coordinate and the momentum operators in the energy basis to be

$$\begin{aligned} \langle n'|X|n\rangle &= \sqrt{\frac{\hbar}{2m\omega}}\langle n'|(a + a^{\dagger})|n\rangle \\ &= \sqrt{\frac{\hbar}{2m\omega}}\left[\sqrt{n}\delta_{n',n-1} + \sqrt{n+1}\delta_{n',n+1}\right], \\ \langle n'|P|n\rangle &= -i\sqrt{\frac{\hbar m\omega}{2}}\langle n'|(a - a^{\dagger})|n\rangle \\ &= -i\sqrt{\frac{\hbar m\omega}{2}}\left[\sqrt{n}\delta_{n',n-1} - \sqrt{n+1}\delta_{n',n+1}\right]. \end{aligned} \tag{5.37}$$

In the energy basis, they have the following explicit off-diagonal matrix forms,

$$X=\sqrt{\frac{\hbar}{2m\omega}}\begin{pmatrix} 0 & \sqrt{1} & 0 & \\ \sqrt{1} & 0 & \sqrt{2} & \\ 0 & \sqrt{2} & 0 & \sqrt{3} \\ 0 & 0 & \sqrt{3} & \ddots \end{pmatrix},$$

$$P=-i\sqrt{\frac{\hbar m\omega}{2}}\begin{pmatrix} 0 & \sqrt{1} & & & \\ -\sqrt{1} & 0 & \sqrt{2} & & \\ 0 & -\sqrt{2} & 0 & \sqrt{3} & \\ 0 & 0 & -\sqrt{3} & 0 & \\ & & & & \ddots \end{pmatrix}. \tag{5.38}$$

These are manifestly Hermitian operators as they should be, but are not diagonal in this basis. The Hamiltonian, on the other hand, is diagonal in the energy basis and has the form

$$H=\frac{\hbar\omega}{2}\begin{pmatrix} 1 & & & \\ & 3 & & \\ & & 5 & \\ & & & \ddots \end{pmatrix}. \tag{5.39}$$

Since the application of $a^\dagger$ gives us a state with a higher energy, we can construct all the higher energy states from the ground state $|0\rangle$. For example,

$$|1\rangle = \frac{a^\dagger}{\sqrt{1}}|0\rangle = a^\dagger|0\rangle,$$

$$|2\rangle = \frac{a^\dagger}{\sqrt{1+1}}|1\rangle = \frac{a^\dagger}{\sqrt{2}}\times\frac{a^\dagger}{\sqrt{1}}|0\rangle = \frac{(a^\dagger)^2}{\sqrt{2!}}|0\rangle. \tag{5.40}$$

In general, we can write

$$|n+1\rangle = \frac{a^\dagger}{\sqrt{n+1}}|n\rangle = \frac{a^\dagger}{\sqrt{n+1}}\frac{a^\dagger}{\sqrt{n}}|n-1\rangle = \cdots$$

$$= \frac{(a^\dagger)^{n+1}}{\sqrt{(n+1)!}}|0\rangle. \tag{5.41}$$

The fact that any higher state can be written as a product of creation operators acting on the ground state and the fact that

$$a|0\rangle = 0 = \langle 0|a^\dagger,$$

greatly simplifies the calculation of matrix elements of operators between different states. For example, consider the matrix element

$$\begin{aligned}\langle 2|X^3|0\rangle &= \langle 0|\frac{a^2}{\sqrt{2!}}\left(\frac{\hbar}{2m\omega}\right)^{\frac{3}{2}}(a+a^\dagger)^3|0\rangle\\ &= \frac{1}{\sqrt{2}}\left(\frac{\hbar}{2m\omega}\right)^{\frac{3}{2}}\langle 0|a^2(a+a^\dagger)^3|0\rangle\\ &= 0,\end{aligned} \tag{5.42}$$

where in the final step we have used the fact that the expectation value of the product of an odd number of a's and $a^\dagger$'s vanishes in the ground state (or in any energy eigenstate). Next, let us calculate

$$\langle 2|X^3|1\rangle = \frac{1}{\sqrt{2}}\left(\frac{\hbar}{2m\omega}\right)^{\frac{3}{2}}\langle 0|a^2(a+a^\dagger)^3 a^\dagger|0\rangle. \tag{5.43}$$

The only terms of $(a+a^\dagger)^3$ that would give a nonzero contribution are (for a non-vanishing expectation value in the ground state, the number of a's must equal that of $a^\dagger$'s in a product)

$$aa^{\dagger 2} + a^\dagger a a^\dagger + a^{\dagger 2}a, \tag{5.44}$$

so that we can obtain

$$\langle 2|X^3|1\rangle = \frac{1}{\sqrt{2}}\left(\frac{\hbar}{2m\omega}\right)^{\frac{3}{2}}\langle 0|a^2(aa^{\dagger 2}+a^\dagger a a^\dagger + a^{\dagger 2}a)a^\dagger|0\rangle. \tag{5.45}$$

Using $[a, a^\dagger] = 1$, we can simplify

$$\begin{aligned}&(aa^\dagger - a^\dagger a) = 1,\\ \text{or,}\quad & a^\dagger a a^\dagger = (aa^\dagger - 1)a^\dagger = aa^{\dagger 2} - a^\dagger,\end{aligned} \tag{5.46}$$

and,

$$\begin{aligned}a^{\dagger 2}a &= a^\dagger(aa^\dagger - 1) = a^\dagger a a^\dagger - a^\dagger\\ &= aa^{\dagger 2} - a^\dagger - a^\dagger = aa^{\dagger 2} - 2a^\dagger.\end{aligned} \tag{5.47}$$

Substituting (5.46) and (5.47) into (5.45), we obtain,

$$\begin{aligned}
\langle 2|X^3|1\rangle &= \frac{1}{\sqrt{2}}\left(\frac{\hbar}{2m\omega}\right)^{\frac{3}{2}} \langle 0|a^2\big(aa^{\dagger 2} + aa^{\dagger 2} - a^\dagger \\
&\qquad\qquad + aa^{\dagger 2} - 2a^\dagger\big)a^\dagger|0\rangle \\
&= \frac{1}{\sqrt{2}}\left(\frac{\hbar}{2m\omega}\right)^{\frac{3}{2}} \langle 0|a^2(3aa^{\dagger 2} - 3a^\dagger)a^\dagger|0\rangle \\
&= \frac{3}{\sqrt{2}}\left(\frac{\hbar}{2m\omega}\right)^{\frac{3}{2}} \Big(\langle 0|a^3a^{\dagger 3}|0\rangle - \langle 0|a^2a^{\dagger 2}|0\rangle\Big) \\
&= \frac{3}{\sqrt{2}}\left(\frac{\hbar}{2m\omega}\right)^{\frac{3}{2}} (3!\langle 3|3\rangle - 2!\langle 2|2\rangle) \\
&= \frac{3}{\sqrt{2}}\left(\frac{\hbar}{2m\omega}\right)^{\frac{3}{2}} (3! - 2!) \\
&= 3\left(\frac{\hbar}{m\omega}\right)^{\frac{3}{2}}.
\end{aligned} \tag{5.48}$$

This shows how the matrix elements of arbitrary operators can be calculated in the energy basis.

Let us next note that the wave function associated with an energy eigenstate is defined as

$$\psi_n(x) = \langle x|n\rangle. \tag{5.49}$$

This measures the probability amplitude for finding the oscillator at the coordinate x with an energy E_n. We know that the ground state satisfies

$$\begin{aligned}
& a|0\rangle = 0, \\
\text{or,}\quad & \langle x|a|0\rangle = 0, \\
\text{or,}\quad & \int \mathrm{d}x'\, \langle x|a|x'\rangle\langle x'|0\rangle = 0.
\end{aligned} \tag{5.50}$$

Using the definition in (5.2)

$$a = \sqrt{\frac{m\omega}{2\hbar}}\left(X + \frac{i}{m\omega}P\right),$$

as well as the basic coordinate representations of X and P,

$$\begin{aligned}\langle x|X|x'\rangle &= x\delta(x-x')\\ \langle x|P|x'\rangle &= -i\hbar\,\frac{\mathrm{d}}{\mathrm{d}x}\,\delta(x-x')\end{aligned}$$

we obtain

$$\langle x|a|x'\rangle = \sqrt{\frac{m\omega}{2\hbar}}\left(x\delta(x-x')+\frac{\hbar}{m\omega}\frac{\mathrm{d}}{\mathrm{d}x}\delta(x-x')\right). \tag{5.51}$$

Using (5.51) in (5.50), we see that the equation for the ground state wave function for the harmonic oscillator becomes

$$\sqrt{\frac{m\omega}{2\hbar}}\int \mathrm{d}x'\left(x\delta(x-x')+\frac{\hbar}{m\omega}\frac{\mathrm{d}}{\mathrm{d}x}\,\delta(x-x')\right)\psi_0(x')=0,$$

$$\text{or,}\quad \sqrt{\frac{m\omega}{2\hbar}}\left(x+\frac{\hbar}{m\omega}\,\frac{\mathrm{d}}{\mathrm{d}x}\right)\psi_0(x)=0. \tag{5.52}$$

This is a simple first order equation, whose solution is easily obtained to be,

$$\begin{aligned}&\frac{\mathrm{d}\psi_0(x)}{\mathrm{d}x} = -\frac{m\omega}{\hbar}x\psi_0(x),\\ \text{or,}\quad &\psi_0(x) = A_0 e^{-\frac{m\omega}{2\hbar}x^2}.\end{aligned} \tag{5.53}$$

Here A_0 is the normalization constant, which can be determined by requiring that

$$\int_{-\infty}^{\infty}\mathrm{d}x\;\psi_0^*(x)\psi_0(x) = A_0^*A_0\int_{-\infty}^{\infty}\mathrm{d}x\;e^{-\frac{m\omega}{\hbar}x^2}=1,$$

$$\text{or,}\quad |A_0|^2\sqrt{\frac{\pi\hbar}{m\omega}}=1,$$

$$\text{or,}\quad |A_0|^2=\sqrt{\frac{m\omega}{\pi\hbar}}. \tag{5.54}$$

Choosing A_0 to be real, we determine

$$A_0 = A_0^* = \left(\frac{m\omega}{\pi\hbar}\right)^{\frac{1}{4}}, \tag{5.55}$$

so that we can write the normalized ground state wave function as

$$\psi_0(x) = \left(\frac{m\omega}{\pi\hbar}\right)^{\frac{1}{4}}e^{-\frac{m\omega}{2\hbar}x^2}. \tag{5.56}$$

To construct the wave functions for higher energy states, we note that in the x basis,

$$a^\dagger = \sqrt{\frac{m\omega}{2\hbar}}\left(X - \frac{i}{m\omega}P\right) \to \sqrt{\frac{m\omega}{2\hbar}}\left(x - \frac{\hbar}{m\omega}\frac{\mathrm{d}}{\mathrm{d}x}\right). \tag{5.57}$$

Furthermore, recall that

$$|n\rangle = \frac{(a^\dagger)^n}{\sqrt{n!}}|0\rangle.$$

Therefore, we determine the wave function for the nth state to be

$$\begin{aligned}\langle x|n\rangle &= \psi_n(x)\\ &= \frac{1}{\sqrt{n!}}\left(\frac{m\omega}{2\hbar}\right)^{\frac{n}{2}}\left(x - \frac{\hbar}{m\omega}\frac{\mathrm{d}}{\mathrm{d}x}\right)^n \psi_0(x)\\ &= \frac{1}{\sqrt{n!}}\left(\frac{m\omega}{\pi\hbar}\right)^{\frac{1}{4}}\left(\frac{m\omega}{2\hbar}\right)^{\frac{n}{2}}\left(x - \frac{\hbar}{m\omega}\frac{\mathrm{d}}{\mathrm{d}x}\right)^n e^{-\frac{m\omega x^2}{2\hbar}}.\end{aligned} \tag{5.58}$$

This completes our investigation of the harmonic oscillator in the matrix (operator) formulation. We have determined the energy eigenvalues and the wave functions for the energy eigenstates (without explicitly solving the Schrödinger equation).

5.3 Solution of the Schrödinger equation

The harmonic oscillator is an important system in quantum mechanics. Most complicated systems can often be split into a part that is of harmonic oscillator type and a part that can be treated as a perturbation on this system. From the studies of black body radiation, we recall that the electromagnetic radiation is treated like a harmonic oscillator system with the quantum of energy $\hbar\omega$. In fact, this is a very general feature. All field theories without any interaction, can be decomposed into harmonic oscillators. Since the study of this system is so significant, we would also study this system by solving the Schrödinger equation for this system.

As we have seen, the Hamiltonian for the system is given by

$$\begin{aligned}H &= \frac{P^2}{2m} + \frac{1}{2}m\omega^2 X^2\\ &\to -\frac{\hbar^2}{2m}\frac{\mathrm{d}^2}{\mathrm{d}x^2} + \frac{1}{2}m\omega^2 x^2.\end{aligned} \tag{5.59}$$

Furthermore, since the Hamiltonian has no explicit time dependence, we are interested in the stationary state solutions. We know that the stationary state solutions have the form (see (3.94))

$$\psi(x,t) = e^{-\frac{i}{\hbar}Et} u_E(x), \tag{5.60}$$

where $u_E(x)$ satisfies the time independent Schrödinger equation,

$$\left(-\frac{\hbar^2}{2m}\frac{\mathrm{d}^2}{\mathrm{d}x^2} + \frac{1}{2}m\omega^2x^2\right) u_E(x) = E u_E(x). \tag{5.61}$$

We recognize that $u_E(x)$ is the wave function associated with the eigenstate of the Hamiltonian corresponding to energy E.

We already know that the energy associated with the oscillator must be positive. It can also be seen here by writing

$$\begin{aligned} E &= \langle H \rangle \\ &= \int \mathrm{d}x\ u_E^*(x) \left(-\frac{\hbar^2}{2m}\frac{\mathrm{d}^2}{\mathrm{d}x^2} + \frac{1}{2}m\omega^2x^2\right) u_E(x) \\ &= \int \mathrm{d}x \left(\frac{\hbar^2}{2m}\left|\frac{\mathrm{d}u_E(x)}{\mathrm{d}x}\right|^2 + \frac{1}{2}m\omega^2x^2\,|u_E(x)|^2\right) > 0, \end{aligned} \tag{5.62}$$

which follows because the right hand side is the integral of a sum of two squares (we are neglecting a surface term which vanishes for well behaved functions).

The time independent Schrödinger equation, (5.61), can be written as

$$\frac{\mathrm{d}^2u_E}{\mathrm{d}x^2} + \frac{2m}{\hbar^2}\left(E - \frac{1}{2}\,m\omega^2x^2\right) u_E = 0. \tag{5.63}$$

First of all, let us note that in solving a differential equation it is always useful to recast the equation in terms of dimensionless variables. This allows us to write down logarithmic or exponential solutions without having to worry about the dimensionality of the arguments. We note that there are three dimensional parameters in our theory and they are

$$\begin{aligned} [\hbar] &= \text{erg-sec} = ML^2T^{-1}, \\ [m] &= \text{gm} = M, \\ [\omega] &= \text{sec}^{-1} = T^{-1}, \end{aligned} \tag{5.64}$$

where M, L, T are three arbitrary units of mass, length and time respectively. We see from (5.64) that

$$\left[\frac{m\omega}{\hbar}\right] = \frac{MT^{-1}}{ML^2T^{-1}} = L^{-2}. \tag{5.65}$$

Consequently, we see that if we define

$$\xi = \left(\frac{m\omega}{\hbar}\right)^{\frac{1}{2}} x, \tag{5.66}$$

then, ξ will be dimensionless. Furthermore, by the chain rule of differentiation, we obtain

$$\frac{\mathrm{d}}{\mathrm{d}x} = \frac{\mathrm{d}\xi}{\mathrm{d}x}\frac{\mathrm{d}}{\mathrm{d}\xi} = \left(\frac{m\omega}{\hbar}\right)^{\frac{1}{2}} \frac{\mathrm{d}}{\mathrm{d}\xi}. \tag{5.67}$$

Substituting (5.67) back into (5.63), we have

$$\left(\frac{m\omega}{\hbar}\right)\frac{\mathrm{d}^2 u_E(\xi)}{\mathrm{d}\xi^2} + \frac{2m}{\hbar^2}\left(E - \frac{1}{2}m\omega^2 \times \frac{\hbar}{m\omega}\xi^2\right) u_E(\xi) = 0,$$

$$\text{or,} \quad \frac{\mathrm{d}^2 u_E(\xi)}{\mathrm{d}\xi^2} + \left(\frac{2E}{\hbar\omega} - \xi^2\right) u_E(\xi) = 0. \tag{5.68}$$

Let us further define

$$\epsilon = \frac{2E}{\hbar\omega}. \tag{5.69}$$

Clearly, ϵ is dimensionless (it measures energy in units of $\frac{\hbar\omega}{2}$). In terms of these dimensionless variables, equation (5.68) becomes

$$\frac{\mathrm{d}^2 u_E(\xi)}{\mathrm{d}\xi^2} + (\epsilon - \xi^2) u_E(\xi) = 0, \tag{5.70}$$

where the independent variable as well as the arbitrary parameter measuring energy are now dimensionless.

Before deriving the solutions, it is useful to find out their asymptotic forms, both in the limit $\xi \to \infty$ and $\xi \to 0$. First of all, for a finite ϵ we see that in the limit $\xi \to \infty$, equation (5.70) becomes

$$\frac{\mathrm{d}^2 u_E}{\mathrm{d}\xi^2} - \xi^2 u_E(\xi) = 0. \tag{5.71}$$

The two independent solutions of this equation have the forms

$$u_E(\xi) \xrightarrow{|\xi|\to\infty} \xi^\ell e^{\pm\frac{1}{2}\xi^2}, \tag{5.72}$$

for any finite integer ℓ. This can be easily checked by noting that

$$\frac{\mathrm{d}u_E}{\mathrm{d}\xi} \xrightarrow{|\xi|\to\infty} e^{\pm\frac{1}{2}\xi^2}\left(\ell\xi^{\ell-1} \pm \xi^{\ell+1}\right),$$

$$\frac{\mathrm{d}^2u_E}{\mathrm{d}\xi^2} \xrightarrow{|\xi|\to\infty} e^{\pm\frac{1}{2}\xi^2}\left(\ell(\ell-1)\xi^{\ell-2} \pm (2\ell+1)\xi^{\ell} + \xi^{\ell+2}\right)$$

$$\longrightarrow \quad e^{\pm\frac{1}{2}\xi^2}\xi^{\ell+2} = \xi^2 u_E(\xi). \tag{5.73}$$

Although both $\xi^{\ell}e^{\pm\frac{1}{2}\xi^2}$ represent asymptotic solutions, a physical solution would only correspond to the one that is normalizable, namely,

$$u_E(\xi) \xrightarrow{|\xi|\to\infty} \xi^{\ell}e^{-\frac{1}{2}\xi^2}. \tag{5.74}$$

Furthermore, in the limit $\xi \to 0$, equation (5.68) reduces to

$$\frac{\mathrm{d}^2u_E}{\mathrm{d}\xi^2} + \epsilon u_E = 0, \tag{5.75}$$

whose solutions are easily seen to be of the form

$$u_E(\xi) \xrightarrow{\xi\to 0} p(\xi), \tag{5.76}$$

where $p(\xi)$ represents a polynomial of positive powers only.

Thus, we can write the general solution of the equation for the harmonic oscillator as

$$u_E(\xi) = f_E(\xi)e^{-\frac{1}{2}\xi^2}, \tag{5.77}$$

with $f_E(\xi)$ representing a polynomial with non-negative powers. We note from (5.77) that

$$\frac{\mathrm{d}u_E}{\mathrm{d}\xi} = \left(f_E'(\xi) - \xi f_E(\xi)\right)e^{-\frac{1}{2}\xi^2},$$

$$\frac{\mathrm{d}^2u_E}{\mathrm{d}\xi^2} = \left(f_E''(\xi) - 2\xi f_E'(\xi) + (\xi^2-1)f_E(\xi)\right)e^{-\frac{1}{2}\xi^2}, \tag{5.78}$$

where primes denote derivatives with respect to ξ. Substituting this back into (5.68), we obtain

$$f_E''(\xi) - 2\xi f_E'(\xi) + (\xi^2-1)f_E(\xi) + (\epsilon - \xi^2)f_E(\xi) = 0,$$

$$\text{or,} \quad f_E''(\xi) - 2\xi f_E'(\xi) + (\epsilon-1)f_E(\xi) = 0. \tag{5.79}$$

This is an equation where the asymptotic behavior at infinity has been factored out. Therefore, we can try a power series solution of the form

$$
\begin{aligned}
f_E(\xi) &= \sum_{n=0}^{\infty} C_n \xi^n, \\
f'_E(\xi) &= \frac{\mathrm{d}f_E(\xi)}{\mathrm{d}\xi} = \sum_{n=1}^{\infty} nC_n\xi^{n-1}, \\
\xi f'_E(\xi) &= \sum_{n=1}^{\infty} nC_n\xi^n = \sum_{n=0}^{\infty} nC_n\xi^n, \\
f''_E(\xi) &= \frac{\mathrm{d}^2 f_E(\xi)}{\mathrm{d}\xi^2} = \sum_{n=2}^{\infty} n(n-1)C_n\xi^{n-2} \\
&= \sum_{n=0}^{\infty}(n+2)(n+1)C_{n+2}\xi^n,
\end{aligned}
\tag{5.80}
$$

where C_n's are constants. Putting these back into (5.79), we have

$$
\sum_{\ell=0}^{\infty}\left((\ell+2)(\ell+1)C_{\ell+2}\xi^\ell - 2\ell C_\ell\xi^\ell + (\epsilon-1)C_\ell\xi^\ell\right) = 0,
$$

$$
\text{or,}\quad \sum_{\ell=0}^{\infty}\xi^\ell\left((\ell+2)(\ell+1)C_{\ell+2} + (\epsilon-1-2\ell)C_\ell\right) = 0. \tag{5.81}
$$

If this has to be true for any arbitrary value of ξ, the coefficients in the parenthesis must vanish, namely,

$$
\begin{aligned}
(\ell+2)(\ell+1)C_{\ell+2} &= -(\epsilon-1-2\ell)C_\ell, \\
C_{\ell+2} &= -\frac{(\epsilon-1-2\ell)}{(\ell+2)(\ell+1)}C_\ell.
\end{aligned}
\tag{5.82}
$$

This defines a recursion relation between the coefficients in the power series in (5.80). It is clear that all of the coefficients can be expressed in terms C_0 and C_1 which are arbitrary. For example,

$$
\begin{aligned}
C_2 &= -\frac{(\epsilon-1)}{2}C_0, \\
C_3 &= -\frac{(\epsilon-3)}{6}C_1, \\
C_4 &= -\frac{(\epsilon-5)}{12}C_2 = \frac{(\epsilon-5)(\epsilon-1)}{24}C_0,
\end{aligned}
\tag{5.83}
$$

and so on. The existence of two arbitrary constants in the solution can be traced to the fact that the Schrödinger equation corresponds to a second order differential equation, (5.79), and a unique solution needs two conditions. However, we also know that this system is invariant under $x \rightarrow -x$ (or, $\xi \rightarrow -\xi$). Therefore, as before, we expect the solutions to be of two types – odd and even. It is clear that if $C_0 = 0$, then, all the even powers in $f_E(\xi)$ would vanish and hence it would be an odd function. On the other hand, if $C_1 = 0$, then, $f_E(\xi)$ would contain only even powers in the expansion and, therefore, would be symmetric. In general, however, it is clear (independent of whether $C_0 = 0$ or $C_1 = 0$) that unless the series terminates at some point, its dominant asymptotic form can be inferred from the ratio

$$\frac{C_{\ell+2}}{C_\ell} \xrightarrow{\ell\rightarrow\infty} \frac{2}{\ell}. \tag{5.84}$$

We recognize that this is the same growth as the asymptotic coefficients of the function $\xi^k e^{\xi^2}$. Therefore, unless the power series terminates, its asymptotic behavior would correspond to that of this function. However, this would lead to an unphysical solution (in the sense that the wave function (5.77) would diverge asymptotically). Therefore, for a physical solution to exist, the series must terminate. The only way this can happen is if the numerator of the recursion relation in (5.82) vanishes for some value $\ell = n$, namely,

$$C_{n+2} = -\frac{(\epsilon - 1 - 2n)}{(n+2)(n+1)} C_n = 0. \tag{5.85}$$

In other words, if for some n,

$$\epsilon - 1 - 2n = 0, \tag{5.86}$$

then, all the higher coefficients would vanish and the series would terminate. This implies that

$$\begin{aligned} &\epsilon_n = (2n+1), \qquad n = 0, 1, 2, \cdots, \\ \text{or,}\quad &\frac{2E_n}{\hbar\omega} = (2n+1), \\ \text{or,}\quad &E_n = \hbar\omega\left(n + \frac{1}{2}\right). \end{aligned} \tag{5.87}$$

Therefore, we see that physical solutions for the system will exist only if the oscillator has the energy values in (5.87), which is what we had obtained earlier in the operator method. (In fact, not only should the above energy values hold for physical solutions, one of the two coefficients, C_0, C_1 should vanish depending on the value n of the physical solution. For example, if n is even, then, $C_1 = 0$ and if n is odd, then, $C_0 = 0$.) And, when a solution is physical, for each value of n, the solution will have the form

$$u_n(\xi) = f_n(\xi)e^{-\frac{1}{2}\xi^2}, \tag{5.88}$$

where

$$f_n(\xi) = \sum_{\ell=0}^{\frac{n}{2}} C_{2\ell}\xi^{2\ell}, \quad \text{or,} \quad f_n(\xi) = \sum_{\ell=0}^{\frac{n-1}{2}} C_{2\ell+1}\xi^{2\ell+1}, \tag{5.89}$$

(depending on whether n is even or odd) with the coefficients satisfying the recursion relation (5.82).

5.4 Hermite polynomials

The polynomials, $f_n(\xi)$, with $\epsilon_n = (2n+1)$ in (5.89) satisfy the differential equation (see (5.79))

$$f_n''(\xi) - 2\xi f_n'(\xi) + 2nf_n(\xi) = 0. \tag{5.90}$$

This is known as the Hermite equation and the solutions, $f_n(\xi)$, are known as the Hermite polynomials of nth order. They are commonly denoted by $H_n(\xi)$. It is obvious from the recursion relations that every Hermite polynomial is completely determined in terms of one arbitrary constant, C_0 or C_1, depending on whether n is even or odd.

We can actually define a function of two variables,

$$S(\xi, s) = e^{\xi^2-(s-\xi)^2} = e^{-s^2+2s\xi} = \sum_{n=0}^{\infty} \frac{H_n(\xi)}{n!} s^n, \tag{5.91}$$

where the coefficients of expansion can be identified with the Hermite polynomials of nth order. To see this, let us calculate

$$\begin{aligned}
\frac{\partial S(\xi, s)}{\partial \xi} &= 2se^{-s^2+2s\xi} = 2s\sum_{n=0}^{\infty} \frac{H_n(\xi)}{n!} s^n \\
&= 2\sum_{n=0}^{\infty} \frac{H_n(\xi)}{n!} s^{n+1} = 2\sum_{n=1}^{\infty} \frac{H_{n-1}(\xi)}{(n-1)!} s^n \\
&= 2\sum_{n=0}^{\infty} \frac{nH_{n-1}(\xi)}{n!} s^n.
\end{aligned} \tag{5.92}$$

On the other hand, differentiating the right hand side of (5.91), we obtain

$$\frac{\partial S(\xi,s)}{\partial \xi} = \sum_{n=0}^{\infty} \frac{H_n'(\xi)}{n!} s^n. \tag{5.93}$$

Thus, comparing (5.92) and (5.93), we have

$$H_n'(\xi) = 2nH_{n-1}(\xi). \tag{5.94}$$

Similarly, taking derivative of (5.91) with respect to s, we obtain

$$\begin{aligned}
\frac{\partial S(\xi,s)}{\partial s} &= (-2s+2\xi)e^{-s^2+2s\xi} = (-2s+2\xi)\sum_{n=0}^{\infty}\frac{H_n(\xi)}{n!}s^n \\
&= 2\xi \sum_{n} \frac{H_n(\xi)}{n!}s^n - 2\sum_{n=0}^{\infty}\frac{H_n(\xi)}{n!}s^{n+1} \\
&= 2\xi \sum_{n} \frac{H_n(\xi)}{n!}s^n - 2\sum_{n=0}^{\infty}\frac{nH_{n-1}}{n!}s^n.
\end{aligned} \tag{5.95}$$

But, differentiating the right hand side of (5.91), we also have

$$\begin{aligned}
\frac{\partial S(\xi,s)}{\partial s} &= \frac{\partial}{\partial s}\sum_{n=0}^{\infty}\frac{H_n(\xi)}{n!}s^n = \sum_{n=1}^{\infty}\frac{H_n(\xi)}{(n-1)!}s^{n-1} \\
&= \sum_{n=0}^{\infty}\frac{H_{n+1}(\xi)}{n!}s^n.
\end{aligned} \tag{5.96}$$

Thus, comparing (5.95) and (5.96), we determine

$$\begin{aligned}
& H_{n+1}(\xi) = 2\xi H_n(\xi) - 2nH_{n-1}(\xi), \\
\text{or,}\quad & 2nH_{n-1}(\xi) = 2\xi H_n(\xi) - H_{n+1}(\xi).
\end{aligned} \tag{5.97}$$

Substituting (5.97) into (5.94), we have

$$\begin{aligned}
& H_n'(\xi) = 2\xi H_n(\xi) - H_{n+1}(\xi), \\
\text{or,}\quad & H_n''(\xi) = 2H_n(\xi) + 2\xi H_n'(\xi) - H_{n+1}'(\xi) \\
& \quad = 2H_n(\xi) + 2\xi H_n'(\xi) - 2(n+1)H_n(\xi) \\
& \quad = 2\xi H_n'(\xi) - 2nH_n(\xi).
\end{aligned} \tag{5.98}$$

Therefore, we see from (5.98) that the functions, $H_n(\xi)$, satisfy the equation

$$H_n''(\xi) - 2\xi H_n'(\xi) + 2nH_n(\xi) = 0, \tag{5.99}$$

which is the Hermite equation. Consequently, the coefficients of expansion in (5.91) are indeed the Hermite polynomials. The relations

$$\begin{aligned} H_n'(\xi) &= 2nH_{n-1}(\xi),\\ H_{n+1}(\xi) &= 2\xi H_n(\xi) - 2nH_{n-1}(\xi), \end{aligned} \tag{5.100}$$

are known as the recursion relations for the Hermite polynomials.

It is clear, from the defining relation (5.91) that we can identify

$$H_n(\xi) = \left.\frac{\partial^n S(\xi,s)}{\partial s^n}\right|_{s=0}. \tag{5.101}$$

This is why $S(\xi,s)$ is also known as the generating function for the Hermite polynomials. Furthermore,

$$\begin{aligned} H_n(\xi) &= \left.\frac{\partial^n S(\xi,s)}{\partial s^n}\right|_{s=0} = \left.\frac{\partial^n}{\partial s^n} e^{\xi^2-(s-\xi)^2}\right|_{s=0}\\ &= e^{\xi^2} \left.\frac{\partial^n}{\partial s^n} e^{-(s-\xi)^2}\right|_{s=0} = e^{\xi^2}\left(-\frac{\partial}{\partial \xi}\right)^n \left. e^{-(s-\xi)^2}\right|_{s=0}\\ &= (-1)^n e^{\xi^2} \frac{\partial^n}{\partial \xi^n} e^{-\xi^2}. \end{aligned} \tag{5.102}$$

This gives a closed form expression for the Hermite polynomials. The first few of these polynomials have the explicit forms

$$\begin{aligned} H_0(\xi) &= 1,\\ H_1(\xi) &= 2\xi,\\ H_2(\xi) &= (4\xi^2 - 2), \end{aligned} \tag{5.103}$$

and so on.

We can also work out the orthogonality relations for the Hermite polynomials by noting that

$$
\begin{aligned}
&\int_{-\infty}^{\infty} d\xi\ H_n(\xi)H_m(\xi)e^{-\xi^2}\\
&= \int_{-\infty}^{\infty} d\xi\ e^{\xi^2}\ \frac{\partial^n}{\partial s^n}e^{-(s-\xi)^2}\Big|_{s=0}\ e^{\xi^2}\frac{\partial^m}{\partial t^m}e^{-(t-\xi)^2}\Big|_{t=0}\ e^{-\xi^2}\\
&= \frac{\partial^n}{\partial s^n}\frac{\partial^m}{\partial t^m}\left(\int_{-\infty}^{\infty} d\xi\ e^{\xi^2}e^{-(s-\xi)^2}e^{-(t-\xi)^2}\right)_{s=t=0}\\
&= \frac{\partial^n}{\partial s^n}\frac{\partial^m}{\partial t^m}\left(\int_{-\infty}^{\infty} d\xi\ e^{-\xi^2-s^2-t^2+2s\xi+2t\xi}\right)_{s=t=0}\\
&= \frac{\partial^n}{\partial s^n}\frac{\partial^m}{\partial t^m}\left(\int_{-\infty}^{\infty} d\xi\ e^{-(\xi-(s+t))^2+2st}\right)_{s=t=0}\\
&= \frac{\partial^n}{\partial s^n}\frac{\partial^m}{\partial t^m}\left(\sqrt{\pi}e^{2st}\right)_{s=t=0}\\
&= \frac{\partial^n}{\partial s^n}\frac{\partial^m}{\partial t^m}\left(\sqrt{\pi}\sum_{p=0}^{\infty}\frac{(2st)^p}{p!}\right)_{s=t=0}\\
&= \begin{cases} 0, & \text{if}\quad n\neq m,\\ \sqrt{\pi}2^n n!, & \text{when}\quad n=m. \end{cases}
\end{aligned}
\tag{5.104}
$$

(Here, we note that the Hermite polynomials are polynomials of positive powers and, consequently, the integral needs a damping factor to be well defined.) Therefore, we can write the orthogonality relation for the Hermite polynomials as

$$
\int_{-\infty}^{\infty} d\xi\, H_n(\xi)H_m(\xi)e^{-\xi^2} = \sqrt{\pi}\,2^n\, n!\,\delta_{mn}. \tag{5.105}
$$

We can now identify the time independent wave functions, (5.88), for the harmonic oscillator as

$$
u_n(\xi) = A_n f_n(\xi)e^{-\frac{1}{2}\xi^2} = A_n H_n(\xi)e^{-\frac{1}{2}\xi^2}, \tag{5.106}
$$

where the constant A_n can be determined from normalization. First of all, we note that in terms of the original variables (see (5.66) and

(5.69)), we can write

$$\xi = \left(\frac{m\omega}{\hbar}\right)^{\frac{1}{2}} x,$$

$$u_n(x) = A_n H_n\left(\left(\frac{m\omega}{\hbar}\right)^{\frac{1}{2}} x\right) e^{-\frac{m\omega}{2\hbar}x^2}. \tag{5.107}$$

Furthermore, we would like the wave function to be normalized, namely,

$$\int_{-\infty}^{\infty} \mathrm{d}x\ |u_n(x)|^2 = 1. \tag{5.108}$$

In terms of the ξ variable, then, this becomes,

$$\int_{-\infty}^{\infty} \frac{\mathrm{d}\xi}{\left(\frac{m\omega}{\hbar}\right)^{\frac{1}{2}}} |u_n(\xi)|^2 = 1,$$

$$\text{or,}\quad |A_n|^2 \left(\frac{\hbar}{m\omega}\right)^{\frac{1}{2}} \int_{-\infty}^{\infty} \mathrm{d}\xi\ H_n^2(\xi) e^{-\xi^2} = 1,$$

$$\text{or,}\quad |A_n|^2 \left(\frac{\hbar}{m\omega}\right)^{\frac{1}{2}} \sqrt{\pi}\ 2^n n! = 1,$$

$$\text{or,}\quad A_n = A_n^* = \left(\left(\frac{m\omega}{\pi\hbar}\right)^{\frac{1}{2}} \frac{1}{2^n n!}\right)^{\frac{1}{2}}. \tag{5.109}$$

Thus, the normalized wave functions for the harmonic oscillator are given by

$$u_n(x) = \left(\frac{m\omega}{\pi\hbar}\right)^{\frac{1}{4}} \frac{1}{\sqrt{2^n\, n!}}\ H_n\left(\sqrt{\frac{m\omega}{\hbar}} x\right) e^{-\frac{m\omega}{2\hbar}x^2}. \tag{5.110}$$

In particular, the ground state wave function is given by

$$u_0(x) = \left(\frac{m\omega}{\pi\hbar}\right)^{\frac{1}{4}} e^{-\frac{m\omega}{2\hbar}x^2}, \tag{5.111}$$

which agrees with the result obtained earlier in the operator method. The complete time dependent wave functions for the harmonic oscillator have the forms

$$\psi_n(x,t) = \left(\frac{m\omega}{\pi\hbar}\right)^{\frac{1}{4}} \frac{1}{\sqrt{2^n\, n!}}\ H_n\left(\sqrt{\frac{m\omega}{\hbar}} x\right) e^{-\frac{m\omega}{2\hbar}x^2 - \frac{i}{\hbar}E_n t},$$

$$(5.112)$$

with

$$E_n = \hbar\omega\left(n + \frac{1}{2}\right). \tag{5.113}$$

We note that these are bound state solutions since they vanish asymptotically.

5.5 Discussion of the results

The harmonic oscillator is a very important system to study, both for physical as well as pedagogical reasons. Physically, of course, one knows, even from classical physics, that small oscillations around a minimum of the potential can be approximated by a harmonic oscillator motion. For example, a potential can be expanded around a minimum as

$$V(x) = V(x_0) + (x - x_0)\, V'(x_0) + \frac{1}{2}(x - x_0)^2\, V''(x_0) + \cdots. \tag{5.114}$$

Furthermore, if x_0 is a minimum of the potential, then, $V'(x_0) = 0$ (and $V''(x_0) > 0$). Therefore, for small $x - x_0$, we have

$$V(x) \approx V(x_0) + \frac{1}{2}(x - x_0)^2\, V''(x_0). \tag{5.115}$$

$V(x_0)$ is a constant and has no effect on the dynamics. In other words, we can properly choose the scale of the energy to get rid of this constant. Then, the potential can be approximated by that of a harmonic oscillator with

$$m\omega^2 = V''(x_0) > 0. \tag{5.116}$$

This is not just a mathematical exercise, but it actually happens in physics. Consider, for example, the case of a crystal where molecules are fixed at definite points as shown in Fig. 5.1. Because the molecules are heavy, they can not move very much. Hence, if disturbed slightly, they execute small oscillations about their positions of equilibrium and behave like a system of harmonic oscillators.

The harmonic oscillator is also significant because it is exactly soluble and we can study various postulates of quantum mechanics in some detail in this system. We have already seen that the motion of the oscillator is bounded and, as emphasized earlier, this leads

Figure 5.1: One dimensional crystal with constituent molecules.

to quantization of energy. Furthermore, the correspondence principle can also be easily checked here, namely, we can analyze the Ehrenfest theorem as well as check that the quantum mechanical system does behave like a classical system for large values of energy. Let us do this in some detail.

First, we note that we can write the solutions of the classical harmonic oscillator as

$$\begin{aligned} x_{\rm cl}(t) &= A \sin(\omega t + \phi), \\ p_{\rm cl}(t) &= m\dot{x}_{\rm cl} = m\omega A \cos(\omega t + \phi), \end{aligned} \tag{5.117}$$

where A represents the amplitude and ϕ a phase angle, both of which can be determined from the initial conditions. In fact, let us next identify

$$\begin{aligned} x_{\rm cl}(0) &= x_0 = A \sin\phi, \\ p_{\rm cl}(0) &= p_0 = m\omega A \cos\phi, \end{aligned} \tag{5.118}$$

so that we can write the solution in terms of the initial conditions as

$$\begin{aligned} x_{\rm cl}(t) &= x_0 \cos\omega t + \frac{p_0}{m\omega} \sin\omega t, \\ p_{\rm cl}(t) &= p_0 \cos\omega t - m\omega x_0 \sin\omega t. \end{aligned} \tag{5.119}$$

Furthermore, the energy associated with the harmonic motion is given by

$$E_{\rm cl} = \frac{p_{\rm cl}^2(t)}{2m} + \frac{1}{2}m\omega^2 x_{\rm cl}^2(t) = \frac{1}{2}m\omega^2 A^2. \tag{5.120}$$

We see from (5.120) that the energy associated with the harmonic motion is constant in time and is proportional to the square of the amplitude which is a continuous variable. Therefore, the energy is also a continuous function and the minimum energy associated with the oscillator is zero, which occurs when the oscillator is sitting at rest at the position of equilibrium.

We can also ask the following probabilistic question. Suppose we randomly try to locate the harmonic oscillator, what is the place where we are most likely to find it. That, of course, would correspond to the place where the oscillator spends most of its time, namely, where the velocity is a minimum. In other words, the classical probability is inversely proportional to the velocity of the oscillator,

$$P(x) \propto \frac{1}{|v(x)|} = \frac{1}{\omega(A^2 - x^2)^{\frac{1}{2}}}. \tag{5.121}$$

Thus, the classical probability peaks around $x = A$ or at the turning points. It has a minimum at the point of equilibrium where the velocity is the largest.

We can now compare these with the properties of the quantum harmonic oscillator. First of all, in the energy basis, states are time independent. This corresponds to the Heisenberg picture of motion where the operators have time dependence and comparison with classical mechanics is the simplest in this picture. Therefore, in this picture, we can ask for the time dependence of various operators. For example (see (3.115) for the case when the operator has no explicit time dependence),

$$\frac{\mathrm{d}a}{\mathrm{d}t} = -\frac{i}{\hbar}\,[a, H] = -\frac{i}{\hbar}\,(\hbar\omega a) = -i\omega a, \tag{5.122}$$

which can be solved yielding

$$a(t) = a(0)e^{-i\omega t}. \tag{5.123}$$

Similarly, for the creation operator, we have

$$\frac{\mathrm{d}a^\dagger}{\mathrm{d}t} = -\frac{i}{\hbar}\,[a^\dagger, H] = -\frac{i}{\hbar}\,(-\hbar\omega a^\dagger) = i\omega a^\dagger, \tag{5.124}$$

which leads to

$$a^\dagger(t) = a^\dagger(0)e^{i\omega t}. \tag{5.125}$$

The creation and the annihilation operators are defined in (5.2) as

$$a = \sqrt{\frac{m\omega}{2\hbar}}\left(X + \frac{i}{m\omega}P\right),$$

$$a^\dagger = \sqrt{\frac{m\omega}{2\hbar}}\left(X - \frac{i}{m\omega}P\right),$$

so that we have

$$X = \sqrt{\frac{\hbar}{2m\omega}}\left(a + a^{\dagger}\right),$$

$$P = -i\sqrt{\frac{\hbar m\omega}{2}}\left(a - a^{\dagger}\right). \tag{5.126}$$

Using (5.123), (5.125) as well as the definition in (5.126), the time dependence of the coordinate and the momentum operators can be determined to be

$$\begin{aligned} X(t) &= \sqrt{\frac{\hbar}{2m\omega}}\left(a(t) + a^{\dagger}(t)\right) \\ &= \sqrt{\frac{\hbar}{2m\omega}}\left(a(0)e^{-i\omega t} + a^{\dagger}(0)e^{i\omega t}\right) \\ &= \sqrt{\frac{\hbar}{2m\omega}}\Big((a(0) + a^{\dagger}(0))\cos\omega t \\ &\qquad\qquad -i(a(0) - a^{\dagger}(0))\sin\omega t\Big) \\ &= \left(X_0\cos\omega t + \frac{P_0}{m\omega}\sin\omega t\right). \end{aligned} \tag{5.127}$$

Similarly, we can show that

$$P(t) = P_0\cos\omega t - m\omega X_0\sin\omega t. \tag{5.128}$$

In general, therefore, in any picture we can write

$$\langle X\rangle(t) = \langle X\rangle(0)\cos\omega t + \frac{1}{m\omega}\langle P\rangle(0)\sin\omega t,$$

$$\langle P\rangle(t) = \langle P\rangle(0)\cos\omega t - m\omega\langle X\rangle(0)\sin\omega t. \tag{5.129}$$

In other words, the expectation values have similar behavior as the classical variables in (5.119) as Ehrenfest's theorem would require.

In the case of the quantum oscillator, however, the energy can not take any continuous value and is quantized

$$E_n = \hbar\omega\left(n + \frac{1}{2}\right), \qquad n = 0, 1, 2, \cdots. \tag{5.130}$$

Furthermore, we see from (5.130) that the minimum of the energy is not zero. This arises basically because of our inability to simultaneously specify both the position as well as the momentum in a quantum mechanical system.

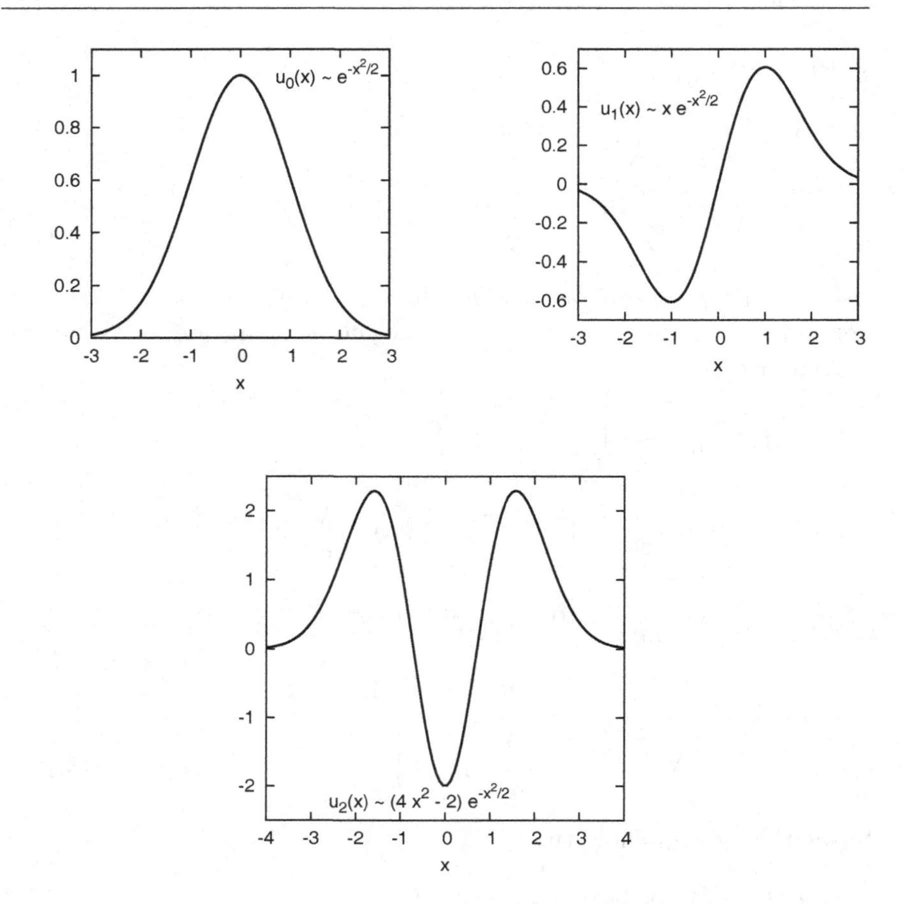

Figure 5.2: The first three wave functions of the harmonic oscillator.

Let us plot a few of the lower order wave functions associated with the motion of the oscillator. We see from Fig. 5.2 that the probability densities for the oscillator (which is the absolute square of the wave functions) seem to behave very differently from that of a classical harmonic oscillator. In particular, for the ground state, we note that the maximum probability is around the point of equilibrium ($x = 0$) and falls off at large distances. This is just the opposite of the classical behavior. However, if we plot the probability density for large values of the quantum numbers n (see Fig. 5.3), the behavior is as follows, as $n \to \infty$, the average value of the probabilities of these plots behaves like the classical oscillator. This is, of course, what the correspondence principle says, namely, when the energy of the system becomes large, the system behaves like a macroscopic system.

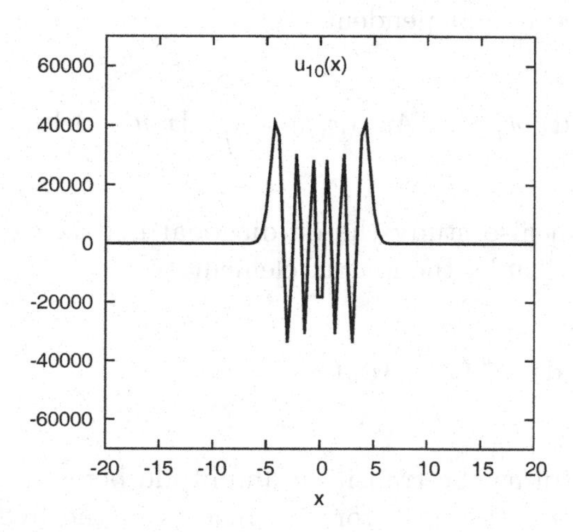

Figure 5.3: The wave function of the oscillator for $n = 10$.

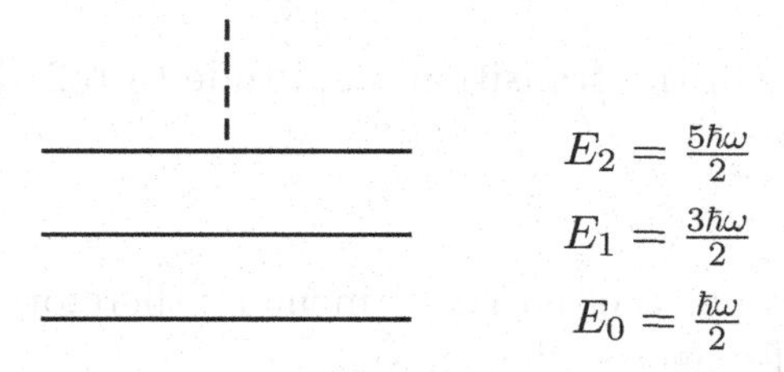

Figure 5.4: Energy levels of the oscillator with uniform spacing of $\hbar\omega$.

Notice that the energy levels of the oscillator are spaced uniformly without any dependence on the mass parameter of the theory. Each level differs from the adjacent by $\hbar\omega$ as shown in Fig. 5.4. This allows us to think as if, with an oscillator of frequency ω, there are associated fictitious particles with quanta of energy $\hbar\omega$. In crystal physics, such particles are known as phonons whereas the interaction of matter with radiation is described in terms of particles known as photons.

We know that the diagonal values of an operator represent expectation values of that operator in a given state. In a stationary state, these are time independent

$$A_{nn} = \int_{-\infty}^{\infty} \mathrm{d}x\ \psi_n^*(x,t) A \psi_n(x,t) = \int_{-\infty}^{\infty} \mathrm{d}x\ u_n^*(x) A u_n(x). \quad (5.131)$$

However, we can also study the off-diagonal matrix elements of an operator. For example, the matrix element

$$A_{mn} = \int_{-\infty}^{\infty} \mathrm{d}x\ \psi_m^*(x,t) A \psi_n(x,t), \quad (5.132)$$

can be thought of as the transition amplitude between the states n and m induced by the operator A. Just as in the hydrogen atom where the electron can drop down from an excited level to a lower level with the emission of a photon, here also any transition between distinct states is accompanied by an emission or absorption of quanta. Furthermore, since

$$\psi_n(x,t) \sim e^{-\frac{i}{\hbar}E_n t}, \quad (5.133)$$

the time dependence of the transition amplitude (5.132) is given by

$$A_{mn} \propto e^{-\frac{i}{\hbar}(E_n - E_m)t}. \quad (5.134)$$

The time dependence of the matrix element is, therefore, nontrivial when $n \neq m$ (for off-diagonal elements).

5.6 Density matrix

Let us next consider not just one isolated system, but an ensemble of identical quantum mechanical systems. Of course, each system in the ensemble can be in a different eigenstate of the Hamiltonian and, consequently, there will be a statistical distribution of systems in various eigenstates of the Hamiltonian. Let us denote by p_n the probability of finding any system in the ensemble in the energy eigenstate $|n\rangle$. Such an ensemble is, in fact, quite physical. For example, we know that in an ensemble which is in thermodynamic equilibrium with a heat reservoir, the probability of finding a system in a specific energy state is given by the Boltzmann distribution law.

Suppose, in such an ensemble we make a measurement of the operator A. It is clear that the value of the measurement has a probability p_n of being $\langle n|A|n\rangle$. The statistical average of these measurements over the entire ensemble is, therefore,

$$\overline{\langle A\rangle} = \sum_n p_n \langle n|A|n\rangle = \sum_n p_n \langle A\rangle_n. \tag{5.135}$$

Actually, there are two kinds of averaging being done in (5.135). First of all, we have the quantum averaging represented by the expectation value $\langle n|A|n\rangle$ and then the classical averaging over systems in different states $|n\rangle$. To distinguish this ensemble average from $\langle A\rangle$, the quantum average, we have put a bar over this expectation value.

Let us define an operator

$$\rho = \sum_n p_n |n\rangle\langle n|, \tag{5.136}$$

where p_n is the probability that a system, picked out randomly from the ensemble, is in the state $|n\rangle$. The p_n's, therefore, satisfy

$$\begin{aligned} &p_n \geq 0, \qquad \text{for all } n,\\ &\sum_n p_n = 1. \end{aligned} \tag{5.137}$$

The operator ρ is called the density operator or the density matrix. Let us now calculate

$$\begin{aligned} \text{Tr}\ \rho A &= \sum_m \langle m|\rho A|m\rangle = \sum_{m,n} p_n \langle m|n\rangle\langle n|A|m\rangle \\ &= \sum_{m,n} p_n \delta_{mn} \langle n|A|m\rangle = \sum_n p_n \langle n|A|n\rangle \\ &= \overline{\langle A\rangle}. \end{aligned} \tag{5.138}$$

Therefore, we see that given any operator Ω, its ensemble average can be defined as

$$\overline{\langle \Omega\rangle} = \text{Tr}\ \rho\Omega. \tag{5.139}$$

In particular, we can choose $\Omega = \mathbb{1}$. Then, we have from (5.139)

$$\text{Tr}\ \rho = \sum_n p_n = 1. \tag{5.140}$$

We can also ask about the statistical average of the probability of obtaining a particular eigenvalue of the operator Ω. We know that

$$P(\omega) = |\langle\omega|\psi\rangle|^2 = \langle\psi|\omega\rangle\langle\omega|\psi\rangle = \langle\psi|P_\omega|\psi\rangle = \langle P_\omega\rangle, \tag{5.141}$$

so that

$$\overline{\langle P_\omega \rangle} = \overline{P}(\omega) = \text{Tr } \rho P_\omega. \tag{5.142}$$

It is clear from the definition of the density matrix that, when all the p_n's are zero except for one, it is a pure ensemble. (Namely, every system, in such an ensemble, is in the same state.) We can derive the following properties of the density matrix quite easily.

1. The density matrix is Hermitian, namely, $\rho^\dagger = \rho$.

2. $\text{Tr}\,\rho = 1 = \sum\limits_n p_n$.

3. The density matrix is positive semi-definite, namely, $\langle u|\rho|u\rangle \geq 0$.

   Proof.

   $$\langle u|\rho|u\rangle = \sum_n p_n \langle u|n\rangle\langle n|u\rangle = \sum_n p_n |\langle u|n\rangle|^2 \geq 0.$$

   ■

4. For a pure ensemble, $\rho^2 = \rho$.

   Proof.

   $$\begin{aligned}
   \rho &= \sum_n p_n |n\rangle\langle n|, \\
   \rho^2 &= \sum_{m,n} p_m p_n |m\rangle\langle m|n\rangle\langle n| = \sum_{m,n} p_m p_n \delta_{mn} |m\rangle\langle n| \\
   &= \sum_n p_n^2 |n\rangle\langle n|.
   \end{aligned}$$

   For a pure ensemble, all the p_n's are zero except for one which is equal to unity. Hence $\rho^2 = \rho$ ■

5. $\text{Tr}\,\rho^n \leq 1$ for $n \geq 2$ and the equality holds for a pure ensemble.

5.7 Planck's law

With these basics for general ensembles, let us turn to a thermodynamic ensemble. In classical thermodynamics, a system in thermodynamic equilibrium at temperature T has a probability of being in an energy state E given by the Boltzmann law

$$p(E) = Ne^{-\frac{E}{kT}}, \tag{5.143}$$

where N is a normalization constant. Let $kT = \beta^{-1}$. If energy is quantized, this probability for a quantum system becomes

$$p_n = N\, e^{-\frac{E_n}{kT}} = \frac{e^{-\frac{E_n}{kT}}}{\sum\limits_n e^{-\frac{E_n}{kT}}} = \frac{e^{-\beta E_n}}{\sum\limits_n e^{-\beta E_n}}, \quad \sum_n p_n = 1. \tag{5.144}$$

Thus, the density matrix, in such a case, takes the form

$$\begin{aligned}\rho = \sum_n p_n |n\rangle\langle n| &= \frac{\sum\limits_n e^{-\beta E_n}|n\rangle\langle n|}{\sum\limits_n e^{-\beta E_n}} = \frac{e^{-\beta H}\sum\limits_n |n\rangle\langle n|}{\sum\limits_n \langle n|e^{-\beta H}|n\rangle}\\ &= \frac{e^{-\beta H}}{\mathrm{Tr}\ e^{-\beta H}} = \frac{e^{-\beta H}}{Z(\beta)},\end{aligned} \tag{5.145}$$

where we have defined the partition function for the system as

$$Z(\beta) = \mathrm{Tr}\ e^{-\beta H} = \sum_n \langle n|e^{-\beta H}|n\rangle = \sum_n e^{-\beta E_n}. \tag{5.146}$$

Our discussion so far has been quite general. Let us next consider an ensemble of harmonic oscillators. For a harmonic oscillator, we know that the energy eigenvalues are given by

$$H|n\rangle = E_n|n\rangle = \hbar\omega\left(n + \frac{1}{2}\right)|n\rangle, \quad n = 0, 1, 2, \cdots . \tag{5.147}$$

Thus, for an ensemble of oscillators, the partition function takes the form

$$\begin{aligned}Z(\beta) &= \sum_{n=0}^{\infty}\langle n|e^{-\beta H}|n\rangle = \sum_{n=0}^{\infty} e^{-\beta\hbar\omega\left(n+\frac{1}{2}\right)}\\ &= e^{-\frac{\beta\hbar\omega}{2}}\sum_{n=0}^{\infty} e^{-\beta\hbar\omega n} = e^{-\frac{\beta\hbar\omega}{2}}\left(\frac{1}{1-e^{-\beta\hbar\omega}}\right)\\ &= \frac{e^{-\frac{\beta\hbar\omega}{2}}}{1-e^{-\beta\hbar\omega}} = \frac{2}{\sinh\frac{\beta\hbar\omega}{2}}.\end{aligned} \tag{5.148}$$

In this ensemble the mean (average) energy of the oscillator is given by (can also be represented as $\overline{\langle H\rangle}$)

$$\begin{aligned}\overline{\langle E\rangle} &= \text{Tr}\ \rho H = \frac{\text{Tr}\ He^{-\beta H}}{\text{Tr}\ e^{-\beta H}} = -\frac{1}{Z(\beta)}\frac{\partial Z(\beta)}{\partial\beta} \\ &= -\frac{\partial \ln Z(\beta)}{\partial\beta},\end{aligned} \tag{5.149}$$

so that using (5.148) we obtain

$$\begin{aligned}\overline{\langle E\rangle} &= -\frac{\partial \ln Z(\beta)}{\partial\beta} = -\frac{\partial}{\partial\beta}\left[-\frac{\beta\hbar\omega}{2} - \ln\left(1 - e^{-\beta\hbar\omega}\right)\right] \\ &= \frac{\hbar\omega}{2} + \frac{\hbar\omega e^{-\beta\hbar\omega}}{1 - e^{-\beta\hbar\omega}} = \frac{\hbar\omega}{2} + \frac{\hbar\omega}{e^{\beta\hbar\omega} - 1}.\end{aligned} \tag{5.150}$$

This is Planck's law. Clearly, for very large $\beta\ (=\frac{1}{kT})$ or very small temperatures, the oscillator remains near the ground state

$$\overline{\langle E\rangle} = \frac{\hbar\omega}{2}. \tag{5.151}$$

For very high temperatures or small β, on the other hand, we have

$$\overline{\langle E\rangle} = \frac{1}{\beta} = kT. \tag{5.152}$$

This is nothing other than the equipartition of energy (for the one dimensional oscillator).

5.8 Oscillator in higher dimensions

Let us next consider a harmonic oscillator in p dimensions ($p \geq 1$). The oscillator has p degrees of freedom and the Hamiltonian, in this case, takes the form

$$H = \sum_{i=1}^{p} H_i, \tag{5.153}$$

where

$$H_i = \frac{P_i^2}{2m} + \frac{1}{2}m\omega^2 X_i^2. \tag{5.154}$$

Namely, the oscillations are harmonic along every direction with the same angular frequency ω. Such an oscillator, whose frequency is the same in every direction, is known as the isotropic oscillator. (In general, the frequency of an oscillator can be different, say ω_i, in different directions.) In this case, we know the basic commutation relations between the coordinates and momenta to be

$$[X_i, X_j] = 0 = [P_i, P_j],$$
$$[X_i, P_j] = i\hbar\delta_{ij}, \tag{5.155}$$

which leads to

$$[H_i, H_j] = 0, \qquad i, j = 1, 2, \cdots, p. \tag{5.156}$$

To solve this problem, we note that we can think of the system as a set of decoupled harmonic oscillators each of which can be solved independently of the others. The Hilbert space of states $\mathcal{E}$ separates, in this case, into a product space as a consequence of (5.156). Thus, we can think of

$$\mathcal{E} = \mathcal{E}_1 \otimes \mathcal{E}_2 \otimes \cdots \otimes \mathcal{E}_p, \tag{5.157}$$

where H_i acts only on the space $\mathcal{E}_i$. We can also define, as before, the annihilation and creation operators (corresponding to every i)

$$a_i = \sqrt{\frac{m\omega}{2\hbar}}\left(X_i + \frac{i}{m\omega}P_i\right),$$
$$a_i^\dagger = \sqrt{\frac{m\omega}{2\hbar}}\left(X_i - \frac{i}{m\omega}P_i\right), \tag{5.158}$$

as well as the number operator

$$N_i = a_i^\dagger a_i, \qquad H_i = \hbar\omega\left(N_i + \frac{1}{2}\right), \tag{5.159}$$

which act only on the states in $\mathcal{E}_i$. The eigenvectors of N_i, which are denoted by $|n_i\rangle$, with $n_i = 0, 1, 2, \ldots, \infty$, therefore, define the basis states of the vector space $\mathcal{E}_i$ and satisfy

$$N_i|n_i\rangle = n_i|n_i\rangle,$$
$$H_i|n_i\rangle = E_{n_i}|n_i\rangle = \hbar\omega\left(n_i + \frac{1}{2}\right)|n_i\rangle. \tag{5.160}$$

Since, the total space is a product of spaces,

$$\mathcal{E} = \mathcal{E}_1 \otimes \mathcal{E}_2 \otimes \cdots \otimes \mathcal{E}_p, \tag{5.161}$$

we can label the states in $\mathcal{E}$ by the quantum numbers of the individual product spaces and define them as

$$|n_1, n_2, \ldots, n_p\rangle = |n_1\rangle \otimes |n_2\rangle \otimes \cdots \otimes |n_p\rangle, \tag{5.162}$$

where

$$n_1, n_2, \ldots, n_p = 0, 1, 2, \ldots, \infty. \tag{5.163}$$

It is easy to show, as in our earlier discussion, that

$$\begin{aligned} [a_i, a_j] &= \left[a_i^\dagger, a_j^\dagger\right] = 0, \\ \left[a_i, a_j^\dagger\right] &= \delta_{ij}. \end{aligned} \tag{5.164}$$

Furthermore, let us define an operator

$$N = \sum_i N_i = \sum_i a_i^\dagger a_i, \tag{5.165}$$

so that

$$\begin{aligned} N|n_1, n_2, \ldots, n_p\rangle \\ &= \sum_i N_i \left(|n_1\rangle \otimes |n_2\rangle \cdots \otimes |n_p\rangle\right) \\ &= (N_1|n_1\rangle) \otimes |n_2\rangle \cdots \otimes |n_p\rangle \\ &\quad + |n_1\rangle \otimes (N_2|n_2\rangle) \cdots \otimes |n_p\rangle \\ &\quad + \ldots \\ &\quad + |n_1\rangle \otimes |n_2\rangle \cdots \otimes (N_p|n_p\rangle) \\ &= (n_1 + n_2 + \ldots n_p)\left(|n_1\rangle \otimes |n_2\rangle \cdots \otimes |n_p\rangle\right) \\ &= (n_1 + n_2 + \ldots n_p)|n_1, n_2, \ldots, n_p\rangle \\ &= n|n_1, n_2, \ldots, n_p\rangle, \end{aligned} \tag{5.166}$$

where we have defined

$$n = n_1 + n_2 + \cdots + n_p. \tag{5.167}$$

This is the total number of quanta in the state and, correspondingly, N is called the total number operator. Similarly,

$$\begin{aligned} H = \sum_i H_i &= \hbar\omega \sum_{i=1}^{p} \left(N_i + \frac{1}{2}\right) \\ &= \hbar\omega \left(N + \frac{p}{2}\right), \end{aligned} \tag{5.168}$$

where we have used the definition in (5.165). We also note that

$$\begin{aligned} H|n_1, n_2, \ldots, n_p\rangle \\ &= \hbar\omega\left(N + \frac{p}{2}\right)|n_1, n_2, \ldots, n_p\rangle \\ &= \hbar\omega\left(n + \frac{p}{2}\right)|n_1, n_2, \ldots, n_p\rangle. \end{aligned} \tag{5.169}$$

Thus, the energy levels of the isotropic oscillator in p dimensions are given by

$$E_n = \hbar\omega\left(n + \frac{p}{2}\right), \qquad n = 0, 1, 2, \ldots. \tag{5.170}$$

Furthermore, the ground state which is denoted by

$$|0, 0, \ldots, 0\rangle = |0\rangle \otimes |0\rangle \otimes \cdots \otimes |0\rangle, \tag{5.171}$$

satisfies

$$a_i|0, 0, \ldots, 0\rangle = 0, \qquad \text{for all } i, \tag{5.172}$$

and is an eigenstate of the Hamiltonian with energy

$$E_0 = \frac{p\hbar\omega}{2}. \tag{5.173}$$

This corresponds to a zero point energy of $\frac{\hbar\omega}{2}$ for every direction (or every degree of freedom). Furthermore, any higher state can be written as

$$|n_1, n_2, \ldots, n_p\rangle = \frac{(a_1^\dagger)^{n_1}(a_2^\dagger)^{n_2}\ldots(a_p^\dagger)^{n_p}}{\sqrt{n_1!n_2!\ldots n_p!}}|0, 0, \ldots, 0\rangle. \tag{5.174}$$

It is clear now that, in higher dimensions, there is degeneracy of states in the spectrum of the oscillator. For example, the state with energy

$$E_1 = \hbar\omega\left(1 + \frac{p}{2}\right), \tag{5.175}$$

is p-fold degenerate. This is easily seen by noting that a state of the form $|1, 0, 0, \ldots, 0\rangle$ has energy E_1. But so does $|0, 1, 0, \ldots, 0\rangle$, $|0, 0, 1, 0, \ldots\rangle$ and so on. There are p-such states.

A state with energy $E_2 = \hbar\omega\left(2 + \frac{p}{2}\right)$ has $\frac{1}{2}(p+1)p$-fold degeneracy. This can be seen by noting that a state of the form $|1, 1, 0, 0, \ldots, 0\rangle$ has energy E_2. There are $\frac{1}{2}p(p-1)$ such states. But, a state of the form $|2,0,0,\ldots,0\rangle$ also has energy E_2. There are p such states. Thus, the total number of states with energy E_2 is

$$\frac{1}{2}p(p-1)+p=\frac{1}{2}p(p+1). \tag{5.176}$$

In general, one can show that in p dimensions a state with energy $E_n = \hbar\omega\left(n+\frac{p}{2}\right)$ has a $\binom{n+p-1}{n}$-fold degeneracy. Let us check this against some known cases. First, in one dimension where $p=1$, the degeneracy formula gives $\binom{n}{n}=1$, namely, there is no degeneracy of states in this case (which we have seen earlier). In p dimensions, for $n=0$, we have the degeneracy $\binom{p-1}{0}=1$, implying that the ground state has no degeneracy. For $n=1$ in p dimensions, we have $\binom{p}{1}=p$, which we have explicitly seen to be the degeneracy of the first excited state. For $n=2$, the formula gives $\binom{p+1}{2}=\frac{1}{2}(p+1)p$, which is the degeneracy of the second excited state as we have seen and so on.

5.9 Selected problems

1. Find $\langle X\rangle, \langle P\rangle, \langle X^2\rangle, \langle P^2\rangle$ and $\Delta X\Delta P$ in the state $|n\rangle$ of the harmonic oscillator. What is the uncertainty relation (for the coordinate and momentum measurements) in the ground state?

2. If $u_n(x)$ and $u_m(x)$ are the eigenfunctions of the harmonic oscillator in one dimension, corresponding to the energy eigenvalues $\hbar\omega(n+\frac{1}{2})$ and $\hbar\omega(m+\frac{1}{2})$ respectively, use the recursion relations for the Hermite polynomials to calculate

$$\langle P\rangle_{nm} = \int_{-\infty}^{\infty} \mathrm{d}x\, u_n^*(x)\left(-i\hbar\,\frac{\mathrm{d}u_m(x)}{\mathrm{d}x}\right). \tag{5.177}$$

3. A particle moving in one dimension has a first excited state eigenfunction associated with the energy eigenvalue E_1 given by

$$\psi_1(x) = x\,\psi_0(x), \tag{5.178}$$

where $\psi_0(x)$ is the ground state wave function associated with the energy eigenvalue E_0. Given that the potential vanishes at $x=0$,

a) Determine the ratio $\frac{E_1}{E_0}$.

b) What is the potential $V(x)$ in which the particle moves?

CHAPTER 6

Symmetries and their consequences

Symmetries play an important role in the study of physical systems – both in classical mechanics as well as in quantum mechanics. In these lectures we will start with a review of classical symmetry transformations before going into a discussion of symmetries in quantum mechanics.

6.1 Symmetries in classical mechanics

Physical objects often possess symmetries. For example, if we look at a circle we say that it is symmetric. That is because if we have not put any distinguishing marks on the circle, any point on the circle is indistinguishable from any other point. Furthermore, if we rotate the circle slightly (about its center) we cannot distinguish it from what it was before the rotation. A deck of (unmarked) playing cards also possesses a symmetry, namely, an up-down symmetry. That is, if we turn the cards upside down we cannot tell it from what it was before turning it.

Symmetries can be classified into two groups – continuous and discrete. We can take the circle and rotate it by any amount (about its center) and it would still look the same. On the other hand, if we are looking at an equilateral triangle, then, it is symmetric only if we rotate it about its center by $120°$ or multiples thereof. In this case, we speak of the equilateral triangle as possessing a discrete symmetry whereas we say that the circle has a continuous symmetry.

Any operation which leaves a system invariant is called a symmetry transformation of the system. Thus, for the circle rotation is a continuous symmetry transformation whereas for the deck of cards reflection (turning the deck of cards upside down) is a discrete symmetry transformation. Furthermore, let us consider the circle and note that we can define a rotation which is infinitesimally close to the original state. The circle would still be invariant. Any finite rotation can be thought of as a series of successive infinitesimal rotations. Therefore, for continuous symmetries, the study of infinitesimal transformations gives all the information about any finite transformation. On the other hand, we note that there does not exist any infinitesimal transformation for discrete symmetries.

Symmetries are not restricted only to physical objects or patterns. We can have functions or theories which also possess symmetries. Thus, for example, consider the simple function

$$f(x,y) = x^2 + y^2. \tag{6.1}$$

Let us note that if we make the change of variables

$$\begin{aligned} x \to x' &= x\cos\theta - y\sin\theta, \\ y \to y' &= x\sin\theta + y\cos\theta, \end{aligned} \tag{6.2}$$

then,

$$f(x,y) = x^2 + y^2 \to x'^2 + y'^2 = x^2 + y^2 = f(x,y). \tag{6.3}$$

We say that the function $f(x,y)$ is invariant or symmetric under the transformation (6.2). Since the parameter θ (of the transformation) can take any value and $f(x,y)$ would still be invariant, this is a continuous symmetry transformation. In fact, we recognize this as the rotation of coordinates x and y (around the z axis) and the function $f(x,y)$ as representing the length of a two dimensional vector which we know to be invariant under a rotation. (Alternatively, we note that $f(x,y) = a^2$ defines a circle and the invariance we are discussing can be thought of as a mathematical description of rotations as symmetries of a circle.) In this case, we can also define an infinitesimal transformation from (6.2) by identifying

$$\theta = \epsilon = \text{ infinitesimally small}. \tag{6.4}$$

Then, the transformation (infinitesimal rotation) in (6.2) takes the form

$$\begin{aligned} x' &= x - \epsilon y, \\ y' &= \epsilon x + y. \end{aligned} \tag{6.5}$$

A physical theory defined by the Hamiltonian $H = H(x^i, p_i)$ is said to possess a certain symmetry if the Hamiltonian is invariant under the corresponding transformations. Let us consider the following infinitesimal canonical transformation (a canonical transformation is one which preserves the fundamental Poisson bracket relations)

$$\begin{aligned} x^i \to x'^i &= x^i + \epsilon \frac{\partial g}{\partial p_i} = x^i + \delta x^i, \\ p_i \to p'_i &= p_i - \epsilon \frac{\partial g}{\partial x^i} = p_i + \delta p_i, \end{aligned} \tag{6.6}$$

where ϵ represents the infinitesimal parameter of transformation and $g = g(x^i, p_i)$ is called the generator of the infinitesimal transformations.

If the Hamiltonian is invariant under the above transformations, then, $g(x^i, p_i)$ – the generator of the transformations – is conserved or is a constant of motion. To see this, let us note that the change in the Hamiltonian under the transformations (6.6) is given by

$$\begin{aligned} \delta H(x^i, p_i) &= \sum_i \left(\frac{\partial H}{\partial x^i} \delta x^i + \frac{\partial H}{\partial p_i} \delta p_i \right) \\ &= \sum_i \left(\frac{\partial H}{\partial x^i} \epsilon \frac{\partial g}{\partial p_i} + \frac{\partial H}{\partial p_i} \left(-\epsilon \frac{\partial g}{\partial x^i} \right) \right) \\ &= \epsilon \sum_i \left(\frac{\partial H}{\partial x^i} \frac{\partial g}{\partial p_i} - \frac{\partial H}{\partial p_i} \frac{\partial g}{\partial x^i} \right) \\ &= \epsilon \{H, g\}. \end{aligned} \tag{6.7}$$

If H is invariant, this implies that for all values of the parameter ϵ,

$$\delta H = 0, \quad \text{or,} \quad \{H, g\} = 0. \tag{6.8}$$

But, by Hamilton's equation, (1.54), this leads to

$$\frac{\mathrm{d}g}{\mathrm{d}t} = \{g, H\} = 0. \tag{6.9}$$

In other words, $g(x^i, p_i)$ is a constant of motion. Conversely, every conserved quantity generates a continuous symmetry of the Hamiltonian.

Note that any dynamical variable $\omega(x^i, p_i)$ has the following transformation properties under the transformations (6.6),

$$\begin{aligned}\delta\omega(x^i,p_i) &= \sum_i\left(\frac{\partial\omega}{\partial x^i}\,\delta x^i + \frac{\partial\omega}{\partial p_i}\delta p_i\right)\\ &= \epsilon\sum_i\left(\frac{\partial\omega}{\partial x^i}\frac{\partial g}{\partial p_i} - \frac{\partial\omega}{\partial p_i}\frac{\partial g}{\partial x^i}\right)\\ &= \epsilon\{\omega,g\}.\end{aligned} \tag{6.10}$$

This is why g is known as the generator of the symmetry transformation. In particular, choosing $\omega(x^i,p_i)=x^i$, we have

$$\delta x^i = \epsilon\{x^i,g\} = \epsilon\frac{\partial g}{\partial p_i}, \tag{6.11}$$

and, similarly, for $\omega(x^i,p_i)=p_i$, we have

$$\delta p_i = \epsilon\{p_i,g\} = -\epsilon\frac{\partial g}{\partial x^i}, \tag{6.12}$$

which coincides with (6.6). Let us next examine a few classical symmetries.

► **Example.** Let us consider a theory in one dimension described by the Hamiltonian $H(x,p)$. Let us choose

$$g(x,p)=p. \tag{6.13}$$

In this case, we obtain the transformations explicitly from (6.10) to be

$$\begin{aligned}&\delta x = \epsilon\{x,g\} = \epsilon\{x,p\} = \epsilon, &&\Rightarrow\ x\to x' = x+\epsilon,\\ &\delta p = \epsilon\{p,g\} = \epsilon\{p,p\} = 0, &&\Rightarrow\ p\to p' = p,\end{aligned} \tag{6.14}$$

which we recognize as an infinitesimal translation of the coordinate x. We see that momentum is the generator of infinitesimal translations and it follows that momentum is conserved in theories which are invariant under translations.

Physically what this means is that since for a single particle,

$$H = \frac{p^2}{2m} + V(x) = T + V,$$

and since the momentum is unaffected by translations of the coordinate, the kinetic energy is invariant. Furthermore, if the potential is such that $V(x) = V(x+\epsilon)$, i.e., if it is a constant, then, the Hamiltonian will be invariant under translations. In this case, the force acting on the particle is zero and, consequently, momentum is conserved. ◄

► **Example.** Let us next consider the same theory described by the Hamiltonian $H(x,p)$ and identify

$$g(x,p) = H(x,p). \tag{6.15}$$

In this case, we see that

$$\{H, g\} = \{H, H\} = 0, \tag{6.16}$$

and

$$\begin{aligned}
&\delta x = \epsilon\{x, H\} = \epsilon \dot{x},\\
\text{or,}\quad &x'(t) = x(t) + \epsilon \frac{\mathrm{d}x(t)}{\mathrm{d}t} = x(t+\epsilon),\\
&\delta p = \epsilon\{p, H\} = \epsilon \dot{p},\\
\text{or,}\quad &p'(t) = p(t) + \epsilon \frac{\mathrm{d}p(t)}{\mathrm{d}t} = p(t+\epsilon).
\end{aligned} \tag{6.17}$$

Thus, it is clear that these transformations correspond to a translation of time and if the Hamiltonian does not depend on time explicitly, then, from (6.16) we note that it is a symmetry of the theory and the total energy is a constant. Hamiltonian is the generator of infinitesimal time translations. ◀

► **Example.** As another example, let us consider a two dimensional theory parameterized by (x, y, p_x, p_y) and described by the Hamiltonian $H(x, y, p_x, p_y)$. Let us further identify $g(x, y, p_x, p_y) = xp_y - yp_x = \ell_z$ = angular momentum about the z-axis. In this case, we have

$$\begin{aligned}
\delta x &= \epsilon\{x, g\} = \epsilon\{x, xp_y - yp_x\} = -\epsilon y\{x, p_x\} = -\epsilon y,\\
\delta y &= \epsilon\{y, g\} = \epsilon\{y, xp_y - yp_x\} = \epsilon x\{y, p_y\} = \epsilon x,\\
\delta p_x &= \epsilon\{p_x, g\} = \epsilon\{p_x, xp_y - yp_x\} = \epsilon p_y\{p_x, x\} = -\epsilon p_y,\\
\delta p_y &= \epsilon\{p_y, g\} = \epsilon\{p_y, xp_y - yp_x\} = -\epsilon p_x\{p_y, y\} = \epsilon p_x.
\end{aligned} \tag{6.18}$$

It is clear, therefore, that under this transformation,

$$\begin{aligned}
x' &= x + \delta x = x - \epsilon y,\\
y' &= y + \delta y = y + \epsilon x,
\end{aligned} \tag{6.19}$$

and momenta also transform in an analogous manner.

As we have seen earlier in (6.5), this is precisely an infinitesimal rotation about the z-axis and we conclude that angular momentum is the generator of rotations. Furthermore, angular momentum is conserved if the Hamiltonian is invariant under rotations. Hamiltonians of the form of the isotropic harmonic oscillator

$$H = \frac{\mathbf{p}^2}{2m} + \frac{1}{2} m\omega^2 \mathbf{x}^2,$$

in higher dimensions would have conservation of angular momentum since both $\mathbf{p}^2$ and $\mathbf{x}^2$ are invariant under rotations. ◀

Theorem. *If the Hamiltonian of a system is invariant under some transformation $(x, p) \to (x', p')$ which is not necessarily infinitesimal, then, if $(x(t), p(t))$ denotes a solution of Hamilton's equations of motion, then, so does $(x'(t), p'(t))$.*

Furthermore, symmetries of a theory can be viewed in two different but equivalent ways. First of all, we can think of a fixed coordinate system in which the physical system is being transformed. Thus, for example, we can think of a particle at x being displaced by an amount 'a'. On the other hand, the same phenomenon can also be viewed as the object being undisturbed, rather the coordinate system being displaced by an amount '$-a$' along the x axis. The former view of the transformation where the system undergoes a change is called the active transformation. The second description where the coordinate system undergoes a change is known as the passive transformation.

6.2 Symmetries in quantum mechanics

Let us now try to investigate how symmetries are realized in quantum mechanics and what are their consequences. First of all, we note that in quantum mechanics the position of a particle or its momentum are not always well defined. Thus, to extend even a simple transformation like an infinitesimal translation (ϵ is infinitesimal)

$$x \to x + \epsilon, \qquad p \to p, \tag{6.20}$$

to quantum mechanics, we have to invoke Ehrenfest's theorem. We know that the expectation values of operators behave like classical quantities. Therefore, the natural generalization of the classical transformation, (6.20), is

$$\langle X\rangle \to \langle X\rangle + \epsilon, \qquad \langle P\rangle \to \langle P\rangle, \tag{6.21}$$

where we have denoted

$$\langle \Omega\rangle = \langle\psi|\Omega|\psi\rangle. \tag{6.22}$$

Of course, one of the ways to look at this is to assume that under a translation the states change as

$$|\psi\rangle \to |\psi_\epsilon\rangle = |\psi'\rangle = T(\epsilon)|\psi\rangle, \tag{6.23}$$

such that $\langle\psi|X|\psi\rangle \to \langle\psi'|X|\psi'\rangle$ where

$$\begin{aligned} \langle\psi'|X|\psi'\rangle &= \langle\psi_\epsilon|X|\psi_\epsilon\rangle = \langle\psi|X|\psi\rangle + \epsilon, \\ \text{or,}\quad \langle\psi|T^\dagger(\epsilon)XT(\epsilon)|\psi\rangle &= \langle\psi|X|\psi\rangle + \epsilon. \end{aligned} \tag{6.24}$$

Here $T(\epsilon)$ denotes the operator which implements an infinitesimal translation on the Hilbert space of states. We also have $\langle\psi|P|\psi\rangle \to \langle\psi'|P|\psi'\rangle$ such that

$$\begin{aligned}
&\langle\psi'|P|\psi'\rangle = \langle\psi_\epsilon|P|\psi_\epsilon\rangle = \langle\psi|P|\psi\rangle,\\
\text{or,}\quad &\langle\psi|T^\dagger(\epsilon)PT(\epsilon)|\psi\rangle = \langle\psi|P|\psi\rangle. \qquad (6.25)
\end{aligned}$$

This point of view is known as the active transformation. For here the state of the system directly undergoes the change.

The passive view, of course, is that the state of the system remains unaltered. Rather, the operators change as

$$X \to T^\dagger(\epsilon)XT(\epsilon), \qquad P \to T^\dagger(\epsilon)PT(\epsilon), \qquad (6.26)$$

such that

$$T^\dagger(\epsilon)XT(\epsilon) = X + \epsilon\mathbb{1}, \qquad (6.27)$$

(which implies that $T^\dagger(\epsilon)XT(\epsilon)$ also measures position but with respect to an origin shifted by ϵ to the left) and

$$T^\dagger(\epsilon)PT(\epsilon) = P. \qquad (6.28)$$

Let us first consider translations from the active point of view. To understand how an arbitrary state transforms under translations, let us recall how the x-basis vectors transform. We know that under a translation,

$$T(\epsilon)|x\rangle = |x+\epsilon\rangle. \qquad (6.29)$$

Namely, the effect of $T(\epsilon)$ is to displace x to $x+\epsilon$. Note that

$$\langle x'|T^\dagger(\epsilon)T(\epsilon)|x\rangle = \langle x'+\epsilon|x+\epsilon\rangle = \delta(x'-x) = \langle x'|x\rangle. \qquad (6.30)$$

Since the states are normalized, it follows that

$$T^\dagger(\epsilon)T(\epsilon) = \mathbb{1}. \qquad (6.31)$$

This means that in quantum mechanics translations are represented by operators which are unitary .

To understand how an arbitrary state transforms under a translation, we note that

$$\begin{aligned}
|\psi'\rangle = |\psi_\epsilon\rangle &= T(\epsilon)|\psi\rangle = T(\epsilon)\int \mathrm{d}x\ |x\rangle\langle x|\psi\rangle\\
&= \int \mathrm{d}x\ T(\epsilon)|x\rangle\psi(x) = \int \mathrm{d}x\ |x+\epsilon\rangle\psi(x). \qquad (6.32)
\end{aligned}$$

Let us define $x' = x + \epsilon$, so that we have

$$|\psi'\rangle = |\psi_\epsilon\rangle = \int \mathrm{d}x'\ |x'\rangle\psi(x'-\epsilon) = \int \mathrm{d}x\ |x\rangle\psi(x-\epsilon),$$

$$\text{or,}\quad \langle x|\psi'\rangle = \langle x|\psi_\epsilon\rangle = \langle x|T(\epsilon)|\psi\rangle = \psi(x-\epsilon),$$

$$\text{or,}\quad \psi_\epsilon(x) = \langle x|T(\epsilon)|\psi\rangle = \psi(x-\epsilon). \tag{6.33}$$

In other words, if

$$\psi(x) \sim e^{-x^2}, \tag{6.34}$$

then,

$$\psi_\epsilon(x) = \psi(x-\epsilon) \sim e^{-(x-\epsilon)^2}. \tag{6.35}$$

This simply means that if the wave function is a Gaussian centered at the origin $x = 0$, then the transformed wave function is an identical Gaussian centered at $x = \epsilon$ as shown in Fig. 6.1. Namely, the wave function simply gets translated without any change in shape.

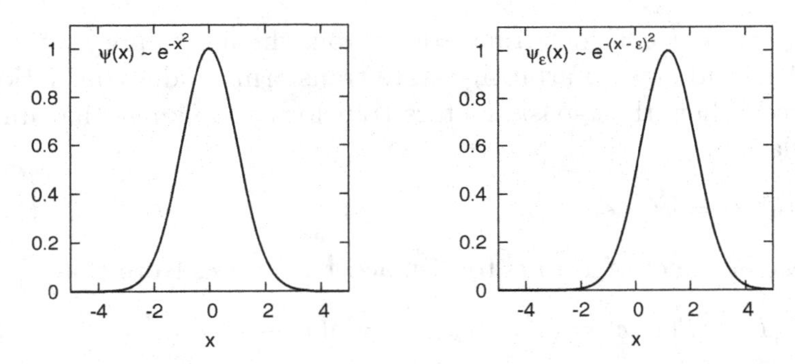

Figure 6.1: A Gaussian wave function and its translated form.

We can also show that under such a transformation,

$$\begin{aligned}
\langle\psi_\epsilon|P|\psi_\epsilon\rangle &= \int \mathrm{d}x\ \psi_\epsilon^*(x)\left(-i\hbar\frac{\mathrm{d}}{\mathrm{d}x}\right)\psi_\epsilon(x)\\
&= \int \mathrm{d}x\ \psi^*(x-\epsilon)\left(-i\hbar\frac{\mathrm{d}}{\mathrm{d}x}\right)\psi(x-\epsilon)\\
&= \int \mathrm{d}x'\ \psi^*(x')\left(-i\hbar\frac{\mathrm{d}}{\mathrm{d}x'}\right)\psi(x')\\
&= \langle\psi|P|\psi\rangle,
\end{aligned} \tag{6.36}$$

as we would expect.

Having defined the translation operator $T(\epsilon)$, let us define the generator of infinitesimal translations as

$$T(\epsilon) = \mathbb{1} - \frac{i\epsilon}{\hbar} G, \tag{6.37}$$

where ϵ is the infinitesimal parameter of transformation and the factor "i" is introduced so that the generator G would be Hermitian, namely,

$$\begin{aligned} T^{\dagger}(\epsilon)T(\epsilon) &= \left(\mathbb{1} + \frac{i\epsilon}{\hbar} G^{\dagger}\right)\left(\mathbb{1} - \frac{i\epsilon}{\hbar} G\right) \\ &= \mathbb{1} + \frac{i\epsilon}{\hbar}\, G^{\dagger} - \frac{i\epsilon}{\hbar} G + O(\epsilon^2). \end{aligned} \tag{6.38}$$

Since $T^{\dagger}(\epsilon)T(\epsilon) = \mathbb{1}$, it follows that $G^{\dagger} = G$, namely, the generator is Hermitian.

To determine the form of the generator of infinitesimal translations, let us note that (see (6.33))

$$\begin{aligned} &\langle x|T(\epsilon)|\psi\rangle = \psi(x-\epsilon), \\ \text{or,}\quad &\langle x|\mathbb{1} - \frac{i\epsilon}{\hbar} G|\psi\rangle = \psi(x) - \epsilon \frac{\mathrm{d}\psi(x)}{\mathrm{d}x} + O(\epsilon^2), \\ \text{or,}\quad &\psi(x) - \frac{i\epsilon}{\hbar}\langle x|G|\psi\rangle = \psi(x) - \epsilon\frac{\mathrm{d}\psi}{\mathrm{d}x}, \\ \text{or,}\quad &-\frac{i\epsilon}{\hbar}\langle x|G|\psi\rangle = -\epsilon\frac{\mathrm{d}\psi}{\mathrm{d}x}. \end{aligned} \tag{6.39}$$

This determines $G = P$ and, therefore, the operator implementing infinitesimal translations in quantum mechanics has the form

$$T(\epsilon) = \mathbb{1} - \frac{i\epsilon}{\hbar} G = \mathbb{1} - \frac{i\epsilon}{\hbar} P. \tag{6.40}$$

In other words, we recover the familiar result that the momentum operator is the generator of infinitesimal translations.

Furthermore, a quantum mechanical theory would be invariant under translations if

$$\begin{aligned} &\langle\psi_\epsilon|H|\psi_\epsilon\rangle = \langle\psi|H|\psi\rangle, \\ \text{or,}\quad &\langle\psi|T^{\dagger}(\epsilon)HT(\epsilon)|\psi\rangle = \langle\psi|H|\psi\rangle, \\ \text{or,}\quad &\langle\psi|\left(\mathbb{1} + \frac{i\epsilon}{\hbar}P\right) H \left(\mathbb{1} - \frac{i\epsilon}{\hbar}P\right)|\psi\rangle = \langle\psi|H|\psi\rangle, \end{aligned}$$

$$\begin{aligned}
\text{or,}\quad & \frac{i\epsilon}{\hbar}\langle\psi|[P,H]|\psi\rangle = 0,\\
\text{or,}\quad & \langle\psi|[P,H]|\psi\rangle = 0.
\end{aligned}\tag{6.41}$$

(Incidentally, since this must be true for any state, this also implies that $[P,H] = 0$ for translation invariance to hold.) By Ehrenfest's theorem this has the consequence that

$$\frac{\mathrm{d}}{\mathrm{d}t}\langle P\rangle = 0,\tag{6.42}$$

namely, in a translation invariant theory the expectation value of momentum in any quantum state is a constant in time.

Let us next discuss translations in the passive picture where we know that states do not change. Rather, the operators change as

$$\begin{aligned}
& T^{\dagger}(\epsilon)XT(\epsilon) = X + \epsilon\mathbb{1},\\
& T^{\dagger}(\epsilon)PT(\epsilon) = P.
\end{aligned}\tag{6.43}$$

Substituting $T(\epsilon) = \mathbb{1} - \frac{i\epsilon}{\hbar}G$ into the first relation in (6.42), we have

$$\begin{aligned}
& \left(\mathbb{1} + \frac{i\epsilon}{\hbar}G\right)X\left(\mathbb{1} - \frac{i\epsilon}{\hbar}G\right) = X + \epsilon\mathbb{1},\\
\text{or,}\quad & \frac{i\epsilon}{\hbar}[G,X] = \epsilon\mathbb{1},\\
\text{or,}\quad & G = P + f(X).
\end{aligned}\tag{6.44}$$

On the other hand, upon using (6.44), the second relation in (6.43) leads to

$$\begin{aligned}
& T^{\dagger}(\epsilon)PT(\epsilon) = P,\\
\text{or,}\quad & \frac{i\epsilon}{\hbar}[G,P] = 0,\\
\text{or,}\quad & \frac{i\epsilon}{\hbar}[P + f(X),P] = 0,
\end{aligned}\tag{6.45}$$

which implies that $f(X)$ is at best a constant which we choose to be zero (since the identity operator does not generate any transformation). As a result, we determine

$$T(\epsilon) = \mathbb{1} - \frac{i\epsilon}{\hbar}G = \mathbb{1} - \frac{i\epsilon}{\hbar}P.\tag{6.46}$$

Invariance of the Hamiltonian, in this picture, implies that

$$T^\dagger(\epsilon)HT(\epsilon) = H, \quad \text{or}, \quad [P, H] = 0. \tag{6.47}$$

Let us also note that in the passive picture,

$$\delta X = T^\dagger(\epsilon)XT(\epsilon) - X = \frac{i\epsilon}{\hbar}[P, X] = \epsilon\mathbb{1},$$
$$\delta P = T^\dagger(\epsilon)PT(\epsilon) - P = \frac{i\epsilon}{\hbar}[P, P] = 0. \tag{6.48}$$

In general, if $\Omega(X, P)$ is an observable, it is easy to check that it will transform under an infinitesimal translation as

$$\Omega(X, P) \to T^\dagger(\epsilon)\Omega(X, P)T(\epsilon),$$
$$\text{or,} \quad \delta\Omega(X, P) = \frac{i\epsilon}{\hbar}[P, \Omega(X, P)]. \tag{6.49}$$

Clearly, the passive picture is more analogous to classical mechanics.

In general, if G is the generator of an infinitesimal transformation, then, in the passive picture any dynamical variable (observable) transforms as

$$\delta\Omega(X, P) = \frac{i\epsilon}{\hbar}\,[G, \Omega(X, P)]. \tag{6.50}$$

Such relations are quite useful when one studies complicated symmetries in quantum field theory.

Finite translations. Once we understand infinitesimal translations, we can ask what is the form of the operator which implements a finite translation in a quantum mechanical system. First of all, let us recall that any finite translation can be thought of as a series of successive infinitesimal translations. For example, let 'a' represent a finite translation. Then, we can define $\epsilon = \frac{a}{N}$ as the parameter of an infinitesimal translation where N is large. (In other words, a finite translation by a is achieved by N successive infinitesimal translations by an amount ϵ as defined above.) Therefore, clearly,

$$\begin{aligned} T(a) &= \lim_{N\to\infty} (T(\epsilon))^N = \lim_{N\to\infty}\left(\mathbb{1} - \frac{i\epsilon P}{\hbar}\right)^N \\ &= \lim_{N\to\infty}\left(\mathbb{1} - \frac{iaP}{\hbar N}\right)^N \\ &= e^{-\frac{i}{\hbar}aP}. \end{aligned} \tag{6.51}$$

Physically we know that a translation by an amount 'a' followed by a translation by an amount 'b' is equivalent to a translation by an amount $(a + b)$. Mathematically, this implies that

$$T(b)T(a) = T(a+b),$$

$$\text{or,}\quad e^{-\frac{i}{\hbar}bP}e^{-\frac{i}{\hbar}aP} = e^{-\frac{i}{\hbar}(a+b)P}, \tag{6.52}$$

leading to the fact that (or this is true because)

$$[P,P] = 0. \tag{6.53}$$

In other words, the algebra of the generators of infinitesimal transformations defines the combination of two finite transformations.

► **Example (Bloch function).** Let us consider a simple one dimensional quantum mechanical system with a periodic potential. Such potentials arise in the study of electronic properties of solids. For example, in a simple solid we can think of the positively charged ions of atoms fixed in a one dimensional lattice of lattice spacing a and the valence electrons moving in the background potential of these ions. Clearly the potential in this case will be periodic with a period of the lattice spacing, namely,

$$V(x) = V(x+a). \tag{6.54}$$

To solve the time independent Schrödinger equation for such a system, let us note that the Hamiltonian describing the dynamics

$$H = \frac{P^2}{2m} + V(X), \tag{6.55}$$

will also be periodic. In fact, denoting the operator for a finite translation by an amount a by $T(a)$ (see (6.51)), the Hamiltonian, in this case, is invariant under the finite translation

$$T^\dagger(a)HT(a) = H, \tag{6.56}$$

as a consequence of (6.54). The translation operator, as we have seen (see (6.31)), is unitary

$$T^\dagger(a)T(a) = \mathbb{1} = T(a)T^\dagger(a). \tag{6.57}$$

The unitarity relation (6.57) allows us to write the invariance condition (6.56) also in the equivalent form

$$HT(a) = T(a)H, \tag{6.58}$$

which can also be thought of as a consequence of (6.47) (see also (6.51). As a result, we note that if $|\psi_E\rangle$ denotes an eigenstate of the Hamiltonian with the energy eigenvalue E, then

$$\begin{aligned} & H|\psi_E\rangle = E|\psi_E\rangle, \\ \text{or,} \quad & HT(a)|\psi_E\rangle = T(a)H|\psi_E\rangle = ET(a)|\psi_E\rangle, \end{aligned} \tag{6.59}$$

so that $|\psi_{E,a}\rangle = T(a)|\psi_E\rangle$ is also an eigenstate of the Hamiltonian with energy E. On the other hand, as we have discussed earlier, there can be no degeneracy of states in one dimension. Therefore, the two states must be proportional to each other, namely,

$$\begin{aligned} & |\psi_{E,a}\rangle = T(a)|\psi_E\rangle = \lambda_a|\psi_E\rangle, \\ \text{or,} \quad & \langle\psi_E|T^\dagger(a)T(a)|\psi_E\rangle = |\lambda_a|^2\langle\psi_E|\psi_E\rangle. \end{aligned} \tag{6.60}$$

Furthermore, since the finite translation operator is unitary (see (6.57)), it follows that

$$|\lambda_a|^2 = 1, \qquad \lambda_a = e^{i\delta_a}, \tag{6.61}$$

with δ_a a real constant. In terms of wave functions, we can write

$$\psi_{E,a}(x) = \langle x|T(a)|\psi_E\rangle = e^{i\delta_a}\psi_E(x). \tag{6.62}$$

A form of the wave function that satisfies all these requirements can be written in the form

$$\psi_E(x) = e^{ikx}\, u_E(x), \tag{6.63}$$

with

$$u_E(x-a) = u_E(x), \tag{6.64}$$

so that we have (see, for example, (6.33))

$$\psi_{E,a}(x) = \psi_E(x-a) = e^{ik(x-a)}\, u_E(x-a) = e^{i\delta_a} e^{ikx}\, u_E(x), \tag{6.65}$$

where we have identified $\delta_a = -ka$. Wave functions satisfying (6.63) and (6.64) are known as Bloch functions and arise in the study of systems with a periodic potential. We note that determining the energy eigenfunction $\psi_E(x)$ in such a case reduces to determining the periodic functions $u_E(x)$.

We note that the energy eigenfunction of the system satisfies the coordinate space equation

$$\begin{aligned} & H\psi_E(x) = E\psi_E(x), \\ \text{or,} \quad & \left(-\frac{\hbar^2}{2m}\frac{\mathrm{d}^2}{\mathrm{d}x^2} + V(x)\right)\psi_E(x) = E\psi_E(x). \end{aligned} \tag{6.66}$$

Substituting the form of the wave function from (6.63), we obtain

$$\frac{\mathrm{d}^2 u_E(x)}{\mathrm{d}x^2} + 2ik\frac{\mathrm{d}u_E(x)}{\mathrm{d}x} + \frac{2m}{\hbar^2}\left(E - V(x) - \frac{\hbar^2 k^2}{2m}\right)u_E(x) = 0. \tag{6.67}$$

This can be solved in two neighboring regions with the matching conditions which would determine the energy eigenvalues. In metals, this leads to allowed bands of energy values where electron motion is possible with these bands separated by gaps where motion is not allowed. The value of the gap then determines whether the material is a conductor or an insulator or even a semi-conductor.

◄

6.3 Groups

To understand the properties of symmetry transformations a little better, let us discuss briefly the concept of a group. A group G is a set of elements $\{g_i\}$ with a definite multiplication law such that

1. $g_1 g_2 \in G$, if $g_1, g_2 \in G$.
2. $g_1(g_2 g_3) = (g_1 g_2)g_3$, where $g_1, g_2, g_3 \in G$, namely, multiplication of group elements is associative.
3. there exists an identity element $\mathbb{1} \in G$ which satisfies

$$g_i\mathbb{1} = g_i = \mathbb{1}g_i, \quad \text{for all} \quad g_i \in G.$$

4. for every element g_i in the group, there exists a unique inverse g_i^{-1} also in the group satisfying

$$g_i^{-1}g_i = \mathbb{1} = g_i g_i^{-1}.$$

Let us now define the set of all translations by $\{T(a)\}$ with the range of the parameter $-\infty < a < \infty$.

1. Clearly, $T(a_1)T(a_2) = T(a_1 + a_2)$ and $T(a_1 + a_2) \in \{T(a)\}$ if both $T(a_1), T(a_2) \in \{T(a)\}$.
2. $T(a_1)\left(T(a_2)T(a_3)\right)$ is equal to $\left(T(a_1)T(a_2)\right)T(a_3)$, both products being equal to $T(a_1 + a_2 + a_3)$. This is seen from the fact that

$$T(a_1)\left(T(a_2)T(a_3)\right) = T(a_1)T(a_2 + a_3) = T(a_1 + a_2 + a_3),$$
$$\left(T(a_1)T(a_2)\right)T(a_3) = T(a_1 + a_2)T(a_3) = T(a_1 + a_2 + a_3).$$

3. Clearly, the translation by an amount zero is the identity element since it leaves every element of the group invariant, namely,

$$T(0) = \mathbb{1}.$$

4. For every translation $T(a) \in \{T(a)\}$, there exists a unique translation $T(-a) \in \{T(a)\}$ such that

$$T(a)T(-a) = T(0) = \mathbb{1} = T(-a)T(a).$$

Therefore, we see that the set of all translations form a group. Furthermore, this is a continuous symmetry group called a Lie group. The parameter of translation takes values between $-\infty < a < \infty$. The group is correspondingly called a non-compact group. (It is easy to check that if 'a' takes values over a compact (finite) range, then, translations will not form a group.) Furthermore, it is clear that if we know the generators of infinitesimal transformation and their algebra, the structure of the group is completely known. The algebraic relations satisfied by the generators of infinitesimal transformation are known as the Lie algebra of the group.

It is clear that translation invariance of a physical theory is essential. This is because it implies that an experiment performed at two different places would give the same result. This is quite important because otherwise physical laws would not be unique.

6.4 Parity

In classical mechanics parity is the operation of reflecting vectors through the origin. Thus, in one dimension parity corresponds to

$$x \to -x, \qquad p \to -p. \tag{6.68}$$

Therefore, we recognize that this is not a continuous symmetry. In stead, it is a discrete symmetry.

From our earlier discussions, we recognize that parity transformation can be implemented in quantum mechanics through

$$\langle X \rangle \xrightarrow{\text{Parity}} \langle -X \rangle = -\langle X \rangle,$$

$$\langle P \rangle \xrightarrow{\text{Parity}} \langle -P \rangle = -\langle P \rangle. \tag{6.69}$$

Let us first discuss parity in the active picture. Let $\mathcal{P}$ represent the operator which implements the parity transformation on a quantum mechanical system, namely,

$$|\psi\rangle \to |\psi'\rangle = \mathcal{P}|\psi\rangle. \tag{6.70}$$

To understand how exactly an arbitrary state transforms under parity, let us see how the x-basis transforms under this operation. It is clear that

$$|x\rangle \to |x'\rangle = \mathcal{P}|x\rangle = |-x\rangle. \tag{6.71}$$

This leads to the fact that $\mathcal{P}$ is a unitary operator since

$$\begin{aligned} &\langle y'|x'\rangle = \langle y|\mathcal{P}^\dagger\mathcal{P}|x\rangle = \langle -y|-x\rangle = \delta(y-x) = \langle y|x\rangle, \\ \text{or,}\quad &\mathcal{P}^\dagger\mathcal{P} = \mathbb{1}. \end{aligned} \tag{6.72}$$

Furthermore, since

$$\mathcal{P}|x\rangle = |-x\rangle, \quad \text{or,} \quad \mathcal{P}^2|x\rangle = |x\rangle, \tag{6.73}$$

we conclude that

$$\mathcal{P}^2 = \mathbb{1}. \tag{6.74}$$

Clearly, the eigenvalues of $\mathcal{P}$ are ± 1. The eigenvalues are real and hence $\mathcal{P}$ is a Hermitian operator (Real eigenvalues, of course, do not guarantee that an operator is Hermitian. However, $\mathcal{P}^\dagger\mathcal{P} = \mathbb{1} = \mathcal{P}^2$ determine that the operator $\mathcal{P}$ is Hermitian.),

$$\mathcal{P} = \mathcal{P}^\dagger. \tag{6.75}$$

Furthermore, $\mathcal{P}^2 = \mathbb{1}$ also implies that

$$\mathcal{P} = \mathcal{P}^{-1}. \tag{6.76}$$

Therefore, the parity operator satisfies the following relations.

1. $\mathcal{P} = \mathcal{P}^{-1} = \mathcal{P}^\dagger$.

2. The eigenvalues of $\mathcal{P}$ are ± 1.

► **Example.** A simple example of an operator with these properties is given by the 2×2 matrix

$$\Omega = \begin{pmatrix} 0 & 1 \\ 1 & 0 \end{pmatrix}, \tag{6.77}$$

which is easily seen to satisfy

$$\Omega = \Omega^{-1} = \Omega^\dagger. \tag{6.78}$$

◄

To see how an arbitrary state transforms under parity, we expand the state in the x-basis. Namely,

$$|\psi\rangle = \int_{-\infty}^{\infty} \mathrm{d}x\ \psi(x)|x\rangle, \tag{6.79}$$

so that

$$\begin{aligned}|\psi'\rangle = \mathcal{P}|\psi\rangle &= \mathcal{P} \int_{-\infty}^{\infty} \mathrm{d}x\ |x\rangle\psi(x) = \int_{-\infty}^{\infty} \mathrm{d}x\ \mathcal{P}|x\rangle\psi(x) \\ &= \int_{-\infty}^{\infty} \mathrm{d}x\ |-x\rangle\psi(x) = \int_{-\infty}^{\infty} \mathrm{d}x\ |x\rangle\psi(-x).\end{aligned} \tag{6.80}$$

Therefore, it follows that

$$\langle x|\psi'\rangle = \langle x|\mathcal{P}|\psi\rangle = \psi(-x). \tag{6.81}$$

Hence, we see that under parity,

$$\psi(x) \xrightarrow{\text{Parity}} \psi(-x). \tag{6.82}$$

We know that since the eigenvalues of $\mathcal{P}$ are ± 1, if $|\psi\rangle$ is an eigenstate of parity, then,

$$\begin{aligned}&\mathcal{P}|\psi\rangle = \pm|\psi\rangle, \\ \text{or,}\quad &\langle x|\mathcal{P}|\psi\rangle = \pm\langle x|\psi\rangle, \\ \text{or,}\quad &\psi(-x) = \pm\psi(x).\end{aligned} \tag{6.83}$$

The states with positive eigenvalue are called even parity states and the ones with negative eigenvalue are known as odd parity states (these are the symmetric and the anti-symmetric states that we had talked about earlier). A general state, however, does not have to be an eigenstate of parity.

In the passive description, the quantum mechanical states of a system do not change, rather the operators change under the transformation as

$$\begin{aligned}&X \longrightarrow \mathcal{P}^{\dagger} X \mathcal{P} = -X, \\ &P \longrightarrow \mathcal{P}^{\dagger} P \mathcal{P} = -P.\end{aligned} \tag{6.84}$$

Note that since $\mathcal{P} = \mathcal{P}^\dagger = \mathcal{P}^{-1}$, we can write the above relations also as

$$\begin{aligned}
&\mathcal{P}X = -X\mathcal{P},\\
\text{or,}\quad &\mathcal{P}X + X\mathcal{P} = 0,\\
\text{or,}\quad &\{\mathcal{P}, X\} = 0.
\end{aligned} \tag{6.85}$$

The curly bracket represents the anti-commutator in quantum mechanics (it is not the Poisson bracket of classical mechanics), which is also denoted as $[\cdot,\cdot]_+$. Similarly, we can also show that

$$\{\mathcal{P}, P\} = 0. \tag{6.86}$$

A theory is said to be parity invariant (or invariant under parity) if

$$\begin{aligned}
\mathcal{P}^\dagger H(X,P)\mathcal{P} &= H(\mathcal{P}^\dagger X\mathcal{P}, \mathcal{P}^\dagger P\mathcal{P}) = H(-X,-P)\\
&= H(X,P).
\end{aligned} \tag{6.87}$$

Using the relations derived earlier, namely,

$$\mathcal{P} = \mathcal{P}^\dagger = \mathcal{P}^{-1},$$

we conclude from (6.87) that the theory is invariant under parity if

$$[\mathcal{P}, H] = 0. \tag{6.88}$$

The time evolution operator for a quantum mechanical state is defined to be

$$|\psi(t)\rangle = U(t)|\psi(0)\rangle, \tag{6.89}$$

and, as we have seen earlier (see (3.81) and (3.85)), corresponds to

$$U(t) = \begin{cases} e^{-\frac{i}{\hbar}Ht}, & \text{when } H \text{ is time independent},\\ \mathrm{T}\left(e^{-\frac{i}{\hbar}\int\limits_0^t \mathrm{d}t'\, H(t')}\right), & \text{when } H \text{ is time dependent}. \end{cases} \tag{6.90}$$

Therefore, since H commutes with the parity operator when the theory is parity invariant, it follows that

$$\mathcal{P}U(t) = U(t)\mathcal{P}. \tag{6.91}$$

In other words, we see that

$$\mathcal{P}U(t)|\psi(0)\rangle = U(t)\mathcal{P}|\psi(0)\rangle. \tag{6.92}$$

This implies that in a parity invariant theory, if initially we start with a system in the state $|\psi(0)\rangle$ and another person starts with the parity transformed state of the system $\mathcal{P}|\psi(0)\rangle$, then, after an arbitrary amount of time t, the two states will continue to be parity transforms of each other.

Parity is also commonly known as handedness. If one stands in front of a mirror, in most cases one cannot differentiate between the left hand and the right hand. Namely, our body is left-right symmetric – or parity invariant. Most macroscopic systems or theories possess parity as a symmetry. However, we do find some systems in nature as well as some microscopic phenomena which do not respect this symmetry. For example, certain sugar molecules are known to rotate right circularly polarized light differently from the left circularly polarized light. In quantum mechanics we know of decay processes such as the beta decay

$$n \to p + e^- + \overline{\nu},$$

which are known to violate reflection symmetry (parity).

6.5 Rotations

Let us start with rotations in two dimensions in this lecture. We will follow this up with a discussion of rotations in three dimensions. We know classically that if we rotate a system by an angle α about the z-axis, then, the coordinates of the particle change as (earlier in (6.2) we had called this angle θ, but we do not use it here in order to avoid any confusion with angular coordinates)

$$\begin{aligned} x \to x' &= x\cos\alpha - y\sin\alpha, \\ y \to y' &= x\sin\alpha + y\cos\alpha. \end{aligned} \tag{6.93}$$

Similarly, the momenta transform as

$$\begin{aligned} p_x \to p_x' &= p_x\cos\alpha - p_y\sin\alpha, \\ p_y \to p_y' &= p_x\sin\alpha + p_y\cos\alpha. \end{aligned} \tag{6.94}$$

We can also write the transformations in (6.93) and (6.94) in the matrix form as

$$\begin{pmatrix} x \\ y \end{pmatrix} \longrightarrow \begin{pmatrix} x' \\ y' \end{pmatrix} = \begin{pmatrix} \cos\alpha & -\sin\alpha \\ \sin\alpha & \cos\alpha \end{pmatrix} \begin{pmatrix} x \\ y \end{pmatrix},$$
$$\begin{pmatrix} p_x \\ p_y \end{pmatrix} \longrightarrow \begin{pmatrix} p'_x \\ p'_y \end{pmatrix} = \begin{pmatrix} \cos\alpha & -\sin\alpha \\ \sin\alpha & \cos\alpha \end{pmatrix} \begin{pmatrix} p_x \\ p_y \end{pmatrix}. \tag{6.95}$$

Let us denote by $R(\alpha, \hat{\mathbf{z}})$ the matrix that rotates these vectors. Furthermore, let us denote by $U(R(\alpha, \hat{\mathbf{z}}))$ the operator which implements the effect of the rotation $R(\alpha, \hat{\mathbf{z}})$ on the Hilbert space of states. Then, in the active picture the states will transform as

$$|\psi\rangle \to |\psi_R\rangle = U(R)|\psi\rangle. \tag{6.96}$$

Of course, the rotated state $|\psi_R\rangle$ must be such that

$$\begin{aligned}
\langle\psi_R|X|\psi_R\rangle &= \langle\psi|X|\psi\rangle\cos\alpha - \langle\psi|Y|\psi\rangle\sin\alpha,\\
\langle\psi_R|Y|\psi_R\rangle &= \langle\psi|X|\psi\rangle\sin\alpha + \langle\psi|Y|\psi\rangle\cos\alpha,\\
\langle\psi_R|P_x|\psi_R\rangle &= \langle\psi|P_x|\psi\rangle\cos\alpha - \langle\psi|P_y|\psi\rangle\sin\alpha,\\
\langle\psi_R|P_y|\psi_R\rangle &= \langle\psi|P_x|\psi\rangle\sin\alpha + \langle\psi|P_y|\psi\rangle\cos\alpha,
\end{aligned} \tag{6.97}$$

which can also be written as

$$\begin{aligned}
\langle\psi|U^\dagger(R)XU(R)|\psi\rangle &= \langle\psi|X|\psi\rangle\cos\alpha - \langle\psi|Y|\psi\rangle\sin\alpha,\\
\langle\psi|U^\dagger(R)YU(R)|\psi\rangle &= \langle\psi|X|\psi\rangle\sin\alpha + \langle\psi|Y|\psi\rangle\cos\alpha,\\
\langle\psi|U^\dagger(R)P_xU(R)|\psi\rangle &= \langle\psi|P_x|\psi\rangle\cos\alpha - (\psi|P_y|\psi\rangle\sin\alpha,\\
\langle\psi|U^\dagger(R)P_yU(R)|\psi\rangle &= \langle\psi|P_x|\psi\rangle\sin\alpha + \langle\psi|P_y|\psi\rangle\cos\alpha.
\end{aligned} \tag{6.98}$$

To find out the effect of rotations on an arbitrary state, let us examine the effect of rotation on the coordinate basis states

$$U(R)|x, y\rangle = |x\cos\alpha - y\sin\alpha, x\sin\alpha + y\cos\alpha\rangle. \tag{6.99}$$

As in earlier examples we can show from this that the rotation operator is unitary,

$$U^\dagger(R)U(R) = \mathbb{1}. \tag{6.100}$$

Let us next write the generator for an infinitesimal rotation around the z-axis as

$$U(R(\epsilon, \hat{\mathbf{z}})) = \mathbb{1} - \frac{i\epsilon}{\hbar}G. \tag{6.101}$$

It follows from this that since

$$U^\dagger(R)U(R) = \mathbb{1}, \quad \Rightarrow \quad G^\dagger = G. \tag{6.102}$$

Namely, the generator of infinitesimal rotations around the z-axis is Hermitian.

We note that under an infinitesimal rotation

$$U(R)|x, y\rangle = |x - \epsilon y, y + \epsilon x\rangle, \tag{6.103}$$

so that we obtain

$$\begin{aligned}
|\psi_R\rangle &= U(R)|\psi\rangle = U(R)\int \mathrm{d}x\mathrm{d}y\, |x, y\rangle\langle x, y|\psi\rangle \\
&= \int \mathrm{d}x\mathrm{d}y\, U(R)|x, y\rangle\, \psi(x, y) \\
&= \int \mathrm{d}x\mathrm{d}y\, |x - \epsilon y, y + \epsilon x\rangle\, \psi(x, y) \\
&= \int \mathrm{d}x\mathrm{d}y\, |x, y\rangle\, \psi(x + \epsilon y, y - \epsilon x).
\end{aligned} \tag{6.104}$$

Therefore, we can identify the rotated wave function with

$$\begin{aligned}
&\psi_R(x, y) = \langle x, y|U(R)|\psi\rangle = \psi(x + \epsilon y, y - \epsilon x), \\
\text{or,} \quad &\langle x, y|\mathbb{1} - \frac{i\epsilon}{\hbar}G|\psi\rangle = \psi(x, y) + \epsilon y\frac{\partial\psi(x, y)}{\partial x} - \epsilon x\frac{\partial\psi(x, y)}{\partial y}, \\
\text{or,} \quad &-\frac{i\epsilon}{\hbar}\langle x, y|G|\psi\rangle = \epsilon y\frac{\partial\psi(x, y)}{\partial x} - \epsilon x\frac{\partial\psi(x, y)}{\partial y},
\end{aligned} \tag{6.105}$$

which determines

$$G = XP_y - YP_x = L_z, \quad U(R(\epsilon, \hat{\mathbf{z}})) = \mathbb{1} - \frac{i\epsilon}{\hbar}L_z. \tag{6.106}$$

In other words, the angular momentum operator L_z is the generator of infinitesimal rotations around the z-axis.

In the passive picture, on the other hand, we should have

$$\begin{aligned}
U^\dagger(R)XU(R) &= X\cos\alpha - Y\sin\alpha, \\
U^\dagger(R)YU(R) &= X\sin\alpha + Y\cos\alpha, \\
U^\dagger(R)P_xU(R) &= P_x\cos\alpha - P_y\sin\alpha, \\
U^\dagger(R)P_yU(R) &= P_x\sin\alpha + P_y\cos\alpha.
\end{aligned} \tag{6.107}$$

For infinitesimal rotations we have $\alpha = \epsilon =$ infinitesimal and the first relation in (6.107) yields

$$\left(\mathbb{1} + \frac{i\epsilon}{\hbar}G\right) X \left(\mathbb{1} - \frac{i}{\hbar}\epsilon G\right) = X - \epsilon Y,$$

$$\text{or,} \quad \frac{i\epsilon}{\hbar}[G, X] = -\epsilon Y, \tag{6.108}$$

which determines

$$G = -YP_x + f_1(Y) + f_2(X) + f_3(P_y) + f_4(XY) \\ + f_5(XP_y) + f_6(YP_y) + f_7(XYP_y). \tag{6.109}$$

On the other hand, from the second relation in (6.107) we obtain

$$\left(\mathbb{1} + \frac{i\epsilon}{\hbar}G\right) Y \left(\mathbb{1} - \frac{i}{\hbar}\epsilon G\right) = \epsilon + Y,$$

$$\text{or,} \quad \frac{i\epsilon}{\hbar}[G, Y] = \epsilon X, \tag{6.110}$$

which determines some of the coefficients in (6.109) to be

$$\begin{aligned} f_5 &= XP_y, \\ f_3(P_y) &= 0, \\ f_6(YP_y) &= 0, \\ f_7(XYP_x) &= 0, \end{aligned} \tag{6.111}$$

so that we can write

$$G = XP_y - YP_x + f_1(Y) + f_2(X) + f_4(XY). \tag{6.112}$$

Similarly, from the third relation in (6.107) we obtain

$$\left(\mathbb{1} + \frac{i\epsilon}{\hbar}G\right) P_x \left(\mathbb{1} - \frac{i\epsilon}{\hbar}G\right) = P_x - \epsilon P_y,$$

$$\text{or,} \quad \frac{i\epsilon}{\hbar}[G, P_x] = -\epsilon P_y, \tag{6.113}$$

which determines

$$f_2(X) = f_4(XY) = 0, \tag{6.114}$$

so that we can write

$$G = XP_y - YP_x + f_1(Y). \tag{6.115}$$

Finally, the last relation in (6.107) leads to

$$\left(\mathbb{1}+\frac{i\epsilon}{\hbar}G\right)P_y\left(\mathbb{1}-\frac{i\epsilon}{\hbar}G\right)=\epsilon P_x+P_y,$$

$$\text{or,}\quad \frac{i\epsilon}{\hbar}[G,P_y]=\epsilon P_x, \tag{6.116}$$

which determines $f_1(Y)=0$ so that we have

$$G=XP_y-YP_x=L_z. \tag{6.117}$$

Therefore, we have determined

$$U(R)=\mathbb{1}-\frac{i\epsilon}{\hbar}L_z. \tag{6.118}$$

A quantum mechanical theory is invariant under such a rotation if

$$U^\dagger(R)HU(R)=H. \tag{6.119}$$

Putting in the infinitesimal form for $U(R)$, the invariance condition, (6.119), becomes

$$\frac{i\epsilon}{\hbar}[L_z,H]=0,\quad \text{or,}\quad [L_z,H]=0. \tag{6.120}$$

We can again construct a finite rotation about the z-axis by taking successive infinitesimal rotations which leads to

$$\begin{aligned}U(R(\alpha,\hat{\mathbf{z}}))&=\lim_{N\to\infty}\left(\mathbb{1}-\frac{i\epsilon}{\hbar}L_z\right)^N\\&=\lim_{N\to\infty}\left(\mathbb{1}-\frac{i\alpha}{N\hbar}L_z\right)^N\\&=e^{-\frac{i\alpha}{\hbar}L_z}.\end{aligned} \tag{6.121}$$

Since $[L_z,L_z]=0$, it is follows that

$$U(R(\alpha_1,\hat{\mathbf{z}}))U(R(\alpha_2,\hat{\mathbf{z}}))=U(R((\alpha_1+\alpha_2),\hat{\mathbf{z}})). \tag{6.122}$$

Namely, rotations about the same axis are additive (much like translations).

The two dimensional vectors can equivalently be described in terms of the polar coordinates (r,ϕ). A rotation, of course, does not change the radial coordinate. Rather, it changes the azimuthal angle ϕ. Thus, in this basis,

$$U(R(\alpha,\hat{\mathbf{z}}))|r,\phi\rangle = |r,\phi+\alpha\rangle. \tag{6.123}$$

Furthermore, note that since $0 \leq \phi \leq 2\pi$, the parameter of rotation is also bounded and lies between

$$0 \leq \alpha \leq 2\pi. \tag{6.124}$$

In this basis of circular coordinates,

$$\begin{aligned}
|\psi_R\rangle = U(R)|\psi\rangle &= U(R)\int r\mathrm{d}r\mathrm{d}\phi\, |r,\phi\rangle\langle r,\phi|\psi\rangle \\
&= \int r\mathrm{d}r\mathrm{d}\phi\, U(R)|r,\phi\rangle\ \psi(r,\phi) \\
&= \int r\mathrm{d}r\mathrm{d}\phi\, |r,\phi+\alpha\rangle\ \psi(r,\phi) \\
&= \int r\mathrm{d}r\mathrm{d}\phi\, |r,\phi\rangle\ \psi(r,\phi-\alpha),
\end{aligned} \tag{6.125}$$

so that we obtain

$$\langle r,\phi|\psi_R\rangle = \psi_R(r,\phi) = \psi(r,\phi-\alpha). \tag{6.126}$$

Furthermore, for $\alpha = \epsilon =$ infinitesimal, (6.126) gives

$$\begin{aligned}
&\langle r,\phi|\left(\mathbb{1} - \frac{i\epsilon}{\hbar}L_z\right)|\psi\rangle = \psi(r,\phi) - \epsilon\frac{\partial\psi(r,\phi)}{\partial\phi}, \\
\text{or,}\quad &-\frac{i\epsilon}{\hbar}\langle r,\phi|L_z|\psi\rangle = -\epsilon\frac{\partial\psi(r,\phi)}{\partial\phi}.
\end{aligned} \tag{6.127}$$

In other words, in the (r,ϕ) basis,

$$L_z \to -i\hbar\frac{\partial}{\partial\phi}. \tag{6.128}$$

We can show again that in two dimensions rotations form a group. This is a Lie group (Abelian group like translations). Furthermore, since the parameter of rotation takes only bounded values, $0 \leq \alpha \leq 2\pi$, this is a compact group (unlike translations).

6.6 Selected problems

1. Prove the relation (see (6.51))

$$\lim_{N\to\infty}\left(1-\frac{ix}{N}\right)^{N} = e^{-ix}. \tag{6.129}$$

2. Show that under any unitary transformation U

$$U^{\dagger}f(X,P)U = f(U^{\dagger}XU, U^{\dagger}PU). \tag{6.130}$$

3. Consider the operator describing a general (unitary) infinitesimal transformation of the form

$$U(\epsilon) = \mathbb{1} - \frac{i\epsilon^{i}}{\hbar}G_{i}, \tag{6.131}$$

where G and ϵ^{i} denote respectively the generators and the constant parameters of the infinitesimal transformation. The indices i take values over a range of integers depending on the transformation under consideration. As we have discussed in (6.50), the infinitesimal change of any operator, in this case, can be written as (in higher dimensions)

$$\delta\Omega(\mathbf{X},\mathbf{P}) = \frac{i\epsilon^{i}}{\hbar}\,[G_{i},\Omega(\mathbf{X},\mathbf{P})]. \tag{6.132}$$

Show that if one were to do two successive infinitesimal transformations with parameters ϵ_1 and ϵ_2, then the order of the transformations will not commute in general and the commutator can be written as

$$\begin{aligned}
[\delta_1,\delta_2]\Omega(\mathbf{X},\mathbf{P}) &= \delta_1(\delta_2\Omega(\mathbf{X},\mathbf{P})) - \delta_2(\delta_1\Omega(\mathbf{X},\mathbf{P}))\\
&= -\frac{\epsilon_1^{i}\epsilon_2^{j}}{\hbar^{2}}\left[[G_i,G_j],\Omega(\mathbf{X},\mathbf{P})\right].
\end{aligned} \tag{6.133}$$

This is often useful in determining the algebra of the generators of infinitesimal transformations.

CHAPTER 7

Angular momentum

As we have seen, angular momentum operator is the generator of rotations. In the next few lectures we will study the algebra of rotations in three dimensions by studying the algebraic properties of the angular momentum operators.

7.1 Rotations in three dimensions

Let us generalize the results of rotations in two dimensions to three dimensions. First of all, there are now three different axes about which we can perform rotations. Therefore, there would be three possible generators of infinitesimal rotations which would correspond to the three components of the angular momentum operator. Let us denote them by

$$\begin{aligned} L_x &= YP_z - ZP_y, \\ L_y &= ZP_x - XP_z, \\ L_z &= XP_y - YP_x. \end{aligned} \tag{7.1}$$

Let us next determine various commutators involving these operators. For example, we note that

$$\begin{aligned} [L_x, X] &= [YP_z - ZP_y, X] = 0, \\ [L_y, X] &= [ZP_x - XP_z, X] = Z\,[P_x, X] = -i\hbar Z, \\ [L_z, X] &= [XP_y - YP_x, X] = -Y\,[P_x, X] = i\hbar Y. \end{aligned} \tag{7.2}$$

Similarly, we can derive other commutation relations also. But, to simplify our calculations, let us introduce a compact notation and define

$$\begin{matrix} X = X_1, & P_x = P_1, & L_x = L_1, \\ Y = X_2, & P_y = P_2, & L_y = L_2, \\ Z = X_3, & P_z = P_3, & L_z = L_3. \end{matrix} \tag{7.3}$$

With this notation we can write the three components of the angular momentum operator in (7.1) as (repeated indices are assumed to be summed)

$$L_i = \epsilon_{ijk} X_j P_k, \quad X_i P_j - X_j P_i = \epsilon_{ijk} L_k, \quad i, j, k = 1, 2, 3, \quad (7.4)$$

where ϵ_{ijk} represents the Levi-Civita tensor which is totally anti-symmetric, with $\epsilon_{123} = 1$. In this notation the canonical commutation relations take the form

$$[X_i, P_j] = i\hbar\delta_{ij}, \quad [X_i, X_j] = 0 = [P_i, P_j]. \quad (7.5)$$

Using the notations in (7.3) and (7.4), we can now calculate

$$\begin{aligned}
[L_i, X_j] &= [\epsilon_{ipq} X_p P_q, X_j] = \epsilon_{ipq} X_p [P_q, X_j] \\
&= \epsilon_{ipq} X_p (-i\hbar\delta_{qj}) = -i\hbar\epsilon_{ipj} X_p = i\hbar\epsilon_{ijk} X_k. \quad (7.6)
\end{aligned}$$

Similarly, we can also derive

$$\begin{aligned}
[L_i, P_j] &= [\epsilon_{ipq} X_p P_q, P_j] = \epsilon_{ipq} [X_p, P_j] P_q \\
&= \epsilon_{ipq} (i\hbar\delta_{pj}) P_q = i\hbar\epsilon_{ijq} P_q = i\hbar\epsilon_{ijk} P_k. \quad (7.7)
\end{aligned}$$

The commutators (7.6) and (7.7) merely tell us how the coordinate as well as the momentum (vector) operators transform under an infinitesimal rotation around the i-axis. Furthermore, the commutation relation between two angular momentum operators can now be determined to be

$$\begin{aligned}
[L_i, L_j] &= [L_i, \epsilon_{jmn} X_m P_n] \\
&= \epsilon_{jmn} \left([L_i, X_m] P_n + X_m [L_i, P_n] \right) \\
&= \epsilon_{jmn} \left(i\hbar\epsilon_{imk} X_k P_n + i\hbar\epsilon_{ink} X_m P_k \right) \\
&= i\hbar \left((\delta_{ji}\delta_{nk} - \delta_{jk}\delta_{ni}) X_k P_n - (\delta_{ji}\delta_{mk} - \delta_{jk}\delta_{mi}) X_m P_k \right) \\
&= i\hbar \left(\delta_{ji} X_k P_k - X_j P_i - \delta_{ji} X_k P_k + X_i P_j \right) \\
&= i\hbar (X_i P_j - X_j P_i) \\
&= i\hbar\epsilon_{ijk} L_k, \quad (7.8)
\end{aligned}$$

where we have used (7.6) and (7.7) in the intermediate steps. Thus, we see that the basic commutation relation involving the angular momentum operators is given by

$$[L_i, L_j] = i\hbar\epsilon_{ijk} L_k, \quad (7.9)$$

where repeated indices are assumed to be summed. (Eq. (7.9) basically defines how the angular momentum (vector) operator transforms under an infinitesimal rotation.) This shows that generators of rotation along different directions do not commute. (The algebra is non-Abelian.) However,

$$[L_i, L_i] = 0, \quad \text{for a fixed} \quad i.$$

Let us now define a quadratic operator

$$L^2 = \mathbf{L}^2 = L_i L_i = \sum_{i=1}^{3} L_i L_i. \tag{7.10}$$

This operator is easily seen to commute with all the components of the angular momentum operator,

$$\begin{aligned}[L_i, L^2] &= [L_i, L_j L_j] = L_j [L_i, L_j] + [L_i, L_j] L_j \\ &= L_j (i\hbar\epsilon_{ijk} L_k) + (i\hbar\epsilon_{ijk} L_k) L_j \\ &= i\hbar\epsilon_{ijk} (L_j L_k + L_k L_j) = 0,\end{aligned} \tag{7.11}$$

which follows from the anti-symmetry of the Levi-Civita tensor. In other words, the operator L^2 commutes with all three generators of infinitesimal rotations. In group theory, such an operator is known as the quadratic Casimir operator.

Furthermore, a theory is invariant under rotations if the generators commute with the Hamiltonian. This implies that for all i,

$$[L_i, H] = 0, \tag{7.12}$$

for rotational invariance to hold. Clearly, for such systems it follows that

$$[L^2, H] = [L_i L_i, H] = L_i [L_i, H] + [L_i, H] L_i = 0. \tag{7.13}$$

However, since different components of the angular momentum operator do not commute among themselves, it is clear that H, L^2 and only one component of the angular momentum can be simultaneously diagonalized in a rotationally invariant theory. A simple example of a rotationally invariant theory is, of course, given by

$$H = \frac{\mathbf{P}^2}{2m} + V(R) = \frac{\mathbf{P}^2}{2m} + V(\mathbf{X}^2), \tag{7.14}$$

where the potential depends only on the radial coordinate. For example, the higher dimensional isotropic harmonic oscillator is such a theory.

In such a theory (with rotational invariance) we conventionally choose to diagonalize H, L^2 and L_3 (L_z) simultaneously. This means that they are chosen to have simultaneous eigenstates. To study the eigenvalue spectrum of these operators, let us further define

$$\begin{aligned} L_+ &= L_1 + iL_2, \\ L_- &= L_1 - iL_2, \qquad L_- = (L_+)^\dagger. \end{aligned} \tag{7.15}$$

Since L^2 commutes with all the components L_i it follows that

$$\begin{aligned} \left[L_+, L^2\right] &= 0, \\ \left[L_-, L^2\right] &= 0. \end{aligned} \tag{7.16}$$

On the other hand, we have

$$\begin{aligned} [L_+, L_3] &= [L_1 + iL_2, L_3] = -i\hbar L_2 + i(i\hbar L_1) \\ &= -\hbar(L_1 + iL_2) = -\hbar L_+, \\ [L_-, L_3] &= [L_1 - iL_2, L_3] = -i\hbar L_2 - i(i\hbar L_1) \\ &= \hbar(L_1 - iL_2) = \hbar L_-. \end{aligned} \tag{7.17}$$

Furthermore, it is easy to derive

$$\begin{aligned} [L_+, L_-] &= [L_1 + iL_2, L_1 - iL_2] \\ &= [L_1, -iL_2] + [iL_2, L_1] \\ &= (-i)(i\hbar L_3) + i(-i\hbar L_3) \\ &= 2\hbar L_3, \end{aligned} \tag{7.18}$$

and we know that the Hamiltonian for a rotationally invariant theory commutes with all the components of angular momentum operator. Therefore, we also have

$$[L_+, H] = [L_-, H] = 0. \tag{7.19}$$

Let $|j, m\rangle$ represent the simultaneous eigenstates of the operators L^2 and L_3 such that

$$\begin{aligned} L_3|j, m\rangle &= \hbar m|j, m\rangle, \\ L^2|j, m\rangle &= \hbar^2 j|j, m\rangle. \end{aligned} \tag{7.20}$$

Let us next examine the effect of the operator L_+ on a given eigenstate $|j,m\rangle$. We note that

$$\begin{aligned}
L_3L_+|j,m\rangle &= ([L_3,L_+] + L_+L_3)\,|j,m\rangle \\
&= (\hbar L_+ + L_+L_3)\,|j,m\rangle \\
&= (\hbar + \hbar m)L_+|j,m\rangle = \hbar(m+1)L_+|j,m\rangle. \qquad (7.21)
\end{aligned}$$

Similarly, we have

$$\begin{aligned}
L^2L_+|j,m\rangle &= \left([L^2,L_+] + L_+L^2\right)|j,m\rangle \\
&= L_+L^2|j,m\rangle \\
&= \hbar^2 jL_+|j,m\rangle. \qquad (7.22)
\end{aligned}$$

This shows that the effect of L_+ acting on a given state $|j,m\rangle$ is to take it to a state where the eigenvalue of L_3 is raised by a unit of $\hbar$, while that of L^2 is unchanged. Therefore, we can write

$$L_+|j,m\rangle = \Gamma_+|j,m+1\rangle, \qquad (7.23)$$

with Γ_+ a constant, depending on j and m. We can also show that

$$\begin{aligned}
L_3L_-|j,m\rangle &= ([L_3,L_-] + L_-L_3)\,|j,m\rangle \\
&= (-\hbar L_- + L_-L_3)|j,m\rangle \\
&= (-\hbar + \hbar m)L_-|j,m\rangle = \hbar(m-1)L_-|j,m\rangle, \\
L^2L_-|j,m\rangle &= L_-L^2|j,m\rangle \\
&= \hbar^2 jL_-|j,m\rangle. \qquad (7.24)
\end{aligned}$$

which shows that the operator L_- acting on a state $|j,m\rangle$ lowers the eigenvalue of L_3 by a unit of $\hbar$ while leaving the eigenvalue of L^2 unchanged. Therefore, we conclude that

$$L_-|j,m\rangle = \Gamma_-|j,m-1\rangle. \qquad (7.25)$$

Since the operators L_+ and L_- raise and lower the eigenvalue of L_3 respectively, they are correspondingly known as raising and lowering operators. Furthermore, it follows that given a state $|j,m\rangle$ we can construct a sequence of states $|j,m+1\rangle$, $|j,m+2\rangle,\cdots$ and $|j,m-1\rangle$, $|j,m-2\rangle,\cdots$ respectively by applying raising and lowering operators. However, physically this sequence cannot go on without termination. For, the operator

$$L^2 = L_1^2 + L_2^2 + L_3^2, \quad \text{or,} \quad L^2 - L_3^2 = L_1^2 + L_2^2 \geq 0. \qquad (7.26)$$

This is a positive semi-definite operator so that the eigenvalues must satisfy

$$\hbar^2 j - \hbar^2 m^2 \geq 0, \quad \text{or,} \quad j \geq m^2. \tag{7.27}$$

This implies that there must exist states with a maximum and a minimum value of m such that

$$\begin{aligned}
& L_+|j, m_{\max}\rangle = 0, \\
\text{or,} \quad & \langle j, m_{\max}|L_-L_+|j, m_{\max}\rangle = 0, \\
\text{or,} \quad & \langle j, m_{\max}|\left(L^2 - L_3^2 - \hbar L_3\right)|j, m_{\max}\rangle = 0, \\
\text{or,} \quad & \left(\hbar^2 j - \hbar^2 m_{\max}^2 - \hbar^2 m_{\max}\right)\langle j, m_{\max}|j, m_{\max}\rangle = 0, \\
\text{or,} \quad & j - m_{\max}(m_{\max} + 1) = 0,
\end{aligned} \tag{7.28}$$

where we have used

$$\begin{aligned}
L_-L_+ &= (L_1 - iL_2)(L_1 + iL_2) = L_1^2 + L_2^2 + i[L_1, L_2] \\
&= L^2 - L_3^2 - \hbar L_3.
\end{aligned} \tag{7.29}$$

We can, similarly, show that there must also exist a state with a minimum value of m such that

$$\begin{aligned}
& L_-|j, m_{\min}\rangle = 0, \\
\text{or,} \quad & \langle j, m_{\min}|L_+L_-|j, m_{\min}\rangle = 0, \\
\text{or,} \quad & \langle j, m_{\min}|\left(L^2 - L_3^2 + \hbar L_3\right)|j, m_{\min}\rangle = 0, \\
\text{or,} \quad & \left(\hbar^2 j - \hbar^2 m_{\min}^2 + \hbar^2 m_{\min}\right)\langle j, m_{\min}|j, m_{\min}\rangle = 0, \\
\text{or,} \quad & j - m_{\min}(m_{\min} - 1) = 0.
\end{aligned} \tag{7.30}$$

Comparing the relations in (7.28) and (7.30), it is clear that

$$m_{\min} = -m_{\max}. \tag{7.31}$$

(The other solution, $m_{\max} = m_{\min} - 1$, violates our assumption that $m_{\max} > m_{\min}$ or that $m_{\min}$ denotes the minimum value of m.) Furthermore, let us assume that one goes from the state $|j, m_{\min}\rangle$ to $|j, m_{\max}\rangle$ by applying k times the raising operator L_+. Since every time L_+ is applied to a state $|j, m\rangle$ it shifts $m \to m + 1$, it follows that

$$\begin{aligned}
& m_{\max} - m_{\min} = k, \\
\text{or,} \quad & 2m_{\max} = k, \\
\text{or,} \quad & m_{\max} = -m_{\min} = \frac{k}{2},
\end{aligned} \tag{7.32}$$

where k is an integer.

Therefore, we can write

$$j = m_{\max}(m_{\max} + 1) = \frac{k}{2}\left(\frac{k}{2} + 1\right). \tag{7.33}$$

Let us define $\ell = \frac{k}{2}$ so that ℓ takes only multiples of half integer values. Then, we can write

$$j = \ell(\ell + 1), \tag{7.34}$$

and $-\ell \leq m \leq \ell$ (in steps of unity, namely, m takes $2\ell + 1$ values) where ℓ takes positive multiples of half integer values

$$\ell = 0, \frac{1}{2}, 1, \frac{3}{2}, \cdots, \qquad m = -\ell, -\ell + 1, \cdots, \ell - 1, \ell. \tag{7.35}$$

We can now determine the normalized eigenstates of the angular momentum operators in the following way. We have already determined that

$$\begin{aligned}
& L^2|\ell, m\rangle = \hbar^2 \ell(\ell + 1)|\ell, m\rangle, \\
& L_3|\ell, m\rangle = \hbar m|\ell, m\rangle,
\end{aligned} \tag{7.36}$$

where $-\ell \leq m \leq \ell$. Furthermore, we know that

$$\begin{aligned}
& L_+|\ell, m\rangle = \Gamma_+|\ell, m + 1\rangle, \\
\text{or,} \quad & \langle \ell, m|L_- L_+|\ell, m\rangle = |\Gamma_+|^2, \\
\text{or,} \quad & \langle \ell, m| \left(L^2 - L_3^2 - \hbar L_3\right)|\ell, m\rangle = |\Gamma_+|^2, \\
\text{or,} \quad & \hbar^2 \left(\ell(\ell + 1) - m(m + 1)\right) = |\Gamma_+|^2, \\
\text{or,} \quad & \Gamma_+ = \Gamma_+^* = \hbar\sqrt{\ell(\ell + 1) - m(m + 1)},
\end{aligned} \tag{7.37}$$

so that

$$\begin{aligned}
& L_+|\ell, m\rangle = \hbar\sqrt{\ell(\ell + 1) - m(m + 1)}|\ell, m + 1\rangle, \\
\text{or,} \quad & |\ell, m + 1\rangle = \frac{1}{\hbar\sqrt{\ell(\ell + 1) - m(m + 1)}} L_+|\ell, m\rangle.
\end{aligned} \tag{7.38}$$

Similarly, we can also show that

$$L_-|\ell, m\rangle = \hbar\sqrt{\ell(\ell+1) - m(m-1)}|\ell, m-1\rangle,$$

$$\text{or,}\quad |\ell, m-1\rangle = \frac{1}{\hbar\sqrt{\ell(\ell+1) - m(m-1)}} L_-|\ell, m\rangle. \tag{7.39}$$

This, therefore, defines all the simultaneous eigenstates of L^2 and L_3 for a particular value of ℓ. They define a Hilbert space $\mathcal{E}^{(\ell)}$ which is a subspace of the total Hilbert space of the angular momentum operators. What we mean by this is that the operators L^2, L_3, L_+ and L_- take any vector in this space to another vector in the same space. In other words they leave the space $\mathcal{E}^{(\ell)}$ invariant. The dimensionality of this space is $(2\ell + 1)$.

► **Example.** Let us now look at a few specific examples of the representations of angular momentum.

1. $\ell = 0$: In this case, the dimensionality of the representation is 1 and $m = 0$.
2. $\ell = \frac{1}{2}$: In this case, the dimensionality of the representation is $2\ell + 1 = 2$ and m takes values $\pm\frac{1}{2}$.

   Let the basis states in this space be $|\frac{1}{2}, \frac{1}{2}\rangle$ and $|\frac{1}{2}, -\frac{1}{2}\rangle$. We know from (7.36) that

   $$\langle \ell, m'|L_3|\ell, m\rangle = \hbar m\langle \ell, m'|\ell, m\rangle = \hbar m\delta_{m' m}. \tag{7.40}$$

   This implies that

   $$\begin{aligned} \langle\frac{1}{2}, \frac{1}{2}|L_3|\frac{1}{2}, \frac{1}{2}\rangle &= \frac{\hbar}{2} = -\langle\frac{1}{2}, -\frac{1}{2}|L_3|\frac{1}{2}, -\frac{1}{2}\rangle, \\ \langle\frac{1}{2}, \frac{1}{2}|L_3|\frac{1}{2}, -\frac{1}{2}\rangle &= 0 = \langle\frac{1}{2}, -\frac{1}{2}|L_3|\frac{1}{2}, \frac{1}{2}\rangle. \end{aligned} \tag{7.41}$$

   Therefore, we have the matrix representation,

   $$L_3 = \frac{\hbar}{2}\begin{pmatrix} 1 & 0 \\ 0 & -1 \end{pmatrix}. \tag{7.42}$$

   Similarly, for L^2 we have

   $$\begin{aligned} \langle \ell, m'|L^2|\ell, m\rangle &= \hbar^2\ell(\ell+1)\langle \ell, m'|\ell, m\rangle = \hbar^2\ell(\ell+1)\delta_{m' m} \\ &= \frac{3}{4}\hbar^2\delta_{m' m}, \end{aligned} \tag{7.43}$$

   so that this is a multiple of the identity matrix,

   $$L^2 = \frac{3}{4}\hbar^2\begin{pmatrix} 1 & 0 \\ 0 & 1 \end{pmatrix} = \frac{3}{4}\hbar^2\mathbb{1}. \tag{7.44}$$

   Furthermore, for the raising operator we have from (7.38)

   $$\begin{aligned} \langle \ell, m'|L_+|\ell, m\rangle &= \Gamma_+\langle \ell, m'|\ell, m+1\rangle = \Gamma_+\delta_{m' m+1} \\ &= \hbar\sqrt{\ell(\ell+1) - m(m+1)}\,\delta_{m' m+1} \\ &= \hbar\sqrt{\frac{3}{4} - m(m+1)}\,\delta_{m' m+1}. \end{aligned} \tag{7.45}$$

In the matrix notation, this takes the form

$$L_+ = \hbar \begin{pmatrix} 0 & \sqrt{\frac{3}{4}+\frac{1}{4}} \\ 0 & 0 \end{pmatrix} = \hbar \begin{pmatrix} 0 & 1 \\ 0 & 0 \end{pmatrix}. \tag{7.46}$$

Similarly, (7.39) leads to

$$\begin{aligned} \langle \ell, m'|L_-|\ell, m\rangle &= \Gamma_- \langle \ell, m'|\ell, m-1\rangle = \Gamma_- \delta_{m'\,m-1} \\ &= \hbar\sqrt{\ell(\ell+1) - m(m-1)}\,\delta_{m'\,m-1} \\ &= \hbar\sqrt{\frac{3}{4} - m(m-1)}\,\delta_{m'\,m-1}, \end{aligned} \tag{7.47}$$

which has the matrix form

$$L_- = \hbar \begin{pmatrix} 0 & 0 \\ \sqrt{\frac{3}{4}+\frac{1}{4}} & 0 \end{pmatrix} = \hbar \begin{pmatrix} 0 & 0 \\ 1 & 0 \end{pmatrix} = (L_+)^\dagger . \tag{7.48}$$

But, from the defining relation (7.15) we know that $L_\pm = L_1 \pm iL_2$ so that

$$\begin{aligned} L_1 &= \frac{1}{2}(L_+ + L_-), \\ L_2 &= -\frac{i}{2}(L_+ - L_-). \end{aligned} \tag{7.49}$$

This determines

$$\begin{aligned} L_1 &= \frac{\hbar}{2}\begin{pmatrix} 0 & 1 \\ 1 & 0 \end{pmatrix}, \\ L_2 &= \frac{\hbar}{2}\begin{pmatrix} 0 & -i \\ i & 0 \end{pmatrix}. \end{aligned} \tag{7.50}$$

In other words, the generators of rotation (angular momenta) corresponding to the eigenvalue $\ell = \frac{1}{2}$ are none other than the three Pauli matrices (up to multiplicative constants).

3. $\ell = 1$: In this case, the dimensionality of the representation is $2\ell + 1 = 3$ and m takes values $m = -1, 0, 1$. Let the basis states in this space be $|1,1\rangle, |1,0\rangle, |1,-1\rangle$. Since L_3 is diagonal in this basis,

$$\langle 1, m'|L_3|1, m\rangle = \hbar m \delta_{m'\,m}, \tag{7.51}$$

and we have

$$L_3 = \hbar \begin{pmatrix} 1 & 0 & 0 \\ 0 & 0 & 0 \\ 0 & 0 & -1 \end{pmatrix}. \tag{7.52}$$

Similarly, L^2 is also diagonal with

$$\langle 1, m'|L^2|1, m\rangle = \hbar^2 1(1+1)\delta_{m'\,m} = 2\hbar^2\delta_{m'\,m}, \tag{7.53}$$

which leads to the matrix form

$$L^2 = 2\hbar^2 \begin{pmatrix} 1 & 0 & 0 \\ 0 & 1 & 0 \\ 0 & 0 & 1 \end{pmatrix} = 2\hbar^2 \mathbb{1}. \tag{7.54}$$

Furthermore, for the raising operator we have

$$\langle 1, m'|L_+|1, m\rangle = \hbar\sqrt{2 - m(m+1)}\,\delta_{m'\,m+1}, \tag{7.55}$$

so that the matrix representation has the form

$$L_+ = \hbar \begin{pmatrix} 0 & \sqrt{2} & 0 \\ 0 & 0 & \sqrt{2} \\ 0 & 0 & 0 \end{pmatrix}. \tag{7.56}$$

Similarly, the lowering operator satisfies

$$\langle 1, m'|L_-|1, m\rangle = \hbar\sqrt{2 - m(m-1)}\,\delta_{m'\,m-1}, \tag{7.57}$$

and we have

$$L_- = \hbar \begin{pmatrix} 0 & 0 & 0 \\ \sqrt{2} & 0 & 0 \\ 0 & \sqrt{2} & 0 \end{pmatrix} = (L_+)^\dagger. \tag{7.58}$$

It follows from (7.56) and (7.58), therefore, that

$$L_1 = \frac{1}{2}(L_+ + L_-) = \frac{\hbar}{\sqrt{2}} \begin{pmatrix} 0 & 1 & 0 \\ 1 & 0 & 1 \\ 0 & 1 & 0 \end{pmatrix},$$

$$L_2 = -\frac{i}{2}(L_+ - L_-) = \frac{\hbar}{\sqrt{2}} \begin{pmatrix} 0 & -i & 0 \\ i & 0 & i \\ 0 & i & 0 \end{pmatrix}. \tag{7.59}$$

Thus, we see that the generators of rotation (angular momenta) have different representations in different spaces.

◀

To find out the spatial eigenfunctions (wave functions) associated with angular momentum operators, we note that rotational symmetry is best studied in the spherical coordinates (see Fig. 7.1),

$$x = r\sin\theta\cos\phi,$$

$$y = r\sin\theta\sin\phi,$$

$$z = r\cos\theta. \tag{7.60}$$

In spherical coordinates, the angular momentum operators take the following forms.

$$L_1 = L_x = i\hbar\left(\sin\phi\frac{\partial}{\partial\theta} + \cos\phi\cot\theta\frac{\partial}{\partial\phi}\right),$$

$$L_2 = L_y = i\hbar\left(-\cos\phi\frac{\partial}{\partial\theta} + \sin\phi\cot\theta\frac{\partial}{\partial\phi}\right),$$

$$L_3 = L_z = -i\hbar\frac{\partial}{\partial\phi}. \tag{7.61}$$

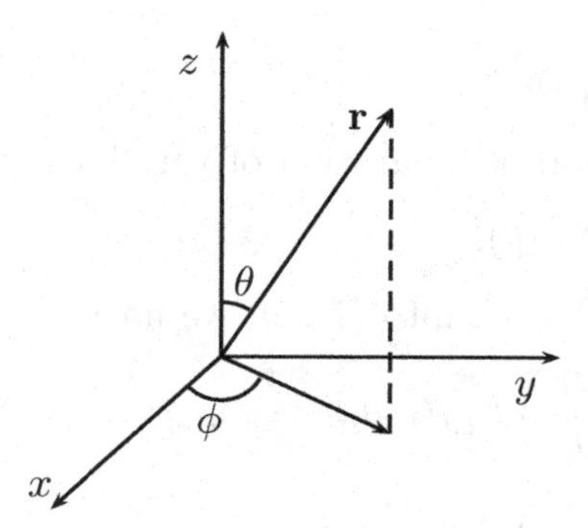

Figure 7.1: Cartesian and spherical coordinate systems.

This determines the forms for the raising and lowering operators to be

$$\begin{aligned} L_{\pm} &= L_1 \pm iL_2 \\ &= i\hbar\left[(\sin\phi \mp i\cos\phi)\frac{\partial}{\partial\theta} + (\cos\phi \pm i\sin\phi)\cot\theta\frac{\partial}{\partial\phi}\right] \\ &= \pm\hbar e^{\pm i\phi}\left[\frac{\partial}{\partial\theta} \pm i\cot\theta\frac{\partial}{\partial\phi}\right]. \end{aligned} \tag{7.62}$$

We know from (7.38) that

$$L_+|\ell,\ell\rangle = 0. \tag{7.63}$$

In the spherical coordinate basis, therefore, this condition becomes

$$\left[\frac{\partial}{\partial\theta} + i\cot\theta\frac{\partial}{\partial\phi}\right]U_{\ell,\ell}(r,\theta,\phi) = 0, \tag{7.64}$$

where

$$U_{\ell,\ell}(r,\theta,\phi) = \langle r,\theta,\phi|\ell,\ell\rangle. \tag{7.65}$$

Furthermore, we also know that

$$\begin{aligned} & L_z|\ell,\ell\rangle = \hbar\ell|\ell,\ell\rangle, \\ \text{or,}\quad & -i\hbar\frac{\partial}{\partial\phi}U_{\ell,\ell}(r,\theta,\phi) = \hbar\ell U_{\ell,\ell}(r,\theta,\phi), \\ \text{or,}\quad & \frac{\partial}{\partial\phi}U_{\ell,\ell}(r,\theta,\phi) = i\ell U_{\ell,\ell}(r,\theta,\phi), \end{aligned} \tag{7.66}$$

so that we can write

$$U_{\ell,\ell}(r,\theta,\phi) = F_{\ell,\ell}(r,\theta)e^{i\ell\phi}. \tag{7.67}$$

Let us now use the method of separation of variables and write

$$F_{\ell,\ell}(r,\theta) = R_{\ell,\ell}(r)\Theta_{\ell,\ell}(\theta). \tag{7.68}$$

Putting (7.67) and (7.68) back into (7.64), we have

$$\begin{aligned}
&\left(\frac{\partial}{\partial\theta} + i\cot\theta\frac{\partial}{\partial\phi}\right)U_{\ell,\ell}(r,\theta,\phi) = 0,\\
\text{or,}\quad &\left(\frac{\mathrm{d}}{\mathrm{d}\theta} + i\cot\theta(i\ell)\right)\Theta_{\ell,\ell}(\theta) = 0,\\
\text{or,}\quad &\frac{\mathrm{d}\Theta_{\ell,\ell}(\theta)}{\mathrm{d}\theta} - \ell\cot\theta\Theta_{\ell,\ell}(\theta) = 0,\\
\text{or,}\quad &\Theta_{\ell,\ell}(\theta) = A\,(\sin\theta)^{\ell}.
\end{aligned} \tag{7.69}$$

Therefore, we have determined

$$U_{\ell,\ell}(r,\theta,\phi) = R_{\ell,\ell}(r)(\sin\theta)^{\ell}e^{i\ell\phi}, \tag{7.70}$$

where we have absorbed the constant A into $R_{\ell,\ell}(r)$.

Furthermore, we note that rotation affects only the angular part of the solution and only the m quantum number changes under a rotation, while ℓ is invariant. Therefore, the radial component of the wave function for a rotationally invariant theory should not depend on the m quantum number. It can at the most depend on the ℓ quantum number and would be determined by the dynamics of the system. Thus, we can write

$$U_{\ell,\ell}(r,\theta,\phi) = R_{\ell}(r)(\sin\theta)^{\ell}e^{i\ell\phi}. \tag{7.71}$$

where we have absorbed the normalization constant into the radial wave function. This is, therefore, the wave function for the state with the highest m quantum number for a given l. The wave function for any other state can be obtained from this by applying the lowering operator. Namely, (see (7.39))

$$\begin{aligned}
|\ell,\ell-1\rangle &= \frac{1}{\hbar\sqrt{\ell(\ell+1)-\ell(\ell-1)}}L_{-}|\ell,\ell\rangle = \frac{1}{\hbar\sqrt{2\ell}}L_{-}|\ell,\ell\rangle,\\
\text{or, } U_{\ell,\ell-1}(r,\theta,\phi) &= -\frac{1}{\hbar\sqrt{2\ell}}\hbar e^{-i\phi}\left[\frac{\partial}{\partial\theta} - i\cot\theta\frac{\partial}{\partial\phi}\right]U_{\ell,\ell}(r,\theta,\phi)\\
&= -\frac{2e^{-i\phi}}{\sqrt{2\ell}}\frac{\partial}{\partial\theta}\,U_{\ell,\ell}(r,\theta,\phi),
\end{aligned} \tag{7.72}$$

where we have used

$$\left(\frac{\partial}{\partial\theta} + i\cot\theta\frac{\partial}{\partial\phi}\right) U_{\ell,\ell}(r,\theta,\phi) = 0,$$

$$\text{or,}\quad i\cot\theta\frac{\partial}{\partial\phi}U_{\ell,\ell}(r,\theta,\phi) = -\frac{\partial}{\partial\theta}U_{\ell,\ell}(r,\theta,\phi). \tag{7.73}$$

Using (7.71) in (7.72), we can write

$$\begin{aligned} U_{\ell,\ell-1}(r,\theta,\phi) &= -\frac{2e^{-i\phi}}{\sqrt{2\ell}}R_\ell(r)e^{i\ell\phi}\frac{\mathrm{d}}{\mathrm{d}\theta}(\sin\theta)^\ell \\ &= -\frac{2\ell}{\sqrt{2\ell}}R_\ell(r)e^{i(\ell-1)\phi}(\sin\theta)^{\ell-1}\cos\theta \\ &= -\sqrt{2\ell}R_\ell(r)(\sin\theta)^{\ell-1}\cos\theta e^{i(\ell-1)\phi}. \end{aligned} \tag{7.74}$$

Similarly, a general wave function, $U_{\ell,m}(r,\theta,\phi)$, can be obtained by applying $(\ell - m)$ times the lowering operator L_- with a suitable normalization.

The interesting conclusions that we can draw from this operator analysis are that the angular momentum operator L^2 has eigenvalues of the form $\hbar^2\ell(\ell+1)$ and that ℓ takes integer as well as half integer values. This is certainly an improvement over the old quantum theory. Let us note that the energy associated with rotations in molecules can be denoted by

$$H = \frac{\mathbf{L}^2}{2I} = \frac{L^2}{2I}, \tag{7.75}$$

where $\mathbf{L}, I$ represent respectively the orbital angular momentum and the moment of inertia of the system. In the old quantum theory, one postulated the eigenvalues of angular momentum to be of the form

$$\text{eigenvalues of } L^2 = \hbar^2\ell^2, \quad \text{where} \quad \ell = 0,1,2,\cdots. \tag{7.76}$$

This, in turn, leads to the separation of energy levels in molecules in the proportions

$$1:3:5:7:\cdots. \tag{7.77}$$

However, this was not observed. The observed separations were in the proportions

$$1:2:3:4:\cdots. \tag{7.78}$$

It is easy to see from (7.75) that this will indeed be the case if

$$\text{eigenvalues of } L^2 = \hbar^2\ell(\ell+1), \quad \text{where } \ell = 0,1,2,3,\cdots. \tag{7.79}$$

This says that the orbital angular momentum eigenvalues take integer values and the extra term arises from the non-commutativity of different components of the angular momentum operators. If we solve the Schrödinger equation, we would obtain only integer values for the angular momentum eigenvalue. The operator method, on the other hand, allows for both integral and half integral eigenvalues.

We will talk about this in more detail later when we solve the Schrödinger equation. But, at the moment, we can understand the difference as follows. The Schrödinger wave function that we normally consider is a scalar function. Thus the effect of a rotation is simply to change the coordinates to a rotated value. Furthermore, a scalar function should have the property that when rotated by an angle 2π around any axis it should come back to its original value. This constraint leads to only integral values for the angular momentum eigenvalue. On the other hand, the wave function may have a nontrivial matrix structure of its own such that a rotation not only changes the coordinates, but also rotates the components of the wave function. Then, in general, we can write

$$\mathbf{J} = \mathbf{L} + \mathbf{S}, \tag{7.80}$$

where the operator $\mathbf{L}$ rotates the spatial coordinates and is known as the orbital angular momentum operator while $\mathbf{S}$ rotates the components of the wave function and is known as the spin angular momentum operator. $\mathbf{J}$ is called the total angular momentum operator. When a wave function has a nontrivial matrix structure, it is not necessary that the wave function remains unchanged under a rotation by 2π. Rather, on physical grounds we would like that the probability density, which is an observable, remains unchanged under a rotation of 2π. This implies that

$$\begin{aligned} &\psi^*\psi \xrightarrow{2\pi} \psi^*\psi, \\ \text{or,} \quad &(\psi,\psi^*) \xrightarrow{2\pi} (\pm\psi,\pm\psi^*). \end{aligned} \tag{7.81}$$

When such a situation occurs where the wave functions change sign under a rotation by 2π, we can have half integral values for the angular momentum. There are a few points to note here.

1. The eigenvalues corresponding to the operator $\mathbf{L}$ are integers. This is because in the absence of $\mathbf{S}$ or any matrix structure, the wave function must be single-valued.

2. $\mathbf{S}$ has both half integer and integer eigenvalues so that $\mathbf{J}$ or the total angular momentum also carries half integer as well as integer eigenvalues.

3. $\mathbf{S}, \mathbf{L}$ and $\mathbf{J}$ satisfy the same commutation relations (same algebra), namely,

$$\begin{aligned} &\mathbf{S} \times \mathbf{S} = i\hbar \mathbf{S}, \\ &\mathbf{L} \times \mathbf{L} = i\hbar \mathbf{L}, \\ &\mathbf{J} \times \mathbf{J} = i\hbar \mathbf{J}. \end{aligned} \tag{7.82}$$

Note that the familiar relation

$$[L_i, L_j] = i\hbar \epsilon_{ijk} L_k, \tag{7.83}$$

can be written as

$$\mathbf{L} \times \mathbf{L} = i\hbar \mathbf{L},$$

generalizing the familiar notation of 3-dimensional vector product

$$\begin{aligned} &\mathbf{C} = \mathbf{A} \times \mathbf{B}, \\ \text{or,} \quad &C_i = \epsilon_{ijk} A_j B_k. \end{aligned} \tag{7.84}$$

7.2 Finite rotations

We have already derived the representations of angular momentum operators corresponding to different eigenvalues. We have noted that for each value of ℓ there exists a representation of dimensionality $2\ell + 1$. In general, however, we can write the matrix elements of various operators as

$$\begin{aligned} \langle \ell', m' | L^2 | \ell, m \rangle &= \hbar^2 \ell(\ell+1) \delta_{\ell',\ell} \delta_{m',m}, \\ \langle \ell', m' | L_3 | \ell, m \rangle &= \hbar m \delta_{\ell',\ell} \delta_{m',m}, \\ \langle \ell', m' | L_+ | \ell, m \rangle &= \hbar \sqrt{\ell(\ell+1) - m(m+1)} \delta_{\ell',\ell} \delta_{m',m+1}, \\ \langle \ell', m' | L_- | \ell, m \rangle &= \hbar \sqrt{\ell(\ell+1) - m(m-1)} \delta_{\ell',\ell} \delta_{m',m-1}, \end{aligned} \tag{7.85}$$

where

$$\begin{aligned} \ell &= 0, \tfrac{1}{2}, 1, \cdots, \infty, \\ m &= -\ell, -\ell+1, \cdots, \ell-1, \ell. \end{aligned} \tag{7.86}$$

Thus, in the complete space, the operators have the following representation.

$$L^2 = \begin{array}{c} (0,0) \\ (\frac{1}{2},\frac{1}{2}) \\ (\frac{1}{2},-\frac{1}{2}) \\ (1,1) \\ (1,0) \\ (1,-1) \\ \vdots \end{array} \left(\begin{array}{cccccc c} \boxed{0} & & & & & & \\ & \frac{3\hbar^2}{4} & 0 & & & & \\ & 0 & \frac{3\hbar^2}{4} & & & & \\ & & & 2\hbar^2 & 0 & 0 & \\ & & & 0 & 0 & 0 & \\ & & & 0 & 0 & 2\hbar^2 & \\ & & & & & & \ddots \end{array}\right). \tag{7.87}$$

Similarly,

$$L_3 = \begin{array}{c} (0,0) \\ (\frac{1}{2},\frac{1}{2}) \\ (\frac{1}{2},-\frac{1}{2}) \\ (1,1) \\ (1,0) \\ (1,-1) \\ \vdots \end{array} \left(\begin{array}{cccccc c} \boxed{0} & & & & & & \\ & \frac{\hbar}{2} & 0 & & & & \\ & 0 & -\frac{\hbar}{2} & & & & \\ & & & \hbar & 0 & 0 & \\ & & & 0 & 0 & 0 & \\ & & & 0 & 0 & -\hbar & \\ & & & & & & \ddots \end{array}\right). \tag{7.88}$$

We also have

$$L_1 = \begin{array}{c} (0,0) \\ (\frac{1}{2},\frac{1}{2}) \\ (\frac{1}{2},-\frac{1}{2}) \\ (1,1) \\ (1,0) \\ (1,-1) \\ \vdots \end{array} \left(\begin{array}{cccccc c} \boxed{0} & & & & & & \\ & 0 & \frac{\hbar}{2} & & & & \\ & \frac{\hbar}{2} & 0 & & & & \\ & & & 0 & \frac{\hbar}{\sqrt{2}} & 0 & \\ & & & \frac{\hbar}{\sqrt{2}} & 0 & \frac{\hbar}{\sqrt{2}} & \\ & & & 0 & \frac{\hbar}{\sqrt{2}} & 0 & \\ & & & & & & \ddots \end{array}\right), \tag{7.89}$$

and similarly for L_2.

First of all, we note that all the operators have infinite dimensional representations. However, the representations are block diagonal. Therefore, any product of the operators would also be block diagonal. In particular, a finite rotation by the angle $\boldsymbol{\alpha}$, generated by

$$U(\boldsymbol{\alpha}) = e^{-\frac{i}{\hbar}\boldsymbol{\alpha}\cdot\mathbf{L}}, \tag{7.90}$$

would again be a block diagonal matrix. Let the $(2\ell+1)$ dimensional block of $U(\boldsymbol{\alpha})$, for a given ℓ, be denoted by $D^{(\ell)}(\boldsymbol{\alpha})$. Then, $D^{(\ell)}(\boldsymbol{\alpha})$ rotates vectors in the space $\mathcal{E}^{(\ell)}$. In other words, if $|\psi_\ell\rangle$ represents an arbitrary vector in the subspace spanned by the $(2\ell+1)$ vectors $|\ell,\ell\rangle$, $|\ell,\ell-1\rangle,\cdots,|\ell,-\ell\rangle$, then, $D^{(\ell)}(\boldsymbol{\alpha})$ is the rotation operator which would act on it. Let us now ask whether $D^{(\ell)}(\boldsymbol{\alpha})$ can be written in a manageable form, at least for small values of ℓ. First of all, note, from the definition, that

$$D^{(\ell)}(\boldsymbol{\alpha}) = e^{-\frac{i}{\hbar}\boldsymbol{\alpha}\cdot\mathbf{L}^{(\ell)}} = \sum_0^\infty \frac{1}{n!}\left(-\frac{i}{\hbar}\right)^n \left(\boldsymbol{\alpha}\cdot\mathbf{L}^{(\ell)}\right)^n. \tag{7.91}$$

It would seem like there is an infinite number of independent terms in this series. However, note that there are only a finite number of independent $(2\ell+1)$ dimensional square matrices. Therefore, after a finite number of terms in the series, we would have linear combinations of known terms. In fact, for a $(2\ell+1)$ dimensional representation, $(\boldsymbol{\alpha}\cdot\mathbf{L}^{(\ell)})^n$, where $n > 2\ell$, can be expressed as a linear combination of $(\boldsymbol{\alpha}\cdot\mathbf{L}^{(\ell)})^k$, where $k \leq 2\ell$. We can easily check this by restricting to rotations around the 3-axis. For example, let

$$\boldsymbol{\alpha}\cdot\mathbf{L}^{(\ell)} \rightarrow \alpha_3 L_3^{(\ell)}. \tag{7.92}$$

Furthermore, we know that the eigenvalues of $L_3^{(\ell)}$ are

$$-\hbar\ell, \hbar(-\ell+1), \cdots, \hbar(\ell-1), \hbar\ell.$$

Let us now look at the operator

$$\begin{aligned}\Omega = &\left(\alpha_3\left(L_3^{(\ell)} - \hbar\ell\mathbb{1}\right)\right)\left(\alpha_3\left(L_3^{(\ell)} - \hbar(\ell-1)\mathbb{1}\right)\right)\cdots\\ &\cdots\left(\alpha_3\left(L_3^{(\ell)} - \hbar(-\ell+1)\mathbb{1}\right)\right)\left(\alpha_3\left(L_3^{(\ell)} + \hbar\ell\mathbb{1}\right)\right).\end{aligned} \tag{7.93}$$

If we now take any vector $|\psi_\ell\rangle$ in this space, it is clear that

$$\Omega|\psi_\ell\rangle = 0. \tag{7.94}$$

This is because this operator annihilates all the $(2\ell+1)$ basis vectors in this space. This, therefore, implies that, in this space, Ω can be identified with the null operator. This, in turn, determines

$$(\alpha_3 L_3^{(\ell)})^{2\ell+1} = \sum_{k=0}^{2\ell} c_k (\alpha_3 L_3^{(\ell)})^k, \tag{7.95}$$

which proves our earlier assertion. Hence, we can write

$$D^{(\ell)}(\boldsymbol{\alpha}) = \sum_{n=0}^{2\ell} f_n(\boldsymbol{\alpha}) \left(\boldsymbol{\alpha}\cdot\mathbf{L}^{(\ell)}\right)^n . \tag{7.96}$$

► **Example.** Let us look at the representation with $\ell = \frac{1}{2}$. We know from (7.42) and (7.50) that the generators of angular momentum in this representation are given by

$$L_i^{(\frac{1}{2})} = \frac{\hbar}{2}\,\sigma_i, \qquad i = 1,2,3, \tag{7.97}$$

where σ_i are the Pauli matrices. They satisfy the commutation relations

$$[\sigma_i, \sigma_j] = 2i\epsilon_{ijk}\sigma_k. \tag{7.98}$$

Furthermore, they also satisfy the anti-commutation relations

$$\{\sigma_i, \sigma_j\} = 2\delta_{ij}\mathbb{1}, \tag{7.99}$$

so that we can write

$$\sigma_i\sigma_j = \frac{1}{2}\{\sigma_i,\sigma_j\} + \frac{1}{2}[\sigma_i,\sigma_j] = \delta_{ij}\mathbb{1} + i\epsilon_{ijk}\sigma_k. \tag{7.100}$$

Multiplying both sides of (7.100) with two arbitrary vectors A_i and B_j we have

$$\begin{aligned}
A_iB_j(\sigma_i\sigma_j) &= A_iB_j(\delta_{ij}\mathbb{1} + i\epsilon_{ijk}\sigma_k),\\
\text{or,}\quad (\sigma_iA_i)(\sigma_jB_j) &= A_iB_i\mathbb{1} + i\epsilon_{ijk}A_iB_j\sigma_k,\\
\text{or,}\quad (\boldsymbol{\sigma}\cdot\mathbf{A})(\boldsymbol{\sigma}\cdot\mathbf{B}) &= (\mathbf{A}\cdot\mathbf{B})\mathbb{1} + i\boldsymbol{\sigma}\cdot(\mathbf{A}\times\mathbf{B}).
\end{aligned} \tag{7.101}$$

Since we have maintained the order of A_i, B_i, this relation also holds for vector operators that commute with the Pauli matrices. Furthermore, this is consistent with our theorem.

In this case, the operator for finite rotations takes the form,

$$\begin{aligned}
D^{(\frac{1}{2})}(\boldsymbol{\alpha}) &= e^{-\frac{i}{\hbar}\boldsymbol{\alpha}\cdot\mathbf{L}^{(\frac{1}{2})}} = e^{-\frac{i}{2}\boldsymbol{\alpha}\cdot\boldsymbol{\sigma}}\\
&= \sum_n \frac{1}{n!}\left(-\frac{i}{2}\right)^n (\boldsymbol{\alpha}\cdot\boldsymbol{\sigma})^n\\
&= \mathbb{1} - \frac{i}{2}\boldsymbol{\alpha}\cdot\boldsymbol{\sigma} - \frac{1}{2!(2)^2}(\boldsymbol{\alpha}\cdot\boldsymbol{\sigma})^2 + \frac{i}{3!2^3}(\boldsymbol{\alpha}\cdot\boldsymbol{\sigma})^3 + \cdots\\
&= \left(1 - \frac{1}{2!}\frac{1}{2^2}\alpha^2 + \frac{1}{4!2^4}\alpha^4\cdots\right)\mathbb{1}\\
&\quad - \frac{i}{2}\boldsymbol{\alpha}\cdot\boldsymbol{\sigma} + \frac{i}{3!2^3}(\boldsymbol{\alpha}\cdot\boldsymbol{\sigma})\,\alpha^2\cdots\\
&= \mathbb{1}\cos\frac{\alpha}{2} - i\,(\hat{\boldsymbol{\alpha}}\cdot\boldsymbol{\sigma})\sin\frac{\alpha}{2},
\end{aligned} \tag{7.102}$$

where $\alpha = |\boldsymbol{\alpha}|$ and we have used the properties of the Pauli matrices following from (7.98) and (7.99). For a rotation by 2π, namely, for $\alpha = 2\pi$, we note that the rotation operator reduces to

$$D^{(\frac{1}{2})}(2\pi) = \mathbb{1}\cos(\pi) - i(\hat{\boldsymbol{\alpha}}\cdot\boldsymbol{\sigma})\sin(\pi) = -\mathbb{1}. \tag{7.103}$$

This shows that two component wave functions belonging to $\ell = \frac{1}{2}$ (also known as spinors) change sign under a rotation by 2π as we have discussed earlier. ◀

7.3 Reducible and irreducible spaces

We have noted earlier that the space $\mathcal{E}^{(\ell)}$ is invariant under rotations. That is, any state in $\mathcal{E}^{(\ell)}$ goes into another state in $\mathcal{E}^{(\ell)}$ under rotation. Such spaces are called invariant spaces and the reason for this invariance is clear. Any state in $\mathcal{E}^{(\ell)}$ has associated with it the angular momentum quantum number ℓ. No rotation will change this value, since $\left[L^2, L_i\right] = 0$ for all i. Thus, a rotated state would continue to have the same ℓ-quantum number associated with it and hence would belong to $\mathcal{E}^{(\ell)}$.

Furthermore, these invariant subspaces are irreducible. That is, they cannot be written as sums of invariant spaces of lower dimensionality. This can be proved by showing that $\mathcal{E}^{(\ell)}$ cannot contain invariant subspaces other than itself. To show this, let us assume that $\overline{\mathcal{E}}^{(\ell)}$ is an invariant subspace of $\mathcal{E}^{(\ell)}$. Let $|\psi\rangle$ be an arbitrary state in $\overline{\mathcal{E}}^{(\ell)}$. We can choose it to be the state $|\ell,\ell\rangle$. (Note that this is always possible by a unitary change of the basis.) Under a rotation

$$|\ell,\ell\rangle \rightarrow D^{(\ell)}(\boldsymbol{\alpha})\,|\ell,\ell\rangle. \tag{7.104}$$

If we are considering an infinitesimal rotation, then ($\boldsymbol{\alpha}$ is infinitesimal)

$$\begin{aligned} D^{(\ell)}(\boldsymbol{\alpha}) &= \mathbb{1} - \frac{i}{\hbar}\boldsymbol{\alpha}\cdot\mathbf{L}^{(\ell)} \\ &= \mathbb{1} - \frac{i}{2\hbar}\left(\alpha_+ L_-^{(\ell)} + \alpha_- L_+^{(\ell)} + 2\alpha_3 L_3^{(\ell)}\right), \end{aligned} \tag{7.105}$$

where we have defined $\alpha_\pm = \alpha_1 \pm i\alpha_2$.

Thus, under an infinitesimal rotation,

$$|\ell,\ell\rangle \rightarrow \left(\mathbb{1} - \frac{i}{2\hbar}\left(\alpha_+ L_-^{(\ell)} + \alpha_- L_+^{(\ell)} + 2\alpha_3 L_3^{(\ell)}\right)\right)|\ell,\ell\rangle. \tag{7.106}$$

Noting that $L_+^{(\ell)}|\ell,\ell\rangle = 0$, we conclude that

$$|\ell,\ell\rangle \rightarrow a_1|\ell,\ell\rangle + b_1|\ell,\ell-1\rangle. \tag{7.107}$$

However, since $\overline{\mathcal{E}}^{(\ell)}$ is an invariant subspace and $|\ell,\ell\rangle$ is a vector in it, this implies that $|\ell,\ell-1\rangle$ must also be in $\overline{\mathcal{E}}^{(\ell)}$. It is useful to note here that a vector $|\ell,m\rangle$, in general, transforms as follows under an infinitesimal rotation.

$$|\ell,m\rangle \to a_1|\ell,m\rangle + a_2|\ell,m+1\rangle + a_3|\ell,m-1\rangle. \tag{7.108}$$

Thus, observing the behavior of successive states under rotation, we can conclude that $\overline{\mathcal{E}}^{(\ell)}$ contains all the states $|\ell,\ell\rangle$, $|\ell,\ell-1\rangle, \cdots,$ $|\ell,-\ell+1\rangle$ and $|\ell,-\ell\rangle$. Therefore, the dimensionality of $\overline{\mathcal{E}}^{(\ell)}$ is $(2\ell+1)$ and hence must be identical to $\mathcal{E}^{(\ell)}$.

On the other hand, a reducible space is one which can be written as a sum of spaces of lower dimensionality, which are themselves invariant spaces. In such a case, the symmetry operators such as $D^{(\ell)}(\boldsymbol{\alpha})$ can be further block diagonalized.

► **Example.** Let us next study a simple example of a reducible space. Let us consider the two element group consisting of $(\mathbb{1},\mathcal{P})$, where $\mathcal{P}$ is the reflection operator (or Parity operator) that we have defined earlier. They satisfy the multiplication law

$$\begin{aligned}
&\mathbb{1}\mathbb{1} = \mathbb{1},\\
&\mathbb{1}\mathcal{P} = \mathcal{P}\mathbb{1} = \mathcal{P},\\
&\mathcal{P}\mathcal{P} = \mathbb{1}.
\end{aligned} \tag{7.109}$$

As operators, they act on scalar functions as (namely, they, of course, act on vectors, but as we have discussed earlier, the effect can be thought of as changing the coefficient scalar functions in an expansion of the state in the coordinate basis)

$$\begin{aligned}
&\mathbb{1}f(x,y,z) = f(x,y,z),\\
&\mathcal{P}f(x,y,z) = f(-x,-y,-z).
\end{aligned} \tag{7.110}$$

Let us consider an arbitrary function $f_1(x,y,z)$ and let

$$f_2(x,y,z) = \mathcal{P}f_1(x,y,z), \qquad \mathcal{P}f_2(x,y,z) = f_1(x,y,z). \tag{7.111}$$

Let us denote by S the set of functions of the form

$$f(x,y,z) = a_1 f_1(x,y,z) + a_2 f_2(x,y,z), \tag{7.112}$$

for all allowed values of the parameter a_1 and a_2. This set of functions in (7.112) defines the space of functions spanned by the basis functions f_1 and f_2. These functions satisfy all the postulates of a linear vector space. The space S is said to be two dimensional because any function, in this space, can be expressed as a combination of the two functions f_1 and f_2. Furthermore, this space is invariant under the operation of the group $(\mathbb{1},\mathcal{P})$. Namely, any vector in S, under the effect of the group, goes into another vector in S.

$$\begin{aligned}
\mathbb{1}f(x,y,z) &= f(x,y,z),\\
\mathcal{P}f(x,y,z) &= \mathcal{P}\left(a_1 f_1(x,y,z) + a_2 f_2(x,y,z)\right)\\
&= a_1 f_2(x,y,z) + a_2 f_1(x,y,z) \in S.
\end{aligned} \tag{7.113}$$

On the other hand, we can define two new functions

$$\tilde{f}_1 = \frac{1}{2}(f_1 + f_2), \qquad \tilde{f}_2 = \frac{1}{2}(f_1 - f_2), \tag{7.114}$$

such that

$$\mathcal{P}\tilde{f}_1 = \tilde{f}_1, \qquad \mathcal{P}\tilde{f}_2 = -\tilde{f}_2. \tag{7.115}$$

Let us denote by S_1 and S_2 the set of functions of the form $a_1\tilde{f}_1$ and $a_2\tilde{f}_2$ respectively, where a_1 and a_2 take all possible allowed values. Clearly, S_1 and S_2 are linear spaces. Moreover, they are also invariant under the action of the group. Namely,

$$\left.\begin{array}{l} \mathbb{1}a_1\tilde{f}_1 = a_1\tilde{f}_1, \\ \mathcal{P}a_1\tilde{f}_1 = a_1\tilde{f}_1, \end{array}\right\} \text{ belongs to } S_1,$$

$$\left.\begin{array}{l} \mathbb{1}a_2\tilde{f}_2 = a_2\tilde{f}_2, \\ \mathcal{P}a_2\tilde{f}_2 = -a_2\tilde{f}_2, \end{array}\right\} \text{ belongs to } S_2. \tag{7.116}$$

Thus S_1 and S_2 are also invariant spaces but they are of lower dimensionality because each of them is spanned by only one function. They are called subspaces of S since any function in S_1 or S_2 is contained in S. Furthermore, the space S is said to be reducible to the two invariant subspaces S_1 and S_2. That is, any function in S can be expressed uniquely as a combination of functions of S_1 and S_2.

$$\begin{aligned} f(x,y,z) &= a_1 f_1 + a_2 f_2 = a_1(\tilde{f}_1 + \tilde{f}_2) + a_2(\tilde{f}_1 - \tilde{f}_2) \\ &= (a_1 + a_2)\,\tilde{f}_1 + (a_1 - a_2)\,\tilde{f}_2. \end{aligned} \tag{7.117}$$

Symbolically, we write

$$S = S_1 \oplus S_2. \tag{7.118}$$

The study of irreducible, invariant spaces is important because any invariant space can be built out of them. ◀

7.4 Selected problems

1. Show that

$$e^{-i\beta L_y} = e^{i\frac{\pi}{2}L_x} e^{-\beta L_z} e^{-i\frac{\pi}{2}L_x}, \tag{7.119}$$

   where β is a real constant.

2. In an eigenstate of L^2 and L_3 denoted by $|\ell, m\rangle$, calculate

$$\langle L_1\rangle, \quad \langle L_2\rangle, \quad \langle L_1^2\rangle, \quad \text{and} \quad \langle L_2^2\rangle. \tag{7.120}$$

3. Consider two harmonic oscillators (in one dimension) with raising and lowering operators $(a, a^\dagger; b, b^\dagger)$ respectively. They satisfy the usual commutation relations

$$[a, a^\dagger] = 1 = [b, b^\dagger], \tag{7.121}$$

with all others vanishing.

From the four operators $a^\dagger b, ab^\dagger, a^\dagger a, b^\dagger b$, show, by taking linear combinations, that one can find operators which have the same commutation relations as the angular momentum operators. What operator plays the role of angular momentum squared?

CHAPTER 8

Schrödinger equation in higher dimensions

In earlier lectures, we solved some quantum mechanical systems in one dimension. In the next few lectures, we will discuss the solutions of some three dimensional systems with spherically symmetric potentials.

8.1 Schrödinger equation in spherically symmetric potential

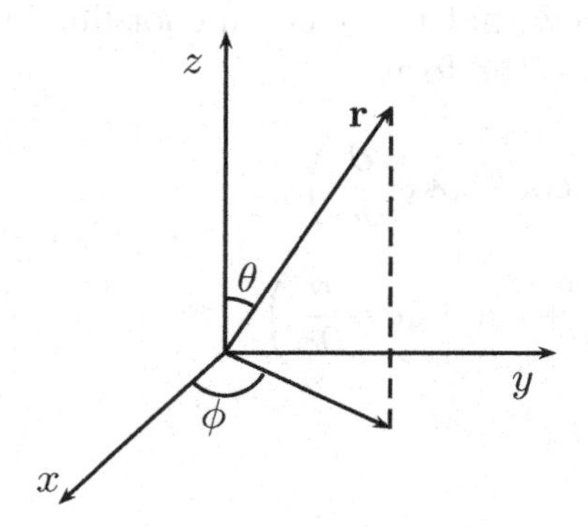

Figure 8.1: Cartesian and spherical coordinates

Let us consider a particle of mass m moving in a three dimensional potential of the form $V = V\left(\mathbf{x}^2\right)$. Thus, the Hamiltonian of the system has the form

$$H = \frac{\mathbf{P}^2}{2m} + V(\mathbf{X}^2) = \frac{\mathbf{P}^2}{2m} + V(R^2). \tag{8.1}$$

In other words, the potential depends only on the magnitude of the vector from the origin. This Hamiltonian is clearly invariant under rotations. Therefore, we expect angular momentum to be conserved. Furthermore, since the potential has spherical symmetry it simplifies to study the problem in spherical coordinates (see Fig. 8.1) where we can identify

$$\begin{aligned}
x &= r\sin\theta\cos\phi,\\
y &= r\sin\theta\sin\phi,\\
z &= r\cos\theta.
\end{aligned} \tag{8.2}$$

The gradient operator in spherical coordinates has the form

$$\boldsymbol{\nabla} = \hat{\mathbf{r}}\frac{\partial}{\partial r} + \frac{\hat{\boldsymbol{\theta}}}{r}\frac{\partial}{\partial\theta} + \frac{\hat{\boldsymbol{\phi}}}{r\sin\theta}\frac{\partial}{\partial\phi}, \tag{8.3}$$

and the Laplacian is given by

$$\boldsymbol{\nabla}^2 = \frac{1}{r^2}\frac{\partial}{\partial r}\left(r^2\frac{\partial}{\partial r}\right) + \frac{1}{r^2\sin\theta}\frac{\partial}{\partial\theta}\left(\sin\theta\frac{\partial}{\partial\theta}\right) + \frac{1}{r^2\sin^2\theta}\frac{\partial^2}{\partial\phi^2}. \tag{8.4}$$

Furthermore, we know that in spherical coordinates the angular momentum operators have the forms

$$\begin{aligned}
L_x &= i\hbar\left(\sin\phi\frac{\partial}{\partial\theta} + \cot\theta\cos\phi\frac{\partial}{\partial\phi}\right),\\
L_y &= i\hbar\left(-\cos\phi\frac{\partial}{\partial\theta} + \cot\theta\sin\phi\frac{\partial}{\partial\phi}\right),\\
L_z &= -i\hbar\frac{\partial}{\partial\phi}.
\end{aligned} \tag{8.5}$$

so that we have

$$\begin{aligned}
L_x^2 &= (i\hbar)^2\left(\sin\phi\frac{\partial}{\partial\theta} + \cot\theta\cos\phi\frac{\partial}{\partial\phi}\right)^2\\
&= (i\hbar)^2\left[\sin^2\phi\frac{\partial}{\partial\theta^2} + \sin\phi\frac{\partial}{\partial\theta}\left(\cot\theta\cos\phi\frac{\partial}{\partial\phi}\right)\right.\\
&\quad + \cot\theta\cos\phi\frac{\partial}{\partial\phi}\left(\sin\phi\frac{\partial}{\partial\theta}\right)\\
&\quad \left. + \cot\theta\cos\phi\frac{\partial}{\partial\phi}\left(\cot\theta\cos\phi\frac{\partial}{\partial\phi}\right)\right]
\end{aligned}$$

$$= (i\hbar)^2\Big[\sin^2\phi\frac{\partial^2}{\partial\theta^2} - \sin\phi\cos\phi\,\mathrm{cosec}^2\theta\frac{\partial}{\partial\phi}$$

$$+\sin\phi\cos\phi\cot\theta\frac{\partial^2}{\partial\theta\partial\phi}$$

$$+\cot\theta\cos^2\phi\frac{\partial}{\partial\theta} + \cot\theta\sin\phi\cos\phi\frac{\partial^2}{\partial\phi\partial\theta}$$

$$-\cot^2\theta\sin\phi\cos\phi\frac{\partial}{\partial\phi} + \cot^2\theta\cos^2\phi\frac{\partial^2}{\partial\phi^2}\Big]. \tag{8.6}$$

Similarly, we have

$$L_y^2 = (i\hbar)^2\Big[\cos^2\phi\frac{\partial^2}{\partial\theta^2} + \sin\phi\cos\phi\,\mathrm{cosec}^2\theta\frac{\partial}{\partial\phi}$$

$$-\sin\phi\cos\phi\cot\theta\frac{\partial^2}{\partial\theta\partial\phi}$$

$$+\cot\theta\sin^2\phi\frac{\partial}{\partial\theta} - \sin\phi\cos\phi\cot\theta\frac{\partial^2}{\partial\theta\partial\phi}$$

$$+\cot^2\theta\sin\phi\cos\phi\frac{\partial}{\partial\phi} + \cot^2\theta\sin^2\phi\frac{\partial^2}{\partial\phi^2}\Big],$$

$$L_z^2 = (-i\hbar)^2\frac{\partial^2}{\partial\phi^2}. \tag{8.7}$$

It follows now, from (8.6) and (8.7), that

$$L^2 = L_x^2 + L_y^2 + L_z^2$$

$$= (i\hbar)^2\left[\frac{\partial^2}{\partial\theta^2} + \cot\theta\frac{\partial}{\partial\theta} + \mathrm{cosec}^2\theta\frac{\partial^2}{\partial\phi^2}\right]$$

$$= (i\hbar)^2\left[\frac{\partial^2}{\partial\theta^2} + \cot\theta\frac{\partial}{\partial\theta} + \frac{1}{\sin^2\theta}\frac{\partial^2}{\partial\phi^2}\right]$$

$$= (i\hbar)^2\left[\frac{1}{\sin\theta}\frac{\partial}{\partial\theta}\left(\sin\theta\frac{\partial}{\partial\theta}\right) + \frac{1}{\sin^2\theta}\frac{\partial^2}{\partial\phi^2}\right]. \tag{8.8}$$

Thus, we see from (8.4) and (8.8) that we can also write

$$\boldsymbol{\nabla}^2 = \frac{1}{r^2}\frac{\partial}{\partial r}\left(r^2\frac{\partial}{\partial r}\right) + \frac{1}{r^2}\frac{L^2}{(i\hbar)^2}$$

$$= \frac{1}{r^2}\frac{\partial}{\partial r}\left(r^2\frac{\partial}{\partial r}\right) - \frac{L^2}{\hbar^2 r^2}. \tag{8.9}$$

The time independent Schrödinger equation for such a system has the form

$$
\begin{aligned}
& H\psi(r,\theta,\phi) = E\psi(r,\theta,\phi), \\
\text{or,}\quad & \left(-\frac{\hbar^2}{2m}\boldsymbol{\nabla}^2 + V(r)\right)\psi = E\psi, \\
\text{or,}\quad & \left[-\frac{\hbar^2}{2m}\left(\frac{1}{r^2}\frac{\partial}{\partial r}\left(r^2\frac{\partial}{\partial r}\right) - \frac{L^2}{\hbar^2 r^2}\right) + V(r)\right]\psi = E\psi. \quad (8.10)
\end{aligned}
$$

Let us now use a separable solution of the form

$$\psi(r,\theta,\phi) = R(r)F(\theta,\phi). \tag{8.11}$$

Substituting this into (8.10), we obtain

$$
\begin{aligned}
& F\left[-\frac{\hbar^2}{2m}\left(\frac{1}{r^2}\frac{\mathrm{d}}{\mathrm{d}r}r^2\frac{\mathrm{d}R}{\mathrm{d}r}\right) + (V(r)-E)R\right] = -R\frac{L^2}{2mr^2}F, \\
\text{or,}\quad & \frac{1}{R}\left[\frac{\mathrm{d}}{\mathrm{d}r}\left(r^2\frac{\mathrm{d}R}{\mathrm{d}r}\right) + \frac{2mr^2}{\hbar^2}(E-V(r))R\right] = \frac{1}{F}\frac{L^2}{\hbar^2}F = \lambda.
\end{aligned}
\tag{8.12}
$$

Here we have used the fact that since the two sides of (8.12) depend on independent coordinates, they can be the same for arbitrary values of these coordinates only if each of the expressions is equal to a constant. Equation (8.12) leads to

$$\frac{\mathrm{d}}{\mathrm{d}r}\left(r^2\frac{\mathrm{d}R}{\mathrm{d}r}\right) + \frac{2mr^2}{\hbar^2}(E-V(r))R = \lambda R, \tag{8.13}$$

$$L^2F(\theta,\phi) = \hbar^2\lambda F(\theta,\phi). \tag{8.14}$$

This shows that $\hbar^2\lambda$ is the eigenvalue of the operator L^2. Furthermore, the solution of the radial equation depends on the form of the potential and, therefore, on the dynamics. Consequently, we will concern ourselves only with the angular equation, for the present, which has the explicit form

$$
\begin{aligned}
& L^2F(\theta,\phi) = \hbar^2\lambda F(\theta,\phi), \\
\text{or,}\quad & -\left[\frac{1}{\sin\theta}\frac{\partial}{\partial\theta}\left(\sin\theta\frac{\partial}{\partial\theta}\right) + \frac{1}{\sin^2\theta}\frac{\partial^2}{\partial\phi^2}\right] = \lambda F(\theta,\phi). \quad (8.15)
\end{aligned}
$$

Let us separate the variables further as

$$F(\theta,\phi) = \Theta(\theta)\Phi(\phi). \tag{8.16}$$

Then, equation (8.15) leads to

$$\Phi\left[\frac{1}{\sin\theta}\frac{\mathrm{d}}{\mathrm{d}\theta}\left(\sin\theta\frac{\mathrm{d}\Theta}{\mathrm{d}\theta}\right)+\lambda\Theta\right]=-\frac{\Theta}{\sin^2\theta}\frac{\mathrm{d}^2\Phi}{\mathrm{d}\phi^2},$$

$$\text{or,}\quad \frac{1}{\Theta}\left[\sin\theta\frac{\mathrm{d}}{\mathrm{d}\theta}\left(\sin\theta\frac{\mathrm{d}\Theta}{\mathrm{d}\theta}\right)+\lambda\sin^2\theta\,\Theta\right]=-\frac{1}{\Phi}\frac{\mathrm{d}^2\Phi}{\mathrm{d}\phi^2}=\alpha, \tag{8.17}$$

so that

$$\sin\theta\frac{\mathrm{d}}{\mathrm{d}\theta}\left(\sin\theta\frac{\mathrm{d}\Theta}{\mathrm{d}\theta}\right)+\lambda\sin^2\theta\,\Theta=\alpha\Theta, \tag{8.18}$$

$$\frac{\mathrm{d}^2\Phi}{\mathrm{d}\phi^2}=\frac{L_z^2}{(-i\hbar)^2}\,\Phi=-\alpha\Phi(\phi). \tag{8.19}$$

The second equation, (8.19), has the simple solution

$$\Phi(\phi)\sim e^{\pm i\sqrt{\alpha}\phi}. \tag{8.20}$$

Since the wave function has to be single valued and continuous, this implies that

$$\alpha=m^2,\quad \text{where}\quad m=0,\pm1,\pm2,\cdots=\ \text{integers}. \tag{8.21}$$

Thus, the normalized eigenstates of L_z can be written as

$$\Phi_m(\phi)=\frac{1}{\sqrt{2\pi}}\,e^{im\phi}, \tag{8.22}$$

where the integer nature of m arises because we require $\Phi_m(\phi)=\Phi_m(\phi+2\pi)$ and the factor of $\frac{1}{\sqrt{2\pi}}$ normalizes the solution in the ϕ space. (Remember $0\leq\phi\leq2\pi$.) It is also clear that $\hbar m$, with m integer, is the eigenvalue of L_z (m should not be confused with the mass).

With (8.21) the θ-equation, (8.18), now becomes

$$\sin\theta\frac{\mathrm{d}}{\mathrm{d}\theta}\left(\sin\theta\frac{\mathrm{d}\Theta}{\mathrm{d}\theta}\right)+\lambda\sin^2\theta\,\Theta=m^2\Theta,$$

$$\text{or,}\quad \sin\theta\frac{\mathrm{d}}{\mathrm{d}\theta}\left(\sin\theta\frac{\mathrm{d}\Theta}{\mathrm{d}\theta}\right)+\left(\lambda\sin^2\theta-m^2\right)\Theta=0. \tag{8.23}$$

Let us define

$$x=\cos\theta, \tag{8.24}$$

so that

$$\frac{\mathrm{d}}{\mathrm{d}\theta} = \frac{\mathrm{d}x}{\mathrm{d}\theta}\frac{\mathrm{d}}{\mathrm{d}x} = -\sin\theta\,\frac{\mathrm{d}}{\mathrm{d}x} = -(1-x^2)^{\frac{1}{2}}\,\frac{\mathrm{d}}{\mathrm{d}x},$$

$$\sin\theta\frac{\mathrm{d}}{\mathrm{d}\theta} = -(1-x^2)\frac{\mathrm{d}}{\mathrm{d}x}.$$

In this variable, equation (8.23) becomes

$$(1-x^2)\frac{\mathrm{d}}{\mathrm{d}x}\left((1-x^2)\frac{\mathrm{d}\Theta}{\mathrm{d}x}\right) + (\lambda(1-x^2) - m^2)\Theta = 0,$$

$$\text{or,}\quad (1-x^2)\frac{\mathrm{d}^2\Theta}{\mathrm{d}x^2} - 2x\frac{\mathrm{d}\Theta}{\mathrm{d}x} + \left(\lambda - \frac{m^2}{1-x^2}\right)\Theta = 0. \tag{8.25}$$

To analyze the asymptotic behavior of the solution of (8.25), let us divide throughout by $(1-x^2)$. Then, the equation becomes

$$\frac{\mathrm{d}^2\Theta}{\mathrm{d}x^2} - \frac{2x}{1-x^2}\frac{\mathrm{d}\Theta}{\mathrm{d}x} + \left(\frac{\lambda}{1-x^2} - \frac{m^2}{(1-x^2)^2}\right)\Theta = 0. \tag{8.26}$$

Since $0 \leq \theta \leq \pi$, it follows that $-1 \leq x \leq 1$ and we see that the coefficients of the terms in (8.26) become singular at $x = \pm 1$.

Near $x = 1$, we can approximate,

$$\frac{2x}{1-x^2} = \frac{2x}{(1-x)(1+x)} \simeq \frac{1}{1-x},$$

$$\frac{m^2}{(1-x^2)^2} = \frac{m^2}{(1-x)^2(1+x)^2} \simeq \frac{m^2}{4(1-x)^2}. \tag{8.27}$$

Thus, for $x \simeq 1$, equation (8.26) reduces to

$$\frac{\mathrm{d}^2\Theta}{\mathrm{d}x^2} - \frac{1}{1-x}\,\frac{\mathrm{d}\Theta}{\mathrm{d}x} - \frac{m^2}{4(1-x)^2}\,\Theta = 0. \tag{8.28}$$

Let us assume a solution of the form ($a_0 \neq 0$)

$$\Theta(x) = (1-x)^{\beta}\left[a_0 + a_1(1-x) + a_2(1-x)^2 + \cdots\right]. \tag{8.29}$$

Putting this back into (8.28), we have for the (leading) term of the form $(1-x)^{\beta-2}$

$$a_0\left[\beta(\beta-1) + \beta - \frac{m^2}{4}\right] = 0,$$

$$\text{or,}\quad \beta = \pm\frac{|m|}{2}. \tag{8.30}$$

since $a_0 \neq 0$. Thus, we have

$$\Theta(x) \xrightarrow{x\to 1} (1-x)^{\frac{|m|}{2}} \left[a_0 + a_1(1-x) + a_2(1-x)^2 + \cdots\right],$$
$$\text{or,}\quad \Theta(x) \xrightarrow{x\to 1} (1-x)^{-\frac{|m|}{2}} \left[a_0 + a_1(1-x) + a_2(1-x)^2 + \cdots\right]. \tag{8.31}$$

It is clear that although the first solution in (8.31) is finite when $x \to 1$ as a physical solution should be, the second solution blows up as $x \to 1$. Thus, we conclude that the physical solution of (8.31) behaves as

$$\Theta(x) \xrightarrow{x\to 1} (1-x)^{\frac{|m|}{2}} \left[a_0 + a_1(1-x) + a_2(1-x)^2 + \cdots\right]. \tag{8.32}$$

Similarly, we can show that the physical solution near $x \simeq -1$ behaves as

$$\Theta(x) \xrightarrow{x\to -1} (1+x)^{\frac{|m|}{2}} \left[a_0' + a_1'(1+x) + a_2'(1+x)^2 + \cdots\right], \tag{8.33}$$

so that we expect a solution of (8.26) to have the general form

$$\Theta(x) = (1-x^2)^{\frac{|m|}{2}} \sum_{k=0}^{\infty} a_k x^k = (1-x^2)^{\frac{|m|}{2}} z(x), \tag{8.34}$$

which leads to

$$\begin{aligned}
\frac{\mathrm{d}\Theta(x)}{\mathrm{d}x} &= (1-x^2)^{\frac{|m|}{2}} \frac{\mathrm{d}z}{\mathrm{d}x} - |m|x(1-x^2)^{\frac{|m|}{2}-1} z,\\
\frac{\mathrm{d}^2\Theta(x)}{\mathrm{d}x^2} &= (1-x^2)^{\frac{|m|}{2}} \frac{\mathrm{d}^2 z}{\mathrm{d}x^2} - 2|m|x(1-x^2)^{\frac{|m|}{2}-1} \frac{\mathrm{d}z}{\mathrm{d}x}\\
&\quad - |m|(1-x^2)^{\frac{|m|}{2}-1} z + |m|(|m|-2)x^2(1-x^2)^{\frac{|m|}{2}-2} z.
\end{aligned} \tag{8.35}$$

Substituting this back into equation (8.26), we have

$$\begin{aligned}
&(1-x^2)\Bigg[(1-x^2)^{\frac{|m|}{2}} \frac{\mathrm{d}^2 z}{\mathrm{d}x^2} - 2|m|x(1-x^2)^{\frac{|m|}{2}-1} \frac{\mathrm{d}z}{\mathrm{d}x}\\
&\quad + \left(-|m|(1-x^2)^{\frac{|m|}{2}-1} + |m|\left(|m|-2\right)x^2(1-x^2)^{\frac{|m|}{2}-2}\right) z\Bigg]\\
&\quad - 2x\left[(1-x^2)^{\frac{|m|}{2}} \frac{\mathrm{d}z}{\mathrm{d}x} - |m|x(1-x^2)^{\frac{|m|}{2}-1} z\right]
\end{aligned}$$

$$+\left(\lambda-\frac{m^2}{(1-x^2)}\right)(1-x^2)^{\frac{|m|}{2}}z=0,$$

$$\text{or,}\quad\left((1-x^2)\frac{\mathrm{d}^2z}{\mathrm{d}x^2}-2(|m|+1)x\frac{\mathrm{d}z}{\mathrm{d}x}+(\lambda-|m|)z\right)(1-x^2)^{\frac{|m|}{2}}$$

$$+(1-x^2)^{\frac{|m|}{2}-1}\left(|m|\left(|m|-2\right)x^2+2|m|x^2-m^2\right)z=0,$$

$$\text{or,}\quad(1-x^2)\frac{\mathrm{d}^2z}{\mathrm{d}x^2}-2(|m|+1)x\frac{\mathrm{d}z}{\mathrm{d}x}+(\lambda-|m|(|m|+1))z=0. \tag{8.36}$$

Let us next substitute a power series solution for z of the form

$$z(x)=\sum_{k=0}^{\infty}a_kx^k, \tag{8.37}$$

where a_i's are arbitrary constant coefficients at this point. With this, equation (8.36) becomes

$$(1-x^2)\sum_{k=2}^{\infty}k(k-1)a_kx^{k-2}-2(|m|+1)x\sum_{k=1}^{\infty}ka_kx^{k-1}$$

$$+(\lambda-|m|(|m|+1))\sum_{k=0}^{\infty}a_kx^k=0,$$

$$\text{or,}\quad\sum_{k=2}^{\infty}k(k-1)a_kx^{k-2}-\sum_{k=2}^{\infty}k(k-1)a_kx^k$$

$$-2(|m|+1)\sum_{k=1}^{\infty}ka_kx^k+(\lambda-|m|(|m|+1))\sum_{k=0}^{\infty}a_kx^k=0,$$

$$\text{or,}\quad\sum_{k=0}^{\infty}(k+2)(k+1)a_{k+2}x^k-\sum_{k=0}^{\infty}k(k-1)a_kx^k$$

$$-2(|m|+1)\sum_{k=0}^{\infty}ka_kx^k+(\lambda-|m|(|m|+1))\sum_{k=0}^{\infty}a_kx^k=0,$$

$$\text{or,}\quad\sum_{k=0}^{\infty}x^k\big[(k+2)(k+1)a_{k+2}$$

$$-\left(k(k-1)+2k(|m|+1)+|m|(|m|+1)-\lambda\right)a_k\big]=0. \tag{8.38}$$

For this to be true we must have

$$\begin{aligned}
a_{k+2} &\\
&= \frac{1}{(k+2)(k+1)}(k^2 + k(2|m|+1) + |m|(|m|+1) - \lambda)a_k \\
&= \frac{1}{(k+2)(k+1)}((k+|m|)(k+|m|+1) - \lambda)a_k. \qquad (8.39)
\end{aligned}$$

Equation (8.39) defines the recursion relation for the coefficients of the power series solution in (8.37). Clearly, for large k

$$a_{k+2} \simeq a_k, \quad \Rightarrow \quad z(x) \sim \frac{1}{1-x^2}, \quad \text{for large } k, \qquad (8.40)$$

which would imply that the solution blows up for some value of m. Therefore, for a physical solution to exist, the series must terminate and we must have

$$\begin{aligned}
&(k+|m|)(k+|m|+1) - \lambda = 0, \\
\text{or,} \quad &\lambda = \ell(\ell+1), \quad \text{where} \quad \ell = k + |m|. \qquad (8.41)
\end{aligned}$$

We recognize here that since both k and m are integers, ℓ also takes integer values. Furthermore, k and $|m|$ are both positive so that,

$$\ell = 0, 1, 2, 3, \cdots \qquad (8.42)$$

and for each value of ℓ, the integer m takes $(2\ell+1)$ values between

$$-\ell \leq m \leq \ell. \qquad (8.43)$$

Thus, we determine that the eigenvalues of L^2 are

$$\hbar^2\lambda = \hbar^2\ell(\ell+1), \quad \text{where} \quad \ell = 0, 1, 2, 3, \cdots \qquad (8.44)$$

and the eigenvalues of L_z are $\hbar m$ where m is an integer lying between $-\ell \leq m \leq \ell$. Note also that if $k = \ell - |m|$ is even, the solution in (8.37) contains only even powers of x. However, if k is odd, then, only the odd terms in the series survive. This is again similar to the solution of the harmonic oscillator.

The power series solution now depends on two quantum numbers ℓ and m and is denoted by

$$z(x) = z_{\ell,m}(x).$$

This is a polynomial of order $\ell - |m|$ and, correspondingly, the θ-solution in (8.34),

$$\Theta_{\ell,m}(x) = (1 - x^2)^{\frac{|m|}{2}} z_{\ell,m}(x), \tag{8.45}$$

is a polynomial of order ℓ and satisfies the equation

$$(1 - x^2)\,\frac{\mathrm{d}^2\Theta_{\ell,m}}{\mathrm{d}x^2} - 2x\frac{\mathrm{d}\Theta_{\ell,m}}{\mathrm{d}x} + \left(\ell(\ell+1) - \frac{m^2}{(1 - x^2)}\right)\Theta_{\ell,m} = 0. \tag{8.46}$$

For $m = 0$, this equation is known as the Legendre equation and the solutions $P_\ell(x)$ of the equation

$$(1 - x^2)\frac{\mathrm{d}^2 P_\ell(x)}{\mathrm{d}x^2} - 2x\frac{\mathrm{d}P_\ell(x)}{\mathrm{d}x} + \ell(\ell+1)P_\ell(x) = 0, \tag{8.47}$$

are known as the Legendre polynomials which are polynomials of order ℓ. The $\Theta_{\ell,m}$'s are related to the Legendre polynomials as

$$\Theta_{\ell,m}(x) = (1 - x^2)^{\frac{|m|}{2}}\,\frac{\mathrm{d}^{|m|}P_\ell(x)}{\mathrm{d}x^{|m|}} = P_{\ell,m}(x), \tag{8.48}$$

for $\ell \geq |m|$ and are known as the associated Legendre polynomials.

The Legendre polynomials have a closed form expression given by

$$P_\ell(x) = \frac{1}{2^\ell \ell!}\,\frac{\mathrm{d}^\ell (x^2 - 1)^\ell}{\mathrm{d}x^\ell}, \tag{8.49}$$

which is known as the Rodrigues' formula. It explicitly shows that the Legendre polynomials are ℓth order polynomials and that the lower order polynomials have the explicit forms

$$\begin{aligned} P_0(x) &= 1, \\ P_1(x) &= x, \\ P_2(x) &= \frac{1}{2}(3x^2 - 1). \end{aligned} \tag{8.50}$$

The associated Legendre polynomials, similarly, have a closed form expression given by

$$P_{\ell,m}(x) = \frac{1}{2^\ell \ell!}\,(1 - x^2)^{\frac{|m|}{2}}\,\frac{\mathrm{d}^{\ell+|m|}(x^2 - 1)^\ell}{\mathrm{d}x^{\ell+|m|}}, \tag{8.51}$$

which are, clearly, also polynomials of order ℓ.

8.2 Generating function for Legendre polynomials

Let us consider the function of two variables

$$T(x,s) = (1-2xs+s^2)^{-\frac{1}{2}} = \sum_{\ell=0}^{\infty} P_\ell(x) s^\ell. \tag{8.52}$$

It can be shown that the coefficient functions, $P_\ell(x)$'s, are the Legendre polynomials of order ℓ. To see this, let us note that

$$\begin{aligned}
\frac{\partial T(x,s)}{\partial x} &= -\frac{1}{2}\frac{1}{(1-2xs+s^2)^{\frac{3}{2}}}(-2s) \\
&= \frac{s}{(1-2xs+s^2)^{\frac{3}{2}}} = \frac{s}{1-2xs+s^2}T(x,s), \\
\frac{\partial T(x,s)}{\partial s} &= -\frac{1}{2}\frac{1}{(1-2xs+s^2)^{\frac{3}{2}}}(-2x+2s) \\
&= \frac{(x-s)}{(1-2xs+s^2)}T(x,s).
\end{aligned} \tag{8.53}$$

Thus, we have

$$(x-s)T(x,s) = (1-2xs+s^2)\,\frac{\partial T(x,s)}{\partial s}, \tag{8.54}$$

$$s\frac{\partial T(x,s)}{\partial s} = (x-s)\frac{\partial T(x,s)}{\partial x}. \tag{8.55}$$

If we substitute the polynomial form of $T(x,s)$ in (8.52) into (8.54), we obtain

$$\begin{aligned}
(x-s)\sum_{\ell=0}^{\infty} P_\ell s^\ell &= (1-2xs+s^2)\sum_{\ell=1}^{\infty} \ell P_\ell s^{\ell-1} \\
&= (1-2xs+s^2)\sum_{\ell=0}^{\infty} \ell P_\ell s^{\ell-1},
\end{aligned} \tag{8.56}$$

which we can also write as

$$\sum_{\ell=0}^{\infty}\left(xP_\ell s^\ell - P_\ell s^{\ell+1} - \ell P_\ell s^{\ell-1} + 2\ell x P_\ell s^\ell - \ell P_\ell s^{\ell+1}\right) = 0. \tag{8.57}$$

Comparing the coefficients of s^{n-1} in (8.57), we obtain

$$xP_{n-1} - P_{n-2} - nP_n + 2(n-1)xP_{n-1} - (n-2)P_{n-2} = 0,$$

$$\text{or,} \quad nP_n - (2n-1)xP_{n-1} + (n-1)P_{n-2} = 0. \tag{8.58}$$

Similarly, putting the polynomial expansion, (8.52), into (8.55), we have

$$s\sum_{\ell=1}^{\infty} \ell P_\ell s^{\ell-1} = (x-s)\sum_{\ell=0}^{\infty} P'_\ell(x)s^\ell,$$

$$\text{or,} \quad \sum_{\ell=0}^{\infty} \ell P_\ell s^\ell = (x-s)\sum_{\ell=0}^{\infty} P'_\ell(x)s^\ell = x\sum_{\ell=0}^{\infty} P'_\ell s^\ell - \sum_{\ell=0}^{\infty} P'_\ell s^{\ell+1}. \tag{8.59}$$

Again comparing the coefficient of s^{n-1} in (8.59), we obtain

$$x\frac{\mathrm{d}P_{n-1}}{\mathrm{d}x} - \frac{\mathrm{d}P_{n-2}}{\mathrm{d}x} = (n-1)P_{n-1},$$

$$\text{or,} \quad x\frac{\mathrm{d}P_n}{\mathrm{d}x} - \frac{\mathrm{d}P_{n-1}}{\mathrm{d}x} = nP_n,$$

$$\text{or,} \quad xP'_n - P'_{n-1} = nP_n. \tag{8.60}$$

Let us next differentiate (8.58) with respect to x, which gives

$$nP'_n - (2n-1)P_{n-1} - (2n-1)xP'_{n-1} + (n-1)P'_{n-2} = 0,$$

$$\text{or,} \quad nP'_n - (2n-1)P_{n-1} - (2n-1)xP'_{n-1}$$
$$+ (n-1)xP'_{n-1} - (n-1)^2P_{n-1} = 0,$$

$$\text{or,} \quad nP'_n - nxP'_{n-1} - n^2P_{n-1} = 0,$$

$$\text{or,} \quad P'_n - xP'_{n-1} - nP_{n-1} = 0, \tag{8.61}$$

where we have used (8.60) in the intermediate steps. Eliminating further P'_{n-1} from (8.61) using (8.60), we have

$$P'_n - x(xP'_n - nP_n) - nP_{n-1} = 0,$$

$$\text{or,} \quad (1-x)^2P'_n + nxP_n - nP_{n-1} = 0. \tag{8.62}$$

Furthermore, differentiating (8.62) with respect to x, we obtain

$$(1-x^2)P''_n - 2xP'_n + nP_n + nxP'_n - nP'_{n-1} = 0,$$

$$\text{or,} \quad (1-x^2)P''_n - 2xP'_n + nP_n + n^2P_n = 0,$$

$$\text{or,} \quad (1-x^2)P''_n - 2xP'_n + n(n+1)P_n = 0, \tag{8.63}$$

where we have used (8.60). This shows that $P_n(x)$'s satisfy the nth order Legendre equation, (8.47), which proves that the function

$$T(x,s) = (1 - 2xs + s^2)^{-\frac{1}{2}} = \sum_{\ell=0}^{\infty} P_\ell(x) s^\ell, \tag{8.64}$$

generates Legendre polynomials. From the relation (8.51), we see that the generating function for the associated Legendre polynomials are given by

$$\begin{aligned} T_m(x,s) &= (1-x^2)^{\frac{|m|}{2}} \frac{\mathrm{d}^{|m|} T(x,s)}{\mathrm{d}x^{|m|}} \\ &= \frac{(2|m|)!(1-x^2)^{\frac{|m|}{2}} s^{|m|}}{2^{|m|}(|m|)!(1-2xs+s^2)^{|m|+\frac{1}{2}}} \\ &= \sum_{\ell=|m|}^{\infty} P_{\ell,m}(x) s^\ell. \end{aligned} \tag{8.65}$$

We can now write the complete angular solution (8.16) of the Schrödinger equation for a spherically symmetric potential as

$$F_{\ell,m}(\theta,\phi) = Y_{\ell,m}(\theta,\phi) = \frac{N_{\ell,m}}{\sqrt{2\pi}} P_{\ell,m}(\cos\theta) e^{im\phi}. \tag{8.66}$$

Here $N_{\ell,m}$'s are normalization constants and $Y_{\ell,m}$'s are called the spherical harmonics. Let us note that we can work out the orthogonality relations for the spherical harmonics in the following way. First of all, we note that

$$\begin{aligned} &\int \sin\theta \mathrm{d}\theta \mathrm{d}\phi \; Y^*_{\ell',m'}(\theta,\phi) \; Y_{\ell,m}(\theta,\phi) \\ &= \frac{N^*_{\ell',m'} N_{\ell,m}}{2\pi} \int \mathrm{d}(\cos\theta) \mathrm{d}\phi \; P_{\ell',m'} P_{\ell,m} e^{-i(m'-m)\phi} \\ &= N^*_{\ell',m'} N_{\ell,m} \delta_{m',m} \int \mathrm{d}(\cos\theta) \; P_{\ell',m'} P_{\ell,m}. \end{aligned} \tag{8.67}$$

Thus the integral in (8.67) vanishes unless the m quantum numbers are equal. Furthermore, it can also be established that even when the m quantum numbers are the same, the integral vanishes unless

$\ell = \ell'$. To see this, let us recall that $P_{\ell,m}$ satisfies the equation

$$(1-x^2)\frac{\mathrm{d}^2 P_{\ell,m}}{\mathrm{d}x^2} - 2x\frac{\mathrm{d}P_{\ell,m}}{\mathrm{d}x} + \left(\ell(\ell+1) - \frac{m^2}{1-x^2}\right)P_{\ell,m} = 0,$$

$$\text{or, } \frac{\mathrm{d}}{\mathrm{d}x}\left((1-x^2)\frac{\mathrm{d}P_{\ell,m}}{\mathrm{d}x}\right) + \left(\ell(\ell+1) - \frac{m^2}{1-x^2}\right)P_{\ell,m} = 0. \tag{8.68}$$

Multiplying (8.68) with $P_{\ell',m}$ and integrating over x, we obtain

$$\int_{-1}^{1} \mathrm{d}x \left[P_{\ell',m}(x)\frac{\mathrm{d}}{\mathrm{d}x}\left((1-x^2)\frac{\mathrm{d}P_{\ell,m}}{\mathrm{d}x}\right) + \left(\ell(\ell+1) - \frac{m^2}{1-x^2}\right)P_{\ell,m}P_{\ell',m}\right] = 0. \tag{8.69}$$

The first term can be integrated by parts and written as

$$(1-x^2)P_{\ell',m}\frac{\mathrm{d}P_{\ell,m}}{\mathrm{d}x}\bigg|_{-1}^{+1} - \int \mathrm{d}x\ (1-x^2)\frac{\mathrm{d}P_{\ell',m}}{\mathrm{d}x}\frac{\mathrm{d}P_{\ell,m}}{\mathrm{d}x}$$
$$= -\int \mathrm{d}x\ (1-x^2)\frac{\mathrm{d}P_{\ell',m}}{\mathrm{d}x}\frac{\mathrm{d}P_{\ell,m}}{\mathrm{d}x}. \tag{8.70}$$

Relation (8.69), therefore, becomes

$$\int \mathrm{d}x \left[-(1-x^2)\frac{\mathrm{d}P_{\ell',m}}{\mathrm{d}x}\frac{\mathrm{d}P_{\ell,m}}{\mathrm{d}x} + \left(\ell(\ell+1) - \frac{m^2}{1-x^2}\right)P_{\ell',m}P_{\ell,m}\right] = 0. \tag{8.71}$$

Similarly, had we started out with the equation for $P_{\ell',m}$ and multiplied by $P_{\ell,m}$, we would have obtained

$$\int \mathrm{d}x \left[-(1-x^2)\frac{\mathrm{d}P_{\ell,m}}{\mathrm{d}x}\frac{\mathrm{d}P_{\ell',m}}{\mathrm{d}x} + \left(\ell'(\ell'+1) - \frac{m^2}{1-x^2}\right)P_{\ell,m}P_{\ell',m}\right] = 0. \tag{8.72}$$

Subtracting the two relations in (8.71) and (8.72), we have

$$\left(\ell(\ell+1) - \ell'(\ell'+1)\right)\int \mathrm{d}x\ P_{\ell',m}P_{\ell,m} = 0. \tag{8.73}$$

However, since ℓ and ℓ' are arbitrary and different, the only way this can be true is if

$$\int \mathrm{d}x\ P_{\ell',m}P_{\ell,m} = 0, \quad \text{for} \quad \ell' \neq \ell. \tag{8.74}$$

Thus, the only nonzero contribution to the normalization in (8.67) comes from

$$\int \mathrm{d}x\ P_{\ell,m}(x)P_{\ell,m}(x), \tag{8.75}$$

and to calculate this, we note from (8.51) (we assume $m > 0$ although we will use $|m|$ keeping the general case in mind) that

$$P_{\ell,m} = (1-x^2)^{\frac{|m|}{2}}\,\frac{\mathrm{d}^{|m|}P_\ell(x)}{\mathrm{d}x^{|m|}}, \tag{8.76}$$

so that

$$\begin{aligned}\frac{\mathrm{d}P_{\ell,m}}{\mathrm{d}x} &= (1-x^2)^{\frac{|m|}{2}}\,\frac{\mathrm{d}^{|m|+1}}{\mathrm{d}x^{|m|+1}}P_\ell(x)\\ &\qquad - |m|x(1-x^2)^{\frac{|m|}{2}-1}\frac{\mathrm{d}^{|m|}P_\ell(x)}{\mathrm{d}x^{|m|}}\\ &= \frac{P_{\ell,m+1}(x)}{(1-x^2)^{\frac{1}{2}}} - |m|x(1-x^2)^{-1}P_{\ell,m}.\end{aligned} \tag{8.77}$$

This determines

$$P_{\ell,m+1}(x) = (1-x^2)^{\frac{1}{2}}\frac{\mathrm{d}P_{\ell,m}}{\mathrm{d}x} + |m|x(1-x^2)^{-\frac{1}{2}}P_{\ell,m}(x). \tag{8.78}$$

Squaring both sides of (8.78) and integrating, we obtain

$$\begin{aligned}&\int_{-1}^{1}\mathrm{d}x\,(P_{\ell,m+1}(x))^2\\ &= \int \mathrm{d}x\left[(1-x^2)\left(\frac{\mathrm{d}P_{\ell,m}}{\mathrm{d}x}\right)^2 + 2|m|x\frac{\mathrm{d}P_{\ell,m}}{\mathrm{d}x}P_{\ell,m}\right.\\ &\qquad\qquad \left. + \frac{|m|^2x^2}{1-x^2}(P_{\ell,m})^2\right]\\ &= \int \mathrm{d}x\ \left[(1-x^2)\left(\frac{\mathrm{d}P_{\ell,m}}{\mathrm{d}x}\right)^2 + |m|x\frac{\mathrm{d}(P_{\ell,m})^2}{\mathrm{d}x}\right.\end{aligned}$$

$$
\begin{aligned}
&\qquad\qquad\qquad\qquad\qquad\qquad + \frac{|m|^2 x^2}{1-x^2}(P_{\ell,m})^2\Big] \\
&= (1-x^2)\frac{\mathrm{d}P_{\ell,m}}{\mathrm{d}x}P_{\ell,m}\Big|_{-1}^{+1} - \int \mathrm{d}x\ P_{\ell,m}\frac{\mathrm{d}}{\mathrm{d}x}\left[(1-x^2)\frac{\mathrm{d}P_{\ell,m}}{\mathrm{d}x}\right] \\
&\quad + |m|x\,(P_{\ell,m}(x))^2\Big|_{-1}^{+1} - |m|\int \mathrm{d}x\ (P_{\ell,m})^2 \\
&\qquad\qquad\qquad\qquad\qquad + \int \mathrm{d}x\frac{|m|^2x^2}{1-x^2}(P_{\ell,m})^2. \qquad (8.79)
\end{aligned}
$$

The first and the third terms in (8.79) give zero, since $P_{\ell,m}$ as well as $(1-x^2)$ vanish at $x=\pm 1$. Furthermore, we can simplify the second term by using the equation for $P_{\ell,m}(x)$, (8.46),

$$
\frac{\mathrm{d}}{\mathrm{d}x}\left((1-x^2)\ \frac{\mathrm{d}P_{\ell,m}}{\mathrm{d}x}\right) = -\left(\ell(\ell+1) - \frac{|m|^2}{1-x^2}\right)P_{\ell,m}, \qquad (8.80)
$$

to obtain

$$
\begin{aligned}
&\int \mathrm{d}x\,(P_{\ell,m+1}(x))^2 \\
&\quad = \int \mathrm{d}x\left(\ell(\ell+1) - \frac{|m|^2}{1-x^2}\right)(P_{\ell,m})^2 \\
&\qquad + \int \mathrm{d}x\left(|m| + \frac{|m|^2x^2}{1-x^2}\right)(P_{\ell,m})^2 \\
&\quad = \int \mathrm{d}x\,\left(\ell(\ell+1) - |m|^2 - |m|\right)(P_{\ell,m}(x))^2 \\
&\quad = \left(\ell(\ell+1) - |m|(|m|+1)\right)\int \mathrm{d}x\ (P_{\ell,m}(x))^2. \qquad (8.81)
\end{aligned}
$$

We can apply the relation in (8.81) repeatedly, which leads to

$$
\begin{aligned}
&\int \mathrm{d}x(P_{\ell,m}(x))^2 \\
&\quad = (\ell - |m| + 1)(\ell + |m|)\int \mathrm{d}x\ (P_{\ell,m-1}(x))^2 \\
&\quad = (\ell - |m| + 1)(\ell + |m|)(\ell - |m| + 2)(\ell + |m| - 1) \\
&\qquad \times \int \mathrm{d}x\ (P_{\ell,m-2}(x))^2 \\
&\quad = (\ell + |m|)(\ell + |m| - 1)\ldots(\ell + 1)
\end{aligned}
$$

$$\times(\ell - |m| + 1)(\ell - |m| + 2) \ldots \ell \int \mathrm{d}x \, (P_\ell(x))^2$$

$$= \frac{(\ell + |m|)!}{\ell!} \times \frac{\ell!}{(\ell - |m|)!} \times \frac{2}{2\ell + 1}$$

$$= \frac{2}{2\ell + 1} \frac{(\ell + |m|)!}{(\ell - |m|)!}. \tag{8.82}$$

Thus, our normalization condition, (8.67), now becomes

$$\int \sin\theta \mathrm{d}\theta \mathrm{d}\phi \; Y^*_{\ell,m} Y_{\ell,m} = 1,$$

$$\text{or,} \quad |N_{\ell,m}|^2 \int \mathrm{d}x (P_{\ell,m}(x))^2 = 1,$$

$$\text{or,} \quad |N_{\ell,m}|^2 \frac{2}{2\ell + 1} \frac{(\ell + |m|)!}{(\ell - |m|)!} = 1, \tag{8.83}$$

which determines the normalization constant to be

$$N_{\ell,m} = N^*_{\ell,m} = \pm \sqrt{\frac{2\ell + 1}{2} \frac{(\ell - |m|)!}{(\ell + |m|)!}}. \tag{8.84}$$

Conventionally, the phase of the spherical harmonics is chosen to be $(-1)^m$ for $m > 0$ and $+1$ for $m \leq 0$.

Therefore, we can write the normalized angular solutions (which are the spherical harmonics) as

$$Y_{\ell,m}(\theta, \phi) = \epsilon \sqrt{\frac{2\ell + 1}{4\pi} \frac{(\ell - |m|)!}{(\ell + |m|)!}} \; P_{\ell,m}(\cos\theta) e^{im\phi}, \tag{8.85}$$

with the phase given by

$$\epsilon = \begin{cases} (-1)^m, & \text{for } m > 0, \\ 1, & \text{for } m \leq 0, \end{cases} \tag{8.86}$$

which we can also write equivalently as $\epsilon = (-1)^{\frac{m+|m|}{2}}$.

The complete solution, (8.11), of the Schrödinger equation is given by

$$\psi_{\ell,m}(r, \theta, \phi) = R(r) Y_{\ell,m}(\theta, \phi), \tag{8.87}$$

where the radial part of the solution is determined from the dynamics of the system. The equation satisfied by the radial part, (8.13), is given by (we represent the mass of the particle as μ to avoid any confusion with the m quantum number)

$$\frac{\mathrm{d}}{\mathrm{d}r}\left(r^2\,\frac{\mathrm{d}R}{\mathrm{d}r}\right)+\frac{2\mu r^2}{\hbar^2}\left(E-V(r)-\frac{\hbar^2\ell(\ell+1)}{2\mu r^2}\right)R=0. \tag{8.88}$$

Thus, we see that a nonzero angular momentum implies the presence of an additional term in the potential. Furthermore, if we differentiate this new potential and calculate the force, we find that it pushes the particle away from the center of the coordinate system and lies along the radial direction. In other words, a nonzero angular momentum gives rise to a centrifugal force (barrier) which is very strong at short distances. In three dimensions, the presence of this repulsive potential prevents the existence of bound states in many cases.

Furthermore, note that the radial solution depends on the quantum number ℓ and the energy eigenvalue. Thus, we can write

$$R(r)=R_{E,\ell}(r). \tag{8.89}$$

However, it does not depend on the azimuthal quantum number m. Of course, there are $(2\ell+1)$ states with the same ℓ value but with different m values. All these states will have the same energy and, therefore, such systems have a $(2\ell+1)$-fold degeneracy as a result of rotational symmetry.

8.3 Parity of spherical harmonics

We can now study the question of parity in the three dimensional case. First of all, parity implies reflecting a vector through the origin (see Fig. 8.2). Thus, in spherical coordinates, the coordinate vector transforms under parity as

$$(r,\theta,\phi)\xrightarrow{\text{Parity}}(r,\pi-\theta,\pi+\phi). \tag{8.90}$$

Since the radial coordinate does not change under reflection, only the angular part of the solution would be effected by such a transformation as

$$Y_{\ell,m}(\theta,\phi)\xrightarrow{\text{Parity}}Y_{\ell,m}(\pi-\theta,\pi+\phi). \tag{8.91}$$

We note that

$$\begin{aligned}&e^{im\phi}\xrightarrow{\text{Parity}}e^{im(\pi+\phi)}=(-1)^{|m|}\,e^{im\phi},\\&x=\cos\theta\xrightarrow{\text{Parity}}\cos(\pi-\theta)=-\cos\theta=-x.\end{aligned} \tag{8.92}$$

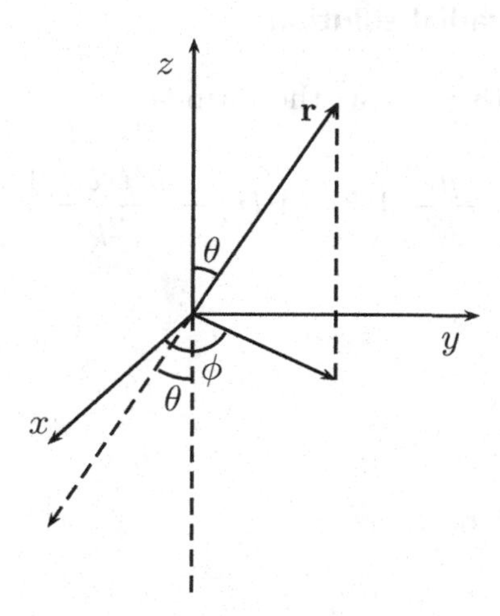

Figure 8.2: Reflection of a vector through the origin.

As a result, we have,

$$\begin{aligned} P_{\ell,m}(x) &\propto (1-x^2)^{\frac{|m|}{2}} \frac{\mathrm{d}^{\ell+|m|}(x^2-1)^\ell}{\mathrm{d}x^{\ell+|m|}} \\ &\xrightarrow{\text{Parity}} (-1)^{\ell+|m|}(1-x^2)^{\frac{|m|}{2}} \frac{\mathrm{d}^{\ell+|m|}(x^2-1)^\ell}{\mathrm{d}x^{\ell+|m|}} \\ &= (-1)^{\ell+|m|}\, P_{\ell,m}(x). \end{aligned} \tag{8.93}$$

Together with (8.92) and (8.93), we obtain

$$\begin{aligned} Y_{\ell,m}(\theta,\phi) &\xrightarrow{\text{Parity}} (-1)^{|m|}(-1)^{\ell+|m|}\, Y_{\ell,m}(\theta,\phi) \\ &= (-1)^\ell\, Y_{\ell,m}(\theta,\phi). \end{aligned} \tag{8.94}$$

In other words, we see that the angular solutions (spherical harmonics) are eigenstates of angular momentum with definite parity eigenvalues. The parity of these states is completely determined by the ℓ-quantum number. It is worth pointing out that all the $(2\ell+1)$ states with different m-quantum numbers have the same parity given by $(-1)^\ell$.

8.4 Behavior of the radial solution

The radial equation, (8.88), has the form

$$\frac{\mathrm{d}}{\mathrm{d}r}\left(r^2\frac{\mathrm{d}R_{E,\ell}}{\mathrm{d}r}\right)+\frac{2\mu r^2}{\hbar^2}\left(E-V(r)-\frac{\hbar^2\ell(\ell+1)}{2\mu r^2}\right)R_{E,\ell}=0. \tag{8.95}$$

Let us now define

$$R_{E,\ell}(r)=\frac{u_{E,\ell}(r)}{r}. \tag{8.96}$$

Then, equation (8.95) becomes

$$\frac{\mathrm{d}}{\mathrm{d}r}\left(r\frac{\mathrm{d}u_{E,\ell}}{\mathrm{d}r}-u_{E,\ell}\right)$$

$$+\frac{2\mu r}{\hbar^2}\left(E-V(r)-\frac{\hbar^2\ell(\ell+1)}{2\mu r^2}\right)u_{E,\ell}(r)=0,$$

$$\text{or,}\quad r\frac{\mathrm{d}^2u_{E,\ell}}{\mathrm{d}r^2}+\frac{2\mu r}{\hbar^2}\left(E-V(r)-\frac{\hbar^2(\ell(\ell+1))}{2\mu r^2}\right)u_{E,\ell}(r)=0,$$

$$\text{or,}\quad \frac{\mathrm{d}^2u_{E,\ell}}{\mathrm{d}r^2}+\frac{2\mu}{\hbar^2}\left(E-V(r)-\frac{\hbar^2\ell(\ell+1)}{2\mu r^2}\right)u_{E,\ell}(r)=0. \tag{8.97}$$

Therefore, we see that, in terms of this new function $u_{E,\ell}(r)$, the equation is the same as the one dimensional Schrödinger equation, but in the presence of an effective potential

$$V_{\text{eff}}(r)=V(r)+\frac{\hbar^2\ell(\ell+1)}{2\mu r^2}. \tag{8.98}$$

There are, however, two important differences. First of all, the radial coordinate takes only nonnegative values,

$$0\leq r<\infty, \tag{8.99}$$

which is to be contrasted with $-\infty<x<\infty$. Furthermore, even though we have the same boundary conditions at $r\to\infty$, namely, for bound states,

$$u(r)\xrightarrow{r\to\infty}e^{-\alpha r},\quad \alpha\text{ real and positive}, \tag{8.100}$$

and, for free particle states or scattering states,

$$u(r)\xrightarrow{r\to\infty}e^{\pm ikr},\quad k\text{ real}, \tag{8.101}$$

there is now a boundary condition to be satisfied at the origin $r = 0$. The normalizability of the radial solution requires

$$\int_0^\infty \mathrm{d}r\ r^2 |R_{E,\ell}(r)|^2 = \text{ finite},$$

$$\text{or,} \quad \int_0^\infty \mathrm{d}r\ r^2 \frac{|u_{E,\ell}(r)|^2}{r^2} = \text{ finite},$$

$$\text{or,} \quad \int_0^\infty \mathrm{d}r\ |u_{E,\ell}(r)|^2 = \text{ finite}. \tag{8.102}$$

Equation (8.102), in principle, allows that

$$u_{E,\ell}(r) \xrightarrow{r\to 0} c, \qquad c = \text{ constant}. \tag{8.103}$$

However, if $c \neq 0$, then, we conclude that near the origin the radial solution will have the form

$$R_{E,\ell}(r) \xrightarrow{r\to 0} \frac{c}{r}, \tag{8.104}$$

and, consequently, will satisfy

$$\boldsymbol{\nabla}^2 R_{E,\ell}(r) = \boldsymbol{\nabla}^2 \left(\frac{c}{r}\right) = -4\pi c\ \delta^3(\mathbf{r}). \tag{8.105}$$

On the other hand, if the potential is a smooth function, this would be hard to satisfy. Thus, for all smooth potentials we have to choose

$$c = 0. \tag{8.106}$$

which determines the boundary condition at the origin to be

$$u_{E,\ell}(r) \xrightarrow{r\to 0} 0. \tag{8.107}$$

Once we keep these two distinctions in mind, we can solve the radial equation, just like the one dimensional Schrödinger equation.

8.5 3-dimensional isotropic oscillator

Let us analyze the three dimensional isotropic oscillator again, but now in spherical coordinates. We know that the Hamiltonian for the system has the form,

$$H = \frac{\mathbf{P}^2}{2\mu} + \frac{1}{2}\mu\omega^2\mathbf{X}^2 = \frac{\mathbf{P}^2}{2\mu} + \frac{1}{2}\mu\omega^2 R^2, \tag{8.108}$$

where the mass is written as μ to distinguish it from the azimuthal quantum number m. Since this is a spherically symmetric system, we can use our earlier separation of variables in the spherical coordinates to write the complete wave function as

$$\psi_{E,\ell,m}(r,\theta,\phi) = R_{E,\ell}(r) Y_{\ell,m}(\theta,\phi). \tag{8.109}$$

Furthermore, defining

$$R_{E,\ell}(r) = \frac{u_{E,\ell}(r)}{r}, \tag{8.110}$$

we recognize that the new function will satisfy the equation

$$\frac{\mathrm{d}^2 u_{E,\ell}(r)}{\mathrm{d}r^2} + \frac{2\mu}{\hbar^2}\left(E - \frac{1}{2}\mu\omega^2 r^2 - \frac{\hbar^2 \ell(\ell+1)}{2\mu r^2}\right) u_{E,\ell}(r) = 0. \tag{8.111}$$

Let us first determine the asymptotic behavior of the solutions. We note that as $r \to \infty$, the equation takes the form

$$\begin{aligned}
&\frac{\mathrm{d}^2 u_{E,\ell}(r)}{\mathrm{d}r^2} - \frac{\mu^2\omega^2 r^2}{\hbar^2} u_{E,\ell}(r) = 0,\\
\text{or,}\quad & u_{E,\ell}(r) \sim e^{-\frac{1}{2}\frac{\mu\omega r^2}{\hbar}} = e^{-\frac{1}{2}y^2},
\end{aligned} \tag{8.112}$$

where we have defined a dimensionless variable

$$y = \left(\frac{\mu\omega}{\hbar}\right)^{\frac{1}{2}} r. \tag{8.113}$$

Equation (8.111), written in terms of this dimensionless variable, takes the form

$$\begin{aligned}
&\left(\frac{\mu\omega}{\hbar}\right)\frac{\mathrm{d}^2 u_{E,\ell}}{\mathrm{d}y^2} + \frac{2\mu}{\hbar^2}\left(E - \frac{1}{2}\mu\omega^2\frac{\hbar y^2}{\mu\omega} - \frac{\hbar^2\ell(\ell+1)}{2\mu\times\frac{\hbar y^2}{\mu\omega}}\right) u_{E,\ell} = 0,\\
\text{or,}\quad & \frac{\mathrm{d}^2 u_{E,\ell}}{\mathrm{d}y^2} + \frac{2}{\hbar\omega}\left(E - \frac{1}{2}\hbar\omega y^2 - \frac{\hbar\omega\ell(\ell+1)}{2y^2}\right) u_{E,\ell} = 0,\\
\text{or,}\quad & \frac{\mathrm{d}^2 u_{E,\ell}}{\mathrm{d}y^2} + \left(\lambda - y^2 - \frac{\ell(\ell+1)}{y^2}\right) u_{E,\ell} = 0,
\end{aligned} \tag{8.114}$$

where

$$\lambda = \frac{2E}{\hbar\omega}, \tag{8.115}$$

is also dimensionless. For $y \to 0$, equation (8.114) becomes

$$\frac{d^2 u_{E,\ell}}{dy^2} - \frac{\ell(\ell+1)}{y^2}\, u_{E,\ell} = 0. \tag{8.116}$$

The two solutions of (8.116) are easily seen to be

$$u_{E,\ell}(y) \xrightarrow{y\to 0} y^{\ell+1}, \quad \text{or}, \quad y^{-\ell}. \tag{8.117}$$

However, the boundary condition at the origin (see (8.107)), namely, $u_{E,\ell}(y) \xrightarrow{y\to 0} 0$, selects

$$u_{E,\ell}(y) \xrightarrow{y\to 0} y^{\ell+1}, \tag{8.118}$$

as the physical solution.

Equations (8.112) and (8.118) suggest a general solution of the form

$$u_{E,\ell}(y) = e^{-\frac{1}{2}y^2} v(y) = e^{-\frac{1}{2}y^2} \sum_{k=0}^{\infty} a_k y^{k+\ell+1},$$

$$\frac{du_{E,\ell}}{dy} = e^{-\frac{1}{2}y^2} \left[-y v(y) + v'(y)\right],$$

$$\frac{d^2 u_{E,\ell}}{dy^2} = e^{-\frac{1}{2}y^2} \left[(y^2-1) v(y) - 2y v'(y) + v''(y)\right]. \tag{8.119}$$

Putting this back into the equation (8.114), we have

$$e^{-\frac{1}{2}y^2}\Bigg[(y^2-1)v(y) - 2yv'(y) + v''(y) + \left(\lambda - y^2 - \frac{\ell(\ell+1)}{y^2}\right) v(y)\Bigg] = 0,$$

$$\text{or,}\quad v''(y) - 2yv'(y) + \left(\lambda - 1 - \frac{\ell(\ell+1)}{y^2}\right) v(y) = 0. \tag{8.120}$$

Let us next use the power series solution

$$v(y) = \sum_{k=0}^{\infty} a_k y^{k+\ell+1},$$

$$v'(y) = \sum_{k=0}^{\infty} (k+\ell+1) a_k y^{k+\ell},$$

$$v''(y) = \sum_{k=0}^{\infty} (k+\ell+1)(k+\ell) a_k y^{k+\ell-1}. \tag{8.121}$$

With the relations in (8.121), equation (8.120) now becomes

$$\sum_{k=0}^{\infty}\Big[(k+\ell+1)(k+\ell)a_k y^{k+\ell-1} - 2(k+\ell+1)a_k y^{k+\ell+1} + (\lambda-1)a_k y^{k+\ell+1} - \ell(\ell+1)a_k y^{k+\ell-1}\Big] = 0,$$

$$\text{or,}\quad \sum_{k=0}^{\infty}\Big[\left((k+\ell+1)(k+\ell) - \ell(\ell+1)\right) a_k y^{k+\ell-1} + (\lambda - 2k - 2\ell - 3)a_k y^{k+\ell+1}\Big] = 0. \tag{8.122}$$

Looking at the lowest order term in the series in (8.122), namely, $y^{\ell-1}$, we have

$$[(\ell+1)\ell - \ell(\ell+1)]a_0 = 0,$$

$$\text{or,}\quad a_0 = \text{arbitrary}. \tag{8.123}$$

Looking at the coefficient of the next term in the series in (8.122) (namely, the coefficient of (y^ℓ)), we find that

$$[(1+\ell+1)(1+\ell) - \ell(\ell+1)]a_1 = 0,$$

$$\text{or,}\quad (\ell+1)(\ell+2-\ell)a_1 = 0,$$

$$\text{or,}\quad 2(\ell+1)a_1 = 0,$$

$$\text{or,}\quad a_1 = 0. \tag{8.124}$$

Furthermore, from the structure of the equation in (8.122), it is clear that the coefficients a_{k+2} are related to the coefficients a_k. Since $a_1 = 0$, this, therefore, implies that only even terms of the series survive. Thus changing $k \to 2k$ in (8.122), we have

$$\sum_{k=0}^{\infty}\Big([(2k+\ell+1)(2k+\ell) - \ell(\ell+1)]a_{2k} y^{2k+\ell-1} + (\lambda - 4k - 2\ell - 3)a_{2k} y^{2k+\ell+1}\Big) = 0. \tag{8.125}$$

Requiring the coefficient of $y^{2k+\ell+1}$ to vanish we obtain the recursion relation for the coefficients to be

$$[(2k+\ell+3)(2k+\ell+2)-\ell(\ell+1)]a_{2k+2}$$
$$=(4k+2\ell+3-\lambda)a_{2k},$$

$$\text{or,}\quad a_{2k+2}=\frac{(4k+2\ell+3-\lambda)}{(2k+2)(2k+2\ell+3)}a_{2k}. \tag{8.126}$$

Thus, we see that for large k,

$$\frac{a_{2k+2}}{a_{2k}}\longrightarrow\frac{1}{k}. \tag{8.127}$$

At large orders, this is the same behavior as that of the series in e^{y^2} which would lead to an unphysical solution, since it does not fall off at spatial infinity. Thus, for a physical solution to exist the series must terminate which implies

$$4k+2\ell+3-\lambda=0,$$

$$\text{or,}\quad \lambda=\frac{2E}{\hbar\omega}=(2\ell+4k+3),$$

$$\text{or,}\quad E=\hbar\omega\left(\ell+2k+\frac{3}{2}\right)=\hbar\omega\left(n+\frac{3}{2}\right), \tag{8.128}$$

where we have defined $n=\ell+2k$. Since ℓ and k take only nonnegative integer values, it follows that

$$n=0,1,2,3,\cdots,$$
$$\ell=n-2k=n,n-2,n-4,\cdots,1\,(\text{or }0), \tag{8.129}$$

depending on the value of n.

The solutions of (8.114), in this case, are obtained to be

$$u_{n,\ell}(y)=e^{-\frac{1}{2}y^2}\sum_{p=0}^{\frac{(n-\ell)}{2}}a_{2p}y^{2p+\ell+1}, \tag{8.130}$$

and the first few energy eigenvalues are given by

$$\begin{aligned}
E&=\frac{3}{2}\hbar\omega \qquad n=0 \qquad \ell=0 \qquad m=0,\\
E&=\frac{5}{2}\hbar\omega \qquad n=1 \qquad \ell=1 \qquad m=\pm1,0,\\
E&=\frac{7}{2}\hbar\omega \qquad n=2 \qquad \ell=2,0 \qquad m=\pm2,\pm1,0;0,\\
&\ \vdots
\end{aligned} \tag{8.131}$$

It is interesting to note that not only are states with different m values degenerate, but states with different ℓ values are also degenerate, i.e., they have the same energy. Rotational symmetry does not explain why this degeneracy should arise. This kind of a phenomenon is known as accidental degeneracy and we would see later that this is a consequence of a larger symmetry operative in this system.

However, since for each ℓ value there are $(2\ell + 1)$ values of m that are degenerate and that for each n value, there are the states with

$$\ell = \begin{cases} 0, 2, \cdots, n, & \text{for } n \text{ even}, \\ 1, 3, \cdots, n, & \text{for } n \text{ odd}, \end{cases} \tag{8.132}$$

which are all degenerate, the total number of degeneracy for a given n is obtained as follows.

1. Even n:

$$\begin{aligned} \sum_{\ell=0,2}^{n} (2\ell + 1) &= \sum_{p=0,1}^{\frac{n}{2}} (4p + 1) \\ &= 4 \times \frac{1}{2}\left(\frac{n}{2}\right)\left(\frac{n}{2} + 1\right) + \left(\frac{n}{2} + 1\right) \\ &= \frac{1}{2}(n + 1)(n + 2). \end{aligned} \tag{8.133}$$

2. Odd n:

$$\begin{aligned} &\sum_{\ell=1,3}^{n} (2\ell + 1) \\ &\quad = \sum_{p=0,1}^{\frac{n-1}{2}} (2(2p + 1) + 1) \\ &\quad = \sum_{p=0}^{\frac{n-1}{2}} (4p + 3) = 4 \times \frac{1}{2}\frac{(n-1)}{2} \times \frac{(n+1)}{2} + 3 \times \frac{n+1}{2} \\ &\quad = \frac{(n+1)}{2}(n - 1 + 3) = \frac{1}{2}(n + 1)(n + 2). \end{aligned} \tag{8.134}$$

Thus, we see that, independent of whether n is even or odd, the degeneracy of states is given by

$$\frac{1}{2}(n + 1)(n + 2), \tag{8.135}$$

which is the same result we had derived earlier in the Cartesian co-ordinates.

8.6 Square well potential

Let us now investigate the question of bound states in a three dimensional square well potential. The potential depends only on the radial coordinate and has the form (see Fig. 8.3)

$$V(r) = \begin{cases} 0, & \text{for} \quad r > a, \\ -V_0, & \text{for} \quad r < a, \qquad V_0 > 0. \end{cases} \tag{8.136}$$

We are going to assume that $E < 0$ and that $|E| < V_0$ with $V_0 > 0$ for a bound state solution to exist. (Namely, $T = E - V > 0$ for motion to exist. If $E > 0$, we will have free particle motion for $r > a$. On the other hand, we are looking for bound state solutions for which we must have $E < 0$. From $T = E + V_0 > 0$ in side the well, it follows that $V_0 > -E = |E|$.)

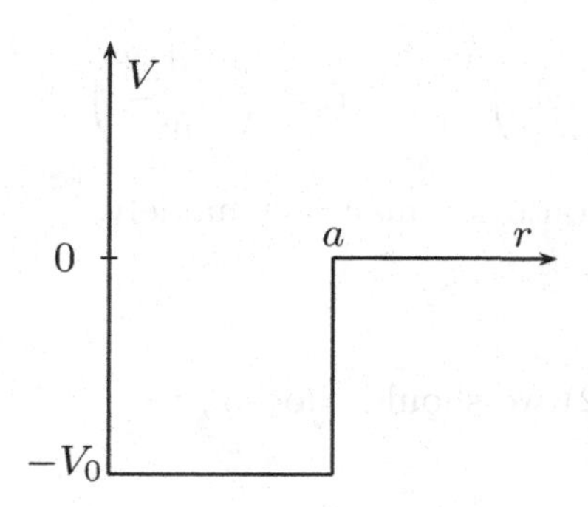

Figure 8.3: Square well potential in three dimensions.

Once again, this is a system with spherical symmetry and, therefore, we can carry out the separation of the angular solution. In the present case, the radial equation in terms of $u_{E,\ell}(r)$ has the form

$$\frac{\mathrm{d}^2 u_{E,\ell}}{\mathrm{d}r^2} + \frac{2\mu}{\hbar^2}\left(E - V(r) - \frac{\hbar^2 \ell(\ell+1)}{2\mu r^2}\right) u_{E,\ell} = 0. \tag{8.137}$$

Thus, for $r < a$, the equation has the form

$$\frac{\mathrm{d}^2 u_{E,\ell}}{\mathrm{d}r^2} + \frac{2\mu}{\hbar^2}\left(E + V_0 - \frac{\hbar^2 \ell(\ell+1)}{2\mu r^2}\right) u_{E,\ell} = 0, \tag{8.138}$$

while, for $r > a$, it has the form

$$\frac{\mathrm{d}^2 u_{E,\ell}}{\mathrm{d}r^2} + \frac{2\mu}{\hbar^2}\left(E - \frac{\hbar^2\ell(\ell+1)}{2\mu r^2}\right) u_{E,\ell} = 0. \tag{8.139}$$

Let us simplify the problem by first looking at the zero angular momentum case. For $\ell = 0$, the equations are given by

$$r < a: \quad \frac{\mathrm{d}^2 u_{E,0}}{\mathrm{d}r^2} + \frac{2\mu(E + V_0)}{\hbar^2} u_{E,0} = 0, \tag{8.140}$$

$$r > a: \quad \frac{\mathrm{d}^2 u_{E,0}}{\mathrm{d}r^2} + \frac{2\mu E}{\hbar^2} u_{E,0} = 0. \tag{8.141}$$

This is identical to the equations for the one dimensional square well potential. Thus, the solutions follow to be

$$\begin{aligned} r < a: &\quad u_{E,0}(r) = A\sin kr + B\cos kr, \\ r > a: &\quad u_{E,0}(r) = Ce^{-\alpha r}, \end{aligned} \tag{8.142}$$

where we have defined

$$k = \left(\frac{2\mu}{\hbar^2}\left(V_0 - |E|\right)\right)^{\frac{1}{2}}, \qquad \alpha = \left(\frac{2\mu|E|}{\hbar^2}\right)^{\frac{1}{2}}. \tag{8.143}$$

The boundary condition at $r = 0$, namely,

$$u_{E,0}(r) \xrightarrow{r\to 0} 0, \tag{8.144}$$

implies that in (8.142) we should choose

$$B = 0.$$

Furthermore, we should match the solution and its derivative at the boundary $r = a$. Matching the solutions at $r = a$, we obtain

$$A\sin ka = Ce^{-\alpha a}. \tag{8.145}$$

Similarly, matching the derivative of the solutions at $r = a$, we find

$$kA\cos ka = -\alpha Ce^{-\alpha a}. \tag{8.146}$$

Thus, dividing the two relations, (8.145) and (8.146), we have

$$\begin{aligned} &ka\cot ka = -\alpha a, \\ \text{or,} \quad &\eta\cot\eta = -\xi, \qquad \eta = ka, \quad \xi = \alpha a. \end{aligned} \tag{8.147}$$

As in the one dimensional case, we have

$$\begin{aligned}
\eta^2 + \xi^2 &= k^2 a^2 + \alpha^2 a^2 = \left(k^2 + \alpha^2\right) a^2 \\
&= \left(\frac{2\mu}{\hbar^2}\left(V_0 - |E|\right) + \frac{2\mu|E|}{\hbar^2}\right) a^2 \\
&= \frac{2\mu V_0 a^2}{\hbar^2}. \qquad (8.148)
\end{aligned}$$

We can carry out the graphical solutions for (8.147) and (8.148) as in the one dimensional case and show that there is no solution if

$$0 \leq V_0 a^2 < \frac{\pi^2 \hbar^2}{8\mu}, \qquad (8.149)$$

and one solution if

$$\frac{\pi^2 \hbar^2}{8\mu} \leq V_0 a^2 < \frac{9\pi^2 \hbar^2}{8\mu}, \qquad (8.150)$$

and so on.

Let us next study the solutions for arbitrary ℓ. We will study the solutions in one region at a time.

Region $r < a$. In this region the equation for arbitrary ℓ is given by (see (8.138))

$$\begin{aligned}
&\frac{\mathrm{d}^2 u_{E,\ell}}{\mathrm{d}r^2} + \frac{2\mu}{\hbar^2}\left(E + V_0 - \frac{\hbar^2 \ell(\ell+1)}{2\mu r^2}\right) u_{E,\ell} = 0, \\
\text{or,}\quad &\frac{\mathrm{d}^2 u_{E,\ell}}{\mathrm{d}r^2} + \left(k^2 - \frac{\ell(\ell+1)}{r^2}\right) u_{E,\ell} = 0. \qquad (8.151)
\end{aligned}$$

Let us now change to the dimensionless variable

$$p = kr. \qquad (8.152)$$

Then, in terms of this variable equation (8.151) becomes

$$\left(\frac{\mathrm{d}^2}{\mathrm{d}p^2} - \frac{\ell(\ell+1)}{p^2}\right) u_{E,\ell} = -u_{E,\ell}. \qquad (8.153)$$

Let us define an operator

$$d_\ell = \frac{\mathrm{d}}{\mathrm{d}p} + \frac{\ell+1}{p}, \qquad (8.154)$$

whose adjoint is given by

$$d_\ell^\dagger = -\frac{\mathrm{d}}{\mathrm{d}p} + \frac{\ell+1}{p}. \qquad (8.155)$$

It follows now that

$$\begin{aligned} d_\ell d_\ell^\dagger &= \left(\frac{\mathrm{d}}{\mathrm{d}p} + \frac{\ell+1}{p}\right)\left(-\frac{\mathrm{d}}{\mathrm{d}p} + \frac{\ell+1}{p}\right) \\ &= -\frac{\mathrm{d}^2}{\mathrm{d}p^2} - \frac{\ell+1}{p^2} + \frac{\ell+1}{p}\frac{\mathrm{d}}{\mathrm{d}p} - \frac{\ell+1}{p}\frac{\mathrm{d}}{\mathrm{d}p} + \frac{(\ell+1)^2}{p^2} \\ &= -\frac{\mathrm{d}^2}{\mathrm{d}p^2} + \frac{\ell(\ell+1)}{p^2}. \end{aligned} \tag{8.156}$$

On the other hand, we note that

$$\begin{aligned} d_\ell^\dagger d_\ell &= \left(-\frac{\mathrm{d}}{\mathrm{d}p} + \frac{(\ell+1)}{p}\right)\left(\frac{\mathrm{d}}{\mathrm{d}p} + \frac{\ell+1}{p}\right) \\ &= -\frac{\mathrm{d}^2}{\mathrm{d}p^2} + \frac{\ell+1}{p^2} - \frac{\ell+1}{p}\frac{\mathrm{d}}{\mathrm{d}p} + \frac{\ell+1}{p}\frac{\mathrm{d}}{\mathrm{d}p} + \frac{(\ell+1)^2}{p^2} \\ &= -\frac{\mathrm{d}^2}{\mathrm{d}p^2} + \frac{(\ell+1)(\ell+2)}{p^2} \\ &= d_{\ell+1}d_{\ell+1}^\dagger. \end{aligned} \tag{8.157}$$

Thus, using the relation (8.156), we can write the dynamical equation

$$\left[\frac{\mathrm{d}^2}{\mathrm{d}p^2} - \frac{\ell(\ell+1)}{p^2}\right] u_{E,\ell}(p) = -u_{E,\ell}(p), \tag{8.158}$$

also as

$$d_\ell d_\ell^\dagger u_{E,\ell}(p) = u_{E,\ell}(p). \tag{8.159}$$

Upon using (8.157), equation (8.159) leads to

$$\begin{aligned} & d_\ell^\dagger\left(d_\ell d_\ell^\dagger u_{E,\ell}(p)\right) = d_\ell^\dagger u_{E,\ell}(p), \\ \text{or,}\quad & d_\ell^\dagger d_\ell\left(d_\ell^\dagger u_{E,\ell}(p)\right) = \left(d_\ell^\dagger u_{E,\ell}(p)\right), \\ \text{or,}\quad & d_{\ell+1}d_{\ell+1}^\dagger\left(d_\ell^\dagger u_{E,\ell}(p)\right) = \left(d_\ell^\dagger u_{E,\ell}(p)\right). \end{aligned} \tag{8.160}$$

From (8.159), we note that $u_{E,\ell}$'s are eigenstates of the operator $d_\ell d_\ell^\dagger$ (corresponding to the value ℓ) with eigenvalue 1. Relation (8.160) implies that $d_\ell^\dagger u_{E,\ell}(p)$ is an eigenstate of $d_{\ell+1}d_{\ell+1}^\dagger$ (corresponding to the value $\ell+1$) with the same eigenvalue 1. Therefore, we must have

$$d_\ell^\dagger u_{E,\ell}(p) = c_\ell u_{E,\ell+1}. \tag{8.161}$$

In other words, $d_\ell^\dagger$'s are like raising operators for ℓ and, for the present, we would omit the normalization constant c_ℓ's, for they can be absorbed into the overall normalization constant of the wave function. Thus, we write

$$u_{E,\ell+1} = d_\ell^\dagger u_{E,\ell}. \tag{8.162}$$

Let us note here parenthetically that this procedure is related to what is known as supersymmetry in quantum mechanics.

The radial function is defined to be (the extra factor of k can be absorbed into the normalization constant)

$$R_{E,\ell} = \frac{u_{E,\ell}}{p}, \tag{8.163}$$

and satisfies

$$\begin{aligned} pR_{E,\ell+1} &= u_{E,\ell+1} = d_\ell^\dagger u_{E,\ell} = d_\ell^\dagger(pR_{E,\ell}) \\ &= \left(-\frac{\mathrm{d}}{\mathrm{d}p} + \frac{\ell+1}{p}\right)(pR_{E,\ell}) \\ &= p\left(-\frac{\mathrm{d}}{\mathrm{d}p} + \frac{\ell}{p}\right)R_{E,\ell}, \end{aligned} \tag{8.164}$$

so that we can write

$$\begin{aligned} R_{E,\ell+1} &= \left(-\frac{\mathrm{d}}{\mathrm{d}p} + \frac{\ell}{p}\right)R_{E,\ell} = p^\ell\left(-\frac{\mathrm{d}}{\mathrm{d}p}\right)\frac{R_{E,\ell}}{p^\ell}, \\ \text{or,}\quad \frac{R_{E,\ell+1}}{p^{\ell+1}} &= \left(-\frac{1}{p}\frac{\mathrm{d}}{\mathrm{d}p}\right)\left(\frac{R_{E,\ell}}{p^\ell}\right) \\ &= \left(-\frac{1}{p}\frac{\mathrm{d}}{\mathrm{d}p}\right)^2\left(\frac{R_{E,\ell-1}}{p^{\ell-1}}\right) \\ &= \quad \cdots\cdots \\ &= \left(-\frac{1}{p}\frac{\mathrm{d}}{\mathrm{d}p}\right)^{\ell+1}\left(\frac{R_{E,0}}{p^0}\right) \\ &= \left(-\frac{1}{p}\frac{\mathrm{d}}{\mathrm{d}p}\right)^{\ell+1} R_{E,0}. \end{aligned} \tag{8.165}$$

Thus, we have the recursion relation

$$R_{E,\ell} = p^\ell\left(-\frac{1}{p}\frac{\mathrm{d}}{\mathrm{d}p}\right)^\ell R_{E,0} = (-p)^\ell\left(\frac{1}{p}\frac{\mathrm{d}}{\mathrm{d}p}\right)^\ell R_{E,0}. \tag{8.166}$$

Thus, we see that if we know one solution (say, for $\ell = 0$), we can construct all the others. Now, for $\ell = 0$, equation (8.158) leads to

$$\frac{\mathrm{d}^2 u_{E,0}(p)}{\mathrm{d}p^2} = -u_{E,0}(p),$$

$$\text{or,} \quad u_{E,\ell}(p) \sim \sin p \quad \text{or} \quad \cos p. \tag{8.167}$$

These are the two independent solutions of the equation. However, the boundary condition at the origin, namely,

$$u_{E,0}(p) \xrightarrow{p\to 0} 0. \tag{8.168}$$

excludes the solution of the form $\cos p$ so that, up to normalization constants, we can write

$$u_{E,0}(p) = \sin p, \tag{8.169}$$

in a region including the origin.

On the other hand, let us note that if we are considering a region where the origin is not included, then, we must allow for the other solution in (8.167) as well. Keeping this in mind, let us denote the two independent solutions of (8.167) by

$$\begin{aligned} u_{E,0}^{(\mathrm{I})} &= \sin p, \\ u_{E,0}^{(\mathrm{II})} &= -\cos p. \end{aligned} \tag{8.170}$$

All other solutions for higher ℓ values can be obtained from these through the use of the relation (8.162) or (8.166).

We note from (8.170) that $R_{E,0} = \frac{u_{E,0}}{p}$ has two independent forms. If

$$R_{E,0}^{(\mathrm{I})} = \frac{u_{E,0}^{(\mathrm{I})}}{p} = \frac{\sin p}{p}, \tag{8.171}$$

then, it follows that

$$R_{E,\ell}^{(\mathrm{I})}(p) = (-p)^{\ell} \left(\frac{1}{p}\frac{\mathrm{d}}{\mathrm{d}p}\right)^{\ell} \left(\frac{\sin p}{p}\right) = j_{\ell}(p), \tag{8.172}$$

which are the spherical Bessel functions of order ℓ.

On the other hand, if

$$R_{E,0}^{(\mathrm{II})} = \frac{u_{E,0}^{(\mathrm{II})}}{p} = -\frac{\cos p}{p}, \tag{8.173}$$

then, we have

$$R_{E,\ell}^{(\mathrm{II})}(p) = (-p)^{\ell}\left(\frac{1}{p}\frac{\mathrm{d}}{\mathrm{d}p}\right)^{\ell}\left(-\frac{\cos p}{p}\right) = \eta_{\ell}(p), \tag{8.174}$$

which are the spherical Neumann functions of order ℓ. The combinations

$$h_{\ell}^{(\pm)} = j_{\ell} \pm i\eta_{\ell}, \tag{8.175}$$

are known as the spherical Hankel functions (of the first and the second kind) and have the asymptotic forms

$$h_{\ell}^{(\pm)}(p) \xrightarrow{p\to\infty} \frac{1}{ip}\, e^{\pm i(p-\frac{1}{2}(\ell+1)\pi)}. \tag{8.176}$$

A few low order spherical Bessel and Neumann functions have the forms

$$\begin{aligned}
j_0(p) &= \frac{\sin p}{p},\\
j_1(p) &= \frac{1}{p}\left(\frac{\sin p}{p} - \cos p\right),\\
j_2(p) &= \left(\frac{3}{p^3} - \frac{1}{p}\right)\sin p - \frac{3\cos p}{p^2},\\
\eta_0(p) &= -\frac{\cos p}{p},\\
\eta_1(p) &= -\frac{1}{p}\left(\frac{\cos p}{p} + \sin p\right),\\
\eta_2(p) &= -\left(\frac{3}{p^3} - \frac{1}{p}\right)\cos p - \frac{3\sin p}{p^2}.
\end{aligned} \tag{8.177}$$

For our problem, of course, we need just the spherical Bessel functions in the region $r < a$, since the solution has to vanish at the origin. Thus, for $r < a$,

$$R_{E,\ell} = \frac{u_{E,\ell}(r)}{r} = a_{\ell} j_{\ell}(kr). \tag{8.178}$$

Region $r > a$. In this region, the equation we have to solve has the form (see (8.139))

$$\frac{\mathrm{d}^2 u_{E,\ell}}{\mathrm{d}r^2} + \frac{2\mu}{\hbar^2}\left(E - \frac{\hbar^2\ell(\ell+1)}{2\mu r^2}\right)u_{E,\ell} = 0,$$

$$\text{or,}\quad \frac{\mathrm{d}^2 u_{E,\ell}}{\mathrm{d}r^2} - \left(\alpha^2 + \frac{\ell(\ell+1)}{r^2}\right)u_{E,\ell} = 0. \tag{8.179}$$

We can now define $q = i\alpha r$, in terms of which equation (8.179) becomes

$$\left(-\frac{\mathrm{d}^2}{\mathrm{d}q^2} + \frac{\ell(\ell+1)}{q^2}\right) u_{E,\ell}(q) = u_{E,\ell}(q),$$

$$\text{or,} \quad \left(\frac{\mathrm{d}^2}{\mathrm{d}q^2} - \frac{\ell(\ell+1)}{q^2}\right) u_{E,\ell}(q) = -u_{E,\ell}(q). \tag{8.180}$$

We note that equation (8.180) is the same as in (8.151), but in the variable $q = i\alpha r$. Furthermore, since now the solution does not include the origin, it can be a combination of both the spherical Bessel and Neumann functions. Thus, for $r > a$, we have

$$R_{E,\ell}(q) = A_\ell j_\ell(q) + B_\ell \eta_\ell(q). \tag{8.181}$$

On the other hand, the solution must fall off exponentially at spatial infinity. This, therefore, selects for us

$$\begin{aligned} R_{E,\ell}(q) &= A_\ell h_\ell^{(+)}(q) = A_\ell(j_\ell(q) + i\eta_\ell(q)) \\ &\xrightarrow{r\to\infty} \frac{A_\ell}{iq}\, e^{i(q-\frac{1}{2}(\ell+1)\pi)} \\ &= -\frac{A_\ell}{\alpha r}\, e^{-(\alpha r+\frac{i}{2}(\ell+1)\pi)}. \end{aligned} \tag{8.182}$$

The first few Hankel functions have the explicit forms

$$\begin{aligned} h_0^{(+)}(i\alpha r) &= -\frac{1}{\alpha r}\, e^{-\alpha r}, \\ h_1^{(+)}(i\alpha r) &= i\left(\frac{1}{\alpha r} + \frac{1}{\alpha^2 r^2}\right) e^{-\alpha r}, \\ h_2^{(+)}(i\alpha r) &= \left(\frac{1}{\alpha r} + \frac{3}{\alpha^2 r^2} + \frac{3}{\alpha^3 r^3}\right) e^{-\alpha r}. \end{aligned} \tag{8.183}$$

We can now match solutions and the derivatives at $r = a$ to determine the energy eigenvalues. We have already done this for $\ell = 0$. For arbitrary ℓ, we can apply numerical methods or analyze graphs to determine the existence of solutions. For $\ell = 1$, for example, the equation becomes

$$\frac{\cot\eta}{\eta} - \frac{1}{\eta^2} = \frac{1}{\xi} + \frac{1}{\xi^2}, \quad \text{where} \quad \eta^2 + \xi^2 = \frac{2\mu V_0 a^2}{\hbar^2}. \tag{8.184}$$

We can show that, in this case, there is no bound state for $V_0 a^2 < \frac{\pi^2\hbar^2}{2\mu}$. For $\frac{\pi^2\hbar^2}{2\mu} \leq V_0 a^2 < \frac{(2\pi)^2\hbar^2}{2\mu}$ there is one bound state and so on.

Thus, we see that the minimum value of V_0a^2 for p-wave binding ($\ell = 1$) is larger than that for s-wave binding ($\ell = 0$), namely, for bound states,

$$\begin{aligned} \ell = 0: &\quad V_0a^2 \geq \tfrac{\pi^2\hbar^2}{8\mu}, \\ \ell = 1: &\quad V_0a^2 \geq \tfrac{\pi^2\hbar^2}{2\mu}. \end{aligned} \tag{8.185}$$

Physically, the meaning of this is very clear. In the case of $\ell = 1$, there exists a centrifugal barrier and, therefore, a particle requires stronger attraction for binding. In fact, it can be shown that the strength of the square well potential, V_0a^2, required to bind a particle of arbitrary ℓ increases monotonically with ℓ. This system does not show any degeneracy in the ℓ quantum number.

8.7 Selected problems

1. Find the energy eigenvalues and eigenfunctions of a particle in a two dimensional circular box that has perfectly rigid walls.

2. A one dimensional square well potential has a bound state for any positive V_0a^2. The three dimensional square well has a bound state only if $V_0a^2 \geq \frac{\pi^2\hbar^2}{8\mu}$. What is the analogous condition for a two dimensional circularly symmetric square well potential?

3. Consider the three dimensional isotropic oscillator described by

$$H_0 = \frac{\mathbf{p}^2}{2\mu} + \frac{1}{2}\,\mu\omega_0^2\mathbf{r}^2. \tag{8.186}$$

Assume that the particle has a charge $e > 0$ and is in a uniform magnetic field $\mathbf{B}$ along the z-axis. Writing $\omega_L = -\frac{eB}{2\mu}$ and choosing

$$\mathbf{A}(\mathbf{r}) = -\frac{1}{2}\mathbf{r} \times \mathbf{B}, \tag{8.187}$$

where $\mathbf{A}$ is the vector potential, we can write

$$H = H_0 + H_I(\omega_L). \tag{8.188}$$

Here, $H_I(\omega_L)$ is a sum of two terms - one which is linear in ω_L (paramagnetic) and the other quadratic in ω_L (diamagnetic).

Determine the exact stationary states of the (complete) system and their degeneracy. How does the energy of the ground state vary with ω_L? Is the ground state an eigenstate of $\mathbf{L}^2$? of L_z? of L_x? (This is the Zeeman effect for the harmonic oscillator.)

4. A particle moves in a potential in three dimensions of the form

$$V(\mathbf{r}) = \begin{cases} \infty, & z^2 \geq a^2, \\ \frac{1}{2}m\omega^2(x^2+y^2), & \text{otherwise}. \end{cases} \tag{8.189}$$

Determine the energy eigenvalues for this system, the degeneracy of each level as well as the eigenfunctions associated with them.

5. A particle moves, in three dimensions, in an *anisotropic* oscillator potential

$$V(x,y,z) = \frac{1}{2}m\omega^2\left(x^2+4y^2+9z^2\right). \tag{8.190}$$

a) What is the general expression for the energy eigenvalues for such an oscillator?

b) What are the associated energy eigenfunctions (they need not be normalized)?

c) What are the degeneracies of the three lowest energy eigenvalues?

6. The range of the strong force which binds a neutron and a proton in deuteron in a s-wave state ($\ell = 0$) is $a = 2\times 10^{-13}$cm. The binding can be explained quite well with a three dimensional square well potential of depth $(-V_0)$ and range a.

a) Given that the ground state has the binding energy 2.23MeV and that the potential strength V_0 is much larger than the binding energy, what is the (approximate) value of the potential?

b) Explain qualitatively, from the behavior of the solutions, whether this model would predict a mean radius of deuteron shorter or larger than the range of the force.

CHAPTER 9

Hydrogen atom

The hydrogen atom is an important physical system that can be exactly solved. In the next few lectures, we will study this system in detail to bring out various quantum mechanical features. We note that, unlike the systems we have studied so far, the hydrogen atom is a two particle system. It consists of a negatively charged electron interacting electromagnetically with a positively charged proton. Thus, let us first set up the formalism to study such systems.

9.1 Relative motion of two particles

Consider an isolated system of two interacting particles of masses m_1 and m_2 at positions $\mathbf{r}_1$ and $\mathbf{r}_2$. Let us suppose that they interact through a potential which depends only on the relative separation of the two particles. Then, the motion of the system can be separated into two distinct and decoupled parts – a part that describes the motion of the center of mass of the system and another which describes the relative motion of the two particles. Let us analyze this both in the classical as well as in the quantum mechanical descriptions.

Classical. The Lagrangian for such a system can be written as

$$L = \frac{1}{2}m_1\dot{\mathbf{r}}_1^2 + \frac{1}{2}m_2\dot{\mathbf{r}}_2^2 - V\left(\mathbf{r}_1 - \mathbf{r}_2\right). \tag{9.1}$$

Let us now define the new set of coordinates

$$\begin{aligned} \mathbf{r} &= \mathbf{r}_1 - \mathbf{r}_2, \\ \mathbf{R} &= \frac{m_1\mathbf{r}_1 + m_2\mathbf{r}_2}{(m_1 + m_2)}, \end{aligned} \tag{9.2}$$

where $\mathbf{r}, \mathbf{R}$ denote respectively the relative coordinate and the center of mass coordinate of the two particle system. The relations in (9.2) can be inverted to give

$$
\begin{aligned}
\mathbf{r}_1 &= \mathbf{R} + \frac{m_2}{m_1+m_2}\,\mathbf{r},\\
\mathbf{r}_2 &= \mathbf{R} - \frac{m_1}{m_1+m_2}\,\mathbf{r}.
\end{aligned} \tag{9.3}
$$

In terms of the variables in (9.2), the Lagrangian, (9.1), becomes

$$
\begin{aligned}
L &= \frac{1}{2}m_1\left(\dot{\mathbf{R}} + \frac{m_2}{m_1+m_2}\dot{\mathbf{r}}\right)^2 + \frac{1}{2}m_2\left(\dot{\mathbf{R}} - \frac{m_1}{m_1+m_2}\dot{\mathbf{r}}\right)^2 - V(\mathbf{r})\\
&= \frac{1}{2}(m_1+m_2)\dot{\mathbf{R}}^2 + \frac{1}{2}\left(\frac{m_1m_2^2}{(m_1+m_2)^2} + \frac{m_2m_1^2}{(m_1+m_2)^2}\right)\dot{\mathbf{r}}^2 - V(\mathbf{r})\\
&= \frac{1}{2}(m_1+m_2)\dot{\mathbf{R}}^2 + \frac{1}{2}\frac{m_1m_2}{m_1+m_2}\dot{\mathbf{r}}^2 - V(\mathbf{r})\\
&= \frac{1}{2}M\dot{\mathbf{R}}^2 + \frac{1}{2}\mu\dot{\mathbf{r}}^2 - V(\mathbf{r}),
\end{aligned} \tag{9.4}
$$

where we have identified

$$
\begin{aligned}
M &= m_1 + m_2 = \text{total mass of the system},\\
\mu &= \frac{m_1m_2}{m_1+m_2} = \text{reduced mass of the system}.
\end{aligned} \tag{9.5}
$$

The conjugate momenta corresponding to these two new coordinates can be obtained to be

$$
\begin{aligned}
\mathbf{P} &= \frac{\partial L}{\partial \dot{\mathbf{R}}} = M\dot{\mathbf{R}} = (m_1+m_2)\frac{m_1\dot{\mathbf{r}}_1 + m_2\dot{\mathbf{r}}_2}{m_1+m_2}\\
&= m_1\dot{\mathbf{r}}_1 + m_2\dot{\mathbf{r}}_2 = \mathbf{p}_1 + \mathbf{p}_2\\
&= \text{total momentum},\\
\mathbf{p} &= \frac{\partial L}{\partial \dot{\mathbf{r}}} = \mu\dot{\mathbf{r}} = \frac{m_1m_2}{m_1+m_2}\,(\dot{\mathbf{r}}_1 - \dot{\mathbf{r}}_2)\\
&= \frac{m_2\mathbf{p}_1 - m_1\mathbf{p}_2}{m_1+m_2}.
\end{aligned} \tag{9.6}
$$

Thus, the Hamiltonian for the system can be written as

$$
H = \mathbf{P}\dot{\mathbf{R}} + \mathbf{p}\dot{\mathbf{r}} - L = \frac{\mathbf{P}^2}{2M} + \frac{\mathbf{p}^2}{2\mu} + V(\mathbf{r}). \tag{9.7}
$$

We see that the motion of the system can be equivalently described by the motion of two fictitious particles – one with the total mass and the coordinates of the center of mass of the system and the other with the reduced mass and with the relative coordinates. Furthermore, the motion of these two fictitious particles are uncoupled and, since the variable $\mathbf{R}$ is cyclic, it follows that (see (1.52))

$$\dot{\mathbf{P}} = 0. \tag{9.8}$$

The total momentum of the system is constant and, consequently, we can go to the center of mass frame, in which case we have

$$\mathbf{P} = 0. \tag{9.9}$$

and in this frame the Hamiltonian (9.7) becomes

$$H = \frac{\mathbf{p}^2}{2\mu} + V(\mathbf{r}). \tag{9.10}$$

The problem of two interacting particles (with a potential depending on the relative separation) in the center of mass frame, therefore, reduces to that of a single particle with a reduced mass and with the relative coordinates.

Quantum mechanical. The Hamiltonian for this system can be written in terms of the new variables as

$$\begin{aligned} H &= \frac{\mathbf{p}_1^2}{2m_1} + \frac{\mathbf{p}_2^2}{2m_2} + V(\mathbf{r}_1 - \mathbf{r}_2) \\ &= \frac{\mathbf{P}^2}{2M} + \frac{\mathbf{p}^2}{2\mu} + V(\mathbf{r}). \end{aligned} \tag{9.11}$$

We see that the Hamiltonian of two particles with an interaction which depends only on the relative separation can be equivalently written as a sum of two uncoupled terms. Quantum mechanically, of course, we know that the coordinate and the momentum operators would satisfy the following commutation relations:

$$\begin{aligned} &[r_{1i}, r_{1j}] = [r_{2i}, r_{2j}] = 0 = [p_{1i}, p_{1j}] = [p_{2i}, p_{2j}]\,, \\ &[r_{1i}, p_{2j}] = [r_{2i}, p_{1j}] = 0 = [r_{1i}, r_{2j}] = [p_{1i}, p_{2j}]\,, \\ &[r_{1i}, p_{1j}] = i\hbar\delta_{ij}\mathbb{1} = [r_{2i}, p_{2j}]\,. \end{aligned} \tag{9.12}$$

Using these as well as the definitions in (9.2) and (9.6), we can derive that

$$\begin{aligned} &[R_i, R_j] = [r_i, r_j] = 0 = [P_i, P_j] = [p_i, p_j]\,, \\ &[R_i, r_j] = [R_i, p_j] = 0 = [r_i, P_j] = [P_i, p_j]\,, \end{aligned} \tag{9.13}$$

and

$$[R_i, P_j] = i\hbar\delta_{ij}\mathbb{1} = [r_i, p_j]. \tag{9.14}$$

Thus, $(\mathbf{r}, \mathbf{p})$ and $(\mathbf{R}, \mathbf{P})$ behave like two pairs of conjugate variables. In the coordinate basis, therefore, we can write these operators as

$$\mathbf{p} = -i\hbar\boldsymbol{\nabla} = -i\hbar\,\frac{\partial}{\partial\mathbf{r}},$$

$$\mathbf{P} = -i\hbar\boldsymbol{\nabla}_{R} = -i\hbar\,\frac{\partial}{\partial\mathbf{R}}. \tag{9.15}$$

Furthermore, since the two sets $(\mathbf{r}, \mathbf{p})$ and $(\mathbf{R}, \mathbf{P})$ commute with each other, the Hamiltonian can be written as the direct sum of two Hamiltonians

$$H = H_R \oplus H_r, \tag{9.16}$$

where H_R is the Hamiltonian associated with the motion of the center of mass and H_r is associated with the relative motion of the two particles. Since,

$$[H_r, H_R] = 0, \tag{9.17}$$

they can be simultaneously diagonalized. The Hilbert space, in fact, becomes a product space of two Hilbert spaces

$$\mathcal{E} = \mathcal{E}_R \otimes \mathcal{E}_r, \tag{9.18}$$

where $(H_R, \mathbf{R}, \mathbf{P})$ act only on the space $\mathcal{E}_R$ while $(H_r, \mathbf{r}, \mathbf{p})$ act only on $\mathcal{E}_r$. A general state of $\mathcal{E}$, (9.18), can be written as

$$|\psi_R, \psi_r\rangle = |\psi_R\rangle \otimes |\psi_r\rangle. \tag{9.19}$$

The situation here is exactly like the higher dimensional oscillator in the Cartesian coordinates that we have discussed earlier. Thus, a general wave function

$$\langle \mathbf{r}, \mathbf{R}|\psi_R, \psi_r\rangle = \psi_R(\mathbf{R})\psi_r(\mathbf{r}), \tag{9.20}$$

becomes a product of two wave functions. Consequently, we know that the Schrödinger equation will separate into two equations (because the wave function is separable)

$$-\frac{\hbar^2}{2M}\boldsymbol{\nabla}_R^2\psi_R(\mathbf{R}) = E_R\psi_R(\mathbf{R}),$$

$$\left(-\frac{\hbar^2}{2\mu}\boldsymbol{\nabla}^2 + V(\mathbf{r})\right)\psi_r(\mathbf{r}) = E_r\psi_r(\mathbf{r}), \tag{9.21}$$

where $E_r + E_R = E =$ total energy of the system. The first equation in (9.21) is easy to solve and yields

$$\psi_R(\mathbf{R}) = e^{-\frac{i}{\hbar}\mathbf{P}\cdot\mathbf{R}}, \qquad E_R = \frac{\mathbf{P}^2}{2M}, \tag{9.22}$$

which simply represents a free particle motion. Namely, the center of mass behaves like a free particle independent of the form of the interaction potential. The interesting dynamics lies in the second equation in (9.21) describing the relative motion of the two particles.

9.2 Hydrogen atom

The hydrogen atom is a two particle system consisting of an electron and a proton. In a simplified picture, we can think of the system as describing the motion of an electron in the Coulomb potential of a proton since the proton is very much heavier than the electron. In fact, in the case of the hydrogen atom we know that

$$\begin{aligned} m_1 &= m_P \simeq 1000 \text{ MeV}/c^2, \\ m_2 &= m_e \simeq .5 \text{ MeV}/c^2. \end{aligned} \tag{9.23}$$

The Coulomb potential of the proton is $\frac{e}{r}$ (in CGS units) and, therefore, the potential energy of the system is given by

$$V = V(r) = -\frac{e^2}{r}, \tag{9.24}$$

where $r = |\mathbf{r}_1 - \mathbf{r}_2|$ is the separation between the proton and the electron. Thus, we see that our earlier discussion of reducing the problem to a single particle motion in the center of mass frame can be applied here and we can identify the Hamiltonian associated with the relative motion of the two particles with

$$H = -\frac{\hbar^2}{2\mu}\boldsymbol{\nabla}^2 + V(r) = -\frac{\hbar^2}{2\mu}\boldsymbol{\nabla}^2 - \frac{e^2}{r}. \tag{9.25}$$

Note that, in this case,

$$\begin{aligned} M &= m_1 + m_2 \simeq m_1, \\ \mu &= \frac{m_1 m_2}{m_1 + m_2} \simeq m_2. \end{aligned} \tag{9.26}$$

Therefore, we can identify the proton with the center of mass which is stationary and the motion of the electron as describing the relative motion.

This Hamiltonian has rotational symmetry and, therefore, following our earlier discussions, we can write the solution in the separable form as

$$\psi_{E,\ell,m}(r,\theta,\phi) = R_{E,\ell}(r)Y_{\ell,m}(\theta,\phi). \tag{9.27}$$

Furthermore, as before, defining $R_{E,\ell}(r) = \frac{u_{E,\ell}(r)}{r}$, the radial equation for $u_{E,\ell}$ becomes (see (8.97))

$$\frac{\mathrm{d}^2 u_{E,\ell}}{\mathrm{d}r^2} + \frac{2\mu}{\hbar^2}\left(E + \frac{e^2}{r} - \frac{\hbar^2\ell(\ell+1)}{2\mu r^2}\right)u_{E,\ell} = 0. \tag{9.28}$$

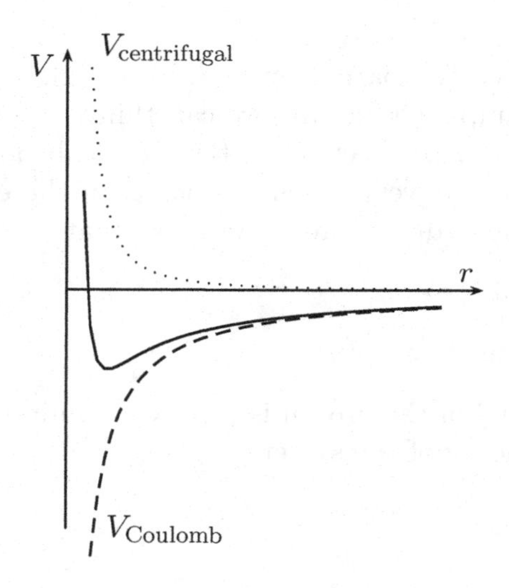

Figure 9.1: Effective potential for the reduced particle in the Hydrogen atom.

For a nonzero value of ℓ, the effective potential has the shape as shown in Fig. 9.1. This shows that the effective potential, in this case, supports both free particle as well as bound state solutions and we are interested in studying the bound state solutions, namely, solutions with $E < 0$. Therefore, writing

$$E = -|E|, \tag{9.29}$$

we see that equation (9.28) has the asymptotic form ($r \to \infty$)

$$\frac{\mathrm{d}^2 u_{E,\ell}}{\mathrm{d}r^2} - \frac{2\mu|E|}{\hbar^2}\,u_{E,\ell} = 0,$$

$$\text{or,}\quad u_{E,\ell}(r) \sim e^{-\left(\frac{2\mu|E|}{\hbar^2}\right)^{\frac{1}{2}}r} = e^{-\frac{1}{2}y}. \tag{9.30}$$

Here, we have defined the dimensionless variable

$$y = 2\left(\frac{2\mu|E|}{\hbar^2}\right)^{\frac{1}{2}} r,$$

$$\frac{\mathrm{d}}{\mathrm{d}r} = 2\left(\frac{2\mu|E|}{\hbar^2}\right)^{\frac{1}{2}} \frac{\mathrm{d}}{\mathrm{d}y}, \tag{9.31}$$

and have discarded the other solution with the positive exponent since it is not normalizable.

In terms of the y variable, equation (9.28) becomes

$$\frac{\mathrm{d}^2 u_{E,\ell}}{\mathrm{d}y^2} + \frac{1}{4}\left(-1 + \frac{e^2}{|E|r} - \frac{\hbar^2\ell(\ell+1)}{2\mu|E|r^2}\right) u_{E,\ell} = 0,$$

$$\text{or,}\quad \frac{\mathrm{d}^2 u_{E,\ell}}{\mathrm{d}y^2} + \left(-\frac{1}{4} + \frac{\lambda}{y} - \frac{\ell(\ell+1)}{y^2}\right) u_{E,\ell} = 0, \tag{9.32}$$

where we have defined another dimensionless variable

$$\lambda = \left(\frac{\mu}{2\hbar^2|E|}\right)^{\frac{1}{2}} e^2 = \left(\frac{\mu e^4}{2\hbar^2|E|}\right)^{\frac{1}{2}}. \tag{9.33}$$

Near the origin, $r \to 0$ $(y \to 0)$, equation (9.32) leads to

$$\frac{\mathrm{d}^2 u_{E,\ell}}{\mathrm{d}y^2} - \frac{\ell(\ell+1)}{y^2} u_{E,\ell} = 0,$$

$$\text{or,}\quad u_{E,\ell}(y) \sim y^{\ell+1}, \tag{9.34}$$

where we have discarded the other solution, $y^{-\ell}$, since it does not satisfy the boundary condition at the origin, (8.107).

The asymptotic forms of the solution in (9.30) and (9.34), therefore, suggest a general solution of the form

$$u_{E,\ell}(y) = e^{-\frac{1}{2}y} v(y) = e^{-\frac{1}{2}y} y^{\ell+1} \sum_{k=0}^{\infty} a_k y^k = e^{-\frac{1}{2}y} \sum_{k=0}^{\infty} a_k y^{k+\ell+1},$$

$$\frac{\mathrm{d}u_{E,\ell}}{\mathrm{d}y} = e^{-\frac{1}{2}y}\left(-\frac{1}{2}v + v'\right),$$

$$\frac{\mathrm{d}^2 u_{E,\ell}}{\mathrm{d}y^2} = e^{-\frac{1}{2}y}\left(\frac{1}{4}v - v' + v''\right). \tag{9.35}$$

Substituting (9.35) into (9.32), we obtain the equation for v to be

$$e^{-\frac{1}{2}y}\left[v''-v'+\frac{1}{4}v+\left(-\frac{1}{4}+\frac{\lambda}{y}-\frac{\ell(\ell+1)}{y^2}\right)v\right]=0,$$

$$\text{or,}\quad \frac{\mathrm{d}^2v}{\mathrm{d}y^2}-\frac{\mathrm{d}v}{\mathrm{d}y}+\left(\frac{\lambda}{y}-\frac{\ell(\ell+1)}{y^2}\right)v=0. \tag{9.36}$$

Let us now use the power series expansion for $v(y)$ given in (9.35),

$$v(y)=\sum_{k=0}^{\infty}a_k y^{k+\ell+1},$$

$$\frac{\mathrm{d}v}{\mathrm{d}y}=\sum_{k=0}^{\infty}(k+\ell+1)a_k y^{k+\ell},$$

$$\frac{\mathrm{d}^2v}{\mathrm{d}y^2}=\sum_{k=0}^{\infty}(k+\ell+1)(k+\ell)a_k y^{k+\ell-1}. \tag{9.37}$$

With this, equation (9.36) becomes

$$\sum_{k=0}^{\infty}\Big[(k+\ell+1)(k+\ell)a_k y^{k+\ell-1}-(k+\ell+1)a_k y^{k+\ell}$$
$$+\lambda a_k y^{k+\ell}-\ell(\ell+1)a_k y^{k+\ell-1}\Big]=0,$$

$$\text{or,}\quad \sum_{k=0}^{\infty}\Big[\{(k+\ell+1)(k+\ell)-\ell(\ell+1)\}\,a_k y^{k+\ell-1}$$
$$-(k+\ell+1-\lambda)a_k y^{k+\ell}\Big]=0. \tag{9.38}$$

Looking at the coefficient of the lowest power of y in (9.38) (namely, $y^{\ell-1}$), we obtain

$$[\ell(\ell+1)-\ell(\ell+1)]a_0=0,\quad \text{or,}\quad a_0=\text{arbitrary}. \tag{9.39}$$

The coefficient of the next term in the series (namely, y^{ℓ}) gives

$$[(\ell+2)(\ell+1)-\ell(\ell+1)]a_1-(\ell+1-\lambda)a_0=0, \tag{9.40}$$

which implies that $a_1\neq 0$ if $a_0\neq 0$.

In general, the recursion relation would connect a_{k+1} to a_k. Thus, looking at the coefficient of $y^{k+\ell}$ in (9.38), we have

$$[(k+1+\ell+1)(k+1+\ell)-\ell(\ell+1)]a_{k+1}=(k+\ell+1-\lambda)a_k,$$

which leads to

$$a_{k+1} = \frac{(k+\ell+1-\lambda)}{(k+1)(k+2\ell+2)}\,a_k, \tag{9.41}$$

so that for large k, we have

$$\frac{a_{k+1}}{a_k} \to \frac{1}{k}.$$

This is the behavior of the series e^y for large orders and, therefore, unless the series terminates, this would lead to an unphysical solution. The series would terminate if

$$\begin{aligned}
&k+\ell+1-\lambda = 0,\\
\text{or,}\quad &\lambda = \left(\frac{\mu e^4}{2\hbar^2|E|}\right)^{\frac{1}{2}} = k+\ell+1 = n,\\
\text{or,}\quad &|E_n| = \frac{\mu e^4}{2\hbar^2 n^2},\\
\text{or,}\quad &E_n = -|E_n| = -\frac{\mu e^4}{2\hbar^2 n^2},
\end{aligned} \tag{9.42}$$

which determines the energy eigenvalues for the hydrogen atom. Now, since both k and ℓ take positive integer values, n also takes positive integer values. Even when ℓ and k are both equal to zero, $n = 1$. Thus, the allowed values for n are

$$n = 1, 2, 3, \cdots. \tag{9.43}$$

Furthermore, the allowed values of the orbital angular momentum, for a fixed n-value, are given by

$$\ell = n-k-1 = n-1, n-2, \cdots, 0. \tag{9.44}$$

Thus, the solution of the differential equation, (9.32), is obtained to be

$$\begin{aligned}
u_{n,\ell}(y) &= e^{-\frac{1}{2}y}\sum_{k=0}^{n-\ell-1} a_k y^{k+\ell+1}\\
&= e^{-\frac{1}{2}y}v_{n,\ell}(y) = e^{-\frac{1}{2}y}y^{\ell+1}w_{n,\ell}(y).
\end{aligned} \tag{9.45}$$

This leads to

$$\frac{dv_{n,\ell}}{dy} = y^{\ell+1}\frac{dw_{n,\ell}}{dy} + (\ell+1)y^{\ell}w_{n,\ell},$$
$$\frac{d^2v_{n,\ell}}{dy^2} = y^{\ell+1}\frac{d^2w_{n,\ell}}{dy^2} + 2(\ell+1)y^{\ell}\frac{dw_{n,\ell}}{dy} + \ell(\ell+1)y^{\ell-1}w_{n,\ell},$$
$$(9.46)$$

so that the differential equation that $w_{n,\ell}(y)$ satisfies follows from (9.36) to be

$$y\,\frac{d^2w}{dy^2} + (2\ell+2-y)\,\frac{dw}{dy} + (n-\ell-1)w(y) = 0. \qquad (9.47)$$

We know that the equation

$$y\frac{d^2L_q}{dy^2} + (1-y)\frac{dL_q}{dy} + qL_q = 0, \qquad (9.48)$$

is known as the Laguerre equation and L_q's are known as the Laguerre polynomials of order q. The functions $L_q^p(y)$, related to the L_q's by the relation,

$$L_q^p(y) = \frac{d^pL_q(y)}{dy^p}, \qquad q \geq p, \qquad (9.49)$$

are known as the associated Laguerre polynomials. (This is one way of defining the associated Laguerre polynomials and there is an alternate way in the literature as well. The two, however, are related to each other and as long as we follow one definition consistently, there is no possibility for confusion.) They are polynomials of order $(q-p)$ and satisfy the differential equation

$$y\frac{d^2L_q^p}{dy^2} + (p+1-y)\frac{dL_q^p}{dy} + (q-p)L_q^p(y) = 0. \qquad (9.50)$$

Comparing equation (9.50) with the one satisfied by the $w_{n,\ell}$'s, (9.47), we can identify

$$w_{n,\ell}(y) = L_{n+\ell}^{2\ell+1}(y). \qquad (9.51)$$

The generating function for the Laguerre polynomials is

$$S(y,t) = \frac{e^{-\frac{yt}{1-t}}}{1-t} = \sum_{n=0}^{\infty}\frac{L_n(y)}{n!}\,t^n, \qquad t<1, \qquad (9.52)$$

where the coefficient of expansion on the right hand side are the Laguerre polynomials. To see that this actually generates the Laguerre polynomials, we note that

$$\frac{\partial S}{\partial y} = -\frac{t}{1-t} S,$$

$$\text{or,} \quad (1-t)\frac{\partial S}{\partial y} = -tS,$$

$$\text{or,} \quad (1-t)\sum_{n=0}^{\infty} \frac{L_n'}{n!} t^n = -t \sum_{n=0}^{\infty} \frac{L_n}{n!} t^n,$$

$$\text{or,} \quad \sum_{n=0}^{\infty} \frac{L_n'}{n!} t^n - \sum_{n=0}^{\infty} \frac{L_n'}{n!} t^{n+1} = -\sum_{n=0}^{\infty} \frac{L_n}{n!} t^{n+1},$$

$$\text{or,} \quad \sum_{n=0}^{\infty} \frac{L_n'}{n!} t^n - \sum_{n=1}^{\infty} \frac{L_{n-1}'}{(n-1)!} t^n = -\sum_{n=1}^{\infty} \frac{L_{n-1}}{(n-1)!} t^n,$$

$$\text{or,} \quad \sum_{n=0}^{\infty} \frac{L_n'}{n!} t^n - \sum_{n=0}^{\infty} n \frac{L_{n-1}'}{n!} t^n = -\sum_{n=0}^{\infty} \frac{L_{n-1}}{n!} t^n,$$

$$\text{or,} \quad L_n' - nL_{n-1}' = -nL_{n-1}, \tag{9.53}$$

$$\text{or,} \quad L_{n+1}' - (n+1)L_n' = -(n+1)L_n. \tag{9.54}$$

Furthermore, differentiating with respect to t, we obtain

$$\frac{\partial S}{\partial t} = -y\left[\frac{1}{1-t} + \frac{t}{(1-t)^2}\right] S + \frac{1}{1-t} S$$

$$= \left[-\frac{y}{(1-t)^2} + \frac{1}{1-t}\right] S = \frac{(1-y-t)}{(1-t)^2} S,$$

which leads to

$$(1-2t+t^2)\frac{\partial S}{\partial t} = (1-y-t)S,$$

$$\text{or,} \quad (1-2t+t^2)\sum_{n=1}^{\infty} \frac{L_n}{(n-1)!} t^{n-1} = (1-y-t)\sum_{n=0}^{\infty} \frac{L_n}{n!} t^n,$$

$$\text{or,} \quad \sum_{n=1}^{\infty} \left[\frac{L_n}{(n-1)!} t^{n-1} - \frac{2L_n}{(n-1)!} t^n + \frac{L_n}{(n-1)!} t^{n+1}\right]$$

$$= (1-y)\sum_{n=0}^{\infty} \frac{L_n}{n!} t^n - \sum_{n=0}^{\infty} \frac{L_n}{n!} t^{n+1},$$

$$\text{or,}\quad \sum_{n=0}^{\infty}\frac{L_{n+1}}{n!}t^n - 2\sum_{n=0}^{\infty}n\frac{L_n}{n!}t^n + \sum_{n=0}^{\infty}n\frac{L_n}{n!}t^{n+1}$$

$$= (1-y)\sum_{n=0}^{\infty}\frac{L_n}{n!}t^n - \sum_{n=0}^{\infty}\frac{L_n}{n!}t^{n+1},$$

$$\text{or,}\quad \sum_{n=0}^{\infty}\left[L_{n+1} - 2nL_n - (1-y)L_n\right]\frac{t^n}{n!}$$

$$= \sum_{n=0}^{\infty}-(n+1)\frac{L_n}{n!}t^{n+1} = -\sum_{n=1}^{\infty}n\frac{L_{n-1}}{(n-1)!}t^n$$

$$= -\sum_{n=0}^{\infty}n^2\frac{L_{n-1}}{n!}t^n. \tag{9.55}$$

Thus, comparing coefficients, we obtain

$$L_{n+1} = (2n+1-y)L_n - n^2L_{n-1}. \tag{9.56}$$

Differentiating (9.56) with respect to y we have

$$L'_{n+1} = (2n+1-y)L'_n - L_n - n^2L'_{n-1},$$

$$\text{or,}\quad n^2L'_{n-1} = (2n+1-y)L'_n - L'_{n+1} - L_n. \tag{9.57}$$

Furthermore, multiplying (9.53) throughout by n and eliminating $n^2L'_{n-1}$ using (9.57), we have

$$nL'_n + L'_{n+1} - (2n+1-y)L'_n + L_n = -n^2L_{n-1},$$

$$\text{or,}\quad L'_{n+1} - (n+1-y)L'_n + L_n = -n^2L_{n-1}. \tag{9.58}$$

If we now use (9.54), equation (9.58) becomes

$$-(n+1)L_n + yL'_n + L_n = -n^2L_{n-1},$$

$$\text{or,}\quad yL'_n - nL_n = -n^2L_{n-1}. \tag{9.59}$$

Differentiating (9.59) with respect to y we obtain,

$$yL''_n + L'_n - nL'_n = -n^2L'_{n-1},$$

$$\text{or,}\quad yL''_n + (1-n)L'_n = -n^2L'_{n-1}, \tag{9.60}$$

which upon eliminating the $n^2L'_{n-1}$ term using (9.57) leads to

$$yL''_n + (1-n)L'_n + (2n+1-y)L'_n - L'_{n+1} - L_n = 0. \tag{9.61}$$

Furthermore, eliminating L'_{n+1} through the relation (9.54), we finally obtain

$$\begin{aligned} & yL''_n + (1-n)L'_n + (n-y)L'_n + (n+1)L_n - L_n = 0, \\ \text{or,} \quad & yL''_n + (1-y)L'_n + nL_n = 0. \end{aligned} \tag{9.62}$$

Thus, we see that the L_n's satisfy the Laguerre equation. It follows now, from (9.49), that the generating function for the associated Laguerre polynomials are given by

$$\begin{aligned} S_p(y,t) &= \frac{\partial^p S(y,t)}{\partial y^p} = \frac{(-t)^p}{(1-t)^{p+1}}\, e^{-\frac{yt}{1-t}} \\ &= \sum_{n=p}^{\infty} \frac{L_n^p(y)}{n!} t^n. \end{aligned} \tag{9.63}$$

We can now identify the radial wave function for the hydrogen atom with

$$R_{n,\ell} = N_{n,\ell}\, e^{-\frac{1}{2}y} y^{\ell} L_{n+\ell}^{2\ell+1}(y), \tag{9.64}$$

where

$$y = 2\left(\frac{2\mu|E_n|}{\hbar^2}\right)^{\frac{1}{2}} r = \frac{2\mu e^2}{\hbar^2 n} r, \tag{9.65}$$

and the total wave function is given by

$$\psi_{n,\ell,m}(r,\theta,\phi) = R_{n,\ell} Y_{\ell,m}(\theta,\phi). \tag{9.66}$$

The normalization constant $N_{n,\ell}$ can be obtained from the orthonormality relations for the associated Laguerre polynomials, which can be derived using the generating functions in the following way. The orthogonality of the $Y_{\ell,m}$'s tells us that a nonzero contribution is obtained only if $\ell = \ell'$ and $m = m'$ in

$$\int \mathrm{d}^3 r\ \psi^*_{n',\ell',m'} \psi_{n,\ell,m}, \tag{9.67}$$

and when these quantum numbers are the same, the angular integral becomes unity. Thus, we only have to look at the radial part of the solution. Furthermore, we can also show, using the equations of motion, that if $n \neq n'$, this integral vanishes. Therefore, let us look at

$$\int r^2 \mathrm{d}r\ R_{n,\ell}R_{n,\ell}$$

$$= \left(\frac{\hbar^2 n}{2\mu e^2}\right)^3 |N_{n,\ell}|^2 \int \mathrm{d}y\ y^2 e^{-y} y^{2\ell} L_{n+\ell}^{2\ell+1}(y) L_{n+\ell}^{2\ell+1}(y)$$

$$= \left(\frac{\hbar^2 n}{2\mu e^2}\right)^3 |N_{n,\ell}|^2 \int \mathrm{d}y\ y^{2\ell+2} e^{-y} L_{n+\ell}^{2\ell+1}(y) L_{n+\ell}^{2\ell+1}(y). \quad (9.68)$$

We now write the associated Laguerre polynomials in terms of their generating functions as (see (9.63))

$$L_{n+\ell}^{2\ell+1}(y) = \frac{\partial^{n+\ell}}{\partial t^{n+\ell}} \left[(-1)^{2\ell+1} \frac{t^{2\ell+1}}{(1-t)^{2\ell+2}} e^{-\frac{yt}{(1-t)}}\right]_{t=0}. \quad (9.69)$$

Using this in (9.68), we obtain

$$\int \mathrm{d}y\ y^{2\ell+2} e^{-y} L_{n+\ell}^{2\ell+1}(y) L_{n+\ell}^{2\ell+1}(y)$$

$$= \frac{\partial^{n+\ell}}{\partial t^{n+\ell}} \frac{\partial^{n+\ell}}{\partial x^{n+\ell}} \left[\frac{(tx)^{2\ell+1}}{(1-t)^{2\ell+2}(1-x)^{2\ell+2}} \times \int \mathrm{d}y\ y^{2\ell+2} e^{-y} e^{-\frac{yt}{(1-t)}} e^{-\frac{yx}{(1-x)}}\right]_{t,x=0}$$

$$= \frac{\partial^{n+\ell}}{\partial t^{n+\ell}} \frac{\partial^{n+\ell}}{\partial x^{n+\ell}} \left[\frac{(tx)^{2\ell+1}}{(1-t)^{2\ell+2}(1-x)^{2\ell+2}} \times \int \mathrm{d}y\ y^{2\ell+2} e^{-\frac{y(1-xt)}{(1-t)(1-x)}}\right]_{t,x=0}. \quad (9.70)$$

Changing the variable of integration as

$$\frac{y(1-xt)}{(1-t)(1-x)} \to y, \quad (9.71)$$

the integral in (9.70) becomes

$$\frac{\partial^{n+\ell}}{\partial t^{n+\ell}} \frac{\partial^{n+\ell}}{\partial x^{n+\ell}} \left[\frac{(tx)^{2\ell+1}}{(1-t)^{2\ell+2}(1-x)^{2\ell+2}} \times \frac{(1-t)^{2\ell+3}(1-x)^{2\ell+3}}{(1-xt)^{2\ell+3}} \int \mathrm{d}y\ y^{2\ell+2} e^{-y}\right]_{t,x=0}$$

$$= \frac{\partial^{n+\ell}}{\partial t^{n+\ell}} \frac{\partial^{n+\ell}}{\partial x^{n+\ell}} \left[\frac{(tx)^{2\ell+1}(1-t)(1-x)}{(1-xt)^{2\ell+3}} \Gamma(2\ell+3)\right]_{t,x=0}. \quad (9.72)$$

It is clear that terms with different powers of t and x inside the square bracket vanish since these variables are set to zero at the end. Thus, the integral becomes

$$\frac{\partial^{n+\ell}}{\partial t^{n+\ell}}\frac{\partial^{n+\ell}}{\partial x^{n+\ell}}\left[\frac{(tx)^{2\ell+1}(1+tx)}{(1-tx)^{2\ell+3}}\right]_{t,x=0}\Gamma(2\ell+3)$$

$$= \left[(2\ell+1)!\binom{n+\ell}{2\ell+1}(2\ell+3)(2\ell+4)\cdots\right.$$

$$\cdots(2\ell+3+n+\ell-2\ell-2)(n+\ell)!$$

$$+(2\ell+2)!\binom{n+\ell}{2\ell+2}(2\ell+3)(2\ell+4)\cdots$$

$$\left.\times(2\ell+3+n+\ell-2\ell-3)(n+\ell)!\right]\Gamma(2\ell+3)$$

$$= \left[(2\ell+1)!\frac{((n+\ell)!)^2}{(2\ell+1)!(n-\ell-1)!}\right.$$

$$\times\ (2\ell+3)(2\ell+4)\ldots(n+\ell+1)$$

$$\left.+(2\ell+2)!\frac{((n+\ell)!)^2}{(2\ell+2)!(n-\ell-2)!}(2\ell+3)\cdots(n+\ell)\right]$$

$$\times(2\ell+2)!$$

$$= ((n+\ell)!)^2(2\ell+2)!\left[\frac{(n+\ell+1)!}{(2\ell+2)!(n-\ell-1)!}\right.$$

$$\left.+\frac{(n+\ell)!}{(2\ell+2)!(n-\ell-2)!}\right]$$

$$= ((n+\ell)!)^2\left[\frac{(n+\ell)!(n+\ell+1)}{(n-\ell-1)!}+\frac{(n+\ell)!(n-\ell-1)}{(n-\ell-1)!}\right]$$

$$= 2n\frac{((n+\ell)!)^3}{(n-\ell-1)!}. \tag{9.73}$$

(Basically, the origin of the combinatoric factors is as follows. Let us look at the first term and note that, $(2\ell+1)$ derivatives of x have to act on the numerator, since we are setting $x=0$. This can act in $\binom{n+\ell}{2\ell+1}$ ways and brings out a factor of $(2\ell+1)!$ from the derivatives. The rest of the $(n-\ell-1)$ derivatives act on the denominator and bring out the factors following these. When we set $x=0$, the denominator becomes unity and the numerator has a power of $t^{n+\ell}$. The t derivatives now give the additional factor of $(n+\ell)!$.)

Thus, the normalization relation becomes

$$\int r^2 \mathrm{d}r \; R_{n,\ell}(r) R_{n,\ell}(r) = 1,$$

$$\text{or,} \quad \left(\frac{\hbar^2 n}{2\mu e^2}\right)^3 |N_{n,\ell}|^2 \int \mathrm{d}y \; y^{2\ell+2} e^{-y} L_{n+\ell}^{2\ell+1} L_{n+\ell}^{2\ell+1} = 1,$$

$$\text{or,} \quad |N_{n,\ell}|^2 \left(\frac{\hbar^2 n}{2\mu e^2}\right)^3 2n \frac{((n+\ell)!)^3}{(n-\ell-1)!} = 1,$$

$$\text{or,} \quad N_{n,\ell} = -\left[\left(\frac{2\mu e^2}{\hbar^2 n}\right)^3 \frac{(n-\ell-1)!}{2n((n+\ell)!)^3}\right]^{\frac{1}{2}} = N_{n,\ell}^*. \tag{9.74}$$

The normalized radial wave functions can now be written as

$$R_{n,\ell} = -\left[\left(\frac{2\mu e^2}{\hbar^2 n}\right)^3 \frac{(n-\ell-1)!}{2n((n+\ell)!)^3}\right]^{\frac{1}{2}} \times e^{-\frac{\mu e^2}{\hbar^2 n} r} \left(\frac{2\mu e^2 r}{\hbar^2 n}\right)^\ell L_{n+\ell}^{2\ell+1}\left(\tfrac{2\mu e^2 r}{\hbar^2 n}\right), \tag{9.75}$$

and the first three radial functions have the explicit forms,

$$R_{1,0} = 2\left(\frac{\mu e^2}{\hbar^2}\right)^{\frac{3}{2}} e^{-\frac{\mu e^2}{\hbar^2} r},$$

$$R_{2,0} = \left(\frac{\mu e^2}{2\hbar^2}\right)^{\frac{3}{2}} \left(2 - \frac{\mu e^2 r}{\hbar^2}\right) e^{-\frac{\mu e^2}{2\hbar^2} r},$$

$$R_{2,1} = \left(\frac{\mu e^2}{2\hbar^2}\right)^{\frac{3}{2}} \frac{\mu e^2 r}{\hbar^2 \sqrt{3}} e^{-\frac{\mu e^2}{2\hbar^2} r}. \tag{9.76}$$

9.3 Fundamental quantities associated with the hydrogen atom

Looking at the wave functions for the hydrogen atom in (9.75) (or (9.76)), we notice that there is a fundamental length scale that enters the solutions,

$$a_0 = \frac{\hbar^2}{\mu e^2}. \tag{9.77}$$

This is known as the Bohr radius for the hydrogen atom. In terms of this quantity we can write down the radial solution, (9.75), as

$$R_{n,\ell}(r) \sim e^{-\frac{r}{n a_0}} \left(\frac{2r}{n a_0}\right)^\ell L_{n+\ell}^{2\ell+1}\left(\tfrac{2r}{n a_0}\right). \tag{9.78}$$

Remembering that $L_{n+\ell}^{2\ell+1}$ is a polynomial of order $(n-\ell-1)$, the leading behavior of the wave function for large r is given by $(r \gg a_0)$

$$R_{n,\ell}(r) \sim r^{n-1} e^{-\frac{r}{na_0}}, \tag{9.79}$$

independent of the value of ℓ. Therefore, the dominant (leading) behavior of the probability for finding the electron (in the nth state) in a spherical shell of radius r and thickness dr, when r is large, can be obtained to be

$$\begin{aligned}\int_{\Omega} r^2 \mathrm{d}r\mathrm{d}\Omega\, \psi^*_{n,\ell,m}\psi_{n,\ell,m} &= r^2\mathrm{d}r\, R^2_{n,\ell}(r)\\ &\approx e^{-\frac{2r}{na_0}} r^{2n}\mathrm{d}r = P_n(r)\mathrm{d}r.\end{aligned} \tag{9.80}$$

We can, therefore, determine the radius of maximum probability for finding the electron in the nth state as

$$\begin{aligned}&\frac{\mathrm{d}P_n(r)}{\mathrm{d}r} = \frac{\mathrm{d}}{\mathrm{d}r}\left(r^{2n} e^{-\frac{2r}{na_0}}\right) = 0,\\ \text{or,}\quad & 2nr^{2n-1}e^{-\frac{2r}{na_0}} - \frac{2}{na_0} r^{2n} e^{-\frac{2r}{na_0}} = 0,\\ \text{or,}\quad & \left(r - n^2 a_0\right) r^{2n-1} e^{-\frac{2r}{na_0}} = 0,\\ \text{or,}\quad & r_{\max} = n^2 a_0.\end{aligned} \tag{9.81}$$

Thus, we see that the Bohr radius, a_0, is the most probable value of r in the ground state $(n = 1)$ and, therefore, defines the natural size of the hydrogen atom. We also see that $r_{\max}$ grows as n^2 for higher states.

The theory also possesses a natural energy scale. Let us define

$$\mathrm{Ry} = \frac{\mu e^4}{2\hbar^2} = \text{Rydberg}. \tag{9.82}$$

In terms of the Rydberg, the energy levels of the hydrogen atom, (9.42), can be written as

$$E_n = -\frac{\mathrm{Ry}}{n^2}. \tag{9.83}$$

Numerical estimates. Let us estimate the values of these length and energy scales. From the definition of the Bohr radius, (9.77), we note that

$$a_0 = \frac{\hbar^2}{\mu e^2} = \frac{(\hbar c)^2}{\mu c^2 e^2} = \frac{\hbar c}{\mu c^2}\frac{\hbar c}{e^2}. \tag{9.84}$$

We know that $\hbar c \simeq 2000$ eV-A° (1A° $= 10^{-8}$ cm). Furthermore, as we have seen in (9.23), $\mu c^2 \simeq .5$ MeV $= 5\times 10^5$ eV. We also know that

$$\alpha = \frac{e^2}{\hbar c} = \text{ fine structure constant } \simeq \frac{1}{137}. \tag{9.85}$$

The fine structure constant is a dimensionless constant that measures the strength of electromagnetic interactions. It follows from these that

$$a_0 \simeq \frac{2000 \text{ eV-A}^\circ}{5\times 10^5 \text{ eV}} \times 137 \simeq .5 \text{ A}^\circ, \tag{9.86}$$

which determines the size of the hydrogen atom. Furthermore,

$$\begin{aligned} \text{Ry} &= \frac{\mu e^4}{2\hbar^2} = \frac{\mu c^2 e^4}{2(\hbar c)^2} = \frac{\mu c^2}{2} \times \left(\frac{e^2}{\hbar c}\right)^2 \\ &\simeq \frac{5\times 10^5}{2} \text{ eV} \times \left(\frac{1}{137}\right)^2 \simeq 13.3 \text{ eV}. \end{aligned} \tag{9.87}$$

A more accurate value for the Rydberg is 13.6 eV (namely, if we use $\mu c^2 = .511$ MeV and so on). So using this more accurate value, we can write the energy levels of hydrogen, (9.83), as

$$E_n = -\frac{13.6}{n^2} \text{ eV}. \tag{9.88}$$

Thus, the ground state of hydrogen, which is the most tightly bound, has an energy -13.6 eV and, therefore, it would take 13.6 eV to release the electron from its ground state. Consequently, this is also known as the binding energy of the hydrogen atom.

9.3.1 Comparison with experiment. Quantum mechanics predicts explicit energy levels for the hydrogen atom and, in principle, we can measure these. However, in practice, we only measure the relative separation between the energy levels. For example, if an atom is in the state characterized by the quantum numbers (n,ℓ,m) with energy E_n, it would remain there forever – since that is a stationary state. However, if we disturb the system, it may make a transition to another state (n',ℓ',m') with energy $E_{n'} \neq E_n$. Furthermore, if $E_{n'} < E_n$, the atom would emit a photon with energy $(E_n - E_{n'})$. Thus, the frequency of the emitted photon would be

$$\omega_{n,n'} = \frac{1}{\hbar}\left(E_n - E_{n'}\right), \tag{9.89}$$

and can be measured in the laboratory. Quantum mechanically, it follows from (9.88) that in the hydrogen atom we must have

$$\omega_{n,n'} = \frac{\mathrm{Ry}}{\hbar}\left(\frac{1}{n'^2} - \frac{1}{n^2}\right). \tag{9.90}$$

For a fixed value of n' we get a family of lines (spectra) as we vary n (for transitions to a given level). Thus, for example,

$$\omega_{n,1} = \frac{\mathrm{Ry}}{\hbar}\left(1 - \frac{1}{n^2}\right), \tag{9.91}$$

is called the Lyman series. Similarly,

$$\omega_{n,2} = \frac{\mathrm{Ry}}{\hbar}\left(\frac{1}{2^2} - \frac{1}{n^2}\right), \tag{9.92}$$

is known as the Balmer series and so on. These lines are observed experimentally and agree with the quantum mechanical predictions. However, there are slight discrepancies between the measurements and the theoretical predictions. But these can all be explained as limitations in the form in which we have used quantum mechanics. For example, we have to correct for the fact that the proton is not really immobile, i.e., it does not have an infinite mass.

Furthermore, we have treated the electron as a non-relativistic particle whereas in reality one finds that the relativistic effects are not completely negligible. These are known as fine structure corrections and are calculable. (We will derive these later when we discuss perturbation methods.) However, we must remember that all such corrections are extremely small and that the non-relativistic Schrödinger equation describes the hydrogen atom extremely well.

Degeneracy of states in the hydrogen atom. The hydrogen atom possesses rotational symmetry. This implies that

$$[H, L_i] = 0. \tag{9.93}$$

In particular, this implies that

$$[H, L_\pm] = 0. \tag{9.94}$$

However, $L_\pm$ change the m-quantum numbers for a given ℓ. Since $L_\pm$ commute with the Hamiltonian, this implies that all the $2\ell + 1$ states with different m-values have the same energy. Thus, rotational invariance implies degeneracy in the m-quantum numbers. On the other hand, we have noted in (9.44) that in the hydrogen atom the orbital angular momentum ℓ takes integer values $0, 1, \ldots, n-1$ for a given value of n. And, furthermore, since the energy levels are characterized by the n-quantum number only, all these states with different ℓ-quantum numbers also have the same energy. Thus, for example, the degeneracy in the first few levels are given by

$$\begin{array}{llll} n & E_n & \ell & m \\ 1 & -13.6 \text{ eV} & 0 & 0 \\ 2 & -\frac{13.6}{4} \text{ eV} & 0,1 & 0; \pm 1, 0 \\ 3 & -\frac{13.6}{9} \text{ eV} & 0,1,2 & 0; \pm 1, 0; \pm 2, \pm 1, 0. \end{array} \tag{9.95}$$

In general, the total number of degenerate states for a given n in the case of the hydrogen atom is obtained as

$$\begin{aligned} \sum_{\ell=0}^{n-1} (2\ell+1) &= 2 \times \frac{1}{2}(n-1)n + n \\ &= n(n-1+1) = n^2. \end{aligned} \tag{9.96}$$

We had seen a similar degeneracy in the ℓ quantum number earlier in the study of the 3-dimensional isotropic harmonic oscillator and had characterized this as accidental degeneracy. Such a degeneracy is not explained by the rotational symmetry since there is no operator within the rotation group which would change the ℓ-quantum number. In fact, this degeneracy is a consequence of the special form of the potential which gives rise to a larger symmetry in the system under study as we will see next.

9.4 Dynamical symmetry in hydrogen

As we have noted earlier, a special form of the potential can sometimes enhance the geometrical symmetry that we would expect in a system. Consider, for example, the classical Keplerian problem where

$$H = \frac{\mathbf{p}^2}{2\mu} - \frac{\kappa}{r}, \tag{9.97}$$

where κ is a constant. For the case of the hydrogen, we can identify $\kappa = e^2$, but let us leave it arbitrary for the present.

This is a rotationally invariant system and hence angular momentum is conserved. In this case, we know the classical trajectory of the particle to be elliptical. Symmetry considerations alone (that is, rotational symmetry) tell us that the motion must be planar. Namely, the motion must lie in a plane so that the angular momentum vector does not change. However, rotational symmetry alone does not require the orbit to be closed. For example, if we perturb the orbit slightly from its closed form, it can, in principle, precess in the same plane without violating angular momentum conservation. On the other hand, the fact that closed orbits are stable implies that there must be another quantity which is conserved and lies in the plane of motion so that the orbit remains closed. In fact, we know of such a vector in this classical problem, which is known as the Runge-Lenz vector and is defined as

$$\mathbf{M} = \frac{\mathbf{p}\times\mathbf{L}}{\mu} - \frac{\kappa}{r}\mathbf{r}. \tag{9.98}$$

It is obvious from the definition in (9.98) that

$$\begin{aligned}
&\mathbf{L}\cdot\mathbf{M} = 0,\\
&\mathbf{M}^2 = M_i^2 = \left(\frac{1}{\mu}\epsilon_{ijk}p_jL_k - \frac{\kappa}{r}r_i\right)^2\\
&\qquad = \frac{1}{\mu^2}\mathbf{p}^2\mathbf{L}^2 - \frac{2\kappa}{\mu r}\mathbf{L}^2 + \kappa^2\\
&\qquad = \frac{2H\mathbf{L}^2}{\mu} + \kappa^2.
\end{aligned} \tag{9.99}$$

It is straightforward to show that $\mathbf{M}$ is a conserved quantity, simply by calculating its Poisson bracket with the Hamiltonian. (It is, of course, obvious that $\mathbf{M}^2$ is conserved.) Furthermore, this is a vector that is orthogonal to the angular momentum and, therefore, would lie in the plane of motion. The conservation of the Runge-Lenz vector is the reason that closed orbits are stable.

Quantum mechanically, for the hydrogen atom, we define the Runge-Lenz operator as ($\kappa = e^2$)

$$M_i = \frac{1}{2\mu}\epsilon_{ijk}\left(p_jL_k - L_jp_k\right) - \frac{e^2}{r}r_i, \quad L_iM_i = \mathbf{L}\cdot\mathbf{M} = 0, \tag{9.100}$$

where symmetrization of products has been used. Using the fundamental canonical commutators, we can show that

$$[M_i, H] = 0. \tag{9.101}$$

Therefore, these are conserved in the quantum system as is the case classically. Furthermore,

$$\mathbf{M}^2 = M_i^2 = \frac{2H}{\mu}\left(\mathbf{L}^2 + \hbar^2\right) + e^4, \tag{9.102}$$

where the $\hbar^2$ term arises from the non-commutativity of quantum operators.

We already know that the angular momentum operators satisfy the commutation relations,

$$[L_i, L_j] = i\hbar\epsilon_{ijk}L_k. \tag{9.103}$$

We can also calculate in a simple manner that

$$\begin{aligned} [M_i, L_j] &= i\hbar\epsilon_{ijk}M_k, \\ [M_i, M_j] &= -\frac{2i\hbar}{\mu}\epsilon_{ijk}HL_k. \end{aligned} \tag{9.104}$$

The L_i's, of course, generate rotations and define a closed algebra. But, L_i's and M_i's do not form a closed algebra since the last relation involves the Hamiltonian. However, remembering that the Hamiltonian is independent of time, we can work in the subspace of the Hilbert space that corresponds to a particular energy value, say, E. In this subspace, the last relation in (9.104) becomes

$$[M_i, M_j] = -\frac{2i\hbar E}{\mu}\epsilon_{ijk}L_k. \tag{9.105}$$

If we are interested in bound states, we note that $E < 0$ and if we scale the generators as

$$M_i \to \left(-\frac{\mu}{2E}\right)^{\frac{1}{2}} M_i, \tag{9.106}$$

then, the commutation relations in (9.103) and (9.104) take the simple form (in a subspace of constant energy)

$$\begin{aligned} [L_i, L_j] &= i\hbar\epsilon_{ijk}L_k, \\ [M_i, L_j] &= i\hbar\epsilon_{ijk}M_k, \\ [M_i, M_j] &= i\hbar\epsilon_{ijk}L_k. \end{aligned} \tag{9.107}$$

This defines a larger symmetry algebra that is operative in this system. This algebra is isomorphic (equivalent) to the algebra of rotations in 4-dimensions (or the O(4) group). Since our system is

3-dimensional whereas the symmetry is four dimensional, this symmetry is called a dynamical symmetry as opposed to a geometrical symmetry. Let us note from the above algebra that

$$[M_i, \mathbf{L}^2] \neq 0. \tag{9.108}$$

Therefore, in this larger algebra there exist operators that do not commute with $\mathbf{L}^2$ and, therefore, can change the ℓ quantum number.

To understand the representations of this algebra a little better, let us define two sets of new generators as

$$\begin{aligned} I_i &= \frac{1}{2}(L_i + M_i), \\ K_i &= \frac{1}{2}(L_i - M_i). \end{aligned} \tag{9.109}$$

It is then straightforward to show that

$$\begin{aligned} [I_i, I_j] &= i\hbar\epsilon_{ijk} I_k, \\ [K_i, K_j] &= i\hbar\epsilon_{ijk} K_k, \\ [I_i, K_j] &= 0. \end{aligned} \tag{9.110}$$

Furthermore, it is obvious that $[I_i, H] = [K_i, H] = 0$ so that these operators are also conserved.

As we see from (9.110), in this basis the algebra becomes equivalent to that of two decoupled algebras of angular momenta. It follows from our earlier discussion of the angular momentum algebra that the operators $\mathbf{I}^2$ and $\mathbf{K}^2$ will have the eigenvalues

$$\mathbf{I}^2 : \hbar^2 i(i+1), \qquad \mathbf{K}^2 : \hbar^2 k(k+1), \tag{9.111}$$

where $i, k = 0, \frac{1}{2}, 1, \ldots$. We recognize that $\mathbf{I}^2$ and $\mathbf{K}^2$ are the two quadratic Casimir operators of the algebra. Equivalently, we can also define two operators as linear combinations

$$\begin{aligned} C &= \mathbf{I}^2 + \mathbf{K}^2 = \frac{1}{2}\left(\mathbf{L}^2 + \mathbf{M}^2\right), \\ C' &= \mathbf{I}^2 - \mathbf{K}^2 = \mathbf{L}\cdot\mathbf{M} = 0, \end{aligned} \tag{9.112}$$

where the second relation is the quantum analog of the orthogonality of $\mathbf{L}$ and $\mathbf{M}$ in (9.99) (see (9.100)). Furthermore, the second relation implies that, for any representation, we must have $i = k$. Correspondingly, the allowed values of C, for any representation, are

$$C: \quad 2\hbar^2 k(k+1). \tag{9.113}$$

We know that, in this subspace (remember the scaling in (9.106), $M_i \to (-\frac{\mu}{2E})^{\frac{1}{2}} M_i$),

$$\mathbf{M}^2 = -\left(\mathbf{L}^2 + \hbar^2\right) - \frac{\mu e^4}{2E}, \tag{9.114}$$

so that we can write

$$\begin{aligned} C &= \frac{1}{2}\left(\mathbf{L}^2 + \mathbf{M}^2\right) \\ &= \frac{1}{2}\left(\mathbf{L}^2 - \mathbf{L}^2 - \hbar^2 - \frac{\mu e^4}{2E}\right) \\ &= \left(-\frac{\mu e^4}{4E} - \frac{1}{2}\hbar^2\right)\mathbb{1} = 2\hbar^2 k(k+1)\mathbb{1}, \\ \text{or,}\quad E &= -\frac{\mu e^4}{2\hbar^2(2k+1)^2} = -\frac{\mu e^4}{2\hbar^2 n^2}, \end{aligned} \tag{9.115}$$

where $n = 2k+1 = 1, 2, 3, \ldots$ (remember that $k = 0, \frac{1}{2}, 1, \frac{3}{2}, \cdots$ for angular momentum algebra).

Thus, we see that we get the right energy levels for the hydrogen atom from this operator analysis of the symmetries in the theory. Let us note that it is not objectionable for the eigenvalue k to be half integer. We only have the physical requirement that the eigenvalues of the $\mathbf{L}$ operator take integer values. Since $\mathbf{L} = \mathbf{I} + \mathbf{K}$ (see (9.109)) we see that it takes values (we are assuming here the composition of angular momentum which we will discuss later)

$$\ell = (i+k, i+k-1, \cdots, |i-k|) = (n-1, n-2, \ldots, 0). \tag{9.116}$$

All these levels would be degenerate in energy. This explains the peculiar degeneracy noticed in this system.

9.5 Selected problems

1. a) Prove the Thomas-Reiche-Kuhn sum rule

$$\sum_{n'} (E_{n'} - E_n)\, |\langle n'|X|n\rangle|^2 = \frac{\hbar^2}{2m}, \tag{9.117}$$

where X is the coordinate operator and $|n\rangle$ represent the eigenstates of the Hamiltonian

$$H = \frac{p^2}{2m} + V(X). \tag{9.118}$$

b) Test the sum rule on the n-th state of the harmonic oscillator.

2. Consider the Runge-Lenz operator defined in (9.100)

$$M_i = \frac{1}{2\mu}\epsilon_{ijk}\left(p_j L_k - L_j p_k\right) - \frac{e^2}{r} r_i, \tag{9.119}$$

where L_i denotes the three orbital angular momentum operators. Given that the Hamiltonian for the Hydrogen atom has the form

$$H = \frac{\mathbf{p}^2}{2\mu} - \frac{e^2}{r}, \tag{9.120}$$

show that

a) $[M_i, H] = 0$.

b) $[M_i, L_j] = i\hbar\epsilon_{ijk}M_k$.

c) $[M_i, M_j] = -\frac{2i\hbar}{\mu}\epsilon_{ijk}HL_k$.

3. Find the lowest energy eigenstate (ground state) of the hydrogen atom in the coordinate representation starting from the operator formalism of $O(4)$ symmetry discussed in the last section.

4. What are the generators of the larger symmetry group for the 3-dimensional isotropic oscillator? What is the algebra they satisfy? (The larger symmetry in the oscillator case corresponds to the group $SU(3)$ which has eight generators.)

CHAPTER 10

Approximate methods

So far, we have studied quantum mechanical systems that can be exactly solved. Most often, however, we are confronted with systems which are very difficult (if not impossible) to solve exactly. That is, the Hamiltonians for such systems cannot be diagonalized exactly in a simple way. In such a case, we look for approximate methods for finding the eigenvalues and eigenstates of the Hamiltonian. And sometimes, these approximate methods give results which are amazingly close to the true experimental values. There are various approximate methods which one can apply to different physical problems. We will discuss all these methods systematically starting with the variational method.

10.1 Variational method

The variational method is an excellent approximate method when we are interested in estimates of the ground state (or higher) energy of a complicated physical system. The basic idea behind the variational method is contained in the following two theorems.

Theorem. *The expectation value of the Hamiltonian of a physical system is stationary in the neighborhood of its eigenstates.*

This theorem is also known as the Ritz theorem.

Proof. Let us assume that $|\psi\rangle$ is a state in which we are evaluating the expectation value of the Hamiltonian. The expectation value of the Hamiltonian, in this state, is defined to be

$$\begin{aligned} \langle H \rangle &= \frac{\langle\psi|H|\psi\rangle}{\langle\psi|\psi\rangle}, \\ \text{or,} \quad \langle\psi|\psi\rangle\langle H\rangle &= \langle\psi|H|\psi\rangle. \end{aligned} \tag{10.1}$$

Here we are assuming that $|\psi\rangle$, in general, is not normalized.

Let us next modify the state infinitesimally to

$$|\psi\rangle \to |\psi\rangle + |\delta\psi\rangle. \tag{10.2}$$

The infinitesimal change in the expectation value of the Hamiltonian, introduced by changing the state, can be calculated as follows (keeping terms up to linear order in the change). First, we note that

$$\delta(\langle\psi|\psi\rangle\langle H\rangle) = \langle\delta\psi|\psi\rangle\langle H\rangle + \langle\psi|\delta\psi\rangle\langle H\rangle + \langle\psi|\psi\rangle\delta\langle H\rangle. \tag{10.3}$$

On the other hand, we see that since the Hamiltonian is unchanged,

$$\delta\langle\psi|H|\psi\rangle = \langle\delta\psi|H|\psi\rangle + \langle\psi|H|\delta\psi\rangle. \tag{10.4}$$

Thus, comparing (10.3) and (10.4), we obtain from (10.1),

$$\langle\psi|\psi\rangle\delta\langle H\rangle = \langle\delta\psi|(H - \langle H\rangle)|\psi\rangle + \langle\psi|(H - \langle H\rangle)|\delta\psi\rangle. \tag{10.5}$$

Let us next define

$$(H - \langle H\rangle)|\psi\rangle = |\phi\rangle. \tag{10.6}$$

Thus, the relation (10.5) can be written as

$$\langle\psi|\psi\rangle\delta\langle H\rangle = \langle\delta\psi|\phi\rangle + \langle\phi|\delta\psi\rangle. \tag{10.7}$$

As a result, we conclude that the expectation value of H, in the state $|\psi\rangle$, will be stationary (namely, $\delta\langle H\rangle = 0$) if

$$\begin{aligned} &\langle\delta\psi|\phi\rangle + \langle\phi|\delta\psi\rangle = 0,\\ \text{or,}\quad &\text{Real } \langle\phi|\delta\psi\rangle = 0. \end{aligned} \tag{10.8}$$

This last relation must be true for any $|\delta\psi\rangle$ in order that $\langle H\rangle$ is stationary. In particular, it must be true if (ϵ is an infinitesimal real parameter)

$$|\delta\psi\rangle = \epsilon|\phi\rangle. \tag{10.9}$$

However, in this case, we will have from (10.8)

$$\langle\phi|\delta\psi\rangle = \epsilon\langle\phi|\phi\rangle = 0. \tag{10.10}$$

In other words, the norm of the state $|\phi\rangle$ must be zero, which, in turn, implies that $|\phi\rangle$ must be the null vector. Thus,

$$\begin{aligned} &|\phi\rangle = (H - \langle H\rangle)|\psi\rangle = 0,\\ \text{or,}\quad &H|\psi\rangle = \langle H\rangle|\psi\rangle. \end{aligned} \tag{10.11}$$

This shows that $|\psi\rangle$ must be an eigenstate of the Hamiltonian for its expectation value to be stationary and this proves the theorem that the expectation value of the Hamiltonian is stationary near its eigenstates. (An alternate, simple way to see this is to note that, for all infinitesimal changes satisfying $\langle\delta\psi|\phi\rangle^* = \langle\delta\psi|\phi\rangle$, the expectation value of the Hamiltonian can be stationary only if $\langle\delta\psi|\phi\rangle = 0$, which implies that $|\phi\rangle = 0$.) ∎

Let us now assume that the state in which the expectation value is evaluated is not an exact eigenstate but differs infinitesimally from one. Namely, let

$$|\psi\rangle = |\psi_n\rangle + \epsilon|\psi_{n+1}\rangle, \tag{10.12}$$

where ϵ is an infinitesimal, real parameter and $|\psi_n\rangle$ and $|\psi_{n+1}\rangle$ are eigenstates of the Hamiltonian with eigenvalues E_n and E_{n+1} respectively. Let us now calculate

$$\begin{aligned}
\langle H\rangle &= \frac{\langle\psi|H|\psi\rangle}{\langle\psi|\psi\rangle} \\
&= \frac{(\langle\psi_n| + \epsilon\langle\psi_{n+1}|)H(|\psi_n\rangle + \epsilon|\psi_{n+1}\rangle)}{(\langle\psi_n| + \epsilon\langle\psi_{n+1}|)(|\psi_n\rangle + \epsilon|\psi_{n+1}\rangle)} \\
&= \frac{(\langle\psi_n| + \epsilon\langle\psi_{n+1}|)(E_n|\psi_n\rangle + \epsilon E_{n+1}|\psi_{n+1}\rangle)}{\langle\psi_n|\psi_n\rangle + \epsilon^2\langle\psi_{n+1}|\psi_{n+1}\rangle} \\
&= \frac{E_n + \epsilon^2 E_{n+1}}{1+\epsilon^2} \approx E_n(1-\epsilon^2) + \epsilon^2 E_{n+1} \\
&= E_n + \epsilon^2(E_{n+1} - E_n) = E_n + O(\epsilon^2).
\end{aligned} \tag{10.13}$$

This shows that if a wave function differs from an energy eigenstate by order ϵ terms, then, the expectation value of the Hamiltonian in this state will be different from the corresponding energy eigenvalue by $O(\epsilon^2)$ terms. Therefore, by cleverly choosing a wave function, we can come very close to the true eigenvalue of the Hamiltonian.

Exercise. Define $|\psi'\rangle = |\psi\rangle + |\delta\psi\rangle$. Then,

$$\delta\langle H\rangle = \frac{\langle\psi'|H|\psi'\rangle}{\langle\psi'|\psi'\rangle} - \frac{\langle\psi|H|\psi\rangle}{\langle\psi|\psi\rangle}.$$

Show from this definition that if $|\delta\psi\rangle$ is infinitesimally small, we obtain the result derived in (10.7).

Theorem. *The expectation value of the Hamiltonian in an arbitrary state is greater than or equal to the ground state energy.*

$$\frac{\langle\psi|H|\psi\rangle}{\langle\psi|\psi\rangle} \geq E_0. \tag{10.14}$$

Proof. To prove this, let us expand the state in the basis of the eigenstates of the Hamiltonian. Namely, let

$$|\psi\rangle = \sum_n c_n |\psi_n\rangle, \tag{10.15}$$

where $|\psi_n\rangle$'s denote eigenstates of the Hamiltonian with eigenvalues E_n. (Here we are assuming the eigenvalues to be discrete. This can always be achieved by quantizing in a box. However, the theorem is true otherwise also.) Thus,

$$\langle\psi|\psi\rangle = \sum_n |c_n|^2,$$

$$\langle\psi|H|\psi\rangle = \sum_n E_n |c_n|^2,$$

$$\frac{\langle\psi|H|\psi\rangle}{\langle\psi|\psi\rangle} = \frac{\sum\limits_n E_n |c_n|^2}{\sum\limits_n |c_n|^2}. \tag{10.16}$$

By definition, $E_n \geq E_0$ for all n (where E_0 represents the ground state energy), so that

$$\frac{\langle\psi|H|\psi\rangle}{\langle\psi|\psi\rangle} \geq E_0 \frac{\sum\limits_n |c_n|^2}{\sum\limits_n |c_n|^2} = E_0. \tag{10.17}$$

This proves that the expectation value of the Hamiltonian in any arbitrary state is greater than or equal to the ground state energy. The equality in (10.17) holds when $|\psi\rangle$ happens to coincide with the ground state of the system. ■

The variational method makes use of both of these theorems and is mostly used to determine an upper bound on the ground state energy of a system. First of all, we know that the ground state energy is an absolute minimum of the expectation value of the Hamiltonian in any state. Therefore, we can choose a wave function which would resemble the ground state wave function as much as is possible, but allow for a few undetermined parameters in it. The expectation value of the Hamiltonian in such a state, therefore, becomes a function of these parameters. The expectation value is then minimized with respect to these parameters, which determines some or all of the values of these parameters. Furthermore, corresponding to these parameters, the Hamiltonian has an expectation value, which serves as an upper bound on the true ground state energy. We can try to lower this upper bound even further by introducing more and more parameters and using more general wave functions. When there is difficulty in lowering the bound any further, we can think of the expectation value as essentially close to the true ground state energy.

The choice of a good trial wave function is, therefore, a crucial part of this method. We, therefore, try to invoke all the symmetry principles of the system and appeal to all physical intuitions about the ground state of the system in choosing a trial wave function.

The variational method can give a very good estimate of the energy eigenvalues. But, it does not determine the wave function accurately. In other words, we cannot take the wave function and, therefore, the expectation values of other observables calculated with it seriously. This is because other observables may not satisfy any minimum theorem (such as the Ritz theorem).

Furthermore, we can also estimate upper bounds on the energy eigenvalues of the excited states from the variational method. This is done simply by noting that if we choose, as a trial wave function, a state which is orthogonal to the ground state wave function, it would give an upper bound for the energy of the first excited state. By considering a series of orthogonal states we can, therefore, in principle determine upper bounds on all the eigenvalues of the Hamiltonian.

10.2 Harmonic oscillator

We have already studied the harmonic oscillator in one dimension in detail. Let us analyze here the ground state energy of this system from the point of view of the variational method. The Hamiltonian for the system has the form (in the coordinate basis)

$$H = -\frac{\hbar^2}{2m}\,\frac{\mathrm{d}^2}{\mathrm{d}x^2} + \frac{1}{2}\,m\omega^2 x^2. \tag{10.18}$$

We are interested in determining the ground state energy of this system using the variational method and are, therefore, looking for a trial wave function that reflects all the properties of the true ground state wave function. Furthermore, it must have a simple enough form for manipulations (for carrying out computations).

First of all, we recall that the system is invariant under parity, i.e., if $x \leftrightarrow -x$, the Hamiltonian is invariant. Therefore, all solutions of this system can be classified as even or odd under the parity operation. We are looking for the ground state wave function. In one dimension, this would correspond to a function without any node.

Thus, this must be an even function. Furthermore, we are looking for bound state solutions which vanish at spatial infinity in either directions. We can, therefore, choose as a trial wave function

$$\psi_\alpha^{(0)}(x) = e^{-\alpha x^2}, \qquad \alpha > 0, \tag{10.19}$$

which satisfies all of our symmetry requirements and the constant α corresponds to the variational parameter.

Let us next calculate

$$\begin{aligned}
\langle\psi_\alpha^{(0)}|\psi_\alpha^{(0)}\rangle &= \int \mathrm{d}x\ \psi_\alpha^{(0)*}(x)\psi_\alpha^{(0)}(x) \\
&= \int \mathrm{d}x\ e^{-2\alpha x^2} = \left(\frac{\pi}{2\alpha}\right)^{\frac{1}{2}}, \\
\langle\psi_\alpha^{(0)}|V|\psi_\alpha^{(0)}\rangle &= \frac{1}{2}m\omega^2 \int \mathrm{d}x\ x^2 e^{-2\alpha x^2} \\
&= \frac{m\omega^2}{2}\left(-\frac{1}{2}\frac{\mathrm{d}}{\mathrm{d}\alpha}\int \mathrm{d}x\ e^{-2\alpha x^2}\right) \\
&= -\frac{m\omega^2}{4}\frac{\mathrm{d}}{\mathrm{d}\alpha}\left(\frac{\pi}{2\alpha}\right)^{\frac{1}{2}} = \frac{m\omega^2}{8\alpha}\left(\frac{\pi}{2\alpha}\right)^{\frac{1}{2}}, \\
\langle\psi_\alpha^{(0)}|T|\psi_\alpha^{(0)}\rangle &= -\frac{\hbar^2}{2m}\int \mathrm{d}x\ \psi_\alpha^{(0)*}(x)\frac{\mathrm{d}^2}{\mathrm{d}x^2}\psi_\alpha^{(0)}(x) \\
&= -\frac{\hbar^2}{2m}\int \mathrm{d}x\ e^{-\alpha x^2}(-2\alpha + 4\alpha^2 x^2)e^{-\alpha x^2} \\
&= -\frac{\hbar^2}{2m}\int \mathrm{d}x\ (-2\alpha + 4\alpha^2 x^2)e^{-2\alpha x^2} \\
&= \frac{\hbar^2\alpha}{m}\left(\frac{\pi}{2\alpha}\right)^{\frac{1}{2}} - \frac{\hbar^2\alpha}{2m}\left(\frac{\pi}{2\alpha}\right)^{\frac{1}{2}} \\
&= \frac{\hbar^2\alpha}{2m}\left(\frac{\pi}{2\alpha}\right)^{\frac{1}{2}}.
\end{aligned} \tag{10.20}$$

Therefore, we obtain

$$\langle\psi_\alpha^{(0)}|H|\psi_\alpha^{(0)}\rangle = \langle\psi_\alpha^{0}|T+V|\psi_\alpha^{0}\rangle = \left(\frac{\pi}{2\alpha}\right)^{\frac{1}{2}}\left(\frac{\hbar^2\alpha}{2m} + \frac{m\omega^2}{8\alpha}\right),$$

$$\text{or,}\quad \langle H\rangle = \frac{\langle\psi_\alpha^{(0)}|H|\psi_\alpha^{(0)}\rangle}{\langle\psi_\alpha^{(0)}|\psi_\alpha^{(0)}\rangle} = \frac{\hbar^2\alpha}{2m} + \frac{m\omega^2}{8\alpha}. \tag{10.21}$$

The value of the variational parameter α which gives the minimum of the function in (10.21) is determined to be

$$\frac{\mathrm{d}\langle H\rangle}{\mathrm{d}\alpha} = 0,$$

$$\text{or,}\quad \frac{\hbar^2}{2m} - \frac{m\omega^2}{8\alpha^2} = 0,$$

$$\text{or,}\quad \alpha_{\min} = \frac{m\omega}{2\hbar}. \tag{10.22}$$

(The other value of the extremum equation leads to an unphysical solution, namely, a solution that diverges at infinity.) Substituting this into (10.21), we determine the value of the upper bound for the ground state energy to be

$$\langle H\rangle_{\min} = \frac{\hbar\omega}{2},$$

$$\psi^{(0)}_{\alpha_{\min}} = e^{-\frac{m\omega}{2\hbar}x^2}. \tag{10.23}$$

We note that, in this particular example, the ground state energy comes out exact because of our choice of the form of the trial wave function in (10.19).

If we want to calculate the energy for the first excited state of the oscillator using the variational method, we have to choose a wave function that is orthogonal to the ground state. Furthermore, it must have one node and vanish at infinity. This leads to the choice of a wave function of the form

$$\psi^{(1)}_{\beta}(x) = xe^{-\beta x^2}, \quad \beta > 0,$$

$$\int \mathrm{d}x\, \psi^{(1)*}_{\beta}(x)\psi^{(0)}_{\alpha_{\min}}(x) = \int \mathrm{d}x\, xe^{-(\alpha_{\min}+\beta)x^2} = 0. \tag{10.24}$$

This trial wave function leads to

$$\langle\psi^{(1)}_{\beta}|\psi^{(1)}_{\beta}\rangle = \int \mathrm{d}x\, \psi^{(1)*}_{\beta}(x)\psi^{(1)}_{\beta}(x)$$

$$= \int \mathrm{d}x\, x^2 e^{-2\beta x^2} = \frac{1}{4\beta}\left(\frac{\pi}{2\beta}\right)^{\frac{1}{2}},$$

$$\langle\psi^{(1)}_{\beta}|V|\psi^{(1)}_{\beta}\rangle = \frac{1}{2}m\omega^2 \int \mathrm{d}x\, x^4 e^{-2\beta x^2}$$

$$= \frac{m\omega^2}{2}\left(-\frac{1}{2}\frac{\mathrm{d}}{\mathrm{d}\beta}\left(\frac{1}{4\beta}\left(\frac{\pi}{2\beta}\right)^{\frac{1}{2}}\right)\right)$$

$$
\begin{aligned}
&= \frac{m\omega^2}{2} \times \frac{1}{2} \times \frac{1}{4} \times \frac{3}{2\beta^2}\left(\frac{\pi}{2\beta}\right)^{\frac{1}{2}} \\
&= \frac{3m\omega^2}{8\beta}\frac{1}{4\beta}\left(\frac{\pi}{2\beta}\right)^{\frac{1}{2}}, \\
\langle\psi_\beta^{(1)}|T|\psi_\beta^{(1)}\rangle &= -\frac{\hbar^2}{2m}\int \mathrm{d}x\; xe^{-\beta x^2}\frac{\mathrm{d}^2}{\mathrm{d}x^2}(xe^{-\beta x^2}) \\
&= -\frac{\hbar^2}{2m}\int \mathrm{d}x\; (-6\beta x^2 + 4\beta^2 x^4)e^{-2\beta x^2} \\
&= -\frac{\hbar^2}{2m}\left(-6\beta + 4\beta^2 \times \frac{3}{4\beta}\right)\frac{1}{4\beta}\left(\frac{\pi}{2\beta}\right)^{\frac{1}{2}} \\
&= \frac{3\hbar^2\beta}{2m}\frac{1}{4\beta}\left(\frac{\pi}{2\beta}\right)^{\frac{1}{2}}. \qquad (10.25)
\end{aligned}
$$

Therefore, we obtain,

$$
\langle H\rangle = \frac{\langle\psi_\beta^{(1)}|T+V|\psi_\beta^{(1)}\rangle}{\langle\psi_\beta^{(1)}|\psi_\beta^{(1)}\rangle} = \left[\frac{3\hbar^2\beta}{2m} + \frac{3m\omega^2}{8\beta}\right]. \qquad (10.26)
$$

The value of the variational parameter β which gives the minimum of this function is determined to be

$$
\begin{aligned}
&\frac{\mathrm{d}\langle H\rangle}{\mathrm{d}\beta} = 0, \\
\text{or,}\quad &\frac{3\hbar^2}{2m} - \frac{3m\omega^2}{8\beta^2} = 0, \\
\text{or,}\quad &\beta_{\min} = \frac{m\omega}{2\hbar}, \qquad (10.27)
\end{aligned}
$$

which leads, from (10.26), to

$$
\langle H\rangle_{\min} = E_1 = \frac{3\hbar\omega}{2}, \qquad (10.28)
$$

and serves as an upper bound for the energy of the first excited state. We see, again, that we get the exact eigenvalue for the first excited state because of our choice of the trial wave function in (10.24).

Let us now examine how different a bound we would have obtained for the ground state had we chosen a different form for the trial wave function. Let us choose as the trial wave function for the ground state (which satisfies all the symmetry properties)

$$
\psi_\alpha^{(0)}(x) = \frac{1}{x^2+\alpha}, \qquad \alpha > 0. \qquad (10.29)
$$

This leads to

$$
\begin{aligned}
\langle\psi_\alpha^{(0)}|\psi_\alpha^{(0)}\rangle &= \int \mathrm{d}x\ \psi_\alpha^{(0)*}(x)\psi_\alpha^{(0)}(x) = \int \mathrm{d}x\ \frac{1}{(x^2+\alpha)^2} \\
&= 2\pi i \times \frac{1}{4i\alpha\sqrt{\alpha}} = \frac{\pi}{2\alpha\sqrt{\alpha}}, \\
\langle\psi_\alpha^{(0)}|V|\psi_\alpha^{(0)}\rangle &= \frac{1}{2}m\omega^2 \int \mathrm{d}x\ \frac{x^2}{(x^2+\alpha)^2} \\
&= \frac{1}{2}m\omega^2 \int \mathrm{d}x \left[\frac{x^2+\alpha-\alpha}{(x^2+\alpha)^2}\right] \\
&= \frac{1}{2}m\omega^2 \left[\int \mathrm{d}x\ \frac{1}{x^2+\alpha} - \alpha\int \mathrm{d}x\ \frac{1}{(x^2+\alpha)^2}\right] \\
&= \frac{1}{2}m\omega^2 \left[\frac{\pi}{\sqrt{\alpha}} - \alpha\times\frac{\pi}{2\alpha\sqrt{\alpha}}\right] \\
&= \frac{m\omega^2\alpha}{2}\frac{\pi}{2\alpha\sqrt{\alpha}}, \\
\langle\psi_\alpha^{(0)}|T|\psi_\alpha^{(0)}\rangle &= -\frac{\hbar^2}{2m}\int \mathrm{d}x\ \frac{1}{x^2+\alpha}\frac{\mathrm{d}^2}{\mathrm{d}x^2}\left(\frac{1}{x^2+\alpha}\right) \\
&= -\frac{\hbar^2}{2m}\int \mathrm{d}x \left[-2+\frac{8x^2}{x^2+\alpha}\right]\frac{1}{(x^2+\alpha)^3} \\
&= -\frac{\hbar^2}{2m}\int \mathrm{d}x \left[6-\frac{8\alpha}{x^2+\alpha}\right]\frac{1}{(x^2+\alpha)^3} \\
&= -\frac{\hbar^2}{2m}\left[6\times 2\pi i\times\frac{1}{2!}(-3)(-4)\frac{1}{(2i\sqrt{\alpha})^5}\right. \\
&\qquad \left. -8\alpha\times 2\pi i\times\frac{1}{3!}(-4)(-5)(-6)\times\frac{1}{(2i\sqrt{\alpha})^7}\right] \\
&= -\frac{\hbar^2}{2m}\left[\frac{9\pi}{4\alpha^2\sqrt{\alpha}} - \frac{5\pi}{2\alpha^2\sqrt{\alpha}}\right] \\
&= \frac{\hbar^2}{4m\alpha}\frac{\pi}{2\alpha\sqrt{\alpha}}.
\end{aligned}
\tag{10.30}
$$

As a result, we obtain

$$
\langle H\rangle = \frac{\langle\psi_\alpha^{(0)}|T+V|\psi_\alpha^{(0)}\rangle}{\langle\psi_\alpha^{(0)}|\psi_\alpha^{(0)}\rangle} = \frac{\hbar^2}{4m\alpha} + \frac{m\omega^2\alpha}{2}. \tag{10.31}
$$

The minimum of this function occurs at a value of α

$$\frac{\mathrm{d}\langle H\rangle}{\mathrm{d}\alpha}=0,$$

$$\text{or,}\quad -\frac{\hbar^2}{4m\alpha^2}+\frac{m\omega^2}{2}=0,$$

$$\text{or,}\quad \alpha_{\min}=\frac{\hbar}{m\omega\sqrt{2}}, \tag{10.32}$$

which, in turn, gives

$$\begin{aligned}\langle H\rangle_{\min} &= \frac{\hbar^2}{4m}\frac{m\omega\sqrt{2}}{\hbar}+\frac{m\omega^2}{2}\frac{\hbar}{m\omega\sqrt{2}}\\ &= \frac{\hbar\omega}{\sqrt{2}}=\sqrt{2}E_0.\end{aligned} \tag{10.33}$$

As noted earlier, this is larger than the exact value of the ground state energy.

Exercise. Calculate $\langle x\rangle, \langle x^2\rangle$and $\langle x^4\rangle$ in this state and compare with the actual result for these in the ground state of the oscillator.

10.3 Hydrogen atom

Earlier we studied the exact solution of the hydrogen atom. Let us next calculate the ground state energy of the hydrogen atom using the variational method. The Hamiltonian for the hydrogen atom, in the center of mass frame, has the form (see (9.25))

$$H=-\frac{\hbar^2}{2\mu}\boldsymbol{\nabla}^2-\frac{e^2}{r}. \tag{10.34}$$

Here we can think of μ as the mass of the electron if we assume the proton to be infinitely heavy. Otherwise, it can be thought of as the reduced mass of the system. First of all, we note that rotation is a symmetry of the system and, therefore, angular momentum is a conserved quantity. Furthermore, expanding $\boldsymbol{\nabla}^2$ in terms of spherical coordinates (see (8.9)), we note that a non-vanishing angular momentum gives rise to a centrifugal barrier. Therefore, for states with nonzero angular momentum, binding must be weaker. The ground state is, of course, the most tightly bound. Therefore, the angular momentum for this state must be zero so that it must be completely spherically symmetric and hence can only depend on the radial coordinate.

Let us choose as our trial wave function

$$\psi_\alpha^{(0)}(r) = e^{-\alpha r^2}, \qquad \alpha > 0, \quad \text{for bound state.} \tag{10.35}$$

This wave function satisfies all the symmetry requirements of the ground state of the hydrogen atom (and it is different from the true ground state wave function of the system), where α represents a variational parameter. Using the basic integral,

$$\int_0^\infty \mathrm{d}x\, x^n\, e^{-\alpha x^2} = \frac{\Gamma(\frac{n+1}{2})}{2\alpha^{\frac{n+1}{2}}}, \qquad \Gamma(\frac{1}{2}) = \sqrt{\pi}, \tag{10.36}$$

we obtain,

$$\begin{aligned}
\langle\psi_\alpha|\psi_\alpha\rangle &= \int \mathrm{d}^3r\ \psi_\alpha^{(0)*}(r)\psi_\alpha^{(0)}(r) \\
&= 4\pi \int_0^\infty \mathrm{d}r\ r^2 e^{-2\alpha r^2} = 4\pi \frac{\Gamma(\frac{3}{2})}{2(2\alpha)^{\frac{3}{2}}} \\
&= 4\pi \frac{1}{4}\frac{1}{2\alpha}\sqrt{\frac{\pi}{2\alpha}} = \left(\frac{\pi}{2\alpha}\right)^{\frac{3}{2}}, \\
\langle\psi_\alpha|V|\psi_\alpha\rangle &= \int \mathrm{d}^3r\ \psi_\alpha^{(0)*}(r)\left(-\frac{e^2}{r}\right)\psi_\alpha^{(0)}(r) \\
&= -4\pi e^2 \int_0^\infty \mathrm{d}r\, r^2 \frac{1}{r} e^{-2\alpha r^2} \\
&= -4\pi e^2 \int_0^\infty \mathrm{d}r\ r e^{-2\alpha r^2} \\
&= -4\pi e^2 \frac{\Gamma(1)}{2(2\alpha)} = -\frac{\pi e^2}{\alpha} \\
&= -2e^2 \left(\frac{2\alpha}{\pi}\right)^{\frac{1}{2}} \left(\frac{\pi}{2\alpha}\right)^{\frac{3}{2}}, \\
\langle\psi_\alpha|T|\psi_\alpha\rangle &= -\frac{\hbar^2}{2\mu}\int \mathrm{d}^3r\ \psi_\alpha^{(0)*}(r)\left(\frac{1}{r^2}\frac{\mathrm{d}}{\mathrm{d}r}r^2\frac{\mathrm{d}}{\mathrm{d}r}\psi_\alpha^{(0)}(r)\right) \\
&= -\frac{4\pi\hbar^2}{2\mu}\int_0^\infty \mathrm{d}r\ r^2 e^{-\alpha r^2}\left(\frac{1}{r^2}\frac{\mathrm{d}}{\mathrm{d}r}(-2\alpha r^3 e^{-\alpha r^2})\right) \\
&= -\frac{4\pi\hbar^2}{2\mu}\int_0^\infty \mathrm{d}r\ (-6\alpha r^2 + 4\alpha^2 r^4) e^{-2\alpha r^2} \\
&= -\frac{4\pi\hbar^2}{2m}\left(-6\alpha\frac{\Gamma\left(\frac{3}{2}\right)}{2(2\alpha)^{\frac{3}{2}}} + 4\alpha^2\frac{\Gamma\left(\frac{5}{2}\right)}{2(2\alpha)^{\frac{5}{2}}}\right)
\end{aligned}$$

$$= -\frac{4\pi\hbar^2}{2\mu}\left(-\frac{3\alpha}{2}\frac{\sqrt{\pi}}{(2\alpha)^{\frac{3}{2}}} + \frac{3\alpha}{4}\frac{\sqrt{\pi}}{(2\alpha)^{\frac{3}{2}}}\right)$$

$$= \frac{3\hbar^2\alpha}{2\mu}\left(\frac{\pi}{2\alpha}\right)^{\frac{3}{2}}, \tag{10.37}$$

which yields

$$\langle H\rangle = \frac{\langle\psi_\alpha|T+V|\psi_\alpha\rangle}{\langle\psi_\alpha|\psi_\alpha\rangle} = \frac{3\hbar^2\alpha}{2\mu} - 2e^2\sqrt{\frac{2\alpha}{\pi}}. \tag{10.38}$$

The minimum of this occurs for a value of α determined to be

$$\frac{3\hbar^2}{2\mu} - e^2\sqrt{\frac{2}{\pi\alpha}} = 0,$$

$$\text{or,}\quad \alpha_{\text{min}} = \frac{8}{9\pi}\left(\frac{\mu e^2}{\hbar^2}\right)^2 = \frac{8}{9\pi}\frac{1}{a_0^2}, \tag{10.39}$$

where a_0 is the Bohr radius defined in (9.77), so that we have

$$\langle H\rangle_{\text{min}} = \frac{3\hbar^2}{2\mu}\frac{8}{9\pi}\left(\frac{\mu e^2}{\hbar^2}\right)^2 - 2e^2\sqrt{\frac{2}{\pi}}\sqrt{\frac{8}{9\pi}\frac{\mu e^2}{\hbar^2}}$$

$$= \frac{4}{3\pi}\left(\frac{\mu e^4}{\hbar^2}\right) - 2\frac{4}{3\pi}\frac{\mu e^4}{\hbar^2} = -\frac{4}{3\pi}\frac{\mu e^4}{\hbar^2}$$

$$= -\frac{\mu e^4}{2\hbar^2}\frac{8}{3\pi} \simeq -0.85\text{Ry}. \tag{10.40}$$

This again gives us an energy which is slightly higher than the ground state energy. Thus, we see that the variational method gives us a very good approximation to the ground state energy. The drawback of the variational method is that although it gives us an upper bound, we have no way of knowing how close the bound is to the true eigenvalue, in the absence of experimental results. For example, let us consider the one dimensional case of a particle moving in a λx^4 potential, also known as the quartic potential ($\lambda > 0$). The Hamiltonian has the form

$$H = -\frac{\hbar^2}{2m}\frac{\mathrm{d}^2}{\mathrm{d}x^2} + \lambda x^4. \tag{10.41}$$

This system is invariant under parity. The ground state, which has no node, must be an even function. And we choose a trial wave function for the ground state of the form

$$\psi_\alpha^{(0)}(x) = e^{-\alpha x^2}, \qquad \alpha > 0, \tag{10.42}$$

where α is a variational parameter. Using (10.36), we obtain,

$$\begin{aligned}
\langle\psi_\alpha|\psi_\alpha\rangle &= \int \mathrm{d}x\ \psi_\alpha^{(0)*}\psi_\alpha^{(0)} = \int_{-\infty}^{\infty} \mathrm{d}x\, e^{-2\alpha x^2} \\
&= 2\frac{\Gamma(\frac{1}{2})}{2(2\alpha)^{\frac{1}{2}}} = \sqrt{\frac{\pi}{2\alpha}}, \\
\langle\psi_\alpha|V|\psi_\alpha\rangle &= \lambda \int \mathrm{d}x\ x^4 e^{-2\alpha x^2} \\
&= \frac{\lambda\Gamma\left(\frac{5}{2}\right)}{(2\alpha)^{\frac{5}{2}}} = \frac{3}{4}\frac{\lambda}{4\alpha^2}\sqrt{\frac{\pi}{2\alpha}} \\
&= \frac{3\lambda}{16\alpha^2}\sqrt{\frac{\pi}{2\alpha}}, \\
\langle\psi_\alpha|T|\psi_\alpha\rangle &= -\frac{\hbar^2}{2m}\int \mathrm{d}x\, e^{-\alpha x^2}\frac{\mathrm{d}}{\mathrm{d}x}(-2\alpha x e^{-\alpha x^2}) \\
&= -\frac{\hbar^2}{2m}\int \mathrm{d}x\,(-2\alpha + 4\alpha^2 x^2)e^{-2\alpha x^2} \\
&= \frac{\hbar^2\alpha}{2m}\sqrt{\frac{\pi}{2\alpha}},
\end{aligned} \tag{10.43}$$

so that we have

$$\langle H\rangle = \frac{\hbar^2\alpha}{2m} + \frac{3\lambda}{16\alpha^2}. \tag{10.44}$$

The minimum of the expectation value of the Hamiltonian occurs for

$$\frac{\hbar^2}{2m} - \frac{3\lambda}{8\alpha^3} = 0, \quad \text{or,} \quad \alpha_{\text{min}} = \left(\frac{3m\lambda}{4\hbar^2}\right)^{\frac{1}{3}}. \tag{10.45}$$

For this value of the parameter, the expectation value of the Hamiltonian gives

$$\begin{aligned}
\langle H\rangle_{\text{min}} &= \frac{\hbar^2}{2m}\alpha_{\text{min}} + \frac{3\lambda}{16\alpha_{\text{min}}^2} \\
&= \alpha_{\text{min}}\left(\frac{\hbar^2}{2m} + \frac{3\lambda}{16\alpha_{\text{min}}^3}\right) = \frac{3\hbar^2}{4m}\alpha_{\text{min}} \\
&= \frac{3\hbar^2}{4m}\left(\frac{3m\lambda}{4\hbar^2}\right)^{\frac{1}{3}} = \frac{3}{8}\left(\frac{6\lambda\hbar^4}{m^2}\right)^{\frac{1}{3}}.
\end{aligned} \tag{10.46}$$

As we have discussed, this is expected to give an upper bound on the ground state energy of the system. Furthermore, positivity of the energy leads to

$$\frac{3}{8}\left(\frac{6\lambda\hbar^4}{m^2}\right)^{\frac{1}{3}} \geq E_0 \geq 0. \tag{10.47}$$

However, we have no way of knowing how close this bound is to the actual ground state energy.

10.4 Ground state of helium

The nucleus of a helium atom consists of two protons and two neutrons. There are two electrons in orbit around the nucleus. Thus, for all practical purposes, we can neglect the motion of the nucleus, i.e., assume it to be infinitely heavy. An idealized classical picture of the system can be given as shown in Fig. 10.1.

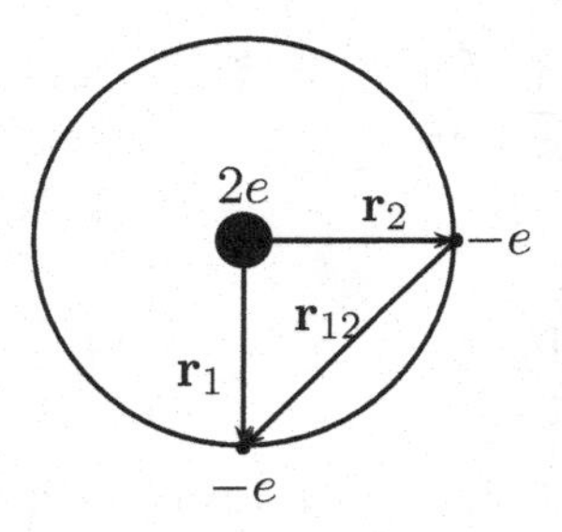

Figure 10.1: A schematic diagram for the two electrons in the ground state of the Helium atom.

Therefore, the Hamiltonian for the system can be written, in the coordinate representation, as

$$H = -\frac{\hbar^2}{2m}\left(\boldsymbol{\nabla}_1^2 + \boldsymbol{\nabla}_2^2\right) - \frac{2e^2}{r_1} - \frac{2e^2}{r_2} + \frac{e^2}{r_{12}}. \tag{10.48}$$

Here $\mathbf{r}_1$ and $\mathbf{r}_2$ are the coordinates of the two electrons (with the nucleus assumed to be at the coordinate origin) and $r_{12} = |\mathbf{r}_1 - \mathbf{r}_2|$ is the distance between them. Thus, in addition to the attractive force between the electrons and the nucleus, there is also repulsion between the two electrons. Here, m denotes the mass of the electron. (Let us note here parenthetically that we are assuming the two electrons to be distinguishable, which they are not. However, as long as we are dealing with the ground state of a two electron system, Pauli exclusion principle tells us that their spins must point in opposite directions. Thus, we can treat them effectively as distinguishable. We are also neglecting the spin-orbit coupling for the present and will talk about it later.)

This atomic system is only slightly more complex than the hydrogen atom and yet it cannot be solved analytically. Consequently, this system has been studied quite extensively by variational techniques so as to draw clues for handling more complex atoms. Let us generalize the system slightly, i.e., rather than assuming the nuclear charge to be $2e$, we consider an arbitrary nucleus of charge Ze so that the Hamiltonian becomes

$$H = -\frac{\hbar^2}{2m}\left(\boldsymbol{\nabla}_1^2 + \boldsymbol{\nabla}_2^2\right) - \frac{Ze^2}{r_1} - \frac{Ze^2}{r_2} + \frac{e^2}{r_{12}}. \tag{10.49}$$

Furthermore, let us ignore the repulsion term for the moment. The system, then, is equivalent to two hydrogenic atoms, i.e., the Hamiltonian is a direct sum of the Hamiltonians for two hydrogenic atoms with nuclear charge Ze. Thus, we can write the ground state wave function for the system to be

$$\psi_0(r_1, r_2) = \psi_0(r_1)\psi_0(r_2). \tag{10.50}$$

We also know from our study of the hydrogen atom that, in this case (The simplest recipe for going from the solution of the hydrogen atom to that of a hydrogenic atom is to let $e^2 \to Ze^2$.),

$$\psi_0(r_1) = \left(\frac{Z^3}{\pi a_0^3}\right)^{\frac{1}{2}} e^{-\frac{Zr_1}{a_0}},$$

$$\psi_0(r_2) = \left(\frac{Z^3}{\pi a_0^3}\right)^{\frac{1}{2}} e^{-\frac{Zr_2}{a_0}}, \tag{10.51}$$

where $a_0 = \frac{\hbar^2}{me^2}$ is the Bohr radius. These are the ground state wave functions for two hydrogenic atoms with nuclear charge Ze and we obtain from (10.50)

$$\psi_0(r_1, r_2) = \psi_0(r_1)\psi_0(r_2) = \frac{Z^3}{\pi a_0^3}\, e^{-\frac{Z}{a_0}(r_1+r_2)}. \tag{10.52}$$

We note that, by construction, the individual states in (10.51) as well as the product state in (10.52) are normalized.

The ground state energy for the system, in this case, is the sum of the ground state energies of the two hydrogenic atoms (in the absence of the mutual repulsion) which can be calculated using the definition of the gamma function

$$\int_0^\infty \mathrm{d}x\, x^n\, e^{-\alpha x} = \frac{\Gamma(n+1)}{\alpha^{n+1}} = \frac{n!}{\alpha^{n+1}}, \tag{10.53}$$

so that (the wave function $\psi_0(r_2)$ integrates to unity since it is normalized)

$$\begin{aligned}
&\int \mathrm{d}^3 r_1\, \psi_0^*(r_1)\left(-\frac{\hbar^2}{2m}\boldsymbol{\nabla}_1^2\psi_0(r_1)\right)\\
&= -\frac{4\pi\hbar^2}{2m}\frac{Z^3}{\pi a_0^3}\int \mathrm{d}r_1\, r_1^2 e^{-\frac{Zr_1}{a_0}}\left(\frac{1}{r_1^2}\frac{\mathrm{d}}{\mathrm{d}r_1}\left(r_1^2\frac{\mathrm{d}}{\mathrm{d}r_1}e^{-\frac{Zr_1}{a_0}}\right)\right)\\
&= -\frac{4\pi\hbar^2}{2m}\frac{Z^3}{\pi a_0^3}\int \mathrm{d}r_1\left[-\frac{2Zr_1}{a_0} + \frac{Z^2}{a_0^2}r_1^2\right]e^{-\frac{2Zr_1}{a_0}}\\
&= -\frac{2\hbar^2 Z^3}{m a_0^3}\left(-\frac{a_0}{2Z} + \frac{a_0}{8Z}\times 2!\right)\\
&= -\frac{2\hbar^2 Z^3}{m a_0^3}\left(-\frac{a_0}{4Z}\right) = \frac{\hbar^2 Z^2}{2m a_0^2} = Z^2\left(\frac{me^4}{2\hbar^2}\right),\\
&\int \mathrm{d}^3 r_1\, \psi_0^*(r_1)\left(-\frac{Ze^2}{r_1}\right)\psi_0(r_1)\\
&= -Ze^2 4\pi\frac{Z^3}{\pi a_0^3}\int \mathrm{d}r_1\, r_1^2\frac{1}{r_1}e^{-\frac{2Zr_1}{a_0}}\\
&= -\frac{4Z^4e^2}{a_0^3}\int \mathrm{d}r_1\, r_1 e^{-\frac{2Zr_1}{a_0}}\\
&= -\frac{4Z^4e^2}{a_0^3}\left(\frac{a_0}{2Z}\right)^2 = -\frac{Z^2e^2}{a_0} = -\frac{Z^2me^4}{\hbar^2}\\
&= -2Z^2\left(\frac{me^4}{2\hbar^2}\right).
\end{aligned} \tag{10.54}$$

The expectation values of the second terms will contribute an identical amount, leading to the ground state energy

$$E_0 = 2\left(Z^2 - 2Z^2\right)\left(\frac{me^4}{2\hbar^2}\right)$$
$$= -2Z^2\left(\frac{me^4}{2\hbar^2}\right) = -2Z^2\text{Ry}. \tag{10.55}$$

Thus, if we take the helium atom and neglect the mutual repulsion of the two electrons, the ground state energy follows from (10.55) to be

$$E_0 = -2(2)^2\text{Ry} = -8\text{Ry} = -108.8 \text{ eV}, \tag{10.56}$$

where we are using $Z = 2$ for helium and $\text{Ry} = 13.6$ eV. However, the measured value of the ground state energy of helium is -78.6 eV. (The ground state energy is the same as the binding energy or ionization energy and can be easily measured.) Therefore, comparing with (10.56), we conclude that the mutual repulsion between the two electrons contributes a significant part to the ground state energy. To calculate its effect, we use the variational technique in the following way. First of all, we would like a trial wave function with no angular momentum, since the ground state has the maximum binding. Furthermore, physical intuition tells us that even though the nucleus has a charge of $2e$, the electrons probably see a much smaller charge because part of the nuclear charge would be screened by the electrons. Thus, we use as a trial wave function

$$\psi_0(r_1, r_2) = \frac{Z^3}{\pi a_0^3}\, e^{-\frac{Z}{a_0}(r_1+r_2)}, \tag{10.57}$$

where we let Z be the variational parameter which would determine the effective charge of the nucleus seen by the electrons. We can calculate the expectation value of the various terms in the Hamiltonian

$$H = -\frac{\hbar^2}{2m}\left(\boldsymbol{\nabla}_1^2 + \boldsymbol{\nabla}_2^2\right) - \frac{2e^2}{r_1} - \frac{2e^2}{r_2} + \frac{e^2}{r_{12}}, \tag{10.58}$$

in this state. We have already calculated, in this state, the expectation values of the kinetic energy terms as well as the attractive potential energy terms in (10.54) (we have to make appropriate changes for the coefficients of the attractive potentials.). Therefore, we only have to evaluate the expectation value of the mutual repulsion term in this state.

$$\begin{aligned}
&\iint \mathrm{d}^3r_1\mathrm{d}^3r_2\ \psi_0^*(r_1,r_2)\frac{e^2}{r_{12}}\psi_0(r_1,r_2) \\
&= \left(\frac{Z^3}{\pi a_0^3}\right)^2 e^2 \iint \mathrm{d}^3r_1\mathrm{d}^3r_2\ e^{-\frac{2Zr_1}{a_0}}\frac{1}{r_{12}}e^{-\frac{2Zr_2}{a_0}}. \qquad (10.59)
\end{aligned}$$

To evaluate this, let us, first of all, scale all coordinates as

$$\begin{aligned}
\mathbf{r}_1 &\to \frac{2Z}{a_0}\mathbf{r}_1, \\
\mathbf{r}_2 &\to \frac{2Z}{a_0}\mathbf{r}_2. \qquad (10.60)
\end{aligned}$$

Then, the integral in (10.59) becomes

$$\begin{aligned}
&= \left(\frac{Z^3}{\pi a_0^3}\right)^2 e^2 \iint \mathrm{d}^3r_1\mathrm{d}^3r_2 \left(\frac{a_0}{2Z}\right)^3\left(\frac{a_0}{2Z}\right)^3 e^{-r_1}\frac{1}{r_{12}}\frac{2Z}{a_0}e^{-r_2} \\
&= \frac{Ze^2}{32\pi^2 a_0}\iint \mathrm{d}^3r_1\mathrm{d}^3r_2\ e^{-r_1}\frac{1}{r_{12}}\ e^{-r_2}. \qquad (10.61)
\end{aligned}$$

The integral can be thought of as the interaction energy of the charge densities $\rho(r_1) = e^{-r_1}$ and $\rho(r_2) = e^{-r_2}$ interacting through a Coulomb potential. This can be evaluated by calculating the potential due to the first distribution by integrating over d^3r_1 and then calculating the energy of the second distribution in the field of the first. The potential due to the first distribution can be calculated in the following way.

Let us consider a spherical shell of radius r_1 and thickness $\mathrm{d}r_1$. The total charge contained in this shell is

$$4\pi r_1^2 e^{-r_1}\mathrm{d}r_1. \qquad (10.62)$$

The potential due to this charge at a point r is given by

$$\begin{aligned}
&4\pi r_1^2 e^{-r_1}\mathrm{d}r_1\,\tfrac{1}{r_1} \quad \text{if} \quad r \leq r_1, \\
&4\pi r_1^2 e^{-r_1}\mathrm{d}r_1\,\tfrac{1}{r} \quad \text{if} \quad r \geq r_1.
\end{aligned} \qquad (10.63)$$

Namely, the potential is a constant within the shell and the charge behaves as if it were at the origin for points outside the shell. Thus, the potential due to the complete distribution can be obtained by integrating over the contributions due to all possible shells,

$$\begin{aligned}\Phi(r) &= \int_0^r 4\pi r_1^2 e^{-r_1} \mathrm{d}r_1 \frac{1}{r} + \int_r^\infty 4\pi r_1^2 e^{-r_1} \mathrm{d}r_1 \frac{1}{r_1} \\ &= \frac{4\pi}{r} \int_0^r \mathrm{d}r_1 r_1^2 e^{-r_1} + 4\pi \int_r^\infty \mathrm{d}r_1 r_1 e^{-r_1} \\ &= \frac{4\pi}{r} \left(-r^2 e^{-r} - 2r e^{-r} - 2e^{-r} + 2\right) \\ &\quad + 4\pi \left(r e^{-r} + e^{-r}\right) \\ &= \frac{4\pi}{r} \left(2 - e^{-r}(r+2)\right). \end{aligned} \tag{10.64}$$

As a result, the mutual interaction energy, (10.61), becomes

$$\begin{aligned} &\frac{Ze^2}{32\pi^2 a_0} \int \mathrm{d}^3 r_2\, \Phi(r_2) e^{-r_2} \\ &= \frac{Ze^2}{32\pi^2 a_0} 4\pi \int_0^\infty r_2^2 \mathrm{d}r_2 \frac{4\pi}{r_2} \left(2 - e^{-r_2}(r_2+2)\right) e^{-r_2} \\ &= \frac{Ze^2}{2a_0} \int_0^\infty \mathrm{d}r_2 \left[2r_2 e^{-r_2} - e^{-2r_2}\left(r_2^2 + 2r_2\right)\right] \\ &= \frac{Ze^2}{2a_0} \left[2 - \frac{1}{4} - \frac{1}{2}\right] = \frac{5Ze^2}{8a_0} = \frac{5Z}{4} \left(\frac{me^4}{2\hbar^2}\right), \end{aligned} \tag{10.65}$$

which represents the expectation value of the repulsive potential in this state. (Another way to evaluate this integral is to expand $\frac{1}{r_{12}}$ in terms of Legendre polynomials and use the orthonormality relations between them.)

Therefore, we can write the expectation value of the total Hamiltonian in this state to be (Note that we are evaluating the expectation value of the helium Hamiltonian in (10.48) (or (10.58)) without the factor of Z in the attractive potential terms which is the reason for the form of the second term.)

$$\langle H \rangle = 2\left(Z^2 - 4Z + \frac{5}{8} Z\right) \frac{me^4}{2\hbar^2}. \tag{10.66}$$

The minimum of this function occurs for

$$2Z - 4 + \frac{5}{8} = 0, \quad \text{or,} \quad Z_{\text{min}} = 2 - \frac{5}{16} = \frac{27}{16}. \tag{10.67}$$

The ground state energy for this value of the variational parameter turns out to be

$$\begin{aligned}\langle H\rangle_{\rm min} &= 2Z_{\rm min}\left(Z_{\rm min} - 4 + \frac{5}{8}\right)\frac{me^4}{2\hbar^2}\\ &= -2Z_{\rm min}^2\,\frac{me^4}{2\hbar^2}\\ &= -2\left(\frac{27}{16}\right)^2\ {\rm Ry}\\ &\simeq -77.5\ {\rm eV}. \end{aligned}\tag{10.68}$$

This serves as an upper bound on the ground state energy of helium and we see that this is quite close to the observed value for the ground state energy. Furthermore, we see that the electrons screen the charge of the nucleus so that the effective charge seen by each of the electrons is $\simeq 1.7e$.

10.5 Selected problems

1. A hydrogen atom is placed in a uniform electric field of strength $\mathcal{E}$ along the z-direction. Choose as a trial wave function

$$\psi_\alpha(x, y, z) = \left(\frac{1}{\pi a_0^3}\right)^{\frac{1}{2}} (1 + \alpha z)e^{-\frac{r}{a_0}},\tag{10.69}$$

   where r is the radial coordinate and a_0, the Bohr radius. Calculate the bound on the ground state energy using this wave function. Can you justify the choice of this wave function? (Neglect higher powers than $\mathcal{E}^2$. Note that this is the second order Stark effect in Hydrogen.)

2. A particle of mass m moves in a one dimensional potential of the form

$$V(x) = \begin{cases} 0 & \text{for } 0 \leq x \leq a,\\ \infty & \text{everywhere else.}\end{cases}\tag{10.70}$$

   Inside the well, it is subjected to another potential of the form $A(x - \frac{a}{2})$ where A is a constant. What is the change in the ground state energy of the particle due to this additional potential? (Keep terms only up to A^2. Can you guess the trial wave function from the previous problem?)

3. Calculate the variational bound on the ground state energy of the Hydrogen atom using the trial wave function

$$\psi_\alpha(r) = \begin{cases} \left(1 - \frac{r}{\alpha}\right) & r \leq \alpha, \\ 0 & r \geq \alpha. \end{cases} \tag{10.71}$$

How is α_{min} related to the Bohr radius?

CHAPTER 11

WKB approximation

In the following lectures, we will study another very powerful approximation method, known as the WKB approximation, that is used in studying quantum mechanical systems subjected to complicated potentials.

11.1 WKB method

WKB approximation is a very powerful method for obtaining approximate solutions to differential equations where the highest derivative term is multiplied by a small parameter. The method was known to Green and Liouville in the early nineteenth century (1837). It was rediscovered in the context of quantum mechanics by Wentzel, Kramers, Brillouin and Jeffreys (1926). Hence the name WKB(J).

The method gives an approximate solution of the Schrödinger equation, no matter how complicated the potential is. It is mostly used in the study of one dimensional systems. However, in higher dimensions, if there is rotational symmetry, then, the method can be applied to the radial equation as well. The idea behind the method is very simple. Consider the Schrödinger equation in one dimension (note that $\hbar$ is a small parameter that multiplies the second derivative term)

$$\left(-\frac{\hbar^2}{2m}\frac{\mathrm{d}^2}{\mathrm{d}x^2} + V\right)\psi = E\psi,$$

$$\text{or,}\quad \frac{\mathrm{d}^2\psi}{\mathrm{d}x^2} + \frac{2m}{\hbar^2}(E-V)\psi = 0. \tag{11.1}$$

Let us assume that $E > V$ and define

$$k^2 = \frac{2m}{\hbar^2}\,(E-V) \equiv \frac{p^2}{\hbar^2}. \tag{11.2}$$

If the potential V is a constant, then, clearly the solutions of equation (11.1),

$$\frac{d^2\psi}{dx^2} + k^2\psi = 0, \tag{11.3}$$

are given by plane waves,

$$\psi(x) = \psi(0)e^{\pm ikx} = \psi(0)e^{\pm \frac{i}{\hbar}px}, \qquad p = \hbar k. \tag{11.4}$$

Thus, the solution, in general, is a superposition of plane waves. We note that the de Broglie wavelength associated with the motion is

$$\lambda = \frac{2\pi\hbar}{p}. \tag{11.5}$$

and the phase change of the solution, (11.4), per unit length is a constant $\frac{p}{\hbar}$.

Let us next suppose that the potential, rather than being a constant, varies slowly. Then, within a region, small compared to the distance over which the potential varies, one can still think of the solution as representing plane waves with wavelength

$$\lambda(x) = \frac{2\pi\hbar}{p(x)} = \frac{2\pi\hbar}{[2m(E - V(x))]^{\frac{1}{2}}}. \tag{11.6}$$

The phase shift per unit length $\frac{p(x)}{\hbar}$ is no longer a constant and the accumulated phase shift between $x = 0$ and x is given by

$$\int_0^x dx'\, \frac{p(x')}{\hbar}. \tag{11.7}$$

Thus, the solution can now be represented by

$$\begin{aligned} \psi(x) &= \psi(0)\, e^{\pm \frac{i}{\hbar}\int_0^x dx' p(x')} \\ &= \psi(x_0)\, e^{\pm \frac{i}{\hbar}\int_{x_0}^x dx' p(x')} . \end{aligned} \tag{11.8}$$

All of this is, of course, valid if the wavelength does not change rapidly. That is, this will be true if the change in the wavelength over a cycle is small compared to the wavelength itself, namely,

$$
\begin{aligned}
& \left|\frac{\delta\lambda}{\lambda}\right| \ll 1, \\
\text{or,} \quad & \left|\frac{\frac{\mathrm{d}\lambda}{\mathrm{d}x}\,\lambda}{\lambda}\right| \ll 1, \\
\text{or,} \quad & \left|\frac{\mathrm{d}\lambda}{\mathrm{d}x}\right| \ll 1. \qquad (11.9)
\end{aligned}
$$

Thus, intuitively, we expect that if the potential, no matter how complicated it is in form, changes very slowly, the general solution will be of the form derived above. Let us now derive things more rigorously.

We are trying to solve the equation

$$\frac{\mathrm{d}^2\psi}{\mathrm{d}x^2} + \frac{p^2}{\hbar^2}\,\psi = 0, \qquad p^2 = 2m(E - V(x)). \qquad (11.10)$$

The general solution of this equation would have the form

$$\psi(x) = f(x)e^{ig(x)}, \qquad (11.11)$$

where $f(x)$ and $g(x)$ are some functions of x (not necessarily real). Noting that, we can write

$$f(x) = e^{\ln f(x)}, \qquad (11.12)$$

for non-negative $f(x)$, we see that we can always write the solution of (11.10) in the form

$$\psi(x) = e^{\frac{i}{\hbar}\phi(x)}, \qquad (11.13)$$

where we have identified

$$\phi(x) = \hbar(g(x) - i\ln f(x)), \qquad (11.14)$$

with $\phi(x)$, in general, assumed to be complex.

If we substitute the form of the solution in (11.14) into the differential equation, (11.10), we have

$$
\begin{aligned}
& \left(\left(\frac{i\phi'}{\hbar}\right)^2 + \frac{i\phi''}{\hbar} + \frac{p^2}{\hbar^2}\right) e^{\frac{i}{\hbar}\phi(x)} = 0, \\
\text{or,} \quad & -\left(\frac{\phi'}{\hbar}\right)^2 + \frac{i\phi''}{\hbar} + \frac{p^2}{\hbar^2} = 0. \qquad (11.15)
\end{aligned}
$$

It is clear now that, for this to be true, $\phi(x)$ must depend on $\hbar$ as well. Therefore, we expand $\phi(x)$ in powers of $\hbar$ as

$$\phi(x) = \phi_0(x) + \hbar\phi_1(x) + \hbar^2\phi_2(x) + \cdots . \tag{11.16}$$

Substituting this back into equation (11.15) and keeping terms only of order $\hbar^{-2}$ and $\hbar^{-1}$, we have

$$\begin{aligned} & -\frac{\phi_0'^2}{\hbar^2} + \frac{p^2(x)}{\hbar^2} + \frac{i\phi_0''}{\hbar} - \frac{2\phi_0'\phi_1'}{\hbar} = 0, \\ \text{or,} \quad & \frac{1}{\hbar^2}\left(-\phi_0'^2 + p^2(x)\right) + \frac{i}{\hbar}\left(\phi_0'' + 2i\phi_0'\phi_1'\right) = 0. \end{aligned} \tag{11.17}$$

Equating the coefficient of the lowest power of $\hbar$ in (11.17) to zero, we obtain

$$-\phi_0'^2 + p^2(x) = 0 \quad \text{or} \quad \phi_0(x) = \pm \int^x \mathrm{d}x'\, p(x'). \tag{11.18}$$

This is, of course, consistent with our intuitive classical result, since in the classical limit, $\hbar \to 0$,

$$\begin{aligned} & \lim_{\hbar\to 0} \phi(x) = \phi_0(x), \\ \text{or,} \quad & \lim_{\hbar\to 0} \psi(x) = e^{\frac{i}{\hbar}\phi_0(x)} = e^{\pm\frac{i}{\hbar}\int^x \mathrm{d}x' p(x')}. \end{aligned} \tag{11.19}$$

However, if we do not set $\hbar = 0$ in the expansion of $\phi(x)$ in (11.16), but rather keep the next order term, then, this is known as the WKB approximation (also known as the semi-classical approximation). Once we have solved for ϕ_0, the next order term, ϕ_1, can be determined by setting the coefficient of terms of order $\hbar^{-1}$ in (11.17) equal to zero. Thus,

$$\begin{aligned} & \phi_0'' + 2i\phi_0'\phi_1' = 0, \\ \text{or,} \quad & \phi_1' = \frac{i}{2}\frac{\phi_0''}{\phi_0'} = \frac{i}{2}(\ln\phi_0')'. \end{aligned} \tag{11.20}$$

Integrating this, we obtain,

$$\phi_1 = \frac{i}{2}\ln\phi_0' + C = i\ln(p(x))^{\frac{1}{2}} + C, \tag{11.21}$$

where we have kept the positive root of (11.18) so that the logarithm is defined. Therefore, to this order, we have

$$\begin{aligned} \phi &= \phi_0(x) + \hbar\phi_1(x) \\ &= \pm\int \mathrm{d}x'\, p(x') + i\hbar\ln(p(x))^{\frac{1}{2}} + \hbar C, \end{aligned} \tag{11.22}$$

so that the solution, to this order, becomes

$$\psi(x) = A\, e^{\pm\frac{i}{\hbar}\int\limits_{x_0}^{x} dx' p(x') - \ln(p(x))^{\frac{1}{2}}}$$

$$= \frac{A}{(p(x))^{\frac{1}{2}}}\, e^{\pm\frac{i}{\hbar}\int\limits_{x_0}^{x} dx' p(x')}. \tag{11.23}$$

The constant A can be determined by noting that

$$\psi(x_0) = \frac{A}{(p(x_0))^{\frac{1}{2}}} \quad \text{or} \quad A = \psi(x_0)(p(x_0))^{\frac{1}{2}}, \tag{11.24}$$

and we can write,

$$\psi(x) = \psi(x_0)\frac{(p(x_0))^{\frac{1}{2}}}{(p(x))^{\frac{1}{2}}}\, e^{\pm\frac{i}{\hbar}\int\limits_{x_0}^{x} dx' p(x')}. \tag{11.25}$$

This is the WKB solution of the Schrödinger equation and is valid only if the potential changes slowly with respect to space. In fact, the WKB solution would be accurate if the successive terms in the expansion in

$$\phi(x) = \phi_0(x) + \hbar\phi_1(x) + \cdots$$

drop off fast. In particular, we should have

$$\left|\frac{\hbar\phi_1}{\phi_0}\right| \ll 1. \tag{11.26}$$

Since $|\phi_0(x)|$ is a monotonically increasing function of x, unless $p(x) = 0$, a small ratio $|\frac{\hbar\phi_1}{\phi_0}|$ also implies that $|\frac{\hbar\phi_1'}{\phi_0'}|$ is small. This, therefore, suggests that

$$\left|\frac{\hbar\phi_1'}{\phi_0'}\right| = \left|\frac{\hbar p'}{2p^2}\right| = \frac{1}{4\pi}\left|\frac{d\lambda}{dx}\right| \ll 1, \tag{11.27}$$

where we have used $\lambda = \frac{2\pi\hbar}{p}$. This, of course, agrees with our earlier intuitive result in (11.9).

In our derivation we assumed that $E > V(x)$. However, if $V(x) > E$, then, everything goes through in a parallel manner. In fact, the WKB solution, in this case, can be obtained from (11.25) simply by letting

$$p(x) \to \pm i|p(x)|, \tag{11.28}$$

so that for, $V(x) > E$, we have

$$\psi(x) = \psi(x_0)\,\frac{|p(x_0)|^{\frac{1}{2}}}{|p(x)|^{\frac{1}{2}}}\,e^{\pm\frac{1}{\hbar}\int\limits_{x_0}^{x} \mathrm{d}x'|p(x')|}, \tag{11.29}$$

where $|p(x)| = [2m(V(x) - E)]^{\frac{1}{2}}$.

We should note, from (11.25) and (11.29), that the WKB approximation breaks down at the classical turning points, i.e., at points where $E = V(x)$. This can be seen simply from the fact that, at such points, $p(x) = 0$ and, as a result, the solution blows up. The physical reason for this breakdown is obvious. In this limit, the de Broglie wavelength becomes infinite and our assumption that the potential changes only slowly over a wavelength is no longer true.

Before, talking about the connection formulae at the turning points, let us see how the WKB method fits into the path integrals (which we discuss in chapter 17). From the path integral formulation, we know that

$$\psi(x,t) = \int \mathrm{d}x' U(x,t;x',t')\psi(x',t'), \tag{11.30}$$

where

$$U(x,t;x',t') = \int \mathcal{D}x \; e^{\frac{i}{\hbar}S[x]}. \tag{11.31}$$

In the classical limit, $\hbar \to 0$ (see chapter 17),

$$\begin{aligned} U(x,t;x',t') &= Ae^{\frac{i}{\hbar}S[x_{cl}]} \\ &= Ae^{\frac{i}{\hbar}\int\limits_{t'}^{t} \mathrm{d}t'' L(x_{cl}(t''),\dot{x}_{cl}(t''))} \\ &= Ae^{\frac{i}{\hbar}\int\limits_{t'}^{t} \mathrm{d}t''(T-V)} \\ &= Ae^{\frac{i}{\hbar}\int\limits_{t'}^{t} \mathrm{d}t''(2T-E)}, \end{aligned} \tag{11.32}$$

where A is a normalization constant.

Since the energy of the system is constant, the second term becomes $(-\frac{i}{\hbar}E(t-t'))$. Furthermore, recalling that

$$T = \frac{p^2}{2m}, \qquad p = m\frac{\mathrm{d}x}{\mathrm{d}t}, \tag{11.33}$$

we obtain,

$$\int_{t'}^{t} \mathrm{d}t'' \, 2T = \frac{1}{m} \int_{t'}^{t} \mathrm{d}t'' \, pm \frac{\mathrm{d}x''}{\mathrm{d}t''} = \int_{x'}^{x} \mathrm{d}x'' \, p, \tag{11.34}$$

so that, we have

$$U(x,t;x',t') = A e^{\frac{i}{\hbar}\int_{x'}^{x} \mathrm{d}x'' p} \; e^{-\frac{i}{\hbar}E(t-t')}. \tag{11.35}$$

Substituting this into (11.30), we get

$$\psi(x,t) = A \int \mathrm{d}x' \; e^{\frac{i}{\hbar}\int_{x'}^{x} \mathrm{d}x'' p} \; \psi(x',t') \; e^{-\frac{i}{\hbar}E(t-t')}. \tag{11.36}$$

Noting that $\psi(x,t) = \psi(x) e^{-\frac{i}{\hbar}Et}$, for stationary states, we obtain

$$\begin{aligned}
\psi(x) &= A \int \mathrm{d}x' \; e^{\frac{i}{\hbar}\int_{x'}^{x} \mathrm{d}x'' p} \; \psi(x') \\
&= A \int \mathrm{d}x' \; e^{\frac{i}{\hbar}\int_{x'}^{x_0} \mathrm{d}x'' p + \frac{i}{\hbar}\int_{x_0}^{x} \mathrm{d}x'' p} \; \psi(x') \\
&= e^{\frac{i}{\hbar}\int_{x_0}^{x} \mathrm{d}x'' p} \; A \int \mathrm{d}x' \; e^{\frac{i}{\hbar}\int_{x'}^{x_0} \mathrm{d}x'' p} \; \psi(x') \\
&= e^{\frac{i}{\hbar}\int_{x_0}^{x} \mathrm{d}x'' p} \; \psi(x_0).
\end{aligned} \tag{11.37}$$

This is, of course, the classical limit of the solution in (11.19). To obtain the terms of the order of $\frac{1}{\sqrt{p}}$ we have to evaluate the transition amplitude keeping terms next to the leading order.

11.2 Connection formulae

When we have a particle moving in a potential, we can divide the entire space into various regions depending on the energy of the particle. For the example of the potential shown in Fig. 11.1, there are three regions. In regions I and III, $V(x) > E$ and hence we should have a damped wave function which vanishes at $x \to \pm\infty$. The WKB approximation gives the wave function, in region I, to be (see (11.29))

$$\psi_{\mathrm{I}}(x) = \frac{A}{|p(x)|^{\frac{1}{2}}} \; e^{-\frac{1}{\hbar}\int_{x}^{x_a} \mathrm{d}x' \, |p(x')|}, \qquad \text{for} \qquad x < x_a. \tag{11.38}$$

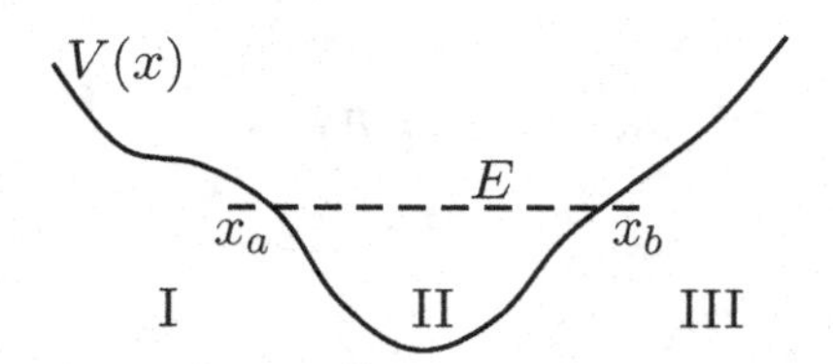

Figure 11.1: A particle with energy E moving in an arbitrary potential.

In region II, $x_a < x < x_b$, on the other hand, we have from (11.25)

$$\psi_{\rm II}(x) = \frac{B}{(p(x))^{\frac{1}{2}}}\, e^{\frac{i}{\hbar}\int\limits_{x_a}^{x} {\rm d}x' p(x')} + \frac{C}{(p(x))^{\frac{1}{2}}}\, e^{-\frac{i}{\hbar}\int\limits_{x_a}^{x} {\rm d}x' p(x')}, \qquad (11.39)$$

while, in the classically inaccessible region III, we have

$$\psi_{\rm III}(x) = \frac{D}{|p(x)|^{\frac{1}{2}}}\, e^{-\frac{1}{\hbar}\int\limits_{x_b}^{x} {\rm d}x' |p(x')|}, \qquad x > x_b. \qquad (11.40)$$

Here, A, B, C, D are normalization constants and x_a, x_b denote the classical turning points. From our study of the Schrödinger equation, we know that the wave function must be continuous across a boundary. This continuity, in addition to giving the physical conservation of particles, also leads to the quantization of energy levels in the case of bound states. If we take the WKB wave functions, however, we see that the wave functions blow up at the classical turning points or at the boundary. Thus, in the present form, there is no way to implement the idea of the matching of solutions in different regions.

Of course, the original Schrödinger equation has smooth solutions at these points as can be seen from the following.

$$\frac{{\rm d}^2\psi}{{\rm d}x^2} + \frac{p^2}{\hbar^2}\,\psi = 0.$$

The coefficients of all the terms in this equation are smooth in the limit $p \to 0$ and hence the solution has to be smooth also. Therefore, the pathology that we encounter in the WKB wave function is a consequence of our approximation scheme. In fact, because

$$p(x) = [2m(E - V(x))]^{\frac{1}{2}},$$

we note that whereas the solutions of the Schrödinger equation are single valued, the WKB solution is multivalued. The difficulty arises because we are trying to approximate a single valued function by multivalued functions. In fact, the question of matching now becomes even more critical because, besides handling the divergence of the wave function, we also have to make sure that the solutions to be matched in the two regions correspond to the same branch. The approach one takes here is to solve the Schrödinger equation exactly in a small region around the turning point and derive the correct prescription for matching.

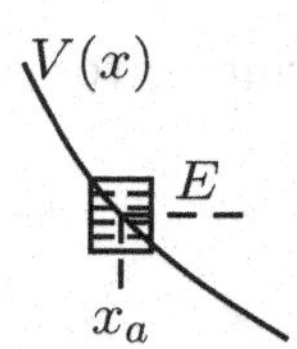

Figure 11.2: An infinitesimal region near the classical turning point $x = x_a$.

Let us consider one of the turning points first and study the Schrödinger equation in the transition region (shaded area) near $x = x_a$ shown in Fig. 11.2. In this region, we would like to solve the equation,

$$\frac{\mathrm{d}^2\psi}{\mathrm{d}x^2} + k^2\psi = 0, \tag{11.41}$$

where $k = \frac{p}{\hbar}$. Let us now define a new function

$$v(x) = \sqrt{k(x)}\psi(x), \tag{11.42}$$

so that

$$\begin{aligned}
\psi(x) &= \frac{1}{\sqrt{k}}v, \\
\frac{\mathrm{d}\psi}{\mathrm{d}x} &= -\frac{1}{2k^{\frac{3}{2}}}\frac{\mathrm{d}k}{\mathrm{d}x}v + \frac{1}{\sqrt{k}}\frac{\mathrm{d}v}{\mathrm{d}x}, \\
\frac{\mathrm{d}^2\psi}{\mathrm{d}x^2} &= \frac{3}{4k^{\frac{5}{2}}}\left(\frac{\mathrm{d}k}{\mathrm{d}x}\right)^2 v - \frac{1}{2k^{\frac{3}{2}}}\frac{\mathrm{d}^2k}{\mathrm{d}x^2}v - \frac{1}{2k^{\frac{3}{2}}}\frac{\mathrm{d}k}{\mathrm{d}x}\frac{\mathrm{d}v}{\mathrm{d}x}
\end{aligned}$$

$$
\begin{aligned}
&- \frac{1}{2k^{\frac{3}{2}}} \frac{\mathrm{d}k}{\mathrm{d}x} \frac{\mathrm{d}v}{\mathrm{d}x} + \frac{1}{\sqrt{k}} \frac{\mathrm{d}^2 v}{\mathrm{d}x^2} \\
&= \frac{1}{\sqrt{k}} \left[\frac{\mathrm{d}^2 v}{\mathrm{d}x^2} - \frac{1}{k} \frac{\mathrm{d}k}{\mathrm{d}x} \frac{\mathrm{d}v}{\mathrm{d}x} + \left(\frac{3}{4k^2} \left(\frac{\mathrm{d}k}{\mathrm{d}x} \right)^2 - \frac{1}{2k} \frac{\mathrm{d}^2 k}{\mathrm{d}x^2} \right) v \right].
\end{aligned} \tag{11.43}
$$

Thus, the equation satisfied by v is obtained from (11.41)-(11.43) to be

$$
\frac{\mathrm{d}^2 v}{\mathrm{d}x^2} - \frac{1}{k} \frac{\mathrm{d}k}{\mathrm{d}x} \frac{\mathrm{d}v}{\mathrm{d}x} + \left[\frac{3}{4k^2} \left(\frac{\mathrm{d}k}{\mathrm{d}x} \right)^2 - \frac{1}{2k} \frac{\mathrm{d}^2 k}{\mathrm{d}x^2} + k^2 \right] v = 0. \tag{11.44}
$$

Furthermore, let us change variables to

$$
x \to y = \int^x \mathrm{d}x' k(x'), \tag{11.45}
$$

so that we have

$$
\begin{aligned}
\frac{\mathrm{d}y}{\mathrm{d}x} &= k(x), \\
\frac{\mathrm{d}v}{\mathrm{d}x} &= \frac{\mathrm{d}y}{\mathrm{d}x} \frac{\mathrm{d}v}{\mathrm{d}y} = k \frac{\mathrm{d}v}{\mathrm{d}y}, \\
\frac{\mathrm{d}^2 v}{\mathrm{d}x^2} &= \frac{\mathrm{d}y}{\mathrm{d}x} \left(\frac{\mathrm{d}k}{\mathrm{d}y} \frac{\mathrm{d}v}{\mathrm{d}y} + k \frac{\mathrm{d}^2 v}{\mathrm{d}y^2} \right) = k \frac{\mathrm{d}k}{\mathrm{d}y} \frac{\mathrm{d}v}{\mathrm{d}y} + k^2 \frac{\mathrm{d}^2 v}{\mathrm{d}y^2}, \\
\frac{\mathrm{d}k}{\mathrm{d}x} &= \frac{\mathrm{d}y}{\mathrm{d}x} \frac{\mathrm{d}k}{\mathrm{d}y} = k \frac{\mathrm{d}k}{\mathrm{d}y}, \\
\frac{\mathrm{d}^2 k}{\mathrm{d}x^2} &= \frac{\mathrm{d}y}{\mathrm{d}x} \left(\left(\frac{\mathrm{d}k}{\mathrm{d}y} \right)^2 + k \frac{\mathrm{d}^2 k}{\mathrm{d}y^2} \right) = k \left(\frac{\mathrm{d}k}{\mathrm{d}y} \right)^2 + k^2 \frac{\mathrm{d}^2 k}{\mathrm{d}y^2}.
\end{aligned} \tag{11.46}
$$

In terms of these variables, then, equation (11.44) becomes

$$
\begin{aligned}
& k^2 \frac{\mathrm{d}^2 v}{\mathrm{d}y^2} + \left[\frac{3}{4} \left(\frac{\mathrm{d}k}{\mathrm{d}y} \right)^2 - \frac{1}{2k} \left(k \left(\frac{\mathrm{d}k}{\mathrm{d}y} \right)^2 + k^2 \frac{\mathrm{d}^2 k}{\mathrm{d}y^2} \right) + k^2 \right] v = 0, \\
\text{or,} \quad & \frac{\mathrm{d}^2 v}{\mathrm{d}y^2} + \left[\frac{1}{4k^2} \left(\frac{\mathrm{d}k}{\mathrm{d}y} \right)^2 - \frac{1}{2k} \frac{\mathrm{d}^2 k}{\mathrm{d}y^2} + 1 \right] v = 0.
\end{aligned} \tag{11.47}
$$

So far, everything has been exact. We now make an approximation. Let us assume that the potential is slowly varying so that, near the turning point, it can be approximated by a linear term. That is, near $x \approx x_a$, let

$$\begin{aligned} V(x) &\simeq V(x_a) - \alpha(x - x_a) \\ &= E - \alpha(x - x_a), \qquad \alpha = -V'(x_a) > 0. \end{aligned} \tag{11.48}$$

Furthermore, we can choose x_a to be the origin of the coordinate system so that x measures the distance from the turning point. Thus, near the turning point $x = x_a$, we can write

$$\begin{aligned} k &= \frac{p}{\hbar} = \left[\frac{2m}{\hbar^2}(E - V(x))\right]^{\frac{1}{2}} \\ &\simeq \begin{cases} cx^{\frac{1}{2}} & \text{for} \quad x > 0 \quad \text{where} \quad E > V, \\ c|x|^{\frac{1}{2}} e^{\frac{i\pi}{2}} & \text{for} \quad x < 0 \quad \text{where} \quad E < V. \end{cases} \end{aligned} \tag{11.49}$$

Here, the constant c is equal to

$$c = \left[\frac{2m\alpha}{\hbar^2}\right]^{\frac{1}{2}}. \tag{11.50}$$

It is easy to see that, near the turning point, (for both positive and negative x)

$$y = \int_0^x \mathrm{d}x' \, k(x') = \frac{2}{3c^2} k^3, \tag{11.51}$$

so that, we have

$$\begin{aligned} \frac{\mathrm{d}y}{\mathrm{d}k} &= \frac{2}{c^2} k^2, \\ \text{or,} \quad \frac{\mathrm{d}k}{\mathrm{d}y} &= \frac{c^2}{2} \frac{1}{k^2}, \end{aligned} \tag{11.52}$$

which leads to

$$\frac{\mathrm{d}^2 k}{\mathrm{d}y^2} = -c^2 \frac{1}{k^3} \frac{\mathrm{d}k}{\mathrm{d}y} = -\frac{c^4}{2} \frac{1}{k^5}. \tag{11.53}$$

Putting the relations in (11.52) and (11.53) back into (11.47), we obtain, in this approximation,

$$\frac{\mathrm{d}^2 v}{\mathrm{d}y^2} + \left[\frac{1}{4k^2}\frac{c^4}{4k^4} - \frac{1}{2k}\left(-\frac{c^4}{2}\frac{1}{k^5}\right) + 1\right] v = 0,$$

$$\text{or,}\quad \frac{\mathrm{d}^2 v}{\mathrm{d}y^2} + \left[\frac{5c^4}{16k^6} + 1\right] v = 0,$$

$$\text{or,}\quad \frac{\mathrm{d}^2 v}{\mathrm{d}y^2} + \left[1 + \frac{5}{36y^2}\right] v = 0. \tag{11.54}$$

This equation is related to the spherical Bessel equation in the following way. We know that the spherical Bessel functions, j_ℓ, satisfy

$$\frac{\mathrm{d}^2 j_\ell}{\mathrm{d}y^2} + \frac{2}{y}\frac{\mathrm{d}j_\ell}{\mathrm{d}y} + \left[1 - \frac{\ell(\ell+1)}{y^2}\right] j_\ell = 0. \tag{11.55}$$

If we define $u_\ell(y) = y j_\ell(y)$, then it follows from (11.55) that

$$\frac{\mathrm{d}^2 u_\ell}{\mathrm{d}y^2} + \left[1 - \frac{\ell(\ell+1)}{y^2}\right] u_\ell = 0. \tag{11.56}$$

Comparing with (11.44), we see that the equation satisfied by v is exactly the same as (11.56) with

$$\ell(\ell+1) + \frac{5}{36} = 0,$$

$$\text{or,}\quad \ell = -\frac{1}{6},\ -\frac{5}{6}. \tag{11.57}$$

Furthermore, recalling that

$$j_\ell(y) = \left(\frac{\pi}{2y}\right)^{\frac{1}{2}} J_{\ell+\frac{1}{2}}(y), \tag{11.58}$$

we have, for $x > 0$ (namely, y real),

$$\psi_{\mathrm{II}}^{\pm} = \frac{1}{\sqrt{k}} v^{\pm} = \frac{A_\ell}{\sqrt{k}} y j_\ell(y),$$

$$= A_\pm \frac{1}{\sqrt{k}} \sqrt{\xi_1} J_{\pm\frac{1}{3}}(\xi_1), \tag{11.59}$$

where we have used (11.50), (11.51) (as well as $\ell = -\frac{1}{6},\ -\frac{5}{6}$) and have defined $\xi_1 = y = \frac{2c}{3}\, x^{\frac{3}{2}}$ which is real for $x > 0$. We have absorbed the factor of $\sqrt{\frac{\pi}{2}}$ into the normalization constant and note, from (11.50), that $k = cx^{\frac{1}{2}}$. The superscripts $\pm$ correspond to the two independent solutions of the Bessel equation. Note that asymptotically, as $y \to \infty$,

$$j_\ell(y) \longrightarrow e^{\pm iy}. \tag{11.60}$$

Similarly, in region I, where y is complex ($x < 0$), we have

$$\psi_{\rm I}^{\pm} = B_{\pm} \frac{1}{\sqrt{|\kappa|}} \sqrt{\xi_2} I_{\pm\frac{1}{3}}(\xi_2), \tag{11.61}$$

where $\xi_2 = |y| = \frac{2c}{3}|x|^{\frac{3}{2}}$, $\kappa = \left[\frac{2m}{\hbar^2}(V(x) - E)\right]^{\frac{1}{2}} = c|x|^{\frac{1}{2}} > 0$ and we have used standard relations like

$$I_n(y) = (i)^{-n} J_n(iy), \tag{11.62}$$

absorbing the factors of (i) into the normalization constant. We note that, for $x \to 0$,

$$\begin{aligned}
&\lim_{\xi_1 \to 0} J_{\pm\frac{1}{3}}(\xi_1) \longrightarrow \frac{\left(\frac{1}{2}\,\xi_1\right)^{\pm\frac{1}{3}}}{\Gamma\left(1 \pm \frac{1}{3}\right)}, \\
&\lim_{\xi_2 \to 0} I_{\pm\frac{1}{3}}(\xi_2) \longrightarrow \frac{\left(\frac{1}{2}\,\xi_2\right)^{\pm\frac{1}{3}}}{\Gamma\left(1 \pm \frac{1}{3}\right)},
\end{aligned} \tag{11.63}$$

so that, as $x \to 0$,

$$\begin{aligned}
\psi_{\rm II}^{+} &\simeq A_{+} \frac{\left(\frac{2c}{3}\right)^{\frac{1}{2}}}{(c)^{\frac{1}{2}}} \frac{\left(\frac{1}{2}\frac{2c}{3}\right)^{\frac{1}{3}}}{\Gamma\left(\frac{4}{3}\right)} x^{-\frac{1}{4}+\frac{3}{4}+\frac{1}{2}} \\
&= A_{+} \left(\frac{2}{3}\right)^{\frac{1}{2}} \left(\frac{c}{3}\right)^{\frac{1}{3}} \frac{1}{\Gamma\left(\frac{4}{3}\right)} x, \\
\psi_{\rm II}^{-} &\simeq A_{-} \frac{\left(\frac{2c}{3}\right)^{\frac{1}{2}}}{(c)^{\frac{1}{2}}} \frac{\left(\frac{1}{2}\frac{2c}{3}\right)^{-\frac{1}{3}}}{\Gamma\left(\frac{2}{3}\right)} x^{-\frac{1}{4}+\frac{3}{4}-\frac{1}{2}} \\
&= A_{-} \left(\frac{2}{3}\right)^{\frac{1}{2}} \left(\frac{c}{3}\right)^{-\frac{1}{3}} \frac{1}{\Gamma\left(\frac{2}{3}\right)}, \\
\psi_{\rm I}^{+} &\simeq B_{+} \frac{\left(\frac{2c}{3}\right)^{\frac{1}{2}}}{(c)^{\frac{1}{2}}} \frac{\left(\frac{1}{2}\frac{2c}{3}\right)^{\frac{1}{3}}}{\Gamma\left(\frac{4}{3}\right)} |x|^{-\frac{1}{4}+\frac{3}{4}+\frac{1}{2}} \\
&= B_{+} \left(\frac{2}{3}\right)^{\frac{1}{2}} \left(\frac{c}{3}\right)^{\frac{1}{3}} \frac{1}{\Gamma\left(\frac{4}{3}\right)} |x|,
\end{aligned}$$

$$\psi_{\rm I}^{-} \simeq B_{-}\frac{\left(\frac{2c}{3}\right)^{\frac{1}{2}}}{(c)^{\frac{1}{2}}}\frac{\left(\frac{1}{2}\frac{2c}{3}\right)^{-\frac{1}{3}}}{\Gamma\left(\frac{2}{3}\right)}|x|^{-\frac{1}{4}+\frac{3}{4}-\frac{1}{2}}$$

$$= B_{-}\left(\frac{2}{3}\right)^{\frac{1}{2}}\left(\frac{c}{3}\right)^{-\frac{1}{3}}\frac{1}{\Gamma\left(\frac{2}{3}\right)}. \qquad (11.64)$$

It is clear from (11.64) that $\psi_{\rm I}^{+}$ joins smoothly on to $\psi_{\rm II}^{+}$ if $A_{+} = -B_{+}$ and $\psi_{\rm I}^{-}$ joins smoothly to $\psi_{\rm II}^{-}$ if $A_{-} = B_{-}$. One should note here that the solutions are well behaved as $x \to 0$ (namely, at the turning points).

We can now use these relations between the coefficients to derive information on the asymptotic forms of the solutions. First of all, we know that

$$J_{\pm\frac{1}{3}}(\xi_1) \xrightarrow{\xi_1\to\text{large}} \left(\frac{1}{2}\pi\xi_1\right)^{-\frac{1}{2}}\cos\left(\xi_1 \mp \frac{\pi}{6} - \frac{\pi}{4}\right),$$

$$I_{\pm\frac{1}{3}}(\xi_2) \xrightarrow{\xi_2\to\text{large}} (2\pi\xi_2)^{-\frac{1}{2}}\left(e^{\xi_2} + e^{-\xi_2}e^{-\left(\frac{1}{2}\pm\frac{1}{3}\right)\pi i}\right). \qquad (11.65)$$

Thus, using the relations for the coefficients in (11.64) already determined, we see from (11.65) that, for large values of the coordinates,

$$\psi_{\rm II}^{+} \longrightarrow \left(\frac{1}{2}\pi k\right)^{-\frac{1}{2}}\cos\left(\xi_1 - \frac{5\pi}{12}\right)$$

$$\longrightarrow -(2\pi|\kappa|)^{-\frac{1}{2}}\left(e^{\xi_2} + e^{-\xi_2-\frac{5\pi i}{6}}\right) = \psi_{\rm I}^{+},$$

$$\psi_{\rm II}^{-} \longrightarrow \left(\frac{1}{2}\pi k\right)^{-\frac{1}{2}}\cos\left(\xi_1 - \frac{\pi}{12}\right)$$

$$\longrightarrow (2\pi|\kappa|)^{-\frac{1}{2}}\left(e^{\xi_2} + e^{-\xi_2-\frac{\pi i}{6}}\right) = \psi_{\rm I}^{-}. \qquad (11.66)$$

In region I, however, we must have a solution which vanishes as $x \to -\infty$. This tells us that the physical solution is the sum of $\psi_{\rm I}^{+}$ and $\psi_{\rm I}^{-}$ (with an appropriate phase), which contains only the term $e^{-\xi_2}$. This leads to our first connection formula

$$\left(\frac{1}{2}\right)(|\kappa|)^{-\frac{1}{2}}e^{-\xi_2} \to (k)^{-\frac{1}{2}}\cos\left(\xi_1 - \frac{\pi}{4}\right),$$

$$\text{or,}\quad \left(\frac{1}{2}\right)(\kappa)^{-\frac{1}{2}}e^{-\int_x^{x_a}\mathrm{d}x'\,\kappa} \to (k)^{-\frac{1}{2}}\cos\left(\int_{x_a}^{x}\mathrm{d}x'\,k - \frac{\pi}{4}\right). \qquad (11.67)$$

Let us note that, since

$$\cos\left(\int \mathrm{d}x'\, k - \frac{\pi}{4}\right) = \sin\left(-\int \mathrm{d}x'\, k - \frac{\pi}{4}\right), \tag{11.68}$$

we can also write the connection formula, (11.67), as

$$\left(\frac{1}{2}\right)(\kappa)^{-\frac{1}{2}} e^{-\int\limits_{x}^{x_a} \mathrm{d}x'\, \kappa} \to (k)^{-\frac{1}{2}} \sin\left(-\int\limits_{x_a}^{x} \mathrm{d}x'\, k - \frac{\pi}{4}\right). \tag{11.69}$$

Similarly, at the other turning point (see Fig. 11.3), we would have a connection formula

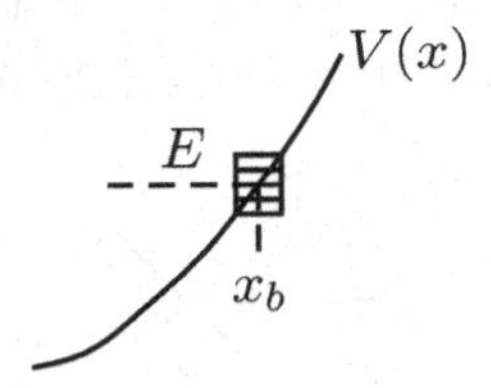

Figure 11.3: An infinitesimal region near the classical turning point $x = x_b$.

$$\left(\frac{1}{2}\right)(\kappa)^{-\frac{1}{2}} e^{-\int\limits_{x_b}^{x} \mathrm{d}x'\, \kappa} \to (k)^{-\frac{1}{2}} \cos\left(\int\limits_{x}^{x_b} \mathrm{d}x'\, k - \frac{\pi}{4}\right). \tag{11.70}$$

Notice that the connection formulae are directional. That is, it is always the solutions in the classically inaccessible regions that are matched on to the classically accessible regions. This is because we know the boundary condition on the wave function in the inaccessible regions – the wave function has to vanish at infinity. The case of the potential where there are several inaccessible regions has to be done extremely carefully. Furthermore, if the form of the potential and the energy of the particle are such that the turning points are not separated sufficiently, for example, as shown in Fig. 11.4, one has to be careful because the region of validity of one solution may overlap with the next.

Furthermore, if one is not interested in the exact form of the wave function in the transition region, there is a simpler way of deriving the connection formulae. The idea is to think of x as a complex

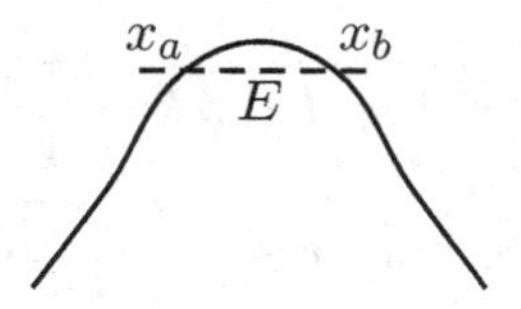

Figure 11.4: A particle moving with energy E in a potential for which the classical turning points may not be well separated.

variable and go, in the complex plane, from the inaccessible region to the accessible region by paths which avoid the turning point. Thus, for example, let us consider the turning point in Fig. 11.5.

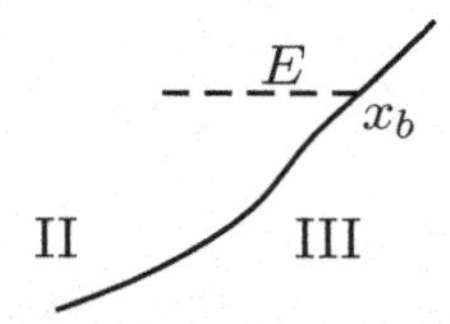

Figure 11.5: The classical turning point at $x = x_b$.

For $x > x_b$ we expect the solution to be of the form

$$\psi_{\text{III}}(x) = \frac{C}{(\kappa)^{\frac{1}{2}}}\, e^{-\int\limits_{x_b}^{x} \mathrm{d}x'\,\kappa}. \tag{11.71}$$

For $x < x_b$ we expect the solution to be of the form

$$\psi_{\text{II}}(x) = \frac{C_1}{(k)^{\frac{1}{2}}}\, e^{i\int\limits_{x_b}^{x} \mathrm{d}x'\,k} + \frac{C_2}{(k)^{\frac{1}{2}}}\, e^{-i\int\limits_{x_b}^{x} \mathrm{d}x'\,k}. \tag{11.72}$$

Let us again choose $x_b = 0$ (to be the origin). Then, in the linear approximation, in the region $x > x_b$

$$\kappa = cx^{\frac{1}{2}}. \tag{11.73}$$

As a complex variable, we can parametrize x as

$$x = \rho e^{i\theta}, \qquad 0 \leq \theta \leq 2\pi. \tag{11.74}$$

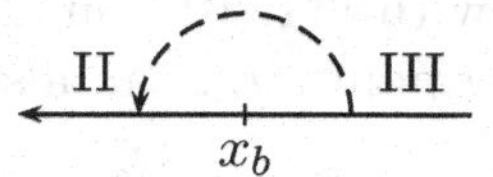

Figure 11.6: The path in the complex x-plane in going from the classically inaccessible to the classically accessible region.

If we follow the path shown in Fig. 11.6 to go from region III to region II, then, in the beginning of the semicircle, $\theta = 0$ and hence κ is real. But at the end of the semicircle ($\theta = \pi$), κ has become purely imaginary. In fact, in crossing the turning point, $\kappa \to ik$, so that

$$e^{-\int_{x_b}^{x} \mathrm{d}x'\, \kappa} \to e^{-i\int_{x_b}^{x} \mathrm{d}x'\, k}, \tag{11.75}$$

so that

$$\begin{aligned} \frac{C}{(\kappa)^{\frac{1}{2}}}\, e^{-\int_{x_b}^{x} \mathrm{d}x'\, \kappa} &\longrightarrow \frac{C}{(ik)^{\frac{1}{2}}}\, e^{-i\int_{x_b}^{x} \mathrm{d}x'\, k} \\ &= \frac{Ce^{-\frac{i\pi}{4}}}{(k)^{\frac{1}{2}}}\, e^{-i\int_{x_b}^{x} \mathrm{d}x'\, k}. \end{aligned} \tag{11.76}$$

Comparing with the form of the solution in region II, (11.72), we see that this determines

$$C_2 = Ce^{-\frac{i\pi}{4}}. \tag{11.77}$$

Similarly, if we had come from region III along a semicircle in the lower half plane, we would have obtained

$$C_1 = Ce^{\frac{i\pi}{4}}. \tag{11.78}$$

Taking the mean of the two ways of coming (namely, (11.77) and (11.78)), we would have derived the connection formula

$$\begin{aligned} \frac{1}{2(\kappa)^{\frac{1}{2}}}\, e^{-\int_{x_b}^{x} \mathrm{d}x'\, \kappa} &\longrightarrow \frac{1}{(k)^{\frac{1}{2}}} \cos\left(\int_{x_b}^{x} \mathrm{d}x'\, k + \frac{\pi}{4}\right) \\ &= \frac{1}{(k)^{\frac{1}{2}}} \cos\left(\int_{x}^{x_b} \mathrm{d}x'\, k - \frac{\pi}{4}\right), \end{aligned} \tag{11.79}$$

which can be compared with (11.70). Similarly, the connection formula at the other turning point, x_a, can also be derived in a similar manner.

11.3 Bohr-Sommerfeld quantization condition

In solutions of the Schrödinger equation, in the case of simple potentials, we have observed that the boundary conditions lead to quantized energy levels for bound states. Since we now have the connection formulae for the WKB solutions, we expect to get quantized energy levels for bound states here also. Let us look at a particle in a general potential well as shown in Fig. 11.7. There are two classical turning points at x_a and x_b.

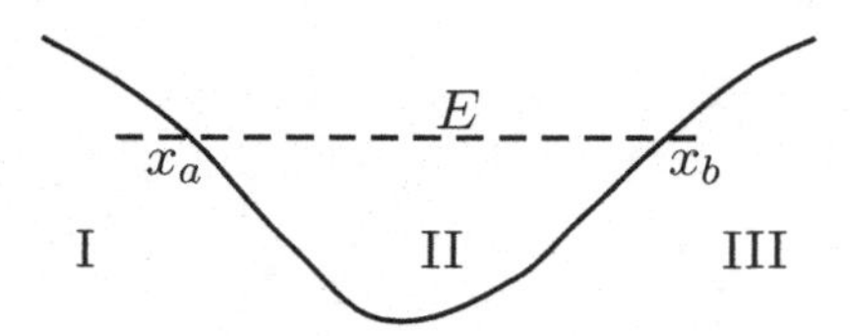

Figure 11.7: A particle with energy E moving in a general potential well.

We know that, in region I, the solution must be of the form

$$\psi_{\text{I}}(x) \simeq \frac{1}{2\kappa^{\frac{1}{2}}}\, e^{-\int\limits_{x}^{x_a} \mathrm{d}x'\,\kappa}, \qquad x < x_a, \tag{11.80}$$

where $\kappa = \left[\frac{2m}{\hbar^2}(V(x) - E)\right]^{\frac{1}{2}}$. Furthermore, through the connection formula, (11.67), this leads to a wave function in region II of the form

$$\psi_{\text{II}} \simeq \frac{1}{k^{\frac{1}{2}}} \cos\left(\int\limits_{x_a}^{x} \mathrm{d}x'\,k - \frac{\pi}{4}\right), \qquad x_a \leq x \leq x_b, \tag{11.81}$$

where $k = \left[\frac{2m}{\hbar^2}(E - V(x))\right]^{\frac{1}{2}}$. On the other hand, we also know that the wave function in region III is of the form

$$\psi_{\text{III}}(x) \simeq \frac{1}{2\kappa^{\frac{1}{2}}}\, e^{-\int\limits_{x_b}^{x} \mathrm{d}x'\,\kappa}, \qquad x > x_b. \tag{11.82}$$

Through the connection formula, (11.70), this leads to a wave function in region II of the form ($x_a \leq x \leq x_b$)

$$\psi_{\rm II}(x) \simeq \frac{1}{k^{\frac{1}{2}}} \cos\left(\int_x^{x_b} dx'\, k - \frac{\pi}{4}\right),$$

$$= \frac{1}{k^{\frac{1}{2}}} \cos\left(\int_{x_b}^{x} dx'\, k + \frac{\pi}{4}\right). \tag{11.83}$$

If both the wave functions in (11.81) and (11.83) are to describe the same particle, then, we expect, at most, their phases to differ by multiples of π (remember that the probability density has to be the same)

$$\int_{x_a}^{x} dx'\, k - \frac{\pi}{4} = \int_{x_b}^{x} dx'\, k + \frac{\pi}{4} + n\pi,$$

$$\text{or,} \quad \int_{x_a}^{x_b} dx\, k = \left(n + \frac{1}{2}\right)\pi,$$

$$\text{or,} \quad \oint dx\, p = \left(n + \frac{1}{2}\right) h, \qquad n = 0, 1, 2, \cdots, \tag{11.84}$$

where the non-negative nature of n arises because $\int_{x_a}^{x_b} dx\, k$ represents the phase between x_a and x_b and cannot be negative. Furthermore, we have used the fact that

$$k = \frac{p}{\hbar}, \qquad h = 2\pi\hbar, \tag{11.85}$$

and $\oint$ = area inside the trajectory of the particle in the phase space. That is, in the phase space of the particle, this is the area enclosed by the particle starting at x_a going to x_b and then coming back to x_a as shown in Fig. 11.8.

The condition, (11.84), is known as the Bohr-Sommerfeld quantization condition with half integer values. It is clear that the wave function vanishes n times inside the potential well, since the total change of phase is $\left(n + \frac{1}{2}\right)\pi$. This is in agreement with the true wave function of the system. However, the WKB method seems to indicate that there will be $\left(\frac{1}{2}n + \frac{1}{4}\right)$ wavelengths inside the well. This does not exactly fit with our solution of the infinite square well, where we

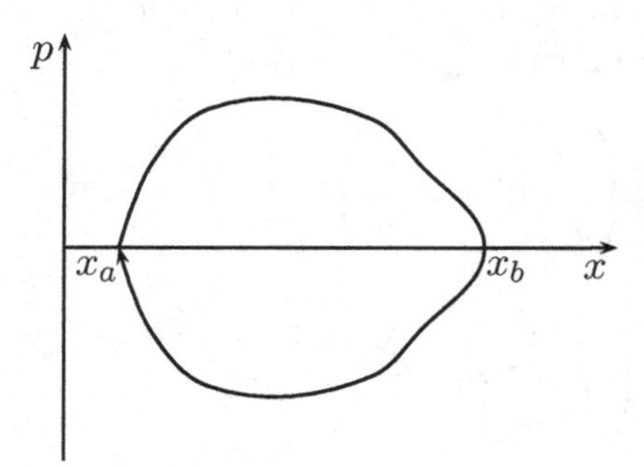

Figure 11.8: Area enclosed by the classical trajectory in phase space in going from one turning point to another and coming back.

have seen that there are $\left(\frac{1}{2}n+\frac{1}{2}\right)$ wavelengths inside the well. The discrepancy arises, for an infinite potential well, because the linear approximation, near the classical turning points, breaks down. We will discuss this case separately.

11.4 Applications of the quantization condition

In what follows, we will discuss several applications of the Bohr-Sommerfeld quantization condition.

1. Let us consider first, a particle moving in an one dimensional potential of the form

$$V(x)=\beta|x|, \qquad \beta>0. \tag{11.86}$$

The potential, in this case, has the form shown in Fig. 11.9.

Thus, we see that the classical turning points are at

$$E=V(x)=\beta|x|, \quad \text{or} \quad x=\pm\frac{E}{\beta}. \tag{11.87}$$

In this case, the quantization condition, (11.84), gives

$$\int\limits_{x_a}^{x_b} \mathrm{d}x\, k=\left(n+\frac{1}{2}\right)\pi,$$

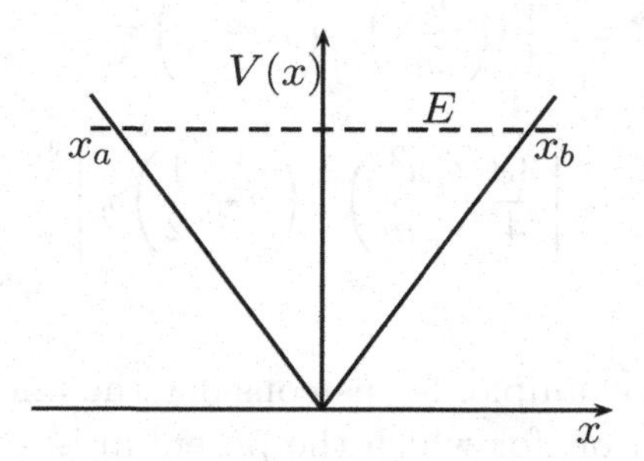

Figure 11.9: Graphical representation of a particle moving in the potential (11.86) with energy E.

$$\text{or,}\quad \int_{-\frac{E}{\beta}}^{\frac{E}{\beta}} \mathrm{d}x \left[\frac{2m}{\hbar^2}(E - V(x))\right]^{\frac{1}{2}} = \left(n + \frac{1}{2}\right)\pi,$$

$$\text{or,}\quad \int_{-\frac{E}{\beta}}^{\frac{E}{\beta}} \mathrm{d}x \left[\frac{2m}{\hbar^2}(E - \beta|x|)\right]^{\frac{1}{2}} = \left(n + \frac{1}{2}\right)\pi,$$

$$\text{or,}\quad 2\left(\frac{2mE}{\hbar^2}\right)^{\frac{1}{2}} \int_{0}^{\frac{E}{\beta}} \mathrm{d}x \left(1 - \frac{\beta x}{E}\right)^{\frac{1}{2}} = \left(n + \frac{1}{2}\right)\pi. \quad (11.88)$$

Let us define

$$y = \frac{\beta x}{E},$$

$$\text{or,}\quad \mathrm{d}y = \frac{\beta}{E}\,\mathrm{d}x. \quad (11.89)$$

This leads to

$$2\left(\frac{2mE}{\hbar^2}\right)^{\frac{1}{2}} \frac{E}{\beta} \int_{0}^{1} \mathrm{d}y\,(1 - y)^{\frac{1}{2}} = \left(n + \frac{1}{2}\right)\pi,$$

$$\text{or,}\quad 2\left(\frac{2mE}{\hbar^2}\right)^{\frac{1}{2}} \frac{E}{\beta}\frac{2}{3} = \left(n + \frac{1}{2}\right)\pi,$$

$$\text{or,}\quad E_n^{\frac{3}{2}} = \frac{3\beta}{4}\left(\frac{\hbar^2}{2m}\right)^{\frac{1}{2}}\left(n+\frac{1}{2}\right)\pi,$$

$$\text{or,}\quad E_n = \left[\frac{3\beta}{4}\left(\frac{\hbar^2}{2m}\right)^{\frac{1}{2}}\left(n+\frac{1}{2}\right)\pi\right]^{\frac{2}{3}}. \tag{11.90}$$

2. As a second example, let us consider the one dimensional harmonic oscillator, for which the potential is given by

$$V(x) = \frac{1}{2}m\omega^2x^2. \tag{11.91}$$

The classical turning points, for this potential, are at

$$E = \frac{1}{2}m\omega^2x^2, \quad \text{or} \quad x = \pm\sqrt{\frac{2E}{m\omega^2}} \tag{11.92}$$

Thus, we have, from (11.84),

$$\int_{x_a}^{x_b} \mathrm{d}x\, k = \left(n+\frac{1}{2}\right)\pi. \tag{11.93}$$

Let us evaluate the left hand side of (11.93).

$$\text{L.H.S.} = \int_{-\sqrt{\frac{2E}{m\omega^2}}}^{\sqrt{\frac{2E}{m\omega^2}}} \mathrm{d}x \left[\frac{2m}{\hbar^2}\left(E - \frac{1}{2}m\omega^2x^2\right)\right]^{\frac{1}{2}}$$

$$= 2\left(\frac{2mE}{\hbar^2}\right)^{\frac{1}{2}} \int_0^{\sqrt{\frac{2E}{m\omega^2}}} \mathrm{d}x \left(1 - \frac{m\omega^2}{2E}x^2\right)^{\frac{1}{2}}. \tag{11.94}$$

If we now make a change of variable,

$$y = \sqrt{\frac{m\omega^2}{2E}}\,x, \quad \Rightarrow \mathrm{d}y = \sqrt{\frac{m\omega^2}{2E}}\,\mathrm{d}x, \tag{11.95}$$

the left hand side of (11.93) in (11.94) becomes

$$\begin{aligned}
\text{L.H.S.} &= 2\left(\frac{2mE}{\hbar^2}\right)^{\frac{1}{2}}\left(\frac{2E}{m\omega^2}\right)^{\frac{1}{2}}\int_0^1 \mathrm{d}y\,(1-y^2)^{\frac{1}{2}}\\
&= 2\left(\frac{2mE}{\hbar^2}\right)^{\frac{1}{2}}\left(\frac{2E}{m\omega^2}\right)^{\frac{1}{2}}\frac{\pi}{4}\\
&= \frac{4E}{\hbar\omega}\frac{\pi}{4}.
\end{aligned}\tag{11.96}$$

Therefore, from (11.93), we obtain,

$$\begin{aligned}
&\frac{4E_n}{\hbar\omega}\frac{\pi}{4} = \left(n+\frac{1}{2}\right)\pi,\\
\text{or,}\quad &E_n = \left(n+\frac{1}{2}\right)\hbar\omega.
\end{aligned}\tag{11.97}$$

In this case, we see that the Bohr-Sommerfeld quantization, following from the WKB approximation, leads to the exact energy eigenvalues.

3. Let us consider next a particle moving in an one dimensional quartic potential, for which

$$V(x) = \lambda x^4,\qquad \lambda > 0.\tag{11.98}$$

The classical turning points, in this case, are determined to be

$$\begin{aligned}
&E = \lambda x^4,\\
\text{or,}\quad &x = \pm\left(\frac{E}{\lambda}\right)^{\frac{1}{4}}.
\end{aligned}\tag{11.99}$$

Thus, from the quantization condition, we have

$$\int_{-(\frac{E}{\lambda})^{\frac{1}{4}}}^{(\frac{E}{\lambda})^{\frac{1}{4}}} \mathrm{d}x\,k = \left(n+\frac{1}{2}\right)\pi.\tag{11.100}$$

We can evaluate the left hand side of (11.100) as

$$\text{L.H.S.} = \left(\frac{2m}{\hbar^2}\right)^{\frac{1}{2}} \int\limits_{-(\frac{E}{\lambda})^{\frac{1}{4}}}^{(\frac{E}{\lambda})^{\frac{1}{4}}} \mathrm{d}x \left(E - \lambda x^4\right)^{\frac{1}{2}}$$

$$= 2\left(\frac{2mE}{\hbar^2}\right)^{\frac{1}{2}} \int\limits_{0}^{(\frac{E}{\lambda})^{\frac{1}{4}}} \mathrm{d}x \left(1 - \frac{\lambda}{E}x^4\right)^{\frac{1}{2}}. \tag{11.101}$$

Let us define

$$y = \left(\frac{\lambda}{E}\right)^{\frac{1}{4}} x, \quad \Rightarrow \mathrm{d}y = \left(\frac{\lambda}{E}\right)^{\frac{1}{4}} \mathrm{d}x. \tag{11.102}$$

With this, equation (11.101) becomes

$$\text{L.H.S.} = 2\left(\frac{2mE}{\hbar^2}\right)^{\frac{1}{2}} \left(\frac{E}{\lambda}\right)^{\frac{1}{4}} \int\limits_0^1 \mathrm{d}y\ (1 - y^4)^{\frac{1}{2}}$$

$$\simeq 2\left(\frac{4m^2}{\hbar^4\lambda}\right)^{\frac{1}{4}} E^{\frac{3}{4}} \frac{9}{10}, \tag{11.103}$$

where we have used

$$\int\limits_0^1 \mathrm{d}y\, (1 - y^4)^{\frac{1}{2}} = \frac{1}{4}B(\frac{1}{4}, \frac{3}{2}) \simeq \frac{9}{10}.$$

Therefore, in this case, the quantization condition, (11.100), gives

$$\frac{9}{5}\left(\frac{4m^2}{\hbar^4\lambda}\right)^{\frac{1}{4}} E_n^{\frac{3}{4}} = \left(n + \frac{1}{2}\right)\pi, \tag{11.104}$$

which determines the energy eigenvalues. Calculating for $n = 0$, we obtain from (11.104)

$$E_0 = \left(\frac{5}{9}\right)^{\frac{4}{3}} \left(\frac{\hbar^4\lambda}{4m^2}\right)^{\frac{1}{3}} \left(\frac{\pi}{2}\right)^{\frac{4}{3}}$$

$$= \left(\frac{5\pi}{18}\right)^{\frac{4}{3}} \left(\frac{1}{24}\right)^{\frac{1}{3}} \left(\frac{6\hbar^4\lambda}{m^2}\right)^{\frac{1}{3}}, \tag{11.105}$$

which can be compared with the value obtained from the variational method in (10.47),

$$\frac{3}{8}\left(\frac{6\hbar^4\lambda}{m^2}\right)^{\frac{1}{3}} \geq E_0 \geq 0.$$

4. As a final application of the quantization condition, let us consider a particle moving in an infinite square well in one dimension (or particle in a box). In this case, the potential has the form shown in Fig. 11.10.

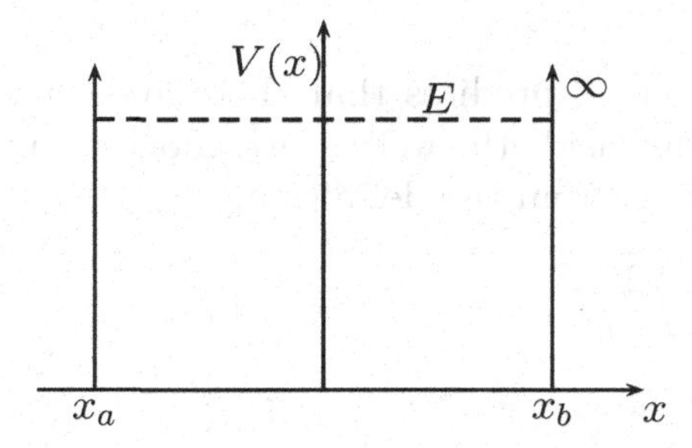

Figure 11.10: A particle with energy E moving in a one dimensional infinite square well potential.

$$\begin{aligned} V(x) &= 0, \qquad & x_a \leq x \leq x_b, \\ &= \infty, \qquad & x \leq x_a \text{ and } x \geq x_b. \end{aligned} \tag{11.106}$$

Contrary to our general analysis, here we cannot, of course, approximate the potential by a linear term in x at the boundaries. Hence the connection formulae breakdown for this case.

However, in this case, we know that, at the boundaries as well as outside the well, the wave function has to vanish identically. This, therefore, suggests the form of the WKB solution, inside the well, to be

$$\psi(x) = \frac{1}{k^{\frac{1}{2}}} \sin\left(\int_{x_a}^{x} \mathrm{d}x'\, k\right), \qquad x_a \leq x \leq x_b,$$

$$\psi(x) = \frac{1}{k^{\frac{1}{2}}} \sin\left(\int_{x_b}^{x} \mathrm{d}x'\, k\right), \qquad x_a \leq x \leq x_b. \tag{11.107}$$

Once again, we note that if they are to describe the same particle, the phase can change by $(n+1)\pi$ (This is because the phase change should, at least, be π since x_a, x_b are distinct points.). Thus, we obtain

$$\int_{x_a}^{x} \mathrm{d}x'\, k = \int_{x_b}^{x} \mathrm{d}x'\, k + (n+1)\pi,$$

$$\text{or,} \quad \int_{x_a}^{x_b} \mathrm{d}x\, k = (n+1)\pi, \qquad n = 0, 1, 2, 3 \cdots . \tag{11.108}$$

This, of course, predicts that there are n nodes and $\left(\frac{1}{2}n + \frac{1}{2}\right)$ wavelengths inside the well, as we know from the exact solution. Furthermore, from the definition,

$$k = \sqrt{\frac{2mE}{\hbar^2}}, \qquad x_a \leq x \leq x_b, \tag{11.109}$$

we determine

$$\sqrt{\frac{2mE_n}{\hbar^2}}\, (x_b - x_a) = (n+1)\pi,$$

$$\text{or,} \quad E_n = \frac{\pi^2 \hbar^2 (n+1)^2}{2ma^2}, \qquad x_b - x_a = a. \tag{11.110}$$

This is, of course, the exact answer that we have determined earlier in (4.23).

As we discussed in the beginning, WKB is a semi-classical approximation. We know that systems tend more towards the classical behavior as the quantum numbers take larger values. Thus, WKB is an extremely good approximation for calculating the behavior of excited states. One should contrast this with the variational method which is extremely good for ground state energy eigenvalues. Thus, the two methods complement each other. (We should, of course, remember that, for certain potentials like the oscillator and the infinite square well, WKB method gives the exact answer.)

11.5 Penetration of a barrier

In (11.67) and (11.70), we derived a set of connection formulae between the classically inaccessible and accessible regions by studying the exact solution of the Schrödinger equation in the transition region with a linear approximation for the potential. There were two independent solutions $\sim J_{\pm\frac{1}{3}}$ and $\sim I_{\pm\frac{1}{3}}$. The formulae we derived corresponded to a particular combination of these solutions. However, one can choose a different combination and derive a second connection formula also. Thus, for the turning point, x_a (see Fig. 11.11), we have

$$\begin{aligned}
\frac{1}{2|\kappa|^{\frac{1}{2}}} e^{-\int_x^{x_a} \mathrm{d}x'\,|\kappa|} &\longleftrightarrow \frac{1}{k^{\frac{1}{2}}} \cos\left(\int_{x_a}^{x} \mathrm{d}x'\,k - \frac{\pi}{4}\right), \\
-\frac{1}{|\kappa|^{\frac{1}{2}}} e^{\int_x^{x_a} \mathrm{d}x'\,|\kappa|} &\longleftrightarrow \frac{1}{k^{\frac{1}{2}}} \sin\left(\int_{x_a}^{x} \mathrm{d}x'\,k - \frac{\pi}{4}\right).
\end{aligned} \tag{11.111}$$

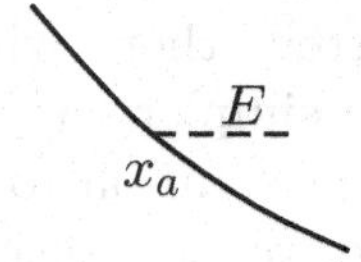

Figure 11.11: The classical turning point $x = x_a$.

Similarly, at the other turning point shown in Fig. 11.12, we have

$$\begin{aligned}
\frac{1}{2|\kappa|^{\frac{1}{2}}} e^{-\int_{x_b}^{x} \mathrm{d}x'\,|\kappa|} &\longleftrightarrow \frac{1}{k^{\frac{1}{2}}} \cos\left(\int_{x}^{x_b} \mathrm{d}x'\,k - \frac{\pi}{4}\right), \\
-\frac{1}{|\kappa|^{\frac{1}{2}}} e^{\int_{x_b}^{x} \mathrm{d}x'\,|\kappa|} &\longleftrightarrow \frac{1}{k^{\frac{1}{2}}} \sin\left(\int_{x}^{x_b} \mathrm{d}x'\,k - \frac{\pi}{4}\right).
\end{aligned} \tag{11.112}$$

Even though the connection formulae work both ways, one has to exercise a certain amount of caution in applying them. That is, one should always start with the region of space where the boundary condition on the wave function can be imposed and, then, match it onto the next region through the connection formula. Thus, for example, in the case of the motion of a particle inside a well, we always match the solution in the classically inaccessible region to that in the classically accessible region.

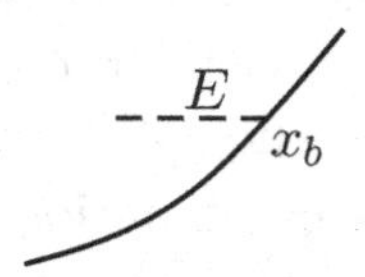

Figure 11.12: The classical turning point at $x = x_b$.

Let us now study the question of penetration of a barrier using these ideas. Classically, of course, if a particle has energy less than the height of the potential barrier, then, it would be completely reflected. On the other hand, if its energy is greater than the potential barrier, then the barrier would behave as if it were completely transparent. Quantum mechanically, however, we know that even when $E < V_{\rm max}$, in addition to the particle being reflected, there will be certain amount of transmission. Similarly, if $E > V_{\rm max}$, in addition to the particle being transmitted, there will be certain amount of reflection. We shall study the simpler case when $E < V_{\rm max}$. It is worth noting here that the exact solution to this barrier penetration problem exists only for square well potentials. However, since not all potentials in nature are so ideal, WKB approximation comes in handy.

Let us again consider a general potential which is slowly varying shown in Fig. 11.13. A particle with energy E is incident from the left. Part of it is reflected and a part is transmitted so that to the right of the barrier there is only a transmitted wave.

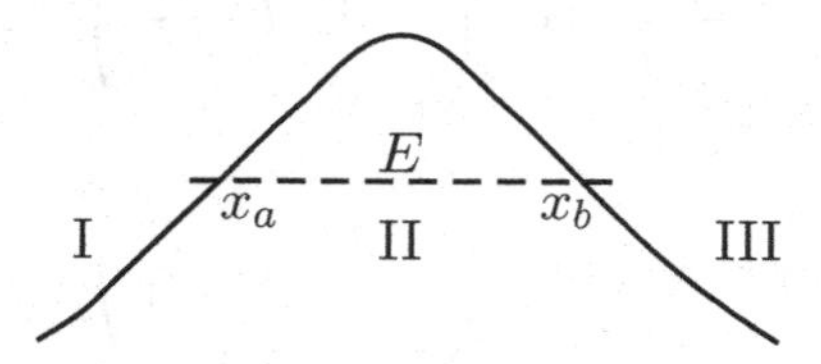

Figure 11.13: A particle with energy E incident from the left on a general potential.

The entire space can be divided into three regions. Region I is to the left of the potential barrier for $x < x_a$ and Region III is to the right of the barrier for $x > x_b$. x_a and x_b are the classical turning points and, therefore, regions I and III are classically accessible regions. The classically inaccessible region is region II for $x_a \leq x \leq x_b$. Furthermore, WKB method is valid if $\int\limits_{x_a}^{x_b} \mathrm{d}x\ \left[\frac{2m}{\hbar^2}(V(x)-E)\right]^{\frac{1}{2}}$ is large. Thus, we are considering the case where the height and the width of the potential are sufficiently large. WKB approximation leads to the wave functions in regions I and III to be of the forms

$$\psi_{\mathrm{I}}(x) = \frac{A_1}{\sqrt{k}}\, e^{i\left(\int\limits_{x}^{x_a} \mathrm{d}x'\, k - \frac{\pi}{4}\right)} + \frac{A_2}{\sqrt{k}}\, e^{-i\left(\int\limits_{x}^{x_a} \mathrm{d}x'\, k - \frac{\pi}{4}\right)}, \qquad x < x_a, \tag{11.113}$$

with

$$k(x) = \left[\frac{2m}{\hbar^2}(E - V(x))\right]^{\frac{1}{2}}, \tag{11.114}$$

and we have included, for later convenience, the phase factors of $\frac{\pi}{4}$ manifestly, which otherwise can always be absorbed into the constants A_1 and A_2. Similarly, we can write

$$\begin{aligned}\psi_{\mathrm{III}}(x) &= \frac{A_3}{\sqrt{k}} e^{i\left(\int\limits_{x_b}^{x} \mathrm{d}x'\, k - \frac{\pi}{4}\right)}, \qquad x > x_b \\ &= \frac{A_3}{\sqrt{k}}\left[\cos\left(\int\limits_{x_b}^{x} \mathrm{d}x'\, k - \frac{\pi}{4}\right) + i\sin\left(\int\limits_{x_b}^{x} \mathrm{d}x'\, k - \frac{\pi}{4}\right)\right].\end{aligned} \tag{11.115}$$

We can now determine the form of the wave function in region II through the connection formulae, (11.111) and (11.112). Thus, starting with region III, we obtain

$$\begin{aligned}\psi_{\mathrm{II}}(x) &= \frac{A_3}{|\kappa|^{\frac{1}{2}}}\left[\frac{1}{2}\, e^{-\int\limits_{x}^{x_b} \mathrm{d}x'\, |\kappa|} - i e^{\int\limits_{x}^{x_b} \mathrm{d}x'\, |\kappa|}\right] \\ &= \frac{A_3}{|\kappa|^{\frac{1}{2}}}\left[\frac{1}{2}\, e^{-\gamma} e^{\int\limits_{x_a}^{x} \mathrm{d}x'\, |\kappa|} - i e^{\gamma} e^{-\int\limits_{x_a}^{x} \mathrm{d}x'\, |\kappa|}\right],\end{aligned} \tag{11.116}$$

where we have defined

$$\gamma = \int\limits_{x_a}^{x_b} \mathrm{d}x\, |\kappa|, \tag{11.117}$$

which is assumed to be large. Thus, the first term is small compared to the second term and we have

$$\psi_{\rm II}(x) \simeq -\frac{iA_3}{|\kappa|^{\frac{1}{2}}}\, e^{\gamma} e^{-\int\limits_{x_a}^{x} \mathrm{d}x'\,|\kappa|}, \qquad x_a \le x \le x_b. \tag{11.118}$$

The connection formula between regions II and I (see (11.111) and (11.112)), namely,

$$\frac{1}{2|\kappa|^{\frac{1}{2}}}\, e^{-\int\limits_{x_a}^{x} \mathrm{d}x'\,|\kappa|} \to \frac{1}{k^{\frac{1}{2}}} \cos\left(\int\limits_{x}^{x_a} \mathrm{d}x'\, k - \frac{\pi}{4}\right),$$

now gives

$$\psi_{\rm I}(x) = -\frac{2iA_3 e^{\gamma}}{k^{\frac{1}{2}}} \cos\left(\int\limits_{x}^{x_a} \mathrm{d}x'\, k - \frac{\pi}{4}\right), \qquad x < x_a. \tag{11.119}$$

Comparing this with (11.113), we determine

$$A_1 = -iA_3 e^{\gamma} = A_2. \tag{11.120}$$

Thus, we have

$$\psi_{\rm I}(x) = -\frac{iA_3 e^{\gamma}}{k^{\frac{1}{2}}} e^{i\left(\int\limits_{x}^{x_a} \mathrm{d}x'\, k - \frac{\pi}{4}\right)} - \frac{iA_3 e^{\gamma}}{k^{\frac{1}{2}}} e^{-i\left(\int\limits_{x}^{x_a} \mathrm{d}x'\, k - \frac{\pi}{4}\right)}, \quad x < x_a,$$

$$\psi_{\rm III}(x) = \frac{A_3}{k^{\frac{1}{2}}} e^{i\left(\int\limits_{x_b}^{x} \mathrm{d}x'\, k - \frac{\pi}{4}\right)}, \qquad x > x_b. \tag{11.121}$$

Recalling the definition of the transmission coefficient, we obtain

$$\begin{aligned} \mathrm{T} &= \frac{v_{\rm TR}|\psi_{\rm TR}|^2}{v_{\rm INC}|\psi_{\rm INC}|^2} = \frac{k_{\rm TR}|\psi_{\rm TR}|^2}{k_{\rm INC}|\psi_{\rm INC}|^2} \\ &\simeq \frac{|A_3|^2}{|A_3|^2 e^{2\gamma}} = e^{-2\gamma} = e^{-2\int\limits_{x_a}^{x_b} \mathrm{d}x\,|\kappa|}. \end{aligned} \tag{11.122}$$

If we calculate the coefficient of reflection, then it is clear that it seems like unity. This is because we approximated the wave function in region II by the second term only. If we had kept both the terms, then, we can show that the reflection coefficient is given by

$$\mathrm{R} = 1 - \mathrm{T} = 1 - e^{-2\int\limits_{x_a}^{x_b} \mathrm{d}x\,|\kappa|}, \tag{11.123}$$

as it should be to conserve probability. We should note here that, for a square well potential with height and width V_0 and a respectively, this is the form of the transmission coefficient except for certain numerical factor. The discrepancy arises because the square well barrier has discontinuities where WKB approximation breaks down. Also note that as the mass of the particle increases, the transmission coefficient in (11.122) decreases. Thus it is difficult for a heavier particle to tunnel through a barrier.

11.6 Applications of tunneling

Let us study next a few applications of tunneling.

1. Cold emission of electrons from a metal:

   The electrons inside a metal move freely in a constant potential. When they reach the surface or the edge of the metal, they are reflected back because of a repulsive potential. This is a very short distance phenomenon (atomic length $\sim 10^{-8}$ cm). By applying an external electric field, the electrons can be given enough energy to overcome this repulsion. The energy W necessary to release an electron from a metal is known as its work function. (Let us note here that if the electron picks up enough energy from thermal motions so that it can go over the potential barrier, then, the process is called thermal emission.)

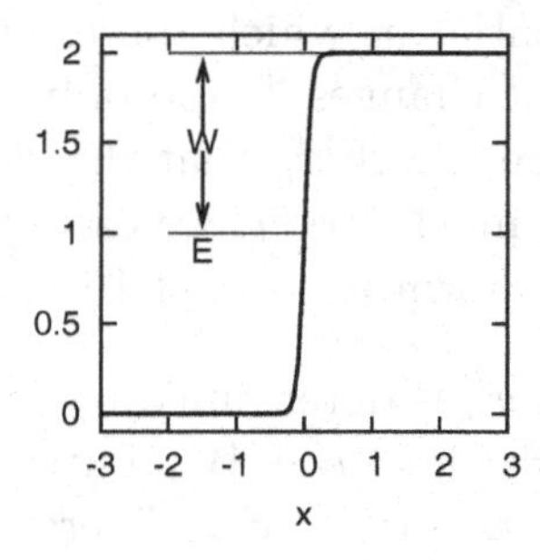

Figure 11.14: The potential in which an electron with energy E moves inside a metal with W denoting the work function.

   The potential in which the electrons move has the shape as shown in Fig. 11.14. Suppose, we now apply a constant electric field $\mathcal{E}$ in a direction (say, x) so as to pull out the electrons. Then, the effective potential seen by the electrons becomes

$$V_{\text{eff}} = V(x) - e\mathcal{E}x, \tag{11.124}$$

and has the form as shown in Fig. 11.15. Because of this applied electric field, the potential curves down for $x > 0$ and an electron with energy $E > 0$ has now a nonzero probability for coming out of the metal.

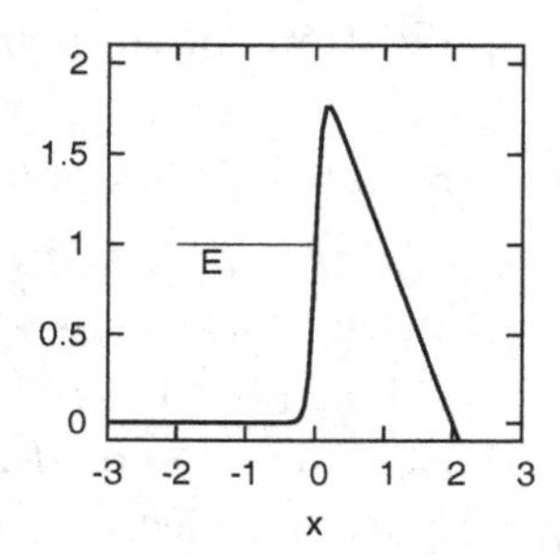

Figure 11.15: Effective potential which the electron sees with an applied electric field.

To calculate this probability we would, of course, need the exact shape of the potential in which the electron moves. However, since the potential changes shape only within an atomic distance and since the length x_a that the electron has to travel to come out is much much larger, we can approximate the potential by the graph shown in Fig. 11.16.

The turning point x_a is determined by (Just to give a few more details, let $V_{\text{max}} = V_0 = E + W$. Furthermore, we know that $V_{\text{eff}} = V_0 - e\mathcal{E}x = E + W - e\mathcal{E}x$. Therefore, the turning points are at $E = V_{\text{eff}} = E + W - e\mathcal{E}x$, which determines the only non-trivial turning point to be as given below.)

$$\begin{aligned} W &= e\mathcal{E}x_a, \\ \text{or,} \quad x_a &= \frac{W}{e\mathcal{E}}. \end{aligned} \tag{11.125}$$

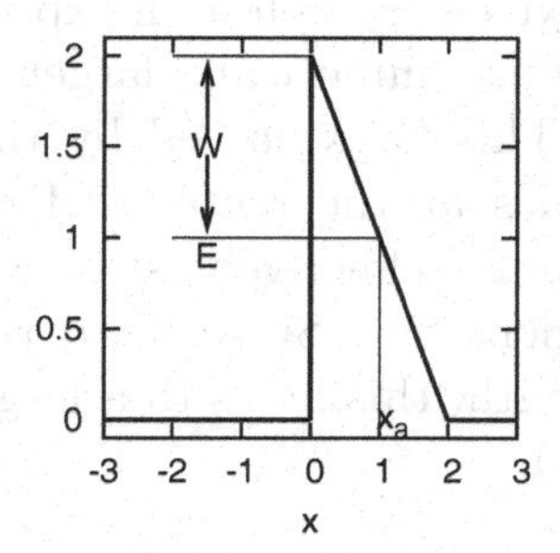

Figure 11.16: Idealized effective potential in which the electron moves.

It follows now that

$$\begin{aligned}\gamma &= \int_0^{x_a} \mathrm{d}x \left[\frac{2m}{\hbar^2}\left(V_{\text{eff}} - E\right)\right]^{\frac{1}{2}} \\ &= \int_0^{x_a} \mathrm{d}x \left[\frac{2m}{\hbar^2}\left(W - e\mathcal{E}x\right)\right]^{\frac{1}{2}} \\ &= \left(\frac{2mW}{\hbar^2}\right)^{\frac{1}{2}} \int_0^{\frac{W}{e\mathcal{E}}} \mathrm{d}x \left(1 - \frac{e\mathcal{E}}{W}\,x\right)^{\frac{1}{2}} \\ &= \left(\frac{2mW}{\hbar^2}\right)^{\frac{1}{2}} \left(\frac{W}{e\mathcal{E}}\right) \int_0^1 \mathrm{d}y\,(1-y)^{\frac{1}{2}} \\ &= \frac{2}{3}\sqrt{2m}\,\frac{W^{\frac{3}{2}}}{\hbar e\mathcal{E}}. \end{aligned} \tag{11.126}$$

Thus, the transmission coefficient, (11.122), which is the probability for transmission, becomes

$$\mathrm{T} = e^{-2\gamma} = e^{-2\int_0^{x_a} \mathrm{d}x\,|\kappa|} = e^{-\frac{4}{3}\sqrt{2m}\frac{W^{\frac{3}{2}}}{\hbar e\mathcal{E}}}. \tag{11.127}$$

From this, one can calculate the current by multiplying this with the number of electrons hitting the metal edge per second.

We note that the current is larger for a larger applied electric field. It is also large for metals with a lower work function. This behavior agrees extremely well with experiments. However, the currents observed are numerically larger than predicted by our WKB analysis. This is explained by the fact that there are strong irregularities at the metal surfaces. As a result, the electric field becomes effectively stronger closer to the metal surface than farther away. Since the current is sensitive to the electric field strength, this leads to a larger current.

2. α decay:

   In heavier nuclei α-particles (namely, the nucleus of the helium atom consisting of two protons and two neutrons) are bound by very strong nuclear forces. If the nucleus has a charge Ze, then outside the nucleus, there is a repulsive Coulomb potential of the type $\frac{Ze}{r}$. The strong force which holds the α-particles inside the nucleus is extremely short ranged. Hence, inside the nucleus, the α-particle can be thought of as free. The potential, therefore, has the shape shown in Fig. 11.17.

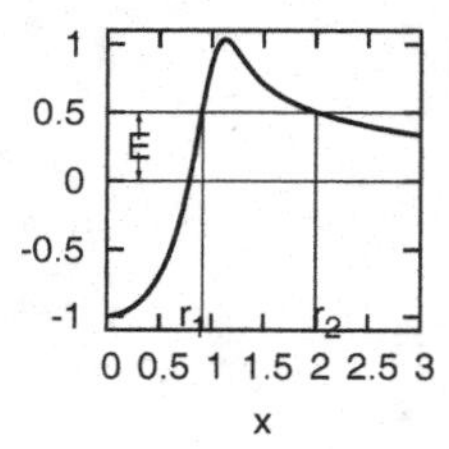

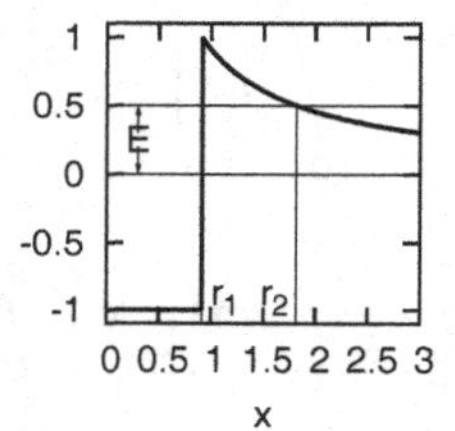

Figure 11.17: The potential in which the alpha particle moves where the potential on the right is a simplified form of the actual potential on the left.

   The α-particle will experience a repulsive force outside the nucleus and would have a (Coulomb) potential energy $\frac{2(Z-2)e^2}{r}$. Since the potential is invariant under rotations, the angular part of the solution can be trivially separated and the problem reduces to an one dimensional problem. Therefore, we can apply the WKB method. To calculate the transmission coefficient, or the tunneling probability, in this case, we again have to know the shape of the potential near r_1 (to determine the turning point). But remembering that the nuclear force is very short ranged and that the distance $r_2 - r_1$ is much larger than nuclear distances we can set

$$r_1 \simeq 2 \times 10^{-13} A^{\frac{1}{3}} \text{ cm}. \tag{11.128}$$

This is the nuclear radius of the system. Furthermore, we can also approximate the strong force by a square well and the turning point r_2 is determined from the fact that (we are assuming that $\ell = 0$)

$$E = V = \frac{2(Z-2)e^2}{r_2}, \quad \text{or} \quad r_2 = \frac{2(Z-2)e^2}{E}. \tag{11.129}$$

Thus, we have

$$\begin{aligned}\gamma &= \int_{r_1}^{r_2} \mathrm{d}r |\kappa| = \int_{r_1}^{r_2} \mathrm{d}r \left[\frac{2m}{\hbar^2}\left(\frac{2(Z-2)e^2}{r} - E\right)\right]^{\frac{1}{2}} \\ &= \left(\frac{2mE}{\hbar^2}\right)^{\frac{1}{2}} \int_{r_1}^{r_2} \mathrm{d}r \left[\left(\frac{r_2}{r} - 1\right)\right]^{\frac{1}{2}}. \end{aligned} \tag{11.130}$$

Let us define

$$\begin{aligned} y &= \left[\frac{r_2}{r} - 1\right]^{\frac{1}{2}}, \\ \text{or,} \quad r &= \frac{r_2}{1+y^2}, \\ \text{or,} \quad \mathrm{d}r &= -\frac{2r_2 y \mathrm{d}y}{(1+y^2)^2}, \end{aligned} \tag{11.131}$$

so that (11.130) becomes

$$\begin{aligned}\gamma &= \left(\frac{2mE}{\hbar^2}\right)^{\frac{1}{2}} (-2r_2) \int_{\sqrt{\frac{r_2}{r_1}-1}}^{0} \mathrm{d}y \, \frac{y^2}{(1+y^2)^2} \\ &= 2r_2 \left(\frac{2mE}{\hbar^2}\right)^{\frac{1}{2}} \int_0^{\sqrt{\frac{r_2}{r_1}-1}} \mathrm{d}y \, \frac{y^2}{(1+y^2)^2} \\ &= 2r_2 \left(\frac{2mE}{\hbar^2}\right)^{\frac{1}{2}} \left[\frac{1}{2}\tan^{-1} y - \frac{1}{2}\frac{y}{1+y^2}\right]_0^{\sqrt{\frac{r_2}{r_1}-1}}. \end{aligned} \tag{11.132}$$

Let us parameterize $\frac{r_1}{r_2} = \cos^2\phi$, $\left(\frac{r_2}{r_1} - 1\right)^{\frac{1}{2}} = \tan\phi$, so that we have,

$$\begin{aligned}\gamma &= 2r_2 \left(\frac{2mE}{\hbar^2}\right)^{\frac{1}{2}} \frac{1}{2} \left[\phi - \frac{\tan\phi}{\sec^2\phi}\right] \\ &= r_2 \left(\frac{2mE}{\hbar^2}\right)^{\frac{1}{2}} \frac{1}{2} \left[2\phi - \sin 2\phi\right]. \end{aligned} \tag{11.133}$$

Substituting, $r_2 = \frac{2(Z-2)e^2}{E}$, $v = \sqrt{\frac{2E}{m}}$, we have, for the transmission coefficient,

$$\mathrm{T} = e^{-2\gamma} = e^{-\frac{4(Z-2)e^2}{\hbar v}[2\phi - \sin 2\phi]}. \tag{11.134}$$

Every time the α-particle collides against the outer wall, this is the probability of escape. Since the α-particle is in a well of depth $-V_0$, its total kinetic energy is

$$\begin{aligned} \text{K.E.} &= E + V_0, \\ v_{\text{inside}} &= \sqrt{\frac{2\text{K.E.}}{m}} = \left[\frac{2(E+V_0)}{m}\right]^{\frac{1}{2}}. \end{aligned} \tag{11.135}$$

The frequency with which it collides against the outer wall is then determined to be

$$f = \frac{v_{\text{inside}}}{2r_1}. \tag{11.136}$$

Thus, the probability of escape per second is

$$\Gamma = fT = fe^{-2\gamma} = \frac{v_{\text{inside}}}{2r_1} e^{-\frac{4(Z-2)e^2}{\hbar v}[2\phi - \sin 2\phi]}, \tag{11.137}$$

and the mean life of the nucleus is obtained to be

$$\tau = \frac{1}{\Gamma} = \frac{2r_1}{v_{\text{inside}}} e^{\frac{4(Z-2)e^2}{\hbar v}[2\phi - \sin 2\phi]}. \tag{11.138}$$

Let us next consider a specific example to get a feeling for some numbers, namely, let us choose Uranium for which we have,

$$Z = 92, \quad r_1 \simeq 10^{-12}\ \text{cm}, \quad E = 4.2\ \text{MeV}. \tag{11.139}$$

Let us choose, for simplicity, $V_0 = 0$. Then, for this nucleus, we have

$$\begin{aligned} r_2 &= \frac{2(Z-2)e^2}{E} = \frac{2\times 90}{4.2\ \text{MeV}}\,\frac{e^2}{\hbar c}\,\hbar c \\ &= \frac{2\times 90}{4.2\ \text{MeV}}\,\frac{1}{137}\times 2\times 10^{-11}\ \text{MeV-cm} \\ &\simeq 7\times 10^{-12}\ \text{cm}. \end{aligned} \tag{11.140}$$

It follows, then, that

$$\begin{aligned} \frac{r_1}{r_2} &= \frac{1}{7} \simeq .14, \\ \phi &= \cos^{-1}\left(\frac{r_1}{r_2}\right)^{\frac{1}{2}} \simeq \cos^{-1}(.4) \simeq \frac{\pi}{2} - .4, \\ \sin 2\phi &= \sin(\pi - .8) \simeq .8. \end{aligned} \tag{11.141}$$

Thus, we have

$$\begin{aligned} 2\phi - \sin 2\phi &= \pi - .8 - .8 \simeq 1.5, \\ v = v_{\text{inside}} &= \sqrt{\frac{2E}{m}} = c\sqrt{\frac{2E}{mc^2}} \\ &= c\sqrt{\frac{2\times 4.2\ \text{MeV}}{4\times 10^3\ \text{MeV}}} \simeq 4.5\times 10^{-2}c. \end{aligned} \tag{11.142}$$

It follows, therefore, that

$$\begin{aligned} &\frac{4(Z-2)e^2}{v\hbar}\,[2\phi - \sin 2\phi] \\ &= \frac{4\times 90}{\frac{v}{c}}\,\frac{e^2}{\hbar c}\times 1.5 \\ &= \frac{4\times 90}{4.5\times 10^{-2}}\times\frac{1}{137}\times 1.5 \\ &\simeq \frac{4\times .7}{3}\times 10^2 \simeq 90. \end{aligned} \tag{11.143}$$

Thus, we determine the life time to be,

$$\begin{aligned} \tau &= \frac{2r_1}{v}\,e^{90} \\ &= \frac{2\times 10^{-12}\ \text{cm}}{4.5\times 10^{-2}\times 3\times 10^{10}\ \text{cm/sec}}\,e^{90} \end{aligned}$$

$$\begin{aligned}
&\simeq 1.5 \times 10^{-21}\, e^{90} \text{ sec} \\
&\simeq 1.5 \times 10^{-21} \times 10^{39} \text{ sec} \simeq 1.5 \times 10^{18} \text{ sec} \\
&\simeq 10^{11} \text{ yrs.}
\end{aligned} \tag{11.144}$$

11.7 Energy splitting due to tunneling

Let us consider a double well potential in one dimension. Consider it to be symmetric in x. Such potentials are frequently used in various branches of physics. A common example is given by

$$V(x) \propto \left(x^2 - b^2\right)^2, \tag{11.145}$$

which is shown in Fig. 11.18.

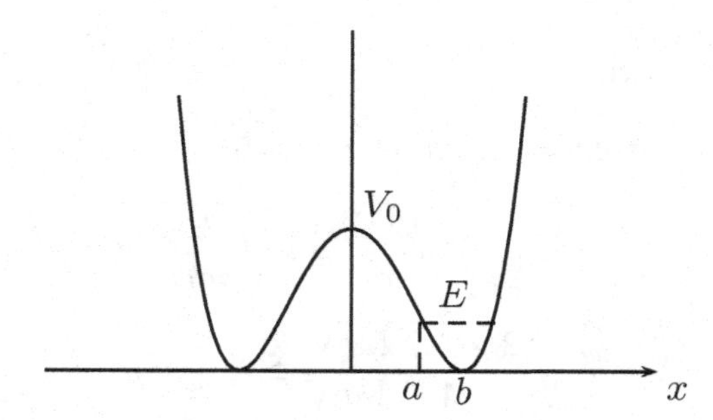

Figure 11.18: A particle with energy E moving in a double well potential in one dimension.

If there is no tunneling between the two wells, then one can solve for the Schrödinger equation in each of the wells separately and would have definite energy eigenvalues. Since the Hamiltonian is parity invariant, we can assume that both $\psi_0(x)$ and $\psi_0(-x)$ belong to the same energy value E_0. Furthermore, the definite parity combinations

$$\begin{aligned}
\psi_1(x) &= \frac{1}{\sqrt{2}}\left(\psi_0(x) + \psi_0(-x)\right), \\
\psi_2(x) &= \frac{1}{\sqrt{2}}\left(\psi_0(x) - \psi_0(-x)\right),
\end{aligned} \tag{11.146}$$

would also belong to the same eigenvalue E_0. Thus, we see that there is a two fold degeneracy in the problem, in the absence of tunneling. In the presence of tunneling, however, the energy levels split and let us calculate the energy splitting in the WKB approximation. We note that at $x = 0$, $\psi_1 = \sqrt{2}\,\psi_0(0)$ and $\psi_1'(0) = 0$ and that similar relations hold for ψ_2.

The Schrödinger equations satisfied by ψ_0 and ψ_1 are

$$\frac{\mathrm{d}^2\psi_0}{\mathrm{d}x^2} + \frac{2m}{\hbar^2}\,(E_0 - V(x))\,\psi_0 = 0,$$
$$\frac{\mathrm{d}^2\psi_1}{\mathrm{d}x^2} + \frac{2m}{\hbar^2}\,(E_1 - V(x))\,\psi_1 = 0, \tag{11.147}$$

where we are assuming that, because of tunneling, the energy of the state ψ_1 changes to E_1.

Let us multiply the first equation in (11.147) by ψ_1 and the second by ψ_0 and integrate from 0 to ∞. Subtracting the two, we have

$$\frac{2m}{\hbar^2}\,(E_1 - E_0)\int_0^\infty \mathrm{d}x\;\psi_0\psi_1 = \psi_1'(0)\psi_0(0) - \psi_1(0)\psi_0'(0)$$
$$= -\psi_1(0)\psi_0'(0). \tag{11.148}$$

But, we know that

$$\int_0^\infty \mathrm{d}x\;\psi_0\psi_1 \simeq \frac{1}{\sqrt{2}}\int_0^\infty \mathrm{d}x\;\psi_0^2 = \frac{1}{\sqrt{2}}, \tag{11.149}$$

and we can use the WKB wave function,

$$\psi_0(0) = \sqrt{\frac{\omega}{2\pi v_0}}\;e^{-\frac{1}{\hbar}\int_0^a \mathrm{d}x|p|},$$
$$\psi_0'(0) = \frac{mv_0}{\hbar}\,\psi_0(0), \tag{11.150}$$

so that $v_0 = \left[\frac{2}{m}\,(V_0 - E_0)\right]^{\frac{1}{2}}$, $\omega = \frac{2\pi}{T}$ with T representing the classical period for bound motion. (The extra factors are due to normalization.) Using these, we obtain from (11.148)

$$\frac{2m}{\hbar^2}\,(E_1 - E_0)\frac{1}{\sqrt{2}} \simeq -\sqrt{2}\frac{mv_0}{\hbar}\,\frac{\omega}{2\pi v_0}\;e^{-\frac{2}{\hbar}\int_0^a \mathrm{d}x|p|},$$

$$\text{or,}\quad (E_1 - E_0) = -\frac{\hbar\omega}{2\pi}\;e^{-\frac{2}{\hbar}\int_0^a \mathrm{d}x|p|}. \tag{11.151}$$

Similarly, we can show

$$E_2 - E_0 \simeq \frac{\hbar\omega}{2\pi}\, e^{-\frac{2}{\hbar}\int_0^a \mathrm{d}x|p|}, \tag{11.152}$$

so that the splitting between the two levels due to tunneling is obtained to be

$$E_2 - E_1 \simeq \frac{\hbar\omega}{\pi}\, e^{-\frac{2}{\hbar}\int_0^a \mathrm{d}x|p|}. \tag{11.153}$$

11.8 Selected problems

1. Consider the WKB solution for a particle bound in a potential well and look at wave functions for large quantum numbers for which the approximation is valid. Assuming that the wave functions are vanishingly small outside the well, derive the normalization constants for the wave functions. Calculate also,

$$x_{nm} = \int_{x_a}^{x_b} dx\, \psi_n^*(x) x \psi_m(x), \tag{11.154}$$

   where x_a, x_b are the classical turning points.

2. What is the WKB (Bohr-Sommerfeld) quantization rule for a particle moving in a potential given by

$$V(x) = \begin{cases} \infty, & \text{for} \quad x \leq 0, \\ \text{continuous and positive}, & x > 0. \end{cases} \tag{11.155}$$

3. Using WKB method, derive the quantization conditions for the energy levels of a particle moving in

   i) the linear potential $V(x) = \beta|x|$, where the real constant $\beta > 0$.

   ii) the oscillator potential $V(x) = \frac{1}{2}m\omega^2 x^2$.

4. Use WKB approximation to show that a spherically symmetric attractive potential, in three dimensions, that falls off like r^{-n} for large r, has an infinite number of bound states if $n \leq 2$.

5. In 1974 two new particles called ψ and ψ' were discovered with rest energies 3.1 GeV and 3.7 GeV respectively (1 GeV $= 10^9$ eV). These are believed to be bound states of a charmed quark of mass 1.5 GeV/c^2 and an anti-quark of the same mass in a linear potential, $V(r) = V_0 + kr$ where V_0, k are constants. By assuming that these are the $n = 0$ and $n = 1$ bound states of zero angular momentum, calculate V_0 using the WKB approximation. What would you predict for the rest mass of ψ'', the $n = 2$ state? (The experimental value is $\simeq 4.2$ GeV/c^2.)

6. For a hydrogen atom, in a constant electric field ($H_1 = -e\mathcal{E}z$), find α and β such that

$$\psi = \left(\frac{1}{\pi a_0^3}\right)^{\frac{1}{2}} (1 + \alpha z + \beta z r) e^{-\frac{r}{a_0}}, \tag{11.156}$$

is the ground state wave function correct to first order in $\mathcal{E}$. Use it to calculate the exact polarizability of the ground state of hydrogen.

CHAPTER 12

Stationary perturbation theory

Perturbation characterizes a series of iterative methods by which one obtains an approximate solution to a difficult problem. The idea is to introduce a small parameter ϵ into the problem and obtain a solution as a series in powers of ϵ. This method is very powerful and is used in all branches of physics.

As an example of perturbation, let us consider the solution of the algebraic equation

$$x^3 - 4.001x + 0.002 = 0. \tag{12.1}$$

The roots of this equation are extremely difficult to obtain exactly. However, we can resort to perturbation theory to find an approximate solution quite easily. First of all, let us introduce a small parameter

$$\epsilon = .001. \tag{12.2}$$

Then, equation (12.1) can be rewritten as

$$x^3 - (4 + \epsilon)x + 2\epsilon = 0. \tag{12.3}$$

Thinking of ϵ as an arbitrary small parameter, we note that the solution of (12.3) would be a function of this parameter. Therefore, we expand the variable x in powers of ϵ, namely,

$$x = x_0 + \epsilon x_1 + \epsilon^2 x_2 + \cdots = \sum_{n=0}^{\infty} \epsilon^n x_n. \tag{12.4}$$

Putting this back into (12.3) and keeping only terms of order ϵ^0, we have

$$x_0^3 - 4x_0 = 0, \quad \text{or} \quad x_0 = 0, \pm 2. \tag{12.5}$$

Thus, to the leading order (in powers of ϵ), the roots of the equation are 0 and ± 2.

If we choose the root $x_0 = 0$, then the expansion in (12.4) becomes

$$x^{(0)} = \epsilon x_1^{(0)} + \epsilon^2 x_2^{(0)} + \cdots . \tag{12.6}$$

Substituting this into (12.3) and keeping terms up to order ϵ, we have

$$-4\epsilon x_1^{(0)} + 2\epsilon = 0, \quad \text{or} \quad x_1^{(0)} = \frac{1}{2} = 0.5, \tag{12.7}$$

so that, to this order, this root is

$$x^{(0)} = x_0^{(0)} + \epsilon x_1^{(0)} = \frac{1}{2}\epsilon = .0005. \tag{12.8}$$

If we choose to work with the root $x_0 = 2$, then, the expansion in (12.4) becomes

$$x^{(2)} = x_0^{(2)} + \epsilon x_1^{(2)} = 2 + \epsilon x_1^{(2)}. \tag{12.9}$$

Putting this into (12.3) and keeping terms up to order ϵ we have

$$\begin{aligned}
&\left(8 + 12\epsilon x_1^{(2)}\right) - 4\left(2 + \epsilon x_1^{(2)}\right) - \epsilon(2) + 2\epsilon = 0,\\
\text{or,}\quad &\epsilon\left(12x_1^{(2)} - 4x_1^{(2)}\right) = 0,\\
\text{or,}\quad &x_1^{(2)} = 0,
\end{aligned} \tag{12.10}$$

so that, to this order, we have

$$x^{(2)} = x_0^{(2)} + \epsilon x_1^{(2)} = 2 + 0 \times \epsilon = 2. \tag{12.11}$$

Similarly, for $x_0 = -2$, the expansion in (12.4) takes the form

$$x^{(-2)} = x_0^{(-2)} + \epsilon x_1^{(-2)} + \cdots = -2 + \epsilon x_1^{(-2)}, \tag{12.12}$$

and, to order ϵ, equation (12.3) gives

$$\begin{aligned}
&\left(-8 + 12\epsilon x_1^{(-2)}\right) - 4\left(-2 + \epsilon x_1^{(-2)}\right) - \epsilon(-2) + 2\epsilon = 0,\\
\text{or,}\quad &\epsilon\left[12x_1^{(-2)} - 4x_1^{(-2)} + 4\right] = 0,\\
\text{or,}\quad &x_1^{(-2)} = -\frac{1}{2}.
\end{aligned} \tag{12.13}$$

Thus, to this order, this root is given by

$$\begin{aligned} x^{(-2)} &= x_0^{(-2)} + \epsilon x_1^{(-2)} \\ &= -2 - .0005 = -2.0005. \end{aligned} \tag{12.14}$$

In fact, we can determine, in a similar manner, that, to the next order, the roots have the forms

$$\begin{aligned} x^{(0)} &= 0 + \frac{1}{2}\epsilon - \frac{1}{8}\epsilon^2 + 0(\epsilon^3), \\ x^{(2)} &= 2 + 0 \times \epsilon + 0 \times \epsilon^2 + 0(\epsilon^3), \\ x^{(-2)} &= -2 - \frac{1}{2}\epsilon + \frac{1}{8}\epsilon^2 + 0(\epsilon^3), \end{aligned} \tag{12.15}$$

where $\epsilon = 0.001$. This gives a solution which is accurate up to one part in 10^9 ($x = 2$ is an exact root.) and shows the power of perturbation theory.

12.1 Non-degenerate perturbation

We now discuss a perturbation approach for physical systems whose Hamiltonians are independent of time. This is why it is also known as stationary perturbation theory. This method is applicable when the complete Hamiltonian is independent of time and can be written as a sum of two parts

$$H = H_0 + H_1, \tag{12.16}$$

where H_0 is the Hamiltonian which we are able to diagonalize. In other words, we can determine the eigenstates and the eigenvalues of H_0. The second term in the Hamiltonian, H_1, is an additional Hamiltonian which is assumed to be small. We would see what smallness means later – for the present, we simply note that the matrix elements of H_1 should be smaller than the differences in the energy levels of the Hamiltonian H_0.

Let $|n_0\rangle$ denote the eigenstates of H_0 with the eigenvalues $E_n^{(0)}$. Thus,

$$H_0|n_0\rangle = E_n^{(0)}|n_0\rangle. \tag{12.17}$$

Furthermore, we assume that all the states are discrete and non-degenerate. The degenerate case has to be discussed separately. Because of the perturbing Hamiltonian H_1, the total Hamiltonian H

would now have a new set of eigenstates $|n\rangle$ with eigenvalues E_n. If the perturbing Hamiltonian is small, then these eigenstates and eigenvalues would be very close to the unperturbed states and eigenvalues. Thus, we can expand these quantities in powers of the effect of the perturbing Hamiltonian. However, since the Hamiltonian is an operator we cannot use it as an expansion parameter and, for book keeping purposes, we introduce a parameter λ to write

$$H = H_0 + \lambda H_1. \tag{12.18}$$

We can now expand various quantities in powers of λ (which measures the power of change due to the perturbing Hamiltonian) and, at the end of our calculations, we can set the constant $\lambda = 1$ to recover the original system. Thus, let

$$\begin{aligned} E_n &= E_n^{(0)} + \lambda E_n^{(1)} + \lambda^2 E_n^{(2)} + \cdots = \sum_{m=0}^{\infty} \lambda^m E_n^{(m)}, \\ |n\rangle &= |n_0\rangle + \lambda|n_1\rangle + \lambda^2|n_2\rangle + \cdots = \sum_{m=0}^{\infty} \lambda^m |n_m\rangle. \end{aligned} \tag{12.19}$$

We are, of course, assuming that such a series expansion converges which, in turn, imposes the condition that each successive term in the expansion must fall off rapidly.

We are now trying to solve the eigenvalue equation

$$H|n\rangle = E_n|n\rangle, \tag{12.20}$$

perturbatively. Putting in the expansion (12.19) into (12.20), we have

$$(H_0 + \lambda H_1) \sum_{m=0}^{\infty} \lambda^m |n_m\rangle = \left(\sum_{m=0}^{\infty} \lambda^m E_n^{(m)}\right) \left(\sum_{m'=0}^{\infty} \lambda^{m'} |n_{m'}\rangle\right). \tag{12.21}$$

Expanding (12.21) to order λ^2, we have

$$\begin{aligned} &H_0|n_0\rangle + \lambda H_1|n_0\rangle + \lambda H_0|n_1\rangle + \lambda^2 H_0|n_2\rangle + \lambda^2 H_1|n_1\rangle + O(\lambda^3) \\ &\quad = E_n^{(0)}|n_0\rangle + \lambda E_n^{(0)}|n_1\rangle + \lambda E_n^{(1)}|n_0\rangle + \lambda^2 E_n^{(0)}|n_2\rangle \\ &\qquad + \lambda^2 E_n^{(2)}|n_0\rangle + \lambda^2 E_n^{(1)}|n_1\rangle + O(\lambda^3). \end{aligned} \tag{12.22}$$

The order λ^0 terms in (12.22) give rise to the relation

$$H_0|n_0\rangle = E_n^{(0)}|n_0\rangle, \tag{12.23}$$

which is the eigenvalue equation for the Hamiltonian H_0 in (12.17) and is automatically satisfied. Terms of order λ give rise to the relation

$$H_1|n_0\rangle + H_0|n_1\rangle = E_n^{(0)}|n_1\rangle + E_n^{(1)}|n_0\rangle, \tag{12.24}$$

$$\text{or,}\quad \langle n_0|H_1|n_0\rangle + \langle n_0|H_0|n_1\rangle = E_n^{(0)}\langle n_0|n_1\rangle + E_n^{(1)}\langle n_0|n_0\rangle,$$

$$\text{or,}\quad E_n^{(1)} = \langle n_0|H_1|n_0\rangle. \tag{12.25}$$

This determines the first order change in the energy eigenvalue and, to this order, we have

$$E_n = E_n^{(0)} + \lambda\langle n_0|H_1|n_0\rangle. \tag{12.26}$$

Furthermore, taking the inner product of (12.24) with $\langle m_0|$ where $m \neq n$, we obtain

$$\langle m_0|H_1|n_0\rangle + \langle m_0|H_0|n_1\rangle = E_n^{(0)}\langle m_0|n_1\rangle + E_n^{(1)}\langle m_0|n_0\rangle. \tag{12.27}$$

But, since $m \neq n$, the orthonormality relation of the energy basis leads to $\langle m_0|n_0\rangle = 0$. Therefore, equation (12.27) determines

$$\langle m_0|H_1|n_0\rangle + E_m^{(0)}\langle m_0|n_1\rangle = E_n^{(0)}\langle m_0|n_1\rangle,$$

$$\text{or,}\quad \langle m_0|n_1\rangle = \frac{\langle m_0|H_1|n_0\rangle}{E_n^{(0)} - E_m^{(0)}}. \tag{12.28}$$

Namely, (12.28) determines all the coefficients of expansion of $|n_1\rangle$ in the basis $|m_0\rangle$ when $m \neq n$. That is, we have

$$|n_1\rangle = \sum c_m^{(n)}|m_0\rangle, \tag{12.29}$$

where

$$c_m^{(n)} = \frac{\langle m_0|H_1|n_0\rangle}{E_n^{(0)} - E_m^{(0)}},\qquad \text{for } m \neq n. \tag{12.30}$$

The coefficient $c_n^{(n)}$ is determined from the fact that the eigenstates are normalized. Thus, to this order,

$$|n\rangle = |n_0\rangle + \lambda|n_1\rangle = |n_0\rangle + \lambda\sum_m c_m^{(n)}|m_0\rangle, \tag{12.31}$$

so that

$$\langle n|n\rangle = \langle n_0|n_0\rangle + 2\lambda\sum_m c_m^{(n)}\langle n_0|m_0\rangle + O(\lambda^2),$$

$$\text{or,}\quad 1 = \langle n_0|n_0\rangle + 2\lambda c_n^{(n)} = 1 + 2\lambda c_n^{(n)},$$

$$\text{or,}\quad c_n^{(n)} = 0. \tag{12.32}$$

This determines that, to this order, we have

$$\begin{aligned}
E_n &= E_n^{(0)} + \langle n_0|H_1|n_0\rangle, \\
|n\rangle &= |n_0\rangle + \sum_m{}' \frac{\langle m_0|H_1|n_0\rangle}{E_n^{(0)} - E_m^{(0)}}|m_0\rangle = |n_0\rangle + |n_1\rangle.
\end{aligned} \tag{12.33}$$

Here, we have set $\lambda = 1$ and the summation with a prime denotes summation over all values of m except $m = n$. We note from (12.31) and (12.32) that $|n_1\rangle$ is orthogonal to $|n_0\rangle$,

$$\langle n_0|n_1\rangle = 0. \tag{12.34}$$

Furthermore, it is obvious from (12.33) that, for the perturbation method to be applicable, not only should the magnitude of perturbation be small, but its off diagonal matrix elements in the unperturbed basis should also be small compared to the level differences of the unperturbed system. If this is not true, this perturbation scheme breaks down. Indeed, this happens if the system has degenerate energy levels.

To order λ^2, the eigenvalue equation, (12.22), gives

$$\begin{aligned}
H_0|n_2\rangle + H_1|n_1\rangle &= E_n^{(0)}|n_2\rangle + E_n^{(1)}|n_1\rangle + E_n^{(2)}|n_0\rangle, \\
\text{or, } \langle n_0|H_0|n_2\rangle + \langle n_0|H_1|n_1\rangle &= E_n^{(0)}\langle n_0|n_2\rangle + E_n^{(1)}\langle n_0|n_1\rangle + E_n^{(2)}.
\end{aligned} \tag{12.35}$$

Using $\langle n_0|n_1\rangle = 0$ and $\langle n_0|H_0 = E_n^{(0)}\langle n_0|$, we obtain, from (12.35),

$$\begin{aligned}
E_n^{(2)} &= \langle n_0|H_1|n_1\rangle \\
&= \langle n_0|H_1 \sum_m{}' \frac{\langle m_0|H_1|n_0\rangle}{E_n^{(0)} - E_m^{(0)}}|m_0\rangle \\
&= \sum_m{}' \frac{\langle m_0|H_1|n_0\rangle\langle n_0|H_1|m_0\rangle}{E_n^{(0)} - E_m^{(0)}} \\
&= \sum_m{}' \frac{|\langle m_0|H_1|n_0\rangle|^2}{E_n^{(0)} - E_m^{(0)}}.
\end{aligned} \tag{12.36}$$

Thus, to second order in the perturbation, the energy eigenvalues are given by

$$E_n = E_n^0 + \langle n_0|H_1|n_0\rangle + \sum_m{}' \frac{|\langle m_0|H_1|n_0\rangle|^2}{E_n^{(0)} - E_m^{(0)}}. \tag{12.37}$$

We note again, from (12.37) that the validity of the perturbation scheme depends on whether or not the off diagonal elements of the perturbing Hamiltonian are small compared to the level splittings of the unperturbed Hamiltonian. We also note that to any order, the corrections of the eigenvalues depend on the eigenstates one order lower and this procedure can be carried out systematically to any order in the perturbation (when the perturbation is small).

► **Example.** Let us consider a one dimensional harmonic oscillator with mass m and charge q moving in an electrostatic potential $(-\mathcal{E}X)$, where $\mathcal{E}$ represents a constant electric field (As we will see, the direction of the electric field is irrelevant for the shift in the energy levels.). Thus, the total Hamiltonian for the system can be written as

$$H = \frac{P^2}{2m} + \frac{1}{2}m\omega^2 X^2 - q\mathcal{E}X = H_0 + H_1, \tag{12.38}$$

where

$$H_0 = \frac{P^2}{2m} + \frac{1}{2}m\omega^2 X^2, \qquad H_1 = -q\mathcal{E}X. \tag{12.39}$$

The eigenstates of H_0 are $|n_0\rangle$, which we have already studied in detail in chapter 5, where we have seen that the operator X can be expressed in terms of the creation and annihilation operators as (see (5.36))

$$X = \sqrt{\frac{\hbar}{2m\omega}}(a + a^\dagger). \tag{12.40}$$

In this case, we know the eigenstates of H_0. The set of vectors $|n_0\rangle$ are such that

$$H_0|n_0\rangle = E_n^{(0)}|n_0\rangle, \qquad E_n^{(0)} = \left(n + \frac{1}{2}\right)\hbar\omega, \tag{12.41}$$

and $\langle n_0|k_0\rangle = \delta_{nk}$. Furthermore, a and $a^\dagger$ are the lowering and raising operators respectively such that

$$\begin{aligned} a|n_0\rangle &= \sqrt{n}|n_0 - 1\rangle, \\ a^\dagger|n_0\rangle &= \sqrt{n+1}|n_0 + 1\rangle. \end{aligned} \tag{12.42}$$

Therefore, the first order change in the energy due to the perturbation, H_1, follows from (12.25) to be

$$\begin{aligned} E_n^{(1)} &= \langle n_0|H_1|n_0\rangle = \langle n_0| - q\mathcal{E}X|n_0\rangle \\ &= -q\mathcal{E}\sqrt{\frac{\hbar}{2m\omega}}\langle n_0|a + a^\dagger|n_0\rangle = 0. \end{aligned} \tag{12.43}$$

To first order in the perturbation, therefore, there is no change in the energy. This result can be understood in simple terms if one works in the coordinate basis rather than the n-basis. Thus,

$$\begin{aligned}
\langle n_0|H_1|n_0\rangle &= \iint\limits_{-\infty}^{\infty} \mathrm{d}x\mathrm{d}x'\ \langle n_0|x\rangle\langle x|H_1|x'\rangle\langle x'|n_0\rangle \\
&= \int\limits_{-\infty}^{\infty} \mathrm{d}x\ \psi_n^*(x)(-q\mathcal{E}x)\psi_n(x) \\
&= -q\mathcal{E}\int\limits_{-\infty}^{\infty} \mathrm{d}x\ x|\psi_n(x)|^2 = 0.
\end{aligned} \tag{12.44}$$

The vanishing of the integral follows because the integrand is an odd function. The physical way of seeing this result is to note that the operator X changes sign under reflection (it has odd parity). Since the eigenstates of the Hamiltonian have definite parity (oscillator states are either even or odd), it follows that the perturbing Hamiltonian cannot connect two states with the same parity.

The change in the eigenstates to first order in the perturbation is obtained from (12.29) and (12.30)

$$|n_1\rangle = \sum_k{}' \frac{\langle k_0|H_1|n_0\rangle}{E_n^{(0)} - E_k^{(0)}}|k_0\rangle = \sum_k{}' \frac{\langle k_0| - q\mathcal{E}X|n_0\rangle}{E_n^{(0)} - E_k^{(0)}}|k_0\rangle. \tag{12.45}$$

We can calculate

$$\begin{aligned}
\langle k_0|X|n_0\rangle &= \sqrt{\frac{\hbar}{2m\omega}}\langle k_0|(a + a^\dagger)|n_0\rangle \\
&= \sqrt{\frac{\hbar}{2m\omega}}(\sqrt{n}\langle k_0|n_0 - 1\rangle + \sqrt{n+1}\langle k_0|n_0 + 1\rangle) \\
&= \sqrt{\frac{\hbar}{2m\omega}}(\sqrt{n}\delta_{k,n-1} + \sqrt{n+1}\delta_{k,n+1}),
\end{aligned} \tag{12.46}$$

so that equation (12.45) gives

$$\begin{aligned}
|n_1\rangle &= \sum_k{}'(-q\mathcal{E})\sqrt{\frac{\hbar}{2m\omega}}\frac{(\sqrt{n}\delta_{k,n-1} + \sqrt{n+1}\delta_{k,n+1})}{E_n^{(0)} - E_k^{(0)}}|k_0\rangle \\
&= -q\mathcal{E}\sqrt{\frac{\hbar}{2m\omega}}\left(\sqrt{n}\frac{|n_0 - 1\rangle}{E_n^{(0)} - E_{n-1}^{(0)}} + \sqrt{n+1}\frac{|n_0 + 1\rangle}{E_n^{(0)} - E_{n+1}^{(0)}}\right) \\
&= -q\mathcal{E}\sqrt{\frac{\hbar}{2m\omega}}\left(\sqrt{n}\frac{|n_0 - 1\rangle}{\hbar\omega} + \sqrt{n+1}\frac{|n_0 + 1\rangle}{-\hbar\omega}\right) \\
&= -\frac{q\mathcal{E}}{\omega}\sqrt{\frac{1}{2m\hbar\omega}}\left(\sqrt{n}|n_0 - 1\rangle - \sqrt{n+1}|n_0 + 1\rangle\right).
\end{aligned} \tag{12.47}$$

Thus, to this order, the eigenstates are

$$|n\rangle = |n_0\rangle - \frac{q\mathcal{E}}{\omega}\sqrt{\frac{1}{2m\hbar\omega}}(\sqrt{n}|n_0 - 1\rangle - \sqrt{n+1}|n_0 + 1\rangle). \tag{12.48}$$

We see that the perturbation, in this case, actually mixes only the adjacent states. We can now calculate the change in the energy to second order in the perturbation from (see (12.37))

$$E_n^{(2)} = \sum_k{}' \frac{|\langle k_0|H_1|n_0\rangle|^2}{E_n^{(0)} - E_k^{(0)}}. \tag{12.49}$$

We know, from (12.46), that

$$\langle k_0|H_1|n_0\rangle = -q\mathcal{E}\sqrt{\frac{\hbar}{2m\omega}}(\sqrt{n}\delta_{k,n-1} + \sqrt{n+1}\delta_{k,n+1}), \tag{12.50}$$

so that

$$\begin{aligned}|\langle k_0|H_1|n_0\rangle|^2 &= q^2\mathcal{E}^2\frac{\hbar}{2m\omega}\left(\sqrt{n}\delta_{k,n-1} + \sqrt{n+1}\delta_{k,n+1}\right)^2 \\ &= \frac{q^2\mathcal{E}^2\hbar}{2m\omega}\left(n\delta_{k,n-1} + (n+1)\delta_{k,n+1}\right).\end{aligned} \tag{12.51}$$

Therefore, we determine the second order change in the energy to be

$$\begin{aligned}E_n^{(2)} &= \sum_k{}' \frac{q^2\mathcal{E}^2\hbar}{2m\omega}\left(\frac{n\delta_{k,n-1} + (n+1)\delta_{k,n+1}}{E_n^{(0)} - E_k^{(0)}}\right) \\ &= \frac{q^2\mathcal{E}^2\hbar}{2m\omega}\left(\frac{n}{E_n^{(0)} - E_{n-1}^{(0)}} + \frac{(n+1)}{E_n^{(0)} - E_{n+1}^{(0)}}\right) \\ &= \frac{q^2\mathcal{E}^2\hbar}{2m\omega}\left(\frac{n}{\hbar\omega} + \frac{(n+1)}{-\hbar\omega}\right) \\ &= \frac{q^2\mathcal{E}^2}{2m\omega^2}(n - (n+1)) = -\frac{q^2\mathcal{E}^2}{2m\omega^2}.\end{aligned} \tag{12.52}$$

Thus, to this order, the energy eigenvalue is

$$E_n = E_n^{(0)} + E_n^{(1)} + E_n^{(2)} = \left(n + \frac{1}{2}\right)\hbar\omega - \frac{q^2\mathcal{E}^2}{2m\omega^2}. \tag{12.53}$$

If we were to calculate the higher order corrections, we will find that, for this system, all the higher order perturbative corrections to the energy vanish. Thus, this seems to be the exact energy eigenvalue. In fact, we can see this in the following way.

$$\begin{aligned}H &= \frac{P^2}{2m} + \frac{1}{2}m\omega^2X^2 - q\mathcal{E}X \\ &= \frac{P^2}{2m} + \frac{1}{2}m\omega^2\left(X^2 - \frac{2q\mathcal{E}X}{m\omega^2}\right) \\ &= \frac{P^2}{2m} + \frac{1}{2}m\omega^2\left(X - \frac{q\mathcal{E}}{m\omega^2}\right)^2 - \frac{1}{2}m\omega^2\left(\frac{q\mathcal{E}}{m\omega^2}\right)^2 \\ &= \frac{P^2}{2m} + \frac{1}{2}m\omega^2\left(X - \frac{q\mathcal{E}}{m\omega^2}\right)^2 - \frac{q^2\mathcal{E}^2}{2m\omega^2}.\end{aligned} \tag{12.54}$$

Let us define a new coordinate operator

$$\tilde{X} = X - \frac{q\mathcal{E}}{m\omega^2}, \tag{12.55}$$

so that we can write

$$H = \frac{P^2}{2m} + \frac{1}{2}m\omega^2\tilde{X}^2 - \frac{q^2\mathcal{E}^2}{2m\omega^2}. \tag{12.56}$$

P and $\tilde{X}$ continue to be conjugate variables. Therefore, we can think of this as a harmonic oscillator whose energy levels are shifted by an amount $(-\frac{q^2\mathcal{E}^2}{2m\omega^2})$. Classically such an oscillator would center around $\tilde{x} = 0$ or $x = \frac{q\mathcal{E}}{m\omega^2}$. Such a system is like a mass attached to a spring hanging under gravity. The effect of the mass is to shift the equilibrium point down. ◀

Exercise. Show that the perturbed eigenstates are centered around $x = \frac{q\mathcal{E}}{m\omega^2}$.

12.2 Ground state of hydrogen and the Stark effect

Degeneracy is a consequence of symmetry. As we have seen earlier, the degeneracy in the m-quantum numbers, in a central potential, is due to the rotational symmetry of the system. Similarly, we have seen earlier, that the degeneracy in the ℓ-quantum numbers in the case of hydrogen atom as well as the isotropic harmonic oscillator is a consequence of an accidental symmetry associated with the system. If one applies a perturbing Hamiltonian to the system which breaks the symmetry, then, the degeneracy would be lifted. Thus, if we apply a constant electric field along the z-axis, the accidental symmetry associated with the hydrogen atom gets broken and the degeneracy in the ℓ-quantum numbers is lifted. But the degeneracy in the m-quantum numbers remains, since there is still a rotational symmetry around the z-axis. If we subject the hydrogen atom further to a constant magnetic field along the z-direction, then even the m-quantum numbers become non-degenerate.

The change in the energy levels due to an external electric field is called the Stark effect. And that due to an external magnetic field is known as the Zeeman effect. Because the higher states of hydrogen are degenerate, we cannot apply non-degenerate perturbation theory to calculate the change in the energy levels, in general. However, since the ground state is non-degenerate, we can calculate the change in its energy due to an electric field, using the perturbation theory developed so far.

Let us assume that the hydrogen atom is in a constant electric field, $\mathcal{E}$, along the negative z-direction (The direction of the electric field, as we will see, is not relevant to the change in the energy up to second order in perturbation.). Thus, one can write down the scalar potential to be

$$\Phi(\mathbf{r}) = \mathcal{E}z = \mathcal{E}r\cos\theta. \tag{12.57}$$

Thus, the total Hamiltonian, in this case, can be written as

$$H = H_0 + H_1 = H_0 - e\mathcal{E}z = H_0 - e\mathcal{E}r\cos\theta, \tag{12.58}$$

where H_0 represents the Hamiltonian for the hydrogen atom and we have used the fact that the electron carries a charge $(-e)$. We know the unperturbed ground state energy of the system to be (see (9.88))

$$H_0|0\rangle = E_0^{(0)}|0\rangle, \quad E_0^{(0)} = -\text{Ry} = -13.6 \text{ eV} = -\frac{e^2}{2a_0}. \tag{12.59}$$

The first order correction to the ground state energy follows to be

$$\begin{aligned} E_0^{(1)} &= \langle 0|H_1|0\rangle = \int \mathrm{d}^3r\ \psi_0^*(r)(-e\mathcal{E}r\cos\theta)\psi_0(r) \\ &= -e\mathcal{E}\int \mathrm{d}^3r\ r\cos\theta|\psi_0(r)|^2 = 0. \end{aligned} \tag{12.60}$$

This is a consequence of the fact that $\psi_0(r)$ has even parity whereas, under parity,

$$\begin{aligned} &r \to r, \\ &\theta \to \pi - \theta, \\ &\cos\theta \to -\cos\theta, \\ &H_1 \to -H_1. \end{aligned} \tag{12.61}$$

Thus, the first order correction to the ground state energy is zero which follows from the symmetry properties of the system. Since the first order correction is proportional to $\mathcal{E}$, one also says that the ground state of hydrogen does not show any linear Stark effect. The second order change in the energy eigenvalue is given by

$$E_0^{(2)} = \sum_k{}' \frac{|\langle k_0|H_1|0\rangle|^2}{E_0^{(0)} - E_k^{(0)}}, \qquad k \neq 0. \tag{12.62}$$

We can, of course, evaluate this integral by brute force. We can also use clever selection rules to restrict the sum to a few terms. We will evaluate this slightly differently.

Let us assume that there exists an operator, Ω (to be determined), such that

$$H_1|0\rangle = (\Omega H_0 - H_0\Omega)|0\rangle. \tag{12.63}$$

Then, it follows that

$$\begin{aligned}\langle k_0|H_1|0\rangle &= \langle k_0|(\Omega H_0 - H_0\Omega)|0\rangle \\ &= E_0^{(0)}\langle k_0|\Omega|0\rangle - E_k^{(0)}\langle k_0|\Omega|0\rangle \\ &= \left(E_0^{(0)} - E_k^{(0)}\right)\langle k_0|\Omega|0\rangle. \end{aligned} \tag{12.64}$$

Thus, the second order correction, in this case, will follow to be

$$\begin{aligned}\sum_k{}' \frac{|\langle k_0|H_1|0\rangle|^2}{E_0^{(0)} - E_k^{(0)}} &= \sum_k{}' \frac{\langle k_0|H_1|0\rangle}{E_0^{(0)} - E_k^{(0)}}\langle 0|H_1|k_0\rangle \\ &= \sum_k{}' \langle k_0|\Omega|0\rangle\langle 0|H_1|k_0\rangle \\ &= \sum_k{}' \langle 0|H_1|k_0\rangle\langle k_0|\Omega|0\rangle \\ &= \langle 0|H_1\Omega|0\rangle - \langle 0|H_1|0\rangle\langle 0|\Omega|0\rangle, \end{aligned} \tag{12.65}$$

where we have used the closure (completeness relation) of the energy eigenstates. Furthermore, we have already shown in (12.60) that

$$\langle 0|H_1|0\rangle = 0.$$

Therefore, it follows that

$$E_0^{(2)} = \langle 0|H_1\Omega|0\rangle. \tag{12.66}$$

Thus, determining the second order correction to the ground state energy depends on finding an operator Ω which satisfies (12.63). We can assume that Ω depends only on the coordinates and, going to the spherical coordinates, we can show that

$$\Omega = \frac{ma_0 e\mathcal{E}}{\hbar^2}\left(\frac{r}{2} + a_0\right) z, \tag{12.67}$$

where a_0 represents the Bohr radius. Thus, we have

$$\begin{aligned}E_0^{(2)} &= -\frac{ma_0 e^2\mathcal{E}^2}{\hbar^2}\langle 0|\left(\frac{r}{2} + a_0\right) z^2|0\rangle \\ &= -\mathcal{E}^2\langle 0|\left(\frac{r}{2} + a_0\right) z^2|0\rangle, \end{aligned} \tag{12.68}$$

where we have used $a_0 = \frac{\hbar^2}{me^2}$. The ground state expectation value in (12.68) can be evaluated simply by noting that the ground state of the hydrogen atom is spherically symmetric and, consequently, in such a state,

$$\langle 0|f(r)x^2|0\rangle = \langle 0|f(r)y^2|0\rangle = \langle 0|f(r)z^2|0\rangle = \frac{1}{3}\langle 0|f(r)r^2|0\rangle, \tag{12.69}$$

which determines

$$E_0^{(2)} = -\frac{9}{4}\, a_0^3 \mathcal{E}^2. \tag{12.70}$$

Here, we have used

$$\begin{aligned}
\langle 0|(\frac{r}{2} + a_0)r^2|0\rangle &= \frac{1}{\pi a_0^3}\int \mathrm{d}^3r\, (\frac{r}{2} + a_0)r^2 e^{-\frac{2}{a_0}r} \\
&= \frac{1}{\pi a_0^3}(4\pi)\int_0^\infty \mathrm{d}r\, (\frac{r}{2} + a_0)r^4 e^{-\frac{2}{a_0}r} \\
&= \frac{a_0^3}{8}\left(\frac{1}{4}\,\Gamma(6) + \Gamma(5)\right) \\
&= \frac{27}{4}\, a_0^3.
\end{aligned} \tag{12.71}$$

Thus, to second order in the perturbation, the ground state energy becomes

$$E_0 = -\frac{e^2}{2a_0} - \frac{9}{4}a_0^3\mathcal{E}^2. \tag{12.72}$$

This is the second order Stark effect or often called the quadratic Stark effect of the ground state of hydrogen. The change in the energy, in this case, is written as

$$E_0^{(2)} = -\frac{1}{2}\alpha\mathcal{E}^2, \tag{12.73}$$

which defines the polarizability of the hydrogen atom and has the value

$$\alpha = \frac{9}{2}a_0^3. \tag{12.74}$$

We note that the effect of the applied electric field is to lower the ground state energy.

Exercise. Compare this result with a variational calculation. Also check explicitly that the operator Ω derived in this example does indeed satisfy the defining relation (12.63).

12.3 Ground state of helium

We have already calculated the ground state energy of helium using the variational method. Let us now see how good perturbation theory is to the lowest order. The Hamiltonian, as we have seen in (10.48), can be written as

$$H = H_0 + H_1, \tag{12.75}$$

where we have identified

$$H_0 = -\frac{\hbar^2}{2m}\nabla_1^2 - \frac{\hbar^2}{2m}\nabla_2^2 - \frac{2e^2}{r_1} - \frac{2e^2}{r_2}, \tag{12.76}$$

and

$$H_1 = \frac{e^2}{r_{12}}. \tag{12.77}$$

As we have discussed earlier in chapter 10, here, 1 and 2 label the two electrons of the system and we are treating the mutual repulsion of the electrons as a perturbation. The ground state of H_0 is given by (see (10.52))

$$\psi_0(r_1, r_2) = \frac{8}{\pi a_0^3}\, e^{-\frac{2}{a_0}(r_1+r_2)}, \quad E_0^{(0)} = -8\frac{e^2}{2a_0} = -8\text{Ry}. \tag{12.78}$$

The first order change in this eigenvalue, due to the perturbation, can be calculated to be

$$\begin{aligned} E_0^{(1)} &= \langle 0|H_1|0\rangle \\ &= \int \mathrm{d}^3r_1\mathrm{d}^3r_2\ \psi_0^*(r_1, r_2)\frac{e^2}{r_{12}}\psi_0(r_1, r_2) \\ &= e^2\int \mathrm{d}^3r_1\mathrm{d}^3r_2\ \frac{1}{r_{12}}\ |\psi_0(r_1, r_2)|^2. \end{aligned} \tag{12.79}$$

We have already evaluated this integral earlier in (10.65) which leads to the result (for $Z = 2$),

$$E_0^{(1)} = \frac{5e^2}{4a_0} = \frac{5}{2}\frac{e^2}{2a_0} = \frac{5}{2}\ \text{Ry} \tag{12.80}$$

Thus, to this order, the ground state energy of helium becomes

$$\begin{aligned} E_0 &= E_0^{(0)} + E_0^{(1)} \\ &= -8\ \text{Ry} + \frac{5}{2}\ \text{Ry} = -\frac{11}{2}\ \text{Ry} \\ &= -\frac{11}{2} \times 13.6\ \text{eV}\ = -74.8\ \text{eV}. \end{aligned} \tag{12.81}$$

We can compare this with the measured value of the ground state energy, which is -78.6 eV. As we have already seen, the variational method gives a value -77.5 eV for the ground state energy of helium. Thus, we see that the variational method gives a much better value than the first order perturbation, which simply means that we have to go to higher orders of perturbation to get closer to the exact energy value.

12.4 Near degenerate systems

If two of the energy levels of a system are very close to each other while all others are far away, the perturbation theory that we have developed so far would not be applicable, since the corrections to the wave functions as well as energy would be large because of the small denominator corresponding to these two levels. So we have to modify our treatment of the problem.

Let the Hamiltonian H_0 have two states $|\psi_1^{(0)}\rangle$ and $|\psi_2^{(0)}\rangle$ with energy eigenvalues $E_1^{(0)}$ and $E_2^{(0)}$ such that they are very close to each other. Thus,

$$E_1^{(0)} - E_2^{(0)} \simeq 0, \qquad E_1^{(0)} > E_2^{(0)}. \tag{12.82}$$

In this case, if we use perturbation theory naively, then, we would obtain

$$\begin{aligned} |\psi_1\rangle &= |\psi_1^{(0)}\rangle + \sum_k{}' \frac{\langle\psi_k^{(0)}|H_1|\psi_1^{(0)}\rangle}{E_1^{(0)} - E_k^{(0)}}\,|\psi_k^{(0)}\rangle, \\ |\psi_2\rangle &= |\psi_2^{(0)}\rangle + \sum_k{}' \frac{\langle\psi_k^{(0)}|H_1|\psi_2^{(0)}\rangle}{E_2^{(0)} - E_k^{(0)}}\,|\psi_k^{(0)}\rangle. \end{aligned} \tag{12.83}$$

Thus, unless

$$\langle\psi_1^{(0)}|H_1|\psi_2^{(0)}\rangle = 0, \tag{12.84}$$

we see that $|\psi_1\rangle$ would contain a large mixture of $|\psi_2^{(0)}\rangle$ and $|\psi_2\rangle$ would contain a large mixture of $|\psi_1^{(0)}\rangle$. Namely, we see that the states $|\psi_1^{(0)}\rangle$ and $|\psi_2^{(0)}\rangle$ will mix a lot. Consequently, let us choose a state of the form

$$|\psi\rangle = a|\psi_1^{(0)}\rangle + b|\psi_2^{(0)}\rangle, \tag{12.85}$$

and try to diagonalize the complete Hamiltonian in this subspace. That is, we are looking for the solutions of the equation

$$\begin{aligned} & H|\psi\rangle = E|\psi\rangle, \\ \text{or,} \quad & (H-E)|\psi\rangle = 0, \\ \text{or,} \quad & (H-E)\left(a|\psi_1^{(0)}\rangle + b|\psi_2^{(0)}\rangle\right) = 0, \end{aligned} \tag{12.86}$$

where $H = H_0 + H_1$. Multiplying (12.86) by $\langle\psi_1^{(0)}|$, we have

$$(H_{11} - E)\,a + H_{12}\,b = 0. \tag{12.87}$$

Similarly, multiplying equation (12.86) by $\langle\psi_2^{(0)}|$, we obtain

$$H_{21}\,a + (H_{22} - E)\,b = 0, \tag{12.88}$$

where we have identified $H_{ij} = \langle\psi_i^{(0)}|H|\psi_j^{(0)}\rangle, i,j = 1,2.$

For simplicity, let us assume that $H_{12} = H_{21}$. Thus, we have two homogeneous equations, (12.87) and (12.88), with two unknown parameters a and b. There would exist a nontrivial solution only if the coefficient matrix has a vanishing determinant, namely,

$$\det\begin{vmatrix} H_{11} - E & H_{12} \\ H_{21} & H_{22} - E \end{vmatrix} = 0, \tag{12.89}$$

which yields

$$E^2 - E\,(H_{11} + H_{22}) + (H_{11}H_{22} - H_{12}H_{21}) = 0. \tag{12.90}$$

The roots of (12.90) determine

$$\begin{aligned} E_{1,2} &= \frac{(H_{11} + H_{22}) \pm \sqrt{(H_{11} + H_{22})^2 - 4(H_{11}H_{22} - H_{12}H_{21})}}{2} \\ &= \frac{1}{2}\left((H_{11} + H_{22}) \pm \sqrt{(H_{11} - H_{22})^2 + 4|H_{12}|^2}\right). \end{aligned} \tag{12.91}$$

This gives the exact energy values of the two levels. Furthermore, it follows from (12.87) that

$$\frac{a}{b} = \frac{H_{12}}{E - H_{11}}. \tag{12.92}$$

Substituting the two roots for E from (12.91) into (12.92) and introducing the parameterization

$$\tan\beta = \frac{2H_{12}}{H_{11} - H_{22}}, \tag{12.93}$$

we have

$$\left(\frac{a}{b}\right)_1 = \cot\frac{\beta}{2}, \qquad \left(\frac{a}{b}\right)_2 = -\tan\frac{\beta}{2}. \tag{12.94}$$

Thus, the normalized states corresponding to the energy values E_1 and E_2 can be written respectively as

$$\begin{aligned}|\psi_1\rangle &= \cos\frac{\beta}{2}\;|\psi_1^{(0)}\rangle + \sin\frac{\beta}{2}\;|\psi_2^{(0)}\rangle,\\ |\psi_2\rangle &= -\sin\frac{\beta}{2}\;|\psi_1^{(0)}\rangle + \cos\frac{\beta}{2}\;|\psi_2^{(0)}\rangle.\end{aligned} \tag{12.95}$$

It is clear now that if

$$|H_{11} - H_{22}| \gg |H_{12}| = |(H_1)_{12}|, \tag{12.96}$$

then, the energy eigenvalues can be expanded as

$$\begin{aligned}E_1 &= H_{11} + \frac{|H_{12}|^2}{H_{11}-H_{22}} + O\left(H_{12}^4\right)\\ &= E_1^{(0)} + \langle\psi_1^{(0)}|H_1|\psi_1^{(0)}\rangle + \frac{|\langle\psi_1^{(0)}|H_1|\psi_2^{(0)}\rangle|^2}{E_1^{(0)} - E_2^{(0)}} + O\left(H_{12}^4\right),\\ E_2 &= H_{22} - \frac{|H_{12}|^2}{H_{11}-H_{22}} + O\left(H_{12}^4\right)\\ &= E_2^{(0)} + \langle\psi_2^{(0)}|H_1|\psi_2^{(0)}\rangle + \frac{|\langle\psi_2^{(0)}|H_1|\psi_1^{(0)}\rangle|^2}{E_2^{(0)} - E_1^{(0)}} + O\left(H_{12}^4\right).\end{aligned} \tag{12.97}$$

We recognize this as nothing other than the usual second order perturbation results (see (12.37)).

Furthermore, in this case,

$$\tan\beta = \frac{2H_{12}}{H_{11}-H_{22}} \simeq 0 \quad \text{or} \quad \beta \simeq 0. \tag{12.98}$$

Therefore, we obtain from (12.95)

$$\begin{aligned}|\psi_1\rangle &\simeq |\psi_1^{(0)}\rangle + \frac{\beta}{2}|\psi_2^{(0)}\rangle\\ &= |\psi_1^{(0)}\rangle + \frac{H_{12}}{H_{11}-H_{22}}|\psi_2^{(0)}\rangle\\ &= |\psi_1^{(0)}\rangle + \frac{\langle\psi_2^{(0)}|H_1|\psi_1^{(0)}\rangle}{E_1^{(0)} - E_2^{(0)}}|\psi_2^{(0)}\rangle + O\left(H_{12}^2\right).\end{aligned} \tag{12.99}$$

Similarly, we have

$$\begin{aligned}|\psi_2\rangle &\simeq -\frac{\beta}{2}|\psi_1^{(0)}\rangle + |\psi_2^{(0)}\rangle \\ &= |\psi_2^{(0)}\rangle + \frac{\langle\psi_1^{(0)}|H_1|\psi_2^{(0)}\rangle}{E_2^{(0)} - E_1^{(0)}}|\psi_1^{(0)}\rangle + O\left(H_{12}^2\right). \qquad (12.100)\end{aligned}$$

This is exactly what we would expect from non-degenerate perturbation theory.

On the other hand, if

$$|H_{11} - H_{22}| \ll |H_{12}|, \qquad (12.101)$$

then, we have

$$\begin{aligned}E_1 &\simeq \frac{1}{2}(H_{11} + H_{22}) + \left(|H_{12}| + \frac{1}{8}\frac{(H_{11} - H_{22})^2}{|H_{12}|}\right), \\ E_2 &\simeq \frac{1}{2}(H_{11} + H_{22}) - \left(|H_{12}| + \frac{1}{8}\frac{(H_{11} - H_{22})^2}{|H_{12}|}\right). \qquad (12.102)\end{aligned}$$

which is very different from the results of non-degenerate perturbation theory. For a fixed value of $|H_{12}|$ we can plot the eigenvalues E_1 and E_2 in (12.102) as a function of $(H_{11} - H_{22})$, which have the forms shown in Fig. 12.1.

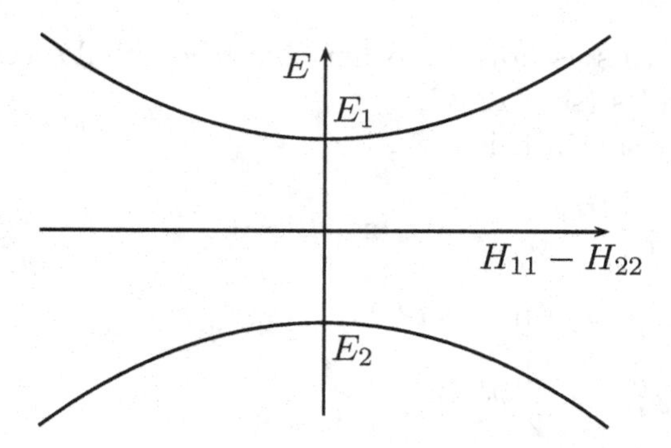

Figure 12.1: Energy levels (12.102) of the two level system as a function of $H_{11} - H_{22}$.

We should note here that the separation between the two levels increases as $(H_{11} - H_{22})$ increases. This is known as the repulsion of the levels.

Furthermore, let us note that, in the limit, $|H_{11} - H_{22}| \ll |H_{12}|$,

$$\tan\beta = \frac{2H_{12}}{H_{11} - H_{22}} \to \infty,$$
$$\text{or,}\quad \beta \simeq \frac{\pi}{2}, \tag{12.103}$$

so that, from (12.95), we obtain

$$\begin{aligned}
|\psi_1\rangle &= \cos\frac{\beta}{2}\;|\psi_1^{(0)}\rangle + \sin\frac{\beta}{2}\;|\psi_2^{(0)}\rangle \\
&\simeq \frac{1}{\sqrt{2}}\left(|\psi_1^{(0)}\rangle + |\psi_2^{(0)}\rangle\right), \\
|\psi_2\rangle &= -\sin\frac{\beta}{2}\;|\psi_1^{(0)}\rangle + \cos\frac{\beta}{2}\;|\psi_2^{(0)}\rangle \\
&\simeq \frac{1}{\sqrt{2}}\left(-|\psi_1^{(0)}\rangle + |\psi_2^{(0)}\rangle\right).
\end{aligned} \tag{12.104}$$

In other words, in this limit, the eigenstates are linear combinations of the unperturbed states where each unperturbed state occurs with equal probability. As a result, the naive perturbation theory developed so far cannot be applied to such a system (where changes in the initial states are not small).

12.5 Degenerate perturbation

Suppose we are dealing with a system where some of the levels are degenerate. Then, in this case, in expressions like

$$|n\rangle = |n_0\rangle + \sum_k{}' \frac{\langle k_0|H_1|n_0\rangle}{E_n^{(0)} - E_k^{(0)}}|k_0\rangle, \tag{12.105}$$

the denominator can vanish for some values of k and hence the expressions become undefined unless, of course,

$$\langle k_0|H_1|n_0\rangle = 0, \tag{12.106}$$

for those values of k. Usually, degeneracy is a consequence of some symmetry of the system. If the perturbing Hamiltonian does not respect the symmetry, degeneracy would be lifted at least partially and hence we do not expect $\langle k_0|H_1|n_0\rangle$ to vanish. Thus, our perturbation scheme breaks down in the sense that whereas theoretically we will predict larger changes in the energy, experimentally the observed splittings are small.

The reason for the breakdown is not hard to see. Suppose an eigenvalue has ℓ-fold degeneracy. Then, $|\psi_{n1}^{(0)}\rangle, |\psi_{n2}^{(0)}\rangle, \cdots, |\psi_{n\ell}^{(0)}\rangle$ or any linear combination of them would have the same energy. We have no information on the unperturbed initial states which are crucial to carry out the perturbation calculations. On the other hand, we know that because of the perturbation, the system would choose to go to a particular state. As we slowly switch off the perturbation, this state would go into a specific linear combination of the ℓ-unperturbed states. That is, therefore, the correct unperturbed state to start with and if we use that, our perturbation calculations will be well behaved. However, if we choose a different starting state with the same energy, then, the terms in our perturbation series become large and, as in the nearly degenerate case, suggest that there is large mixing and hence the states have to be diagonalized further.

Thus, as in the nearly degenerate case, we choose as starting states

$$|\psi_n^{(0)}\rangle = \sum_{i=1}^{\ell} a_i|\psi_{ni}^{(0)}\rangle,$$

$$H = H_0 + H_1, \tag{12.107}$$

where we are assuming that all the states $|\psi_{ni}^{(0)}$ are degenerate with the energy eigenvalue $E_n^{(0)}$. The first order perturbation equation, (12.24), in this case gives

$$\begin{aligned}
&H_0|\psi_n^{(1)}\rangle + H_1|\psi_n^{(0)}\rangle = E_n^{(0)}|\psi_n^{(1)}\rangle + E_n^{(1)}|\psi_n^{(0)}\rangle,\\
\text{or,}\quad &\langle\psi_{nj}^{(0)}|H_0|\psi_n^{(1)}\rangle + \langle\psi_{nj}^{(0)}|H_1|\psi_n^{(0)}\rangle = E_n^{(0)}\langle\psi_{nj}^{(0)}|\psi_n^{(1)}\rangle\\
&\qquad + E_n^{(1)}\langle\psi_{nj}^{(0)}|\psi_n^{(0)}\rangle,\\
\text{or,}\quad &\langle\psi_{nj}^{(0)}|H_1|\psi_n^{(0)}\rangle - E_n^{(1)}\langle\psi_{nj}^{(0)}|\psi_n^{(0)}\rangle = 0,\\
\text{or,}\quad &\sum_{i=1}^{\ell} a_i\left(\langle\psi_{nj}^{(0)}|H_1|\psi_{ni}^{(0)}\rangle - E_n^{(1)}\delta_{ij}\right) = 0.
\end{aligned} \tag{12.108}$$

This set of homogeneous equations has a nontrivial solution if the determinant of the coefficient matrix vanishes. That is,

$$\det\left(\langle\psi_{nj}^{(0)}|H_1|\psi_{ni}^{(0)}\rangle - E_n^{(1)}\delta_{ij}\right) = 0. \tag{12.109}$$

This is known as the secular equation. We have an $\ell \times \ell$ determinant in (12.109). Consequently, it has ℓ roots which are the first order corrections to the energy levels. If H_1 lifts the degeneracy of the levels completely in the first order itself, then, all these roots would be distinct and we can determine all the constants a_i. However, if all the roots are degenerate, then H_1 fails to lift the degeneracy in first order and we have to go to the second order equations to determine a_i's. On the other hand, H_1 may lift the degeneracy only partially. In that case, some of the roots would be distinct and others degenerate. This would correspond to the case where only some of the constants, a_i's, would be determined uniquely – the others remaining arbitrary. If degeneracy is removed in the first order, then, non-degenerate perturbation theory can be applied in higher orders.

12.6 Doubly degenerate level and resonance

Let us consider a Hamiltonian H_0 which has two eigenstates $|\psi_1\rangle$ and $|\psi_2\rangle$ that are degenerate and have the same energy E_0. Let us now perturb the system with a Hamiltonian H' such that

$$\begin{aligned} &\langle\psi_1|H'|\psi_1\rangle = \langle\psi_2|H'|\psi_2\rangle = 0, \\ &\langle\psi_1|H'|\psi_2\rangle = \langle\psi_2|H'|\psi_1\rangle = H'_{12}. \end{aligned} \tag{12.110}$$

Let us choose a linear combination of the two states as our starting unperturbed state.

$$|\psi\rangle = a|\psi_1\rangle + b|\psi_2\rangle. \tag{12.111}$$

Then, as we have seen in (12.109), there exists a nontrivial solution for a and b if

$$\det \begin{vmatrix} H'_{11} - E_0^{(1)} & H'_{12} \\ H'_{12} & H'_{22} - E_0^{(1)} \end{vmatrix} = 0. \tag{12.112}$$

Since $H'_{11} = H'_{22} = 0$, it follows that

$$\det \begin{vmatrix} -E_0^{(1)} & H'_{12} \\ H'_{12} & -E_0^{(1)} \end{vmatrix} = 0, \quad \text{or,} \quad E_0^{(1)} = \pm H'_{12}. \tag{12.113}$$

These are the first order changes in the energy. Clearly, the degeneracy of the two levels is completely lifted. The constants a and b satisfy the condition (see (12.87) and (12.88))

$$\frac{a}{b} = \frac{H'_{12}}{E_0^{(1)}}. \tag{12.114}$$

Thus, corresponding to the two roots in (12.113), we have

$$\left(\frac{a}{b}\right)_+ = 1, \qquad \left(\frac{a}{b}\right)_- = -1. \tag{12.115}$$

This defines the two initial states of perturbation to be

$$\begin{aligned} |\psi^+\rangle &= \frac{1}{\sqrt{2}}\left(|\psi_1\rangle + |\psi_2\rangle\right), \\ |\psi^-\rangle &= \frac{1}{\sqrt{2}}\left(|\psi_1\rangle - |\psi_2\rangle\right). \end{aligned} \tag{12.116}$$

These states have energy values $E_\pm = E_0 \pm H'_{12}$ up to first order so that the states evolve in time as

$$\begin{aligned} \psi^+(x,t) &= \frac{1}{\sqrt{2}}\left(\psi_1(x) + \psi_2(x)\right) e^{-\frac{i}{\hbar}(E_0 + H'_{12})t} \\ \psi^-(x,t) &= \frac{1}{\sqrt{2}}\left(\psi_1(x) - \psi_2(x)\right) e^{-\frac{i}{\hbar}(E_0 - H'_{12})t}. \end{aligned} \tag{12.117}$$

Let us suppose that, at $t = 0$, the system is in the state $|\psi_1\rangle$. Thus, at $t = 0$, the wave function has the form

$$\psi(x) = \psi_1(x) = \frac{1}{\sqrt{2}}\left(\psi^+(x) + \psi^-(x)\right). \tag{12.118}$$

The time evolution of this state is, then, obtained from (12.117) to be

$$\begin{aligned} \psi(x,t) &= \frac{1}{\sqrt{2}}\left[\psi^+(x,t) + \psi^-(x,t)\right] \\ &= \frac{1}{2}\Big[(\psi_1(x) + \psi_2(x))e^{-\frac{i}{\hbar}H'_{12}t} \\ &\qquad + (\psi_1(x) - \psi_2(x))e^{\frac{i}{\hbar}H'_{12}t}\Big]\, e^{-\frac{i}{\hbar}E_0 t}. \end{aligned} \tag{12.119}$$

It is clear that, at $t = 0$, the system is in the state $|\psi_1\rangle$. But, as time grows, it moves more and more into the state $|\psi_2\rangle$. At $t = \frac{\hbar\pi}{2H'_{12}}$, the system is completely in the state $|\psi_2\rangle$. In other words, we see that the system oscillates between the two states $|\psi_1\rangle$ and $|\psi_2\rangle$. This is the quantum resonance phenomenon similar to two springs weakly coupled to each other. Many physical phenomena, such as the $K^0 \leftrightarrow \bar{K}^0$ oscillations, can be described in terms of this simple model.

12.7 Stark effect of the first excited state of hydrogen

The first excited state of hydrogen is four fold degenerate. First of all, this state corresponds to $n = 2$. Furthermore, we know that for every n, the ℓ-quantum number takes values from 0 to $n-1$ in steps of unity. We also know that the m-quantum number for a given ℓ value goes from $-\ell$ to ℓ in steps of one. Thus, the degenerate states, in this case, are

$$|n,\ell,m\rangle: \quad |2,0,0\rangle,\ |2,1,-1\rangle,\ |2,1,0\rangle,\ |2,1,1\rangle. \tag{12.120}$$

All the four states in (12.120) have the same energy eigenvalue (see (9.42))

$$E_2^{(0)} = -\frac{e^2}{2a_0 \times 2^2} = -\frac{e^2}{8a_0}. \tag{12.121}$$

Since the parity of a state, $(-1)^\ell$, depends only on the ℓ-quantum number, we see that the degenerate states do not all have the same parity.

In the presence of a constant electric field along the z direction, our naive non-degenerate perturbation calculation would yield

$$\langle 2,\ell,m|H'|2,\ell,m\rangle = \langle 2,\ell,m| - e\mathcal{E}r\cos\theta|2,\ell,m\rangle = 0, \tag{12.122}$$

since the perturbing Hamiltonian is parity odd and the states have definite parity. This, of course, predicts no first order change in energy. However, this is not true as has been shown in experiments and we would see how degenerate perturbation theory leads to the correct result.

Let us choose, as our initial unperturbed state,

$$|\psi\rangle = a_1|2,0,0\rangle + a_2|2,1,-1\rangle + a_3|2,1,0\rangle + a_4|2,1,1\rangle. \tag{12.123}$$

Then, there exist nontrivial solutions for the a_i's only if

$$\det\left(\langle 2,\ell,m|H'|2,\ell',m'\rangle - E_2^{(1)}\delta_{\ell\ell'}\delta_{mm'}\right) = 0. \tag{12.124}$$

Thus, we have to calculate the matrix elements

$$\langle 2,\ell,m|H'|2,\ell',m'\rangle. \tag{12.125}$$

Note that

$$\begin{aligned} H' &= -e\mathcal{E}Z = -e\mathcal{E}r\cos\theta, \\ L_z &= XP_y - YP_x. \end{aligned} \tag{12.126}$$

As a result, we have

$$\begin{aligned}
&\left[L_z, H'\right] = 0,\\
\text{or,}\quad &\langle n,\ell,m|\left[L_z, H'\right]|n,\ell',m'\rangle = 0,\\
\text{or,}\quad &\langle n,\ell,m|\left[L_zH' - H'L_z\right]|n,\ell',m'\rangle = 0,\\
\text{or,}\quad &\hbar(m-m')\langle n,\ell,m|H'|n,\ell',m'\rangle = 0. \qquad (12.127)
\end{aligned}$$

This tells us that $\langle n,\ell,m|H'|n,\ell',m'\rangle$ has to vanish when $m \neq m'$. Therefore, only the matrix elements of the type $\langle 2,\ell,m|H'|2,\ell',m\rangle$ need to be calculated. Furthermore, note that since H' is odd under parity,

$$\langle n,\ell,m|H'|n,\ell',m\rangle = 0, \qquad (12.128)$$

unless $\ell - \ell' = (2k+1),\ k = 0, \pm1, \pm2, \cdots$. Thus, we see that the only non-vanishing matrix elements are

$$\langle 2,1,0|H'|2,0,0\rangle, \qquad \langle 2,0,0|H'|2,1,0\rangle. \qquad (12.129)$$

Noting that the wave functions for the first excited state of hydrogen have the forms (see (9.27) and (9.76))

$$\begin{aligned}
\psi_{2,0,0}(\mathbf{r}) &= \left(\frac{1}{32\pi a_0^3}\right)^{\frac{1}{2}}\left(2 - \frac{r}{a_0}\right) e^{-\frac{r}{2a_0}},\\
\psi_{2,1,0}(\mathbf{r}) &= \left(\frac{1}{32\pi a_0^3}\right)^{\frac{1}{2}} \frac{r}{a_0}\ \cos\theta\ e^{-\frac{r}{2a_0}}, \qquad (12.130)
\end{aligned}$$

we easily obtain

$$\begin{aligned}
\langle 2,1,0|H'|2,0,0\rangle &= \langle 2,0,0|H'|2,1,0\rangle\\
&= 3ea_0\mathcal{E}. \qquad (12.131)
\end{aligned}$$

Thus, we are looking for the roots of the equation

$$\det\begin{vmatrix} -E_2^{(1)} & 3ea_0\mathcal{E} & 0 & 0\\ 3ea_0\mathcal{E} & -E_2^{(1)} & 0 & 0\\ 0 & 0 & -E_2^{(1)} & 0\\ 0 & 0 & 0 & -E_2^{(1)} \end{vmatrix} = 0.$$

This leads to

$$\begin{aligned}
&\left(E_2^{(1)}\right)^2\left(\left(E_2^{(1)}\right)^2 - (3ea_0\mathcal{E})^2\right) = 0,\\
\text{or,}\quad &E_2^{(1)} = 0, 0, \pm 3ea_0\mathcal{E}. \qquad (12.132)
\end{aligned}$$

These are the first order corrections to the energy of the first excited state in hydrogen. In this example, we see that the degeneracy is lifted only partially. That is, the applied electric field breaks the accidental symmetry and lifts the degeneracy in the ℓ-quantum numbers. But, since the field is applied along the z-direction, there is still a rotational symmetry about the z-axis. This, in turn, implies that the degeneracy in m-quantum numbers still persists. The eigenvalues $E_2^{(1)} = \pm 3ea_0\mathcal{E}$ determine the corresponding states to be

$$\frac{1}{\sqrt{2}}\left(|2,1,0\rangle \pm\ |2,0,0\rangle\right). \tag{12.133}$$

However, the zero eigenvalues allow for any linear combinations of the states $|2,1,1\rangle$ and $|2,1,-1\rangle$. In particular, we can choose as the starting states $|2,1,1\rangle$; $|2,1,-1\rangle$; $\frac{1}{\sqrt{2}}(|2,1,0\rangle \pm |2,0,0\rangle)$ and the perturbation due to the external electric field would be stable.

The eigenvalues can also be given the following interpretation. Since the energy is linear in the electric field, we can think of the first excited state of hydrogen as having a permanent dipole moment of magnitude $3ea_0$ which can be oriented in three different ways – one state parallel to the electric field, one state anti-parallel to the field and two states with zero component along the field. The first excited state of hydrogen, as we see, exhibits linear Stark effect .

12.8 Fine structure of hydrogen levels

The Schrödinger solution gives a very good description of the hydrogen atom. However, as we discussed earlier, there are corrections to these values of the energy. They are known as the fine structure corrections and arise from two sources.

1. Although we treated the electron as a non-relativistic particle, in reality it is not. Note that, in the ground state of hydrogen, we have (for the classical energies)

$$\begin{aligned}
&T = -\frac{1}{2}V = 13.6\,\text{eV},\\
\text{or,}\quad &\left(\frac{v}{c}\right)^2 = \frac{2T}{mc^2} = \frac{27.2\,\text{eV}}{.5\times 10^6\,\text{eV}} = 54.4\times 10^{-6},\\
\text{or,}\quad &\frac{v}{c} \simeq 7\times 10^{-3} \simeq O(\alpha),
\end{aligned} \tag{12.134}$$

where $\alpha = \frac{e^2}{\hbar c} \sim \frac{1}{137}$ represents the fine structure constant. Thus, we have to correct for this discrepancy. We define the kinetic energy as

$$\begin{aligned} T &= E - mc^2 \\ &= (p^2c^2 + m^2c^4)^{\frac{1}{2}} - mc^2 \\ &= mc^2\left(1 + \frac{p^2}{m^2c^2}\right)^{\frac{1}{2}} - mc^2 \\ &= mc^2\left(1 + \frac{1}{2}\frac{p^2}{m^2c^2} - \frac{1}{8}\frac{p^4}{m^4c^4} + \cdots\right) - mc^2 \\ &= \frac{p^2}{2m} - \frac{p^4}{8m^3c^2} + O(p^6). \end{aligned} \tag{12.135}$$

Therefore, under this approximation, the Hamiltonian becomes

$$H = \frac{p^2}{2m} - \frac{e^2}{r} - \frac{p^4}{8m^3c^2} = H_0 + H'. \tag{12.136}$$

First of all, let us note that H' is rotationally invariant. Therefore, it is diagonal in the $|n, \ell, m\rangle$ basis and we have

$$\langle n, \ell, m|H'|n', \ell', m'\rangle = 0, \text{ if } n \neq n', \ \ell \neq \ell', \ m \neq m'. \tag{12.137}$$

As a result, even though the energy levels are degenerate, we can still apply non-degenerate perturbation theory, since the potentially dangerous terms are zero because the numerator, in this case, vanishes.

The first order change to the energy levels can be written as

$$\begin{aligned} E_n^{(1)} &= \langle n, \ell, m|H'|n, \ell, m\rangle \\ &= -\frac{1}{8m^3c^2}\langle n, \ell, m|p^4|n, \ell, m\rangle. \end{aligned} \tag{12.138}$$

Let us note that, since

$$H_0 = \frac{p^2}{2m} - \frac{e^2}{r}, \tag{12.139}$$

we can write

$$\begin{aligned} \frac{p^2}{2m} &= H_0 + \frac{e^2}{r}, \\ p^4 &= 4m^2\left(\frac{p^2}{2m}\right)^2 = 4m^2\left(H_0 + \frac{e^2}{r}\right)^2. \end{aligned} \tag{12.140}$$

Using this, we obtain,

$$
\begin{aligned}
E_n^{(1)} &= -\frac{1}{8m^3c^2} \times 4m^2 \langle n,\ell,m| \left(H_0 + \frac{e^2}{r}\right)^2 |n,\ell,m\rangle \\
&= -\frac{1}{2mc^2}\left[E_n^{(0)2} + 2E_n^{(0)}e^2\langle\frac{1}{r}\rangle_{n\ell m} + e^4\langle\frac{1}{r^2}\rangle_{n\ell m}\right].
\end{aligned}
\tag{12.141}
$$

Let us next develop some tricks for calculating these averages. First of all, we note that the virial theorem applied to hydrogen implies that

$$
\langle T\rangle_{n\ell m} = \left\langle -\frac{1}{2}V\right\rangle_{n\ell m}, \tag{12.142}
$$

so that we have

$$
\begin{aligned}
\langle H_0\rangle_{n\ell m} &= \langle T+V\rangle_{n\ell m} = \left\langle -\frac{1}{2}V + V\right\rangle_{n\ell m} \\
&= \frac{1}{2}\langle V\rangle_{n\ell m} = -\frac{e^2}{2}\left\langle\frac{1}{r}\right\rangle_{n\ell m}.
\end{aligned}
\tag{12.143}
$$

Therefore, we obtain

$$
\begin{aligned}
\left\langle\frac{1}{r}\right\rangle_{n\ell m} &= -\frac{2}{e^2}\langle H_0\rangle_{n\ell m} = -\frac{2}{e^2}\,E_n^{(0)} \\
&= -\frac{2}{e^2}\left(-\frac{e^2}{2a_0n^2}\right) = \frac{1}{a_0n^2}.
\end{aligned}
\tag{12.144}
$$

To calculate $\langle\frac{1}{r^2}\rangle_{n\ell m}$, we note that, if we add a perturbation

$$
H_1 = \frac{\lambda}{r^2}, \tag{12.145}
$$

to the Hamiltonian of the hydrogen atom, then, the first order change in the energy can be written as

$$
\langle H_1\rangle_{n\ell m} = \lambda\left\langle\frac{1}{r^2}\right\rangle_{n\ell m}. \tag{12.146}
$$

On the other hand, with this perturbation, the problem can be exactly solved. For example, the Hamiltonian for the radial equation, in this case, has the form (after factoring out the angular solutions)

$$\begin{aligned} H &= H_0 + H_1 \\ &= -\frac{\hbar^2}{2m}\left[\frac{1}{r^2}\frac{\partial}{\partial r}\left(r^2\frac{\partial}{\partial r}\right) - \frac{\ell(\ell+1)}{r^2}\right] - \frac{e^2}{r} + \frac{\lambda}{r^2} \\ &= -\frac{\hbar^2}{2m}\frac{1}{r^2}\frac{\partial}{\partial r}\left(r^2\frac{\partial}{\partial r}\right) + \frac{\hbar^2\ell'(\ell'+1)}{2mr^2} - \frac{e^2}{r}, \end{aligned} \tag{12.147}$$

where we have defined

$$\ell'(\ell'+1) = \ell(\ell+1) + \frac{2m\lambda}{\hbar^2}. \tag{12.148}$$

In other words, $\ell' = \ell'(\lambda)$ can be thought of as a function of λ. In terms of ℓ', the energy eigenvalues have the form

$$\begin{aligned} E_n &= -\frac{e^2}{2a_0(k+\ell'+1)^2}, \qquad n = k+\ell'+1 \\ &= E_n(\lambda). \end{aligned} \tag{12.149}$$

We can now expand $E_n(\lambda)$ in a Taylor series,

$$E_n(\lambda) = E_n(0) + \lambda\left.\frac{\mathrm{d}E_n}{\mathrm{d}\lambda}\right|_{\lambda=0} + \frac{\lambda^2}{2!}\left.\frac{\mathrm{d}^2E_n}{\mathrm{d}\lambda^2}\right|_{\lambda=0} + \cdots. \tag{12.150}$$

Clearly, $E_n(0)$ is the unperturbed energy (corresponding to $\lambda = 0$), while $\lambda \left.\frac{\mathrm{d}E_n}{\mathrm{d}\lambda}\right|_{\lambda=0}$ is the first order change in the energy. Thus, we have

$$\langle H_1\rangle_{n\ell m} = \lambda\left\langle\frac{1}{r^2}\right\rangle_{n\ell m} = \lambda\left.\frac{dE_n}{d\lambda}\right|_{\lambda=0}, \tag{12.151}$$

which leads to

$$\begin{aligned} \left\langle\frac{1}{r^2}\right\rangle_{n\ell m} &= \left.\frac{\mathrm{d}E_n}{\mathrm{d}\lambda}\right|_{\lambda=0} \\ &= -\frac{e^2}{2a_0}\frac{(-2)}{(k+\ell'+1)^3}\left.\frac{\mathrm{d}\ell'}{\mathrm{d}\lambda}\right|_{\lambda=0 \text{ or } \ell'=\ell}. \end{aligned} \tag{12.152}$$

From the defining relation, (12.148), we note that

$$\begin{aligned}
\ell'^2 + \ell' &= \ell^2 + \ell + \frac{2m\lambda}{\hbar^2}, \\
\text{or,} \quad 2\ell'\frac{\mathrm{d}\ell'}{\mathrm{d}\lambda} + \frac{\mathrm{d}\ell'}{\mathrm{d}\lambda} &= \frac{2m}{\hbar^2}, \\
\text{or,} \quad \frac{\mathrm{d}\ell'}{\mathrm{d}\lambda} &= \frac{2m}{\hbar^2(2\ell'+1)}, \\
\text{or,} \quad \left(\frac{\mathrm{d}\ell'}{\mathrm{d}\lambda}\right)_{\substack{\lambda=0 \\ \text{or, } \ell'=\ell}} &= \frac{2m}{\hbar^2(2\ell+1)}.
\end{aligned} \tag{12.153}$$

Using this, we obtain,

$$\begin{aligned}
\left\langle \frac{1}{r^2} \right\rangle_{n\ell m} &= \frac{e^2}{a_0 n^3} \times \frac{m}{\hbar^2\left(\ell + \frac{1}{2}\right)} = \frac{1}{a_0^2 n^3 \left(\ell + \frac{1}{2}\right)} \\
&= \frac{4nE_n^{(0)2}}{\left(\ell + \frac{1}{2}\right) e^4}.
\end{aligned} \tag{12.154}$$

To calculate $\langle \frac{1}{r^3} \rangle_{n\ell m}$, we use the following trick. Let us define the radial momentum as

$$p_r = -i\hbar \left(\frac{\partial}{\partial r} + \frac{1}{r}\right). \tag{12.155}$$

which is Hermitian and satisfies the canonical commutation relations. It follows now that

$$\begin{aligned}
p_r^2 &= -\hbar^2 \left(\frac{\partial}{\partial r} + \frac{1}{r}\right)\left(\frac{\partial}{\partial r} + \frac{1}{r}\right) \\
&= -\hbar^2 \left(\frac{\partial^2}{\partial r^2} + \frac{1}{r}\frac{\partial}{\partial r} - \frac{1}{r^2} + \frac{1}{r}\frac{\partial}{\partial r} + \frac{1}{r^2}\right) \\
&= -\hbar^2 \left(\frac{\partial^2}{\partial r^2} + \frac{2}{r}\frac{\partial}{\partial r}\right) = -\hbar^2 \frac{1}{r^2}\frac{\partial}{\partial r}\left(r^2 \frac{\partial}{\partial r}\right).
\end{aligned} \tag{12.156}$$

Therefore, we can write

$$H_0 = \frac{1}{2m}\left(p_r^2 + \frac{L^2}{r^2}\right) - \frac{e^2}{r} \tag{12.157}$$

It follows now that

$$\begin{aligned}
[H_0, p_r] &= \frac{L^2}{2m}\left[\frac{1}{r^2}, p_r\right] - e^2\left[\frac{1}{r}, p_r\right] \\
&= \frac{L^2}{2m}(-i\hbar)\left[\frac{1}{r^2}, \left(\frac{\partial}{\partial r} + \frac{1}{r}\right)\right] \\
&\quad - e^2(-i\hbar)\left[\frac{1}{r}, \left(\frac{\partial}{\partial r} + \frac{1}{r}\right)\right] \\
&= -\frac{i\hbar L^2}{2m}\frac{2}{r^3} + i\hbar e^2\frac{1}{r^2} \\
&= i\hbar\left(\frac{e^2}{r^2} - \frac{L^2}{mr^3}\right). \qquad (12.158)
\end{aligned}$$

In an energy eigenbasis, (12.158) would lead to

$$\begin{aligned}
&\langle n, \ell, m|[H_0, p_r]|n, \ell, m\rangle = 0, \\
\text{or,} \quad & i\hbar\left\langle\frac{e^2}{r^2} - \frac{L^2}{mr^3}\right\rangle_{n\ell m} = 0, \\
\text{or,} \quad & \left\langle\frac{e^2}{r^2} - \frac{\hbar^2\ell(\ell+1)}{mr^3}\right\rangle_{n\ell m} = 0, \qquad (12.159)
\end{aligned}$$

which yields

$$\begin{aligned}
\left\langle\frac{1}{r^3}\right\rangle_{n\ell m} &= \frac{me^2}{\hbar^2\ell(\ell+1)}\left\langle\frac{1}{r^2}\right\rangle_{n\ell m} \\
&= \frac{1}{a_0\ell(\ell+1)}\left\langle\frac{1}{r^2}\right\rangle_{n\ell m} \\
&= \frac{1}{a_0^3 n^3\ell(\ell+1)\left(\ell+\frac{1}{2}\right)}, \qquad (12.160)
\end{aligned}$$

where we have used (12.154).

Using (12.144) and (12.154), we can now determine the first order correction to th energy in (12.141) to be

$$\begin{aligned}
E_n^{(1)} &= -\frac{1}{2mc^2}\left(E_n^{(0)^2} + 2E_n^{(0)}e^2\left(-\frac{2}{e^2}E_n^{(0)}\right) + e^4\frac{4nE_n^{(0)^2}}{e^4\left(\ell+\frac{1}{2}\right)}\right) \\
&= -\frac{1}{2mc^2}\left(E_n^{(0)^2} - 4E_n^{(0)^2} + 4n\frac{E_n^{(0)^2}}{\ell+\frac{1}{2}}\right)
\end{aligned}$$

$$
\begin{aligned}
&= -\frac{E_n^{(0)^2}}{2mc^2}\left(-3 + \frac{4n}{\ell+\frac{1}{2}}\right) \\
&= -\frac{1}{2mc^2}\frac{e^4}{4a_0^2n^4}\left(-3 + \frac{4n}{\ell+\frac{1}{2}}\right) \\
&= -\frac{1}{8mc^2}\left(\frac{me^2}{\hbar^2}\right)^2 e^4\left(-\frac{3}{n^4} + \frac{4}{n^3\left(\ell+\frac{1}{2}\right)}\right) \\
&= -\frac{1}{8}(mc^2)\left(\frac{e^2}{\hbar c}\right)^4\left(-\frac{3}{n^4} + \frac{4}{n^3\left(\ell+\frac{1}{2}\right)}\right) \\
&= -\frac{mc^2\alpha^4}{2n^3}\left(-\frac{3}{4n} + \frac{1}{\left(\ell+\frac{1}{2}\right)}\right). \qquad (12.161)
\end{aligned}
$$

2. The other source of correction comes from the spin orbit interaction. It was observed that the electron did possess a magnetic moment which was not due to its orbital motion. Because the electron is not at rest, in its rest frame the proton or the nucleus is moving. Since a moving charge has a magnetic field associated with it, this field would interact with the magnetic moment of the electron. Thus, the spin-orbit interaction Hamiltonian has the form

$$
\begin{aligned}
H'_{\text{S-O}} &= -\boldsymbol{\mu}\cdot\mathbf{B} = -\boldsymbol{\mu}\cdot\left(-\frac{e}{c}\frac{\mathbf{v}\times\mathbf{r}}{r^3}\right) \\
&= -\frac{e}{mcr^3}\boldsymbol{\mu}\cdot(\mathbf{r}\times\mathbf{p}) = -\frac{e}{mc}\frac{\boldsymbol{\mu}\cdot\mathbf{L}}{r^3} \\
&= -\frac{e}{mc}\left(-\frac{e}{mc}\right)\frac{\mathbf{S}\cdot\mathbf{L}}{r^3} = \frac{e^2}{m^2c^2r^3}\,\mathbf{S}\cdot\mathbf{L}. \qquad (12.162)
\end{aligned}
$$

Actually, the correct Hamiltonian is only half of the expression in (12.162). The factor of $\frac{1}{2}$ is due to the fact that the electron motion is not linear and this factor is known as the Thomas factor. In a relativistic theory this factor comes out automatically and we can write the correct spin-orbit interaction as

$$H'_{\text{S-O}} = \frac{e^2}{2m^2c^2r^3}\mathbf{S}\cdot\mathbf{L}. \qquad (12.163)$$

We note that, in this case, neither L_i nor S_i commutes with H. But, we recall that the total angular momentum is given by

$$\mathbf{J} = \mathbf{L} + \mathbf{S}. \qquad (12.164)$$

It is easy to check that J_i commutes with the Hamiltonian even though L_i and S_i do not. Furthermore, we can write (S_i and L_j commute)

$$\mathbf{S}\cdot\mathbf{L} = \frac{1}{2}\left(J^2 - L^2 - S^2\right). \tag{12.165}$$

This allows us to write

$$H'_{\text{S-O}} = \frac{e^2}{4m^2c^2r^3}\left(J^2 - L^2 - S^2\right). \tag{12.166}$$

Let us work in the $|j,m;\ell,s\rangle$ basis for the angular part so that

$$\begin{aligned}&\langle n;j',m';\ell',\frac{1}{2}|H'_{\text{S-O}}|n;j,m;\ell,\frac{1}{2}\rangle\\&= \frac{e^2}{4m^2c^2}\left\langle\frac{1}{r^3}\right\rangle_{n\ell}\hbar^2\left[j(j+1)-\ell(\ell+1)-\frac{3}{4}\right]\delta_{jj'}\delta_{\ell\ell'}\delta_{mm'}.\end{aligned} \tag{12.167}$$

Thus, the first order change in the energy, due to the spin-orbit interaction, is obtained to be

$$E^{(1)}_{n\,\text{S-O}} = \frac{e^2\hbar^2}{4m^2c^2}\left[j(j+1)-\ell(\ell+1)-\frac{3}{4}\right]\left\langle\frac{1}{r^3}\right\rangle_{n\ell}. \tag{12.168}$$

Furthermore, since $\mathbf{J} = \mathbf{L} + \mathbf{S}$ and $s = \frac{1}{2}$, we have

$$j = \ell \pm \frac{1}{2}, \tag{12.169}$$

and this leads to

$$\begin{aligned}E^{(1)}_{n\,\text{S-O}} &= \frac{e^2\hbar^2}{4m^2c^2}\left\langle\frac{1}{r^3}\right\rangle_{n\ell}\begin{cases}\ell & \text{if } j=\ell+\frac{1}{2}\\-(\ell+1) & \text{if } j=\ell-\frac{1}{2}\end{cases}\\&= \frac{e^2\hbar^2}{4m^2c^2}\frac{1}{a_0^3n^3\ell\left(\ell+\frac{1}{2}\right)(\ell+1)}\begin{cases}\ell\\-(\ell+1)\end{cases}\\&= \frac{e^2\hbar^2}{4m^2c^2}\left(\frac{me^2}{\hbar^2}\right)^3\frac{1}{n^3\ell\left(\ell+\frac{1}{2}\right)(\ell+1)}\begin{cases}\ell\\-(\ell+1)\end{cases}\\&= \frac{mc^2}{4}\left(\frac{e^2}{\hbar c}\right)^4\frac{1}{n^3\ell\left(\ell+\frac{1}{2}\right)(\ell+1)}\begin{cases}\ell\\-(\ell+1)\end{cases}\\&= \frac{mc^2\alpha^4}{4}\frac{1}{n^3\ell\left(\ell+\frac{1}{2}\right)(\ell+1)}\begin{cases}\ell,\\-(\ell+1).\end{cases}\end{aligned} \tag{12.170}$$

Thus, adding (12.161) and (12.170), we obtain the total fine structure splitting in hydrogen to be

$$\begin{aligned}
E_{nT}^{(1)} &= E_n^{(1)} + E_{n\,\text{S-O}}^{(1)} \\
&= -\frac{mc^2\alpha^4}{2n^3}\left[-\frac{3}{4n} + \frac{1}{(\ell+\frac{1}{2})}\right] \\
&\quad + \frac{mc^2\alpha^4}{4}\frac{1}{n^3\,(\ell+\frac{1}{2})}\left\{\frac{1}{\ell+1}\text{or} -\frac{1}{\ell}\right\} \\
&= -\frac{mc^2\alpha^2}{2n^2}\frac{\alpha^2}{n}\left[-\frac{3}{4n} + \frac{1}{2\,(\ell+\frac{1}{2})}\right. \\
&\quad \left.\times\left\{2-\left(\frac{1}{\ell+1}\text{ or } -\frac{1}{\ell}\right)\right\}\right] \\
&= -\frac{mc^2\alpha^2}{2n^2}\frac{\alpha^2}{n}\left[-\frac{3}{4n} + \frac{1}{(2\ell+1)}\left\{\frac{2\ell+1}{\ell+1}\text{ or }\frac{2\ell+1}{\ell}\right\}\right] \\
&= -\frac{mc^2\alpha^2}{2n^2}\frac{\alpha^2}{n}\left[-\frac{3}{4n} + \left\{\frac{1}{\ell+1}\text{ or }\frac{1}{\ell}\right\}\right] \\
&= -\frac{mc^2\alpha^2}{2n^2}\frac{\alpha^2}{n}\left[-\frac{3}{4n} + \frac{1}{j+\frac{1}{2}}\right].
\end{aligned} \tag{12.171}$$

for both $j = \ell \pm \frac{1}{2}$. This is the total fine structure splitting of the energy levels in hydrogen.

12.9 Selected problems

1. Brillouin-Wigner perturbation: Assume a completely non-degenerate quantum mechanical system with

$$H_0|u_n\rangle = E_n^{(0)}|u_n\rangle. \tag{12.172}$$

Writing

$$H = H_0 + \lambda H_1, \tag{12.173}$$

derive the perturbation equations by expanding only $|\psi_n\rangle$ (and not E_n) in powers of λ. Solve for the wave function up to first order in λ and the energy eigenvalues up to second order in λ.

2. A two dimensional isotropic oscillator is subjected to a time independent perturbation, H', whose matrix elements vanish between two states which have the same parity in either X or Y (Example: $H' = XY$).

   a) What is the degeneracy of the unperturbed state with energy eigenvalue $E^{(0)} = 3\hbar\omega$?

   b) List, in bra-ket notation, all the matrix elements of H', between the eigenstates belonging to this energy value, which do not vanish from symmetry considerations.

   c) What is the first order change of this energy level in terms of these matrix elements?

3. Because of the finite size of the nucleus in a hydrogenic atom, the potential, in which the electron moves, is of the form

$$V(r) = \begin{cases} -\frac{Ze^2}{r}, & r \geq a, \\ -\frac{Ze^2}{a}\left(\frac{3}{2} - \frac{r^2}{2a^2}\right), & r \leq a, \end{cases} \tag{12.174}$$

   where Z is the nuclear charge and a is the nuclear radius. Assuming that $a \ll \frac{\hbar^2}{me^2} = a_0$ (typically $a \sim 10^{-13}$cm, while $a_0 \sim 10^{-8}$cm), calculate the first order change in the ground state energy from its value for a point nucleus.

CHAPTER 13

Time dependent perturbation theory

Let us consider a system whose Hamiltonian H_0 is time independent and let us assume that we know how to solve for its eigenvalues and eigenfunctions exactly. We know that the eigenstates, in this case, will be stationary states.

$$
\begin{aligned}
& i\hbar \frac{\mathrm{d}}{\mathrm{d}t} |u_n(t)\rangle = H_0 |u_n(t)\rangle, \\
& |u_n(t)\rangle = e^{-\frac{i}{\hbar}E_n t} |u_n(0)\rangle = e^{-\frac{i}{\hbar}E_n t} |u_n\rangle, \qquad (13.1)
\end{aligned}
$$

such that

$$
H_0 |u_n\rangle = E_n |u_n\rangle. \qquad (13.2)
$$

These are stationary states. This simply means that if initially the system is in the state $|u_i\rangle$ (or $|i\rangle$), it remains in that state forever (unless disturbed). Mathematically, this is denoted by

$$
\langle u_f | u_i(t)\rangle = \delta_{if} \times \text{ phase factor}. \qquad (13.3)
$$

That is, the probability amplitude for finding the system in a different state, at a later time, is zero. Another way of saying this is that the system is unable to make a transition to a different state all by itself.

If there is a perturbation, however, things are different. The system can make a transition to a different state because of the perturbation. This is a very physical effect. We may have a system in a stationary state and apply a perturbation for a certain period of time and ask about the state of the system at the end of the perturbation. Clearly it would not necessarily be the same as the initial state. Therefore, one can calculate the transition probabilities for the system going into various states.

Let us now define the problem more precisely. Let us assume that we have a time dependent Hamiltonian of the form

$$
H(t) = H_0 + H'(t). \qquad (13.4)
$$

It is the perturbation Hamiltonian (and, therefore, the total Hamiltonian) which depends on time. Thus, energy is not conserved any more and stationary states are not eigenstates of the total Hamiltonian. We are looking for solutions to the equation

$$i\hbar \frac{\mathrm{d}}{\mathrm{d}t}|\psi(t)\rangle = H(t)|\psi(t)\rangle, \tag{13.5}$$

with the initial condition that

$$|\psi(0)\rangle = |u_i\rangle. \tag{13.6}$$

That is, the system is initially in the ith unperturbed state.

First of all, we note that although the states, $|u_n\rangle$'s, are no longer the eigenstates of the complete Hamiltonian, they still form a complete basis and, therefore, we can expand the state in this basis as

$$|\psi(t)\rangle = \sum_n c_n(t)|u_n\rangle. \tag{13.7}$$

To convince yourself of this, note that for each fixed value of time we can do it and hence we can do it for all times. The only point to note here is that the coefficients of expansion become functions of time. Furthermore, $c_n(t) = \langle u_n|\psi(t)\rangle$ now defines the probability amplitude for finding the system in the nth unperturbed state at time t. Remembering that the system was initially in the ith unperturbed state, this, therefore, measures the transition amplitude from the ith state to the nth state. Thus, we determine the probability of transition at time t to be

$$P_{i\to n} = |c_n(t)|^2 = |\langle u_n|\psi(t)\rangle|^2. \tag{13.8}$$

Furthermore, we would like to define a perturbative expansion for the transition probabilities. Hence we define

$$\begin{aligned} H(t) &= H_0 + \lambda H'(t) \\ |\psi(t)\rangle &= \sum_{n,m} \lambda^m c_n^{(m)}(t)|u_n\rangle, \end{aligned} \tag{13.9}$$

and solve for the time dependent Schrödinger equation, (13.5). This method was developed by Dirac and is known as the method of variation of constants.

Putting the expansions in (13.9) into the Schrödinger equation (13.5), we have

$$i\hbar\frac{\mathrm{d}}{\mathrm{d}t}|\psi(t)\rangle = (H_0 + \lambda H'(t))|\psi(t)\rangle,$$

$$\text{or,}\quad \sum_{n,m}\lambda^m i\hbar\frac{\mathrm{d}c_n^{(m)}(t)}{\mathrm{d}t}|u_n\rangle = \sum_{n,m}\lambda^m c_n^m(t)\left(H_0 + \lambda H'(t)\right)|u_n\rangle. \tag{13.10}$$

Taking the inner product with $\langle u_k|$, we obtain from (13.10)

$$\begin{aligned}\sum_m \lambda^m i\hbar\frac{\mathrm{d}c_k^{(m)}(t)}{\mathrm{d}t} &= \sum_{n,m}\lambda^m c_n^{(m)}(t)\left(E_n\delta_{nk} + \lambda H'_{kn}(t)\right)\\ &= \sum_m \lambda^m\big(E_k c_k^{(m)}(t) + \lambda\sum_n c_n^{(m)}(t)H'_{kn}(t)\big).\end{aligned} \tag{13.11}$$

Matching the lowest power of λ on both sides of (13.11), we obtain,

$$\begin{aligned}& i\hbar\frac{\mathrm{d}c_k^{(0)}(t)}{\mathrm{d}t} = E_k c_k^{(0)}(t),\\ \text{or,}\quad & c_k^{(0)}(t) = c_k^{(0)}(0)e^{-\frac{i}{\hbar}E_k t}.\end{aligned} \tag{13.12}$$

This, of course, tells us that, to zeroth order in the perturbation, the eigenstates are stationary states. Now let us redefine the coefficients of expansion in (13.9) such that

$$c_k^{(m)}(t) = a_k^{(m)}(t)e^{-\frac{i}{\hbar}E_k t}. \tag{13.13}$$

Then, equation (13.11) becomes

$$\sum_m \lambda^m i\hbar\frac{\mathrm{d}c_k^{(m)}(t)}{\mathrm{d}t} = \sum_m \lambda^m\left(E_k c_k^{(m)}(t) + \lambda\sum_n c_n^{(m)}(t)H'_{kn}\right),$$

$$\text{or,}\quad \sum_m \lambda^m i\hbar\left[-\frac{iE_k}{\hbar}e^{-\frac{i}{\hbar}E_k t}a_k^{(m)}(t) + e^{-\frac{i}{\hbar}E_k t}\frac{\mathrm{d}a_k^{(m)}(t)}{\mathrm{d}t}\right]$$

$$= \sum_m \lambda^m\left(E_k e^{-\frac{i}{\hbar}E_k t}a_k^{(m)}(t) + \lambda\sum_n e^{-\frac{i}{\hbar}E_n t}a_n^{(m)}(t)H'_{kn}\right),$$

$$\text{or,}\quad \sum_m \lambda^m e^{-\frac{i}{\hbar}E_k t}i\hbar\frac{\mathrm{d}a_k^{(m)}(t)}{\mathrm{d}t} = \sum_{n,m}\lambda^{m+1}e^{-\frac{i}{\hbar}E_n t}a_n^{(m)}(t)H'_{kn}. \tag{13.14}$$

Defining the Bohr frequency

$$\omega_{kn} = \frac{E_k - E_n}{\hbar}, \tag{13.15}$$

we can write equation (13.14) as

$$\sum \lambda^m i\hbar \frac{\mathrm{d}a_k^{(m)}(t)}{\mathrm{d}t} = \sum_{n,m} \lambda^{m+1} e^{i\omega_{kn}t} a_n^{(m)}(t) H'_{kn}. \tag{13.16}$$

Furthermore, matching powers of λ, we now obtain

$$i\hbar \frac{\mathrm{d}a_k^{(m)}(t)}{\mathrm{d}t} = \sum_n e^{i\omega_{kn}t} a_n^{(m-1)}(t) H'_{kn}. \tag{13.17}$$

For $m = 0$, this leads to

$$i\hbar \frac{\mathrm{d}a_k^{(0)}(t)}{\mathrm{d}t} = 0, \tag{13.18}$$

which implies that

$$a_k^{(0)}(t) = a_k^{(0)} = \text{ constant}. \tag{13.19}$$

Similarly, for $m = 1$, we obtain from (13.17)

$$i\hbar \frac{\mathrm{d}a_k^{(1)}(t)}{\mathrm{d}t} = \sum_n e^{i\omega_{kn}t} a_n^{(0)} H'_{kn}. \tag{13.20}$$

Noting that the system is initially in the state $|u_i\rangle$, we have

$$a_n^{(0)} = \delta_{in}. \tag{13.21}$$

This gives, to first order in the perturbation,

$$i\hbar \frac{\mathrm{d}a_k^{(1)}}{\mathrm{d}t} = \sum_n e^{i\omega_{kn}t} \delta_{in} H'_{kn} = e^{i\omega_{ki}t} H'_{ki},$$

$$\text{or,} \quad a_k^{(1)}(t) = \frac{1}{i\hbar} \int_0^t \mathrm{d}t' \; e^{i\omega_{ki}t'} H'_{ki}(t')$$

$$= \frac{1}{i\hbar} \int_0^t \mathrm{d}t' \; e^{i\omega_{ki}t'} \langle u_k | H'(t') | u_i \rangle. \tag{13.22}$$

Here, we are assuming that the perturbation is switched on at $t = 0$. Otherwise, the lower limit of integration will be different. Therefore, to first order in the perturbation, the probability for transition to a state different from the initial state is obtained to be

$$\begin{aligned} P_{i\to f}(t) &= \left|c_f^{(1)}\right|^2 = \left|a_f^{(1)}\right|^2 \\ &= \frac{1}{\hbar^2}\left|\int_0^t \mathrm{d}t' e^{i\omega_{fi}t'}\langle f|H'(t')|i\rangle\right|^2 . \end{aligned} \tag{13.23}$$

We can, of course, carry through this procedure to higher orders and the perturbation method will be valid only if $P_{i\to f}$'s are small compared to unity.

▶ **Example.** Let us consider an one dimensional oscillator in its ground state at $t = -\infty$, which is subjected to a perturbation

$$H'(t) = -e\mathcal{E}Xe^{-\frac{t^2}{\tau^2}}. \tag{13.24}$$

Here $e, \mathcal{E}$ and τ are constants. (Namely, we have an oscillator in a time dependent electric field.) The perturbation is applied over an infinite time interval. The question we would like to ask is what is the probability that the oscillator will be in the state $|n\rangle$ (of the unperturbed oscillator) as $t \to \infty$.

In this case, we know, from first order perturbation theory, (13.22) that

$$a_k^{(1)}(t) = \frac{1}{i\hbar}\int_{-\infty}^t \mathrm{d}t' e^{i\omega_{ki}t'}\langle u_k|H'(t')|u_i\rangle. \tag{13.25}$$

For the present case,

$$\begin{aligned} |u_i\rangle &= |0\rangle, \\ |u_f\rangle &= |n\rangle, \qquad \omega_{fi} = \frac{E_f - E_i}{\hbar} = n\omega, \\ X &= \sqrt{\frac{\hbar}{2m\omega}}\,(a + a^\dagger). \end{aligned} \tag{13.26}$$

Therefore, we have

$$\langle n|X|0\rangle = \sqrt{\frac{\hbar}{2m\omega}}\langle n|(a + a^\dagger)|0\rangle = \sqrt{\frac{\hbar}{2m\omega}}\delta_{n1}. \tag{13.27}$$

Furthermore, we are interested in the transition probability as $t \to \infty$. Thus, we obtain from (13.25) and (13.27)

$$\begin{aligned}
a_n^{(1)}(\infty) &= \frac{1}{i\hbar}\int_{-\infty}^{\infty} \mathrm{d}t'\ e^{in\omega t'}\left(-e\mathcal{E}e^{-\frac{t'^2}{\tau^2}}\right)\langle n|X|0\rangle \\
&= -\frac{e\mathcal{E}}{i\hbar}\sqrt{\frac{\hbar}{2m\omega}}\delta_{n1}\int_{-\infty}^{\infty}\mathrm{d}t'\ e^{-\frac{t'^2}{\tau^2}+in\omega t'} \\
&= -\frac{e\mathcal{E}}{i\hbar}\sqrt{\frac{\hbar}{2m\omega}}\delta_{n1}\int_{-\infty}^{\infty}\mathrm{d}t'\ e^{-\left(\frac{t'}{\tau}-\frac{i\omega\tau}{2}\right)^2-\frac{\omega^2\tau^2}{4}} \\
&= -\frac{e\mathcal{E}}{i\hbar}\sqrt{\frac{\hbar}{2m\omega}}e^{-\frac{\omega^2\tau^2}{4}}\delta_{n1}\sqrt{\pi}\,\tau.
\end{aligned} \tag{13.28}$$

In other words, because of this perturbation, the oscillator can make a transition only to the first excited state, with the transition probability given by (13.23),

$$P_{0\to 1}(\infty) = \left|a_1^{(1)}(\infty)\right|^2 = \frac{\pi e^2\mathcal{E}^2\tau^2}{2m\hbar\omega}\ e^{-\frac{\omega^2\tau^2}{2}}. \tag{13.29}$$

◀

13.1 Harmonic and constant perturbations

Time dependent perturbations can be of various types, which we discuss in the following.

1. Let us assume that a system is subjected to a perturbation of the form

$$H'(t) = 2\hat{H}\sin\omega t, \tag{13.30}$$

where $\hat{H}$ is constant in time. We assume that the perturbation is turned on between time 0 and t_0. Then, at a later time t ($t > t_0$), we obtain from (13.22)

$$\begin{aligned}
a_f^{(1)}(t) &= \frac{2}{i\hbar}\int_0^{t_0}\mathrm{d}t'\ e^{i\omega_{fi}t'}\left(\frac{e^{i\omega t'}-e^{-i\omega t'}}{2i}\right)\hat{H}_{fi} \\
&= -\frac{\hat{H}_{fi}}{\hbar}\int_0^{t_0}\mathrm{d}t'\left(e^{i(\omega_{fi}+\omega)t'}-e^{i(\omega_{fi}-\omega)t'}\right) \\
&= -\frac{\hat{H}_{fi}}{i\hbar}\left(\frac{1}{\omega_{fi}+\omega}\left(e^{i(\omega_{fi}+\omega)t_0}-1\right)\right.
\end{aligned}$$

$$
\begin{aligned}
&\quad - \frac{1}{\omega_{fi} - \omega}\left(e^{i(\omega_{fi}-\omega)t_0} - 1\right)\Bigg) \\
&= -\frac{2\hat{H}_{fi}}{\hbar}\left(e^{\frac{i(\omega_{fi}+\omega)t_0}{2}}\frac{\sin\frac{(\omega_{fi}+\omega)t_0}{2}}{(\omega_{fi}+\omega)}\right. \\
&\quad \left. - e^{\frac{i(\omega_{fi}-\omega)t_0}{2}}\frac{\sin\frac{(\omega_{fi}-\omega)t_0}{2}}{(\omega_{fi}-\omega)}\right). \qquad (13.31)
\end{aligned}
$$

Looking at the expression in (13.31), it is clear that the dominant contribution to the transition amplitude comes from

$$\omega = \pm\omega_{fi}. \qquad (13.32)$$

From the definition in (13.15), we see that the two cases correspond to

$$E_f = E_i \pm \hbar\omega. \qquad (13.33)$$

The two cases correspond, respectively, to the absorption and emission of a quantum of radiation when electromagnetic interactions are involved. For the present, let us assume that

$$\omega \simeq \omega_{fi}. \qquad (13.34)$$

Then, the dominant contribution in (13.31) has the form

$$a_f^{(1)}(t) = \frac{2\hat{H}_{fi}}{\hbar}\, e^{\frac{i(\omega_{fi}-\omega)t_0}{2}}\, \frac{\sin\frac{(\omega-\omega_{fi})t_0}{2}}{(\omega-\omega_{fi})}, \qquad (13.35)$$

so that the transition probability takes the form

$$P_{i\to f}(t) = \left|a_f^{(1)}(t)\right|^2 = \frac{4|\hat{H}_{fi}|^2}{\hbar^2}\frac{\sin^2\frac{(\omega-\omega_{fi})t_0}{2}}{(\omega-\omega_{fi})^2}. \qquad (13.36)$$

This transition probability for a fixed value of t_0 is shown in Fig. 13.1. There is a peak at $\omega = \omega_{fi}$ with a magnitude $\frac{|\hat{H}_{fi}|^2 t_0^2}{\hbar^2}$. Away from this value of the frequency, the probability oscillates with a very damped amplitude, much like a diffraction pattern. This is, of course, a resonant behavior. The resonance width is defined to be the distance between the first zeros on either side of the resonant frequency. The zeros occur at

$$\frac{(\omega-\omega_{fi})t_0}{2} = \pm\pi. \qquad (13.37)$$

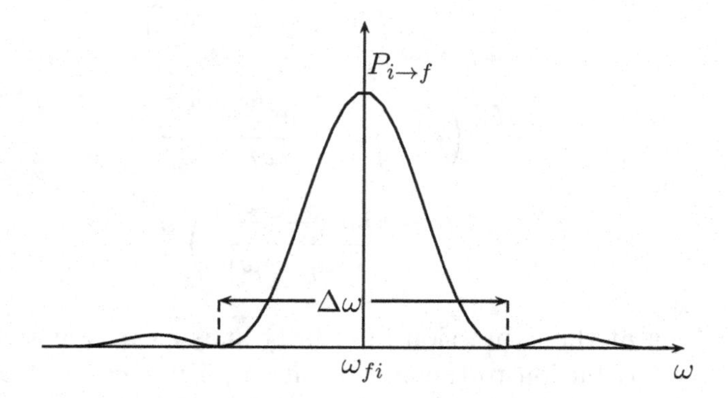

Figure 13.1: Transition probability in (13.36) as a function of frequency ω at a fixed t_0.

Thus, the resonance width is determined to be

$$\Delta\omega = \frac{4\pi}{t_0}. \tag{13.38}$$

This tells us that the longer the perturbation acts, the narrower the resonance would be. This also has a similarity with the uncertainty principle. Let us assume that we wish to measure the energy separation $E_f - E_i$. We can do this by subjecting the system to a harmonic perturbation and looking for a resonance peak. Clearly, the uncertainty in the determination of the energy levels $(E_f - E_i)$ will be of the form

$$\Delta E = \frac{\hbar\Delta\omega}{2} \simeq \frac{2\pi\hbar}{t_0} = \frac{h}{t_0},$$
$$\text{or,} \quad \Delta E t_0 \simeq h. \tag{13.39}$$

Therefore, the product $\Delta E t_0$ cannot be smaller than h.

It is obvious that, had we plotted the total transition probability (without neglecting the contribution from $\omega = -\omega_{fi}$), then, we would have obtained another peak of the same width at $\omega = -\omega_{fi}$. Clearly, we can neglect this resonance while talking about the one at $\omega = \omega_{fi}$ provided

$$2|\omega_{fi}| \gg \Delta\omega,$$
$$\text{or,} \quad t_0 \gg \frac{1}{|\omega_{fi}|} = \frac{1}{\omega}. \tag{13.40}$$

From (13.36), it is clear that, if $\omega = \omega_{fi}$, then,

$$P_{i\to f}(t) = \frac{|\hat{H}_{fi}|^2 t_0^2}{\hbar^2}, \tag{13.41}$$

so that perturbation theory is valid only if

$$t_0 \ll \frac{\hbar}{|\hat{H}_{fi}|}. \tag{13.42}$$

2. Let us next assume that the system is subjected to a constant perturbation between the time interval 0 and t_0. Thus,

$$H'(t) = \hat{H} = \text{ constant}. \tag{13.43}$$

In this case, we have $(t > t_0)$

$$\begin{aligned} a_f^{(1)}(t) &= \frac{1}{i\hbar}\hat{H}_{fi}\int_0^{t_0} \mathrm{d}t'\, e^{i\omega_{fi}t'}, \\ &= -\frac{\hat{H}_{fi}}{\hbar}\frac{\left(e^{i\omega_{fi}t_0} - 1\right)}{\omega_{fi}} \\ &= \frac{2\hat{H}_{fi}}{i\hbar}\frac{\sin\frac{\omega_{fi}t_0}{2}}{\omega_{fi}}\, e^{\frac{i\omega_{fi}t_0}{2}}. \end{aligned} \tag{13.44}$$

Therefore, the transition probability becomes

$$P_{i\to f}(t) = \left|a_f^{(1)}(t)\right|^2 = \frac{4\left|\hat{H}_{fi}\right|^2}{\hbar^2}\frac{\sin^2\frac{\omega_{fi}t_0}{2}}{\omega_{fi}^2}. \tag{13.45}$$

This transition probability shown in Fig. 13.2 is again similar to the case of the harmonic perturbation, except that now we see a resonance occur when (see also Fig. 13.1)

$$\omega_{fi} = 0. \tag{13.46}$$

Namely, resonance phenomenon takes place under constant perturbation if there are degenerate levels. Once again, it is clear, from (13.45), that perturbation theory is applicable at resonance only if

$$t_0 \ll \frac{\hbar}{\left|\hat{H}_{fi}\right|}. \tag{13.47}$$

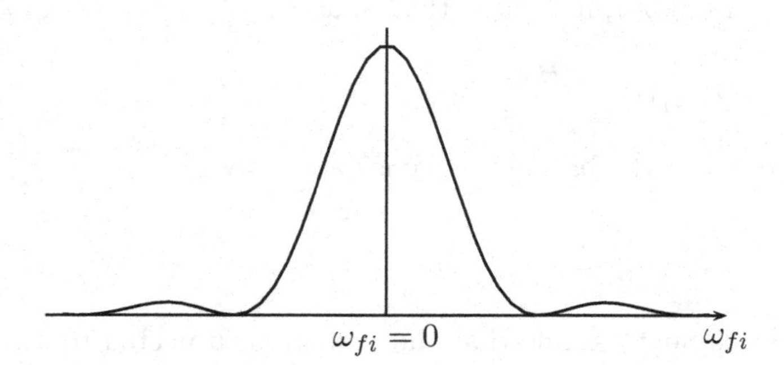

Figure 13.2: Transition probability in (13.45) as a function of ω_{fi}.

13.2 Long perturbation at resonance

As we have seen in (13.42) and (13.47), if the perturbation acts for a longer time at resonance, then, our perturbation scheme breaks down. Here again, as in the case of the nearly degenerate system (in stationary perturbation theory), we solve the system exactly. First of all, let us note that if

$$\omega \simeq \omega_{fi}, \tag{13.48}$$

then, the only state, that would have a dominant probability for transition from the state $|u_i\rangle$, would be the state $|u_f\rangle$. Thus, we neglect all other states and write (namely, we think of this as a two level system)

$$|\psi(t)\rangle = a_i(t)e^{-\frac{i}{\hbar}E_it}|u_i\rangle + a_f(t)e^{-\frac{i}{\hbar}E_ft}|u_f\rangle. \tag{13.49}$$

In this case, we obtain

$$\begin{aligned}
& i\hbar\frac{\mathrm{d}}{\mathrm{d}t}|\psi(t)\rangle = (H_0 + H'(t))|\psi(t)\rangle, \\
\text{or,}\quad & e^{-\frac{i}{\hbar}E_it}\, i\hbar\frac{\mathrm{d}a_i(t)}{\mathrm{d}t}|u_i\rangle + e^{-\frac{i}{\hbar}E_ft}\, i\hbar\frac{\mathrm{d}a_f(t)}{\mathrm{d}t}|u_f\rangle \\
& \quad = e^{-\frac{i}{\hbar}E_it}\, H'(t)|u_i\rangle\, a_i(t) + e^{-\frac{i}{\hbar}E_ft}\, H'(t)|u_f\rangle\, a_f(t).
\end{aligned} \tag{13.50}$$

Taking the inner product of (13.50) with the state $\langle u_i|$, we have

$$i\hbar\frac{\mathrm{d}a_i(t)}{\mathrm{d}t} = (H'(t))_{ii}a_i + e^{-i\omega_{fi}t}(H'(t))_{if}a_f(t). \tag{13.51}$$

Furthermore, choosing $H'(t) = 2\hat{H}\sin\omega t$, we can write (13.51) as

$$i\hbar\frac{\mathrm{d}a_i(t)}{\mathrm{d}t} = -i\left(e^{i\omega t} - e^{-i\omega t}\right)\hat{H}_{ii}a_i(t) - i\left(e^{-i(\omega_{fi}-\omega)t} - e^{-i(\omega_{fi}+\omega)t}\right)\hat{H}_{if}a_f(t). \tag{13.52}$$

Similarly, taking the inner product of (13.50) with $\langle u_f|$, we obtain

$$i\hbar\frac{\mathrm{d}a_f(t)}{\mathrm{d}t} = -i\left(e^{i(\omega_{fi}+\omega)t} - e^{i(\omega_{fi}-\omega)t}\right)\hat{H}_{fi}a_i(t) - i\left(e^{i\omega t} - e^{-i\omega t}\right)\hat{H}_{ff}a_f(t). \tag{13.53}$$

It is clear that, since

$$\omega \simeq \omega_{fi}, \tag{13.54}$$

the terms with $e^{\pm i\omega t}$ and $e^{\pm i(\omega_{fi}+\omega)t}$ will contribute negligibly when integrated. Thus, we can approximate and write

$$i\hbar\frac{\mathrm{d}a_i(t)}{\mathrm{d}t} = -ie^{-i(\omega_{fi}-\omega)t}\hat{H}_{if}a_f(t),$$

$$i\hbar\frac{\mathrm{d}a_f(t)}{\mathrm{d}t} = ie^{i(\omega_{fi}-\omega)t}\hat{H}_{fi}a_i(t). \tag{13.55}$$

If we assume that the system is initially in the ith state, then, the equations in (13.55) have to be solved subject to the initial conditions

$$a_i(0) = 1, \qquad a_f(0) = 0,$$

$$\left.\frac{\mathrm{d}a_i(t)}{\mathrm{d}t}\right|_{t=0} = 0, \qquad \left.\frac{\mathrm{d}a_f(t)}{\mathrm{d}t}\right|_{t=0} = \frac{1}{\hbar}\hat{H}_{fi}. \tag{13.56}$$

This system of coupled equations can be exactly solved for ω near ω_{fi} and the solution gives

$$P_{i\to f} = \frac{4|\hat{H}_{fi}|^2}{4|\hat{H}_{fi}|^2 + \hbar^2(\omega-\omega_{fi})^2} \times \sin^2\left(\frac{\sqrt{4|\hat{H}_{fi}|^2 + \hbar^2(\omega-\omega_{fi})^2}\, t_0}{2\hbar}\right). \tag{13.57}$$

Relation (13.57) is known as the Breit-Wigner formula.

First of all, let us note that this probability lies between 0 and 1, no matter how long the perturbation acts. When the transition probability is zero, the system has oscillated back to the initial state. It is clear that if $\omega = \omega_{fi}$, no matter how small the perturbation is, the system can move from the state $|u_i\rangle$ to $|u_f\rangle$ with a considerable probability. We also see that if t_0 is small (at $\omega = \omega_{fi}$), then, (13.57) gives

$$P_{i\to f} = \frac{|\hat{H}_{if}|^2 t_0^2}{\hbar^2}, \tag{13.58}$$

which is, of course, the first order perturbation result that we have derived earlier in (13.41).

13.3 Transition from a discrete level to continuum

All the perturbation schemes that we have developed so far hold true, no matter whether the spectrum of the Hamiltonian is discrete or continuous. In fact, we can think of a continuous spectrum as a limiting case of a discrete spectrum. Various interesting physical phenomena correspond to a system making a transition from a discrete level to continuum. To name a few such effects, we have ionization of an atom when an electron leaves the atom, photoelectric effect, radioactive beta decay and so on. Let us study such effects in some detail.

First of all, let us remind ourselves of the continuous spectrum. The free Schrödinger equation

$$\boldsymbol{\nabla}^2\psi + k^2\psi = 0, \qquad k^2 = \frac{2mE}{\hbar^2}, \tag{13.59}$$

has solutions of the form

$$\psi(\mathbf{r}) \sim e^{i\mathbf{k}\cdot\mathbf{r}}, \tag{13.60}$$

where the momentum, $\mathbf{p} = \hbar\mathbf{k}$, and hence the energy can take any continuous value. In such a case, we are, of course, not interested in measuring transition to a particular state, rather we would like to study the transition to a group of states lying close together. The reason for this is obvious. There are states lying infinitely close to a particular state. And since our measuring abilities are limited by the uncertainty principle as well as by the capabilities of our measuring devices, we cannot measure a state with appreciable accuracy. Therefore, we would like to study the transition to any one of a number of states that lie within a certain energy interval. This requires the notion of the density of states.

It is not *a priori* clear how we should calculate this for the continuum solutions of the Schrödinger equation. An easy way is to assume that a particle moves in a box of length L. Thus, we look for solutions of

$$\boldsymbol{\nabla}^2\psi + k^2\psi = 0, \qquad |x|, |y|, |z| \leq \frac{L}{2}. \tag{13.61}$$

The natural boundary condition to apply, in this case, is that the wave function vanishes at the walls. However, a more convenient mathematical boundary condition is that the wave function is periodic at the boundaries. That is,

$$\psi\left(-\frac{L}{2}, y, z\right) = \psi\left(\frac{L}{2}, y, z\right), \tag{13.62}$$

and so on for the y, z coordinates. This immediately leads to the condition that

$$k_x = \frac{2\pi n_x}{L}, \quad k_y = \frac{2\pi n_y}{L}, \quad k_z = \frac{2\pi n_z}{L}, \tag{13.63}$$

with the wave function taking the form

$$\psi_{\mathbf{k}}(\mathbf{r}) = \frac{1}{(L)^{\frac{3}{2}}}\, e^{i\mathbf{k}\cdot\mathbf{r}}. \tag{13.64}$$

Here, n_x, n_y and n_z take integer values and

$$E_k = E_n = \frac{\hbar^2 k^2}{2m} = \frac{4\pi^2\hbar^2}{2mL^2}\left(n_x^2 + n_y^2 + n_z^2\right) = \frac{h^2}{2mL^2}n^2. \tag{13.65}$$

Thus, we see that the imposition of periodic boundary condition leads to discrete eigenvalues for momentum as well as energy. As we increase the dimensions of the box, the spacing between the levels decreases and in the limit of an infinite box, the eigenvalues become continuous. Thus, we see that this is a convenient way of looking at the continuous spectrum. Furthermore, it also helps in calculating various quantities more easily.

To calculate the density of states, we go to momentum space and note that each momentum state can be represented as a point in this space with coordinates

$$\left(\frac{2\pi\hbar}{L}n_x, \frac{2\pi\hbar}{L}n_y, \frac{2\pi\hbar}{L}n_z\right). \tag{13.66}$$

Since only one of the n's have to change by unity to give another distinct state or point, the volume associated with each point in this space is $\left(\frac{2\pi\hbar}{L}\right)^3$ or $\left(\frac{h}{L}\right)^3$. This is the volume of each state in momentum space. (Let me emphasize here that, in quantum mechanics, a particle cannot have exact coordinates and momenta simultaneously because of the uncertainty principle. As a result, the state of a system cannot be specified as a point in phase space, as in classical mechanics. Rather, it is assigned a Planck cell of volume $\left(\frac{h}{L}\right)^3$.) It follows now that the number of states in a volume d^3p in momentum space is given by

$$\frac{\mathrm{d}^3 p}{\left(\frac{h}{L}\right)^3} = \frac{L^3}{h^3}\,\mathrm{d}^3 p = \frac{L^3}{h^3}\,p^2 \mathrm{d}p\mathrm{d}\Omega = \left(\frac{L}{2\pi}\right)^3 k^2 \mathrm{d}k\mathrm{d}\Omega. \tag{13.67}$$

On the other hand, a volume d^3p in momentum space corresponds to an energy interval $\mathrm{d}E_p$. Therefore, we can write

$$\rho(E_p)\mathrm{d}E_p = \left(\frac{L}{2\pi}\right)^3 k^2 \mathrm{d}k\mathrm{d}\Omega, \tag{13.68}$$

where $\rho(E_p)$ represents the density of states in the energy space. From the relation for the energy of a free particle, we obtain

$$\begin{aligned} E_p &= E_k = \frac{\hbar^2 k^2}{2m}, \\ \text{or,}\quad \mathrm{d}E_p &= \mathrm{d}E_k = \frac{\hbar^2 k}{m}\,\mathrm{d}k, \end{aligned} \tag{13.69}$$

which determines the density of states, from (13.68), to correspond to

$$\rho(E_p) = \frac{mL^3}{\hbar^2(2\pi)^3} k\mathrm{d}\Omega = \frac{mL^3}{8\pi^3\hbar^2}\,k\mathrm{d}\Omega. \tag{13.70}$$

We note here that, even though L appears in various intermediate steps, physical results are independent of L.

Let us now assume that a system, which is initially in a discrete state $|i\rangle$, is subjected to a harmonic perturbation (the perturbation is assumed to be applied for a long time)

$$H'(t) = 2\hat{H}\sin\omega t. \tag{13.71}$$

As we have seen in (13.36), the probability for transition to a final state, in this case, is ($\omega \simeq \omega_{fi}$)

$$P_{i\to f}(t) = \frac{4|\hat{H}_{fi}|^2}{\hbar^2}\,\frac{\sin^2\frac{(\omega-\omega_{fi})t}{2}}{(\omega-\omega_{fi})^2}. \tag{13.72}$$

Let us further assume that the final state is a state of the continuum, labeled by the wave number k_f (energy E_f). In this case, of course, since there is degeneracy of states, the transition can occur to any nearby state. The relevant question to ask, therefore, is what is the probability of transition to a state with $E = E_f \pm \Delta E$. This is given by

$$\begin{aligned}\tilde{P}_{i\to f}(t) &= \int_{-\Delta E}^{\Delta E} P_{i\to f}(t)\rho_f(E_f)\mathrm{d}E_f \\ &= \int_{-\Delta E}^{\Delta E} \frac{4|\hat{H}_{fi}|^2}{\hbar^2}\frac{\sin^2\frac{(\omega-\omega_{fi})t}{2}}{(\omega-\omega_{fi})^2}\rho_f(E_f)\mathrm{d}E_f. \end{aligned} \tag{13.73}$$

We notice that the only quantity that oscillates appreciably, inside the integrand, is the $\sin^2(\frac{1}{2}(\omega-\omega_{fi})t)/(\omega-\omega_{fi})^2$ term. Furthermore, its behavior is like a delta function, namely, it picks up the dominant contribution when $\omega = \omega_{fi}$. Thus, we can perform the integration by taking out the non-varying terms and extending the limits of integration. In other words ($x = (\omega_{fi} - \omega)t/2$),

$$\begin{aligned}\tilde{P}_{i\to f}(t) &\simeq \frac{4|\hat{H}_{fi}|^2}{\hbar^2}\rho_f \times \frac{\hbar t}{2}\int_{-\infty}^{\infty}\mathrm{d}x\,\frac{\sin^2 x}{x^2} \\ &= \frac{2|\hat{H}_{fi}|^2\rho_f t}{\hbar}\times\pi. \end{aligned} \tag{13.74}$$

Thus, the rate of transition to such a group of states is

$$R_{i\to f}(t) = \frac{\tilde{P}_{i\to f}}{t} = \frac{2\pi\rho_f|\hat{H}_{fi}|^2}{\hbar}. \tag{13.75}$$

This formula is known as Fermi's Golden rule and shows that the rate of transition from a discrete state to a group in the continuum is a constant independent of time. Let us note here that we cannot obtain a constant transition rate if only discrete states are involved in the transition.

13.4 Ionization of hydrogen

Let us consider a hydrogen atom in its ground state subjected to an oscillating electric field. Thus, we can write the perturbing Hamiltonian as

$$H'(t) = -2e\mathbf{r}\cdot\boldsymbol{\mathcal{E}}\sin\omega t = 2\hat{H}\sin\omega t, \tag{13.76}$$

and, in this case, we have

$$\hat{H} = -e\mathbf{r} \cdot \boldsymbol{\mathcal{E}}, \tag{13.77}$$

where we are assuming that $\boldsymbol{\mathcal{E}}$ is constant. We would like to calculate the probability that the hydrogen atom would be ionized because of this field.

We may think of such a perturbation as being affected by a pair of condensers with an alternating voltage. However, this is not realistic since, for ionization to occur,

$$\begin{aligned}
|E_0| &= \hbar\omega, \\
\text{or,}\quad \omega &= \frac{|E_0|}{\hbar}, \\
\text{or,}\quad \nu &= \frac{\omega}{2\pi} = \frac{|E_0|}{2\pi\hbar} = \frac{|E_0|}{h} \\
&\simeq \frac{13.6 \text{ eV}}{6 \times 10^{-21} \text{ MeV-sec}} \\
&\simeq 2 \times 10^{15} \text{ cycles/sec}.
\end{aligned} \tag{13.78}$$

This is a large frequency to achieve in the laboratory. However, the idea is that a traveling electromagnetic wave can have such high frequencies associated with it and hence can cause such transitions.

The calculation of the matrix element now becomes

$$\begin{aligned}
\langle f|\hat{H}|i\rangle &= \langle k|\hat{H}|0\rangle \\
&= -e|\boldsymbol{\mathcal{E}}| \left(\frac{1}{\pi a_0^3 L^3}\right)^{\frac{1}{2}} \int \mathrm{d}^3 r \; e^{-ikr\cos\theta'} r\cos\theta'' e^{-\frac{r}{a_0}},
\end{aligned} \tag{13.79}$$

where $|k\rangle$ denotes the continuum free particle state with wave number k and $|0\rangle$ represents the ground state of hydrogen. Here, θ'' is the angle between the vector $\mathbf{r}$ and the electric field, whereas θ' is the angle between the momentum vector of the electron and the vector $\mathbf{r}$. We can choose the momentum vector $\mathbf{k}$ to be along the z axis and define θ as the angle between $\mathbf{k}$ and $\boldsymbol{\mathcal{E}}$ as shown in Fig. 13.3 ($\mathbf{r}, \boldsymbol{\mathcal{E}}$ and $\mathbf{k}$ need not be coplanar). In this case, we can write

$$\cos\theta'' = \cos\theta' \cos\theta + \sin\theta' \sin\theta \cos(\phi' - \phi). \tag{13.80}$$

Using (13.80), it is clear that the integration over ϕ' gets rid of the second term in (13.79). The integration of the first term is quite

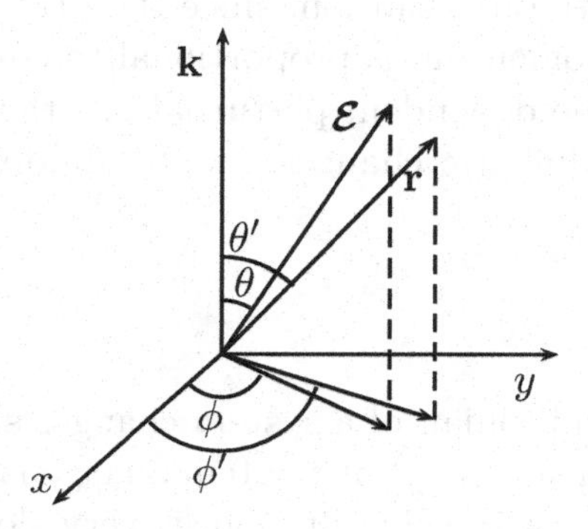

Figure 13.3: The three vectors $\mathbf{k}, \mathbf{r}, \boldsymbol{\mathcal{E}}$ in spherical coordinates.

straightforward and the result is

$$\langle f|\hat{H}|i\rangle = \langle k|\hat{H}|0\rangle = \frac{32i\pi e|\boldsymbol{\mathcal{E}}|ka_0^5 \cos\theta}{\left(\pi a_0^3 L^3\right)^{\frac{1}{2}} \left(1 + k^2 a_0^2\right)^3}. \tag{13.81}$$

Remembering that (see (13.70))

$$\rho_f = \rho(k) = \frac{mL^3}{8\pi^3\hbar^2}\, k\mathrm{d}\Omega, \tag{13.82}$$

we find that the rate of ionization, in this case, takes the form

$$\begin{aligned} R_{i\to f} &= \frac{2\pi}{\hbar} |\langle f|\hat{H}|i\rangle|^2 \rho_f \\ &= \frac{2\pi}{\hbar} \frac{32^2\pi^2 e^2 |\boldsymbol{\mathcal{E}}|^2 k^2 a_0^{10} \cos^2\theta}{\pi a_0^3 L^3 \left(1 + k^2 a_0^2\right)^6} \frac{mL^3}{8\pi^3\hbar^2} k\mathrm{d}\Omega \\ &= \frac{256 m e^2 |\boldsymbol{\mathcal{E}}|^2 a_0^7}{\pi\hbar^3 \left(1 + k^2 a_0^2\right)^6} k^3 \cos^2\theta \mathrm{d}\Omega \\ &= \frac{256 |\boldsymbol{\mathcal{E}}|^2 a_0^6 k^3}{\pi\hbar (1 + k^2 a_0^2)^6} \cos^2\theta \mathrm{d}\Omega. \end{aligned} \tag{13.83}$$

(Note that the physical rate of transition is independent of L as discussed earlier.) The differential ionizability into the solid angle $\mathrm{d}\Omega$, then, becomes

$$\frac{R}{\mathrm{d}\Omega} \propto \cos^2\theta. \tag{13.84}$$

This result is consistent with our intuition, since the driving force in the direction of the final momentum is proportional to $\cos\theta$.

The other kinds of time dependent perturbations that we come across are known as: a) adiabatic changes and b) sudden changes, which we discuss next.

13.5 Adiabatic changes

Let us assume that the Hamiltonian of a system changes slowly from $H(0)$ to $H(t)$. If the system starts out at $t = 0$ in an eigenstate $|n(0)\rangle$ of $H(0)$ and if the variation of the Hamiltonian is very slow in time, then, the adiabatic theorem due to Ehrenfest (Bohr) tells us that the system, at a later time t, would be in the eigenstate $|n(t)\rangle$ of $H(t)$.

The slowness of variation is defined as follows. First of all, let ω be the minimum frequency of motion associated with the system. For example, it can be the minimum of the splitting of the energy levels and so on. Furthermore, the time variation of the Hamiltonian introduces another frequency into the system, namely,

$$\left|\frac{\frac{\partial H'}{\partial t}}{H'}\right|. \tag{13.85}$$

A perturbation is said to be adiabatic, if

$$\left|\frac{\frac{\partial H'}{\partial t}}{H'}\right| \ll \omega. \tag{13.86}$$

That is, the Hamiltonian should not change appreciably during a characteristic cycle of motion, for adiabatic approximation to hold.

An example of adiabatic perturbation is, of course, the case of the one dimensional harmonic oscillator that we have already studied, where

$$H'(t) = -e\mathcal{E}Xe^{-\frac{t^2}{\tau^2}}. \tag{13.87}$$

Here, τ is the time scale associated with the perturbation and the adiabatic approximation applies if τ is large. In this case, we have seen in (13.29) that the oscillator, initially in the ground state, can make a transition only to the first excited state, with the probability of transition given by

$$P_{0\to 1}(\infty) = \frac{\pi e^2\mathcal{E}^2\tau^2}{2m\hbar\omega}\, e^{-\frac{\omega^2\tau^2}{2}}. \tag{13.88}$$

We see from (13.88) that if

$$\omega\tau \gg 1, \tag{13.89}$$

which is the condition of validity for the adiabatic approximation, the transition probability is very small and hence the system would remain in the ground state.

We can make a connection between the time independent perturbation theory and the time dependent one through the adiabatic approximation in the following way. Let us assume that a system, whose Hamiltonian is H_0 at $t = -\infty$, is subjected to a perturbation, which changes its value continuously to $H_0 + H_1$ at $t = 0$. In this case, we can represent

$$H(t) = H_0 + e^{\frac{t}{\tau}} H_1, \qquad -\infty \leq t \leq 0, \quad \tau > 0. \tag{13.90}$$

If τ is large, then, the adiabatic theorem tells us that the eigenstate $|n_0\rangle$ of H_0 would go over into the eigenstate $|n\rangle$ of H at $t = 0$. Thus if we calculate $|n\rangle$ to a given order in the time dependent scheme, and let $\tau \to \infty$, we should get back the time independent result. Let us check this to first order in the perturbation. We know that we can write

$$\begin{aligned}
a_m^{(1)}(0) &= \frac{1}{i\hbar} \int_{-\infty}^{0} \mathrm{d}t \; e^{i\omega_{mn}t} e^{\frac{t}{\tau}} \langle m_0|H_1|n_0\rangle \\
&= \frac{\langle m_0|H_1|n_0\rangle}{i\hbar} \int_{-\infty}^{0} \mathrm{d}t \; e^{t(\frac{1}{\tau} + i\omega_{mn})} \\
&= \frac{\langle m_0|H_1|n_0\rangle}{i\hbar} \frac{1}{\frac{1}{\tau} + i\omega_{mn}},
\end{aligned}$$

$$\text{or,} \quad a_m^{(1)}(0) \xrightarrow{\tau\to\infty} -\frac{\langle m_0|H_1|n_0\rangle}{\hbar\omega_{mn}} = \frac{\langle m_0|H_1|n_0\rangle}{E_n^{(0)} - E_m^{(0)}}. \tag{13.91}$$

On the other hand, these are precisely the coefficients of expansion in the time independent perturbation theory in (13.28), so that we obtain

$$\lim_{\tau\to\infty} a_m^{(1)}(0) = \langle m_0|n_1\rangle. \tag{13.92}$$

13.6 Sudden changes

The other kind of time dependent perturbation that we come across is the other extreme. Here, the Hamiltonian changes very rapidly if not instantaneously. This happens, for example, when

$$\left|\frac{\frac{\partial H'}{\partial t}}{H'}\right| \gg \omega. \tag{13.93}$$

Clearly, a typical example of a sudden change is the system which initially has a Hamiltonian H_0, and at $t = 0$ it suddenly changes to $H_0 + H_1$. These situations often occur in the laboratory. We may change the direction or magnitude of an applied magnetic field suddenly and so on.

In this case, if the system is in a stationary state before $t = 0$, then, it would be in one of the eigenstates of H_0, say, $|u_n\rangle$. After the perturbation is switched on, however, this would no longer be an eigenstate (stationary state) of $H_0 + H_1$. Thus, one expands the initial state in the eigenbasis $|v_m\rangle$ of $H_0 + H_1$, namely,

$$|\psi\rangle = |u_n\rangle = \sum_m c_{mn}|v_m\rangle, \tag{13.94}$$

where

$$c_{mn} = \langle v_m|u_n\rangle. \tag{13.95}$$

Here, the assumption is that both H_0 and $H_0 + H_1$ are soluble, even if only through perturbation theory. In this case, the time evolution of the state can be written as

$$|\psi(t)\rangle = \sum_m c_{mn}|v_m\rangle e^{-\frac{i}{\hbar}E_m t}, \tag{13.96}$$

where E_m is the total energy eigenvalue of the state $|v_m\rangle$.

► **Example.** A tritium atom (H^3), which has two neutrons and one proton in its nucleus, suddenly changes to a helium nucleus (He^3) of two protons and one neutron through beta decay. The emitted electron has a kinetic energy of 16 KeV. We would like to calculate the probability of the ground state electron of H^3 to remain in the ground state of $He^{3\,+}$. We note that, in this case,

$$\begin{aligned}
&\frac{1}{2}mv^2 = 16\ \text{KeV},\\
\text{or,}\quad &\left(\frac{v}{c}\right)^2 = \frac{2\times 16\ \text{KeV}}{mc^2} = \frac{32\ \text{KeV}}{.5\ \text{MeV}} = 64\times 10^{-3},\\
\text{or,}\quad &\frac{v}{c} \simeq .25.
\end{aligned} \tag{13.97}$$

Therefore, the electron, that is emitted, is extremely relativistic. The ground state energy of tritium and He^{3+}, on the other hand, are

$$E_0^{H^3} = E_0^{H} = -13.6 \text{ eV},$$
$$E_0^{He^{3+}} = 4E_0^{H} = -54.4 \text{ eV}. \tag{13.98}$$

Thus, we see the change in the ground state energies is much smaller than the energy of the outgoing electron or the change in the Hamiltonian and, consequently, we can think of this as a sudden change. The emitted electron, of course, has a very large energy and would come out of the atom right away.

In this case, the initial wave function is given by (both H^3 and He^{3+} are hydrogenic atoms)

$$u_0(x) = \frac{1}{(\pi a_0^3)^{\frac{1}{2}}} e^{-\frac{r}{a_0}}, \tag{13.99}$$

while the ground state wave function for He^{3+} has the form

$$v_0(x) = \left(\frac{8}{\pi a_0^3}\right)^{\frac{1}{2}} e^{-\frac{2r}{a_0}}. \tag{13.100}$$

Therefore, the transition amplitude is determined from (13.95) to be

$$\begin{aligned}
c_{00} &= \frac{(8)^{\frac{1}{2}}}{\pi a_0^3} \int \mathrm{d}^3 r\, e^{-\frac{3r}{a_0}} = \frac{\sqrt{8}}{\pi a_0^3} \times 4\pi \int_0^\infty r^2 \mathrm{d}r\, e^{-\frac{3r}{a_0}} \\
&= \frac{4\sqrt{8}}{a_0^3} \left(\frac{a_0}{3}\right)^3 \int_0^\infty \mathrm{d}y\, y^2 e^{-y} = \frac{4\sqrt{8}}{27}\, \Gamma(3) = \frac{8\sqrt{8}}{27} \\
&= \frac{16\sqrt{2}}{27}.
\end{aligned} \tag{13.101}$$

The probability that the system will be in the ground state of the He^{3+} ion, after the beta decay, therefore, follows to be

$$P = |c_{00}|^2 = \frac{2^9}{3^6} \simeq 0.7. \tag{13.102}$$

This calculation shows that, as a consequence of the beta decay, the ground state electron of tritium has a large probability of remaining in the ground state of He^{3+}. However, this also shows that there is a finite probability for it to end up in an excited state. ◀

13.7 Selected problems

1. A one dimensional oscillator, in its ground state, is subjected to a perturbation of the form

$$H'(t) = cpe^{-\alpha|t|} \cos \omega t, \tag{13.103}$$

where c, α, ω are constants with $\alpha > 0$ and p is the momentum operator. What is the probability that, as $t \to \infty$, the oscillator will be found in its first excited state in first order perturbation theory? Discuss the result as a function of α, ω, ω_c, where ω_c represents the oscillator frequency.

2. Consider the problem done in the last example in this chapter, where a tritium atom (H^3) changes to a Helium (He^{3+}) ion by emitting an electron. The out-coming electron has kinetic energy 16 KeV. Assuming that the electron in H^3 was in its ground state before the β decay, what is the probability that it would end up in the $|n = 16, \ell = 3, m = 0\rangle$ eigenstate of the He^{3+} ion?

3. A particle is in the ground state of the one dimensional potential

$$V(x) = \left.\begin{array}{ll} \infty, & x \leq -a, \\ 0, & -a \leq x \leq a, \\ \infty, & x \geq a. \end{array}\right\} \tag{13.104}$$

A time dependent perturbation of the form

$$H'(t) = T_0 V_0 \sin \frac{\pi x}{a}\, \delta(t), \tag{13.105}$$

where T_0, V_0 are constants, acts on the system. What is the probability that the system will be in the first excited state afterwards?

Chapter 14

Spin

From our study of the angular momentum in chapter 7, we have learned so far the following things.

1. The orbital angular momentum, defined as the operator

$$\mathbf{L} = \mathbf{R} \times \mathbf{P}, \tag{14.1}$$

   and studied within the context of the Schrödinger equation, has only integer eigenvalues (for, say, L_z) in units of $\hbar$. In particular, the eigenvalues of the operator $\mathbf{L}^2$ are given by $\hbar^2 \ell(\ell+1)$, $\ell = 0, 1, 2, \ldots$.

2. However, if we treat angular momentum as an abstract operator satisfying the same commutation relations as the orbital angular momentum, namely,

$$[J_i, J_j] = i\hbar\epsilon_{ijk} J_k, \tag{14.2}$$

   and study its representations algebraically, then, we find that the operator $\mathbf{J}^2$ has eigenvalues $\hbar^2 j(j+1)$ where the quantum number j takes multiples of half integer values. This gives theoretical support for the existence of half integer angular momentum. However, let us emphasize here that whenever angular momentum is related to the actual physical motion of the particle, as we have seen earlier, the eigenvalues become restricted to integers.

On the experimental side, on the other hand, there was increasing evidence that somehow half integer angular momentum must be associated with the electron. The simple experiments to suggest this are as follows.

1. Anomalous Zeeman effect:

   As we know, rotational symmetry leads to degeneracy in the m-quantum numbers or the azimuthal quantum numbers. If we apply a magnetic field along the z-direction, rotational symmetry is broken and the degeneracy in the m-quantum numbers is lifted. The azimuthal quantum number, m, takes $2\ell + 1$ values. Thus, each level, under the action of a magnetic field, splits up into $2\ell + 1$ closely spaced levels. Since the orbital angular momentum takes only integer values, the number of such levels is expected to be odd. This phenomenon is known as the normal Zeeman effect. However, for some atoms, particularly the ones with an odd nuclear charge, the original level splits up into an even number of closely spaced levels. This is known as the anomalous Zeeman effect and cannot be explained by an integer value for the angular momentum.

2. Fine structure:

   As we have already seen in chapter 12, even without any external magnetic field, the levels of hydrogen corresponding to a definite energy show fine structure, which can be explained only if a half integer angular momentum is associated with the electron.

3. Stern-Gerlach experiment:

   The definitive proof for half integer angular momentum comes from the Stern-Gerlach experiment. Here a beam of silver atoms (Ag^{47}, which is paramagnetic and has an odd nuclear charge) is sent through an inhomogeneous magnetic field. Since the silver atom is neutral, it does not experience any Lorentz force. However, the atom has a permanent (magnetic) dipole moment because of the outer electron and that can lead to a magnetic energy of the form

   $$H' = -\boldsymbol{\mu} \cdot \mathbf{B}, \tag{14.3}$$

   where $\boldsymbol{\mu}$ denotes the magnetic moment of the electron. The magnetic moment is related to the angular momentum of the electron and hence the behavior of the atoms under the influence of the magnetic field gives information about the angular momentum of the electron. In fact, the result of the experiment shows that the beam of silver atoms splits into two, leading to the fact that the electron must have a half integer angular momentum associated with it.

In 1925, Uhlenbeck and Goudsmit introduced the idea that, in addition to the orbital angular momentum, the electron possesses an intrinsic spin angular momentum of magnitude $\frac{\hbar}{2}$. The name spin was introduced because, in the early days, the picture was that the electron spins around its axis in addition to its motion around the nucleus. In 1927, Pauli introduced the non-relativistic formalism for spin. We would see later, in connection with the Dirac equation, that a relativistic description of particle motion already contains spin in it. But, let us note here that spin is not a consequence of relativity. Furthermore, it does not have any classical analogue. It is completely an intrinsic quantum property of a particle.

This raises a very interesting question. Namely, in developing quantum mechanics, we emphasized that, given a classical observable, we go over to quantum mechanics by promoting that observable to an operator. The commutation relations of the operator can be obtained from its classical Poisson bracket relations through the quantum correspondence principle. On the other hand, if spin has no classical analogue, it is not clear how we can determine the properties of this operator, in particular, its commutation relations. We bypass this problem in the following way. First of all, we note that the total angular momentum of a particle consists of two parts – one due to its orbital motion and the other due to its spin. Thus, we have

$$\mathbf{J} = \mathbf{L} + \mathbf{S}. \tag{14.4}$$

Furthermore, both $\mathbf{L}$ and $\mathbf{J}$ satisfy commutation relations of an angular momentum operator. We also know that $\mathbf{S}$ is an intrinsic operator, i.e., it does not depend on coordinates and momenta. Therefore, $\mathbf{L}$ and $\mathbf{S}$ commute and it follows that

$$\begin{aligned} [J_i, J_j] &= [L_i, L_j] + [S_i, S_j], \\ \text{or,} \quad i\hbar\epsilon_{ijk}J_k &= i\hbar\epsilon_{ijk}L_k + [S_i, S_j]. \end{aligned} \tag{14.5}$$

As a result, for this relation to be true, we must have

$$[S_i, S_j] = i\hbar\epsilon_{ijk}S_k. \tag{14.6}$$

This shows that $\mathbf{S}$ also has the commutation relations of an angular momentum operator. Therefore, the operator $\mathbf{S}^2$ has eigenvalues $\hbar^2 s(s+1)$ where $s = 0, \frac{1}{2}, 1, \frac{3}{2}, \ldots$. Furthermore, the z-component takes $(2s+1)$ values from $-s\hbar$ to $s\hbar$ in steps of unity. Each particle has associated with it a particular s-value. Thus, unlike the orbital angular momentum, whose value can be changed by applying an external field, the spin of a particle is fixed. Furthermore, let us note that, since the spin is a fixed quantity given by $\hbar^2 s(s+1)$, when $\hbar \to 0$ or in the classical limit, spin vanishes. This should be contrasted with orbital angular momentum where the eigenvalues are $\hbar^2\ell(\ell+1)$ with ℓ taking any integer value. Hence, in the classical limit, it does not necessarily vanish, since we can have $\hbar \to 0$, $\ell \to \infty$ such that $\hbar^2\ell(\ell+1)$ is finite.

It can further be shown that particles with integer spin values obey Bose-Einstein statistics and hence are called bosons. On the other hand, particles with half integer spin values obey Fermi-Dirac statistics and are known as fermions. Let us now specialize to the case of electrons. The experiments showed that the electron has spin $s = \frac{1}{2}$. Thus, the spin operators, for the electron, act on a $(2s+1) = 2$ dimensional vector space. The eigenvalues of S_z are $\pm\frac{\hbar}{2}$ and those of $\mathbf{S}^2$ are $\hbar^2 s(s+1) = \frac{3\hbar^2}{4}$. The two basis states, in this two dimensional space, are

$$|s = \frac{1}{2}, s_z = \frac{1}{2}\rangle, \quad |s = \frac{1}{2}, s_z = -\frac{1}{2}\rangle, \tag{14.7}$$

such that

$$\begin{aligned} S_z|\frac{1}{2}, \frac{1}{2}\rangle &= \frac{\hbar}{2}|\frac{1}{2}, \frac{1}{2}\rangle, \\ S_z|\frac{1}{2}, -\frac{1}{2}\rangle &= -\frac{\hbar}{2}|\frac{1}{2}, -\frac{1}{2}\rangle. \end{aligned} \tag{14.8}$$

Sometimes the basis states in (14.7) are also denoted by $|+\rangle$ and $|-\rangle$ respectively, corresponding to the signature of the z-component of the spin value. The states are normalized so that

$$\begin{aligned} \langle +|+\rangle &= \langle -|-\rangle = 1, \\ \langle +|-\rangle &= \langle -|+\rangle = 0. \end{aligned} \tag{14.9}$$

In a two dimensional matrix representation, we can choose,

$$|+\rangle = \begin{pmatrix} 1 \\ 0 \end{pmatrix}, \quad |-\rangle = \begin{pmatrix} 0 \\ 1 \end{pmatrix}. \tag{14.10}$$

Any vector, in this space, can be defined as a linear combination of these two basis states. Thus, we can write

$$|\psi\rangle = \sum_i c_i|i\rangle = c_+|+\rangle + c_-|-\rangle, \tag{14.11}$$

where c_i's represent the components of the vector, $|\psi\rangle$, in this basis. The spin operators have the following matrix representation in this space,

$$S_x = \frac{\hbar}{2}\begin{pmatrix} 0 & 1 \\ 1 & 0 \end{pmatrix}, \quad S_y = \frac{\hbar}{2}\begin{pmatrix} 0 & -i \\ i & 0 \end{pmatrix},$$

$$S_z = \frac{\hbar}{2}\begin{pmatrix} 1 & 0 \\ 0 & -1 \end{pmatrix}, \quad \mathbf{S}^2 = \frac{3\hbar^2}{4}\begin{pmatrix} 1 & 0 \\ 0 & 1 \end{pmatrix}. \tag{14.12}$$

We see, from (14.12), that we can define the spin operators as

$$\mathbf{S} = \frac{\hbar}{2}\,\boldsymbol{\sigma}, \tag{14.13}$$

where $\boldsymbol{\sigma}$'s are known as the Pauli matrices and satisfy the following properties.

1. The Pauli matrices satisfy the commutation relation,

$$[\sigma_i, \sigma_j] = 2i\epsilon_{ijk}\sigma_k. \tag{14.14}$$

2. Distinct Pauli matrices anti-commute (the curly brackets denote anti-commutators).

$$\{\sigma_i, \sigma_j\} = 0, \quad \text{for } i \neq j, \tag{14.15}$$

which implies that

$$\sigma_i\sigma_j = -\sigma_j\sigma_i, \quad \text{for } i \neq j. \tag{14.16}$$

3. From the two relations in (14.14) and (14.15), it follows that, for $i \neq j$,

$$\sigma_i\sigma_j - \sigma_j\sigma_i = 2i\epsilon_{ijk}\sigma_k,$$

$$\text{or,} \quad \sigma_i\sigma_j = i\epsilon_{ijk}\sigma_k. \tag{14.17}$$

4. The Pauli matrices are traceless.

$$\text{Tr}\ \sigma_i = 0. \tag{14.18}$$

Proof. For $i \neq j$, let us note that

$$\text{Tr}\ (\sigma_i\sigma_j) = \text{Tr}\ (\sigma_j\sigma_i) = -\text{Tr}\ (\sigma_i\sigma_j)\,,$$

$$\text{or,} \quad \text{Tr}\ (\sigma_i\sigma_j) = 0, \quad \text{for } i \neq j. \tag{14.19}$$

With this, it follows, from (14.17), that

$$\mathrm{Tr}\ (\sigma_i\sigma_j) = i\epsilon_{ijk}\ \mathrm{Tr}\ (\sigma_k) = 0, \tag{14.20}$$

which proves that

$$\mathrm{Tr}\ (\sigma_k) = 0.$$

■

5. Let $\hat{\mathbf{n}}$ denote an arbitrary unit vector. Then,

$$(\hat{\mathbf{n}}\cdot\boldsymbol{\sigma})^2 = \mathbb{1}, \tag{14.21}$$

where $\mathbb{1}$ represents the two dimensional identity matrix.

Proof. Let us note that the eigenvalues of S_z are $\pm\frac{\hbar}{2}$ so that, in this space, the operator

$$\left(S_z - \frac{\hbar}{2}\ \mathbb{1}\right)\left(S_z + \frac{\hbar}{2}\ \mathbb{1}\right) = 0. \tag{14.22}$$

But, since the z-direction is arbitrary, we can write (14.22) also as

$$\left(\hat{\mathbf{n}}\cdot\mathbf{S} - \frac{\hbar}{2}\mathbb{1}\right)\left(\hat{\mathbf{n}}\cdot\mathbf{S} + \frac{\hbar}{2}\mathbb{1}\right) = 0,$$

$$\text{or,}\quad (\hat{\mathbf{n}}\cdot\mathbf{S})^2 = \frac{\hbar^2}{4}\mathbb{1},$$

$$\text{or,}\quad \left(\frac{\hbar}{2}\hat{\mathbf{n}}\cdot\boldsymbol{\sigma}\right)^2 = \frac{\hbar^2}{4}\mathbb{1},$$

$$\text{or,}\quad (\hat{\mathbf{n}}\cdot\boldsymbol{\sigma})^2 = \mathbb{1}. \tag{14.23}$$

■

In particular, if we choose $\hat{\mathbf{n}}$ to be along a fixed axis, say i, then, (14.23) leads to

$$(\sigma_i)^2 = \mathbb{1}. \tag{14.24}$$

Thus, we see that the square of each Pauli matrix is the identity matrix.

6. We can combine (14.15) and (14.24) to write

$$\{\sigma_i, \sigma_j\} = 2\delta_{ij}\mathbb{1}, \quad \text{for all}\, i \text{ and } j. \tag{14.25}$$

This, in turn, leads to

$$\text{Tr}\ (\sigma_i \sigma_j) = 2\delta_{ij}, \quad \text{for all } i \text{ and } j. \tag{14.26}$$

7. If **A** and **B** are two non-commuting vector operators which, however, commute with the Pauli matrices, then,

$$\begin{aligned}
(\boldsymbol{\sigma}\cdot\mathbf{A})\,(\boldsymbol{\sigma}\cdot\mathbf{B}) &= (\sigma_i A_i)\,(\sigma_j B_j)\\
&= A_i B_j\,(\sigma_i\sigma_j)\\
&= A_i B_j\left(\frac{1}{2}\{\sigma_i,\sigma_j\} + \frac{1}{2}[\sigma_i,\sigma_j]\right)\\
&= A_i B_j\left(\frac{1}{2}\times 2\delta_{ij}\mathbb{1} + \frac{1}{2}\times 2i\epsilon_{ijk}\sigma_k\right)\\
&= A_i B_i\ \mathbb{1} + i\epsilon_{ijk}\sigma_k A_i B_j\\
&= (\mathbf{A}\cdot\mathbf{B})\ \mathbb{1} + i\boldsymbol{\sigma}\cdot(\mathbf{A}\times\mathbf{B})\,.
\end{aligned} \tag{14.27}$$

8. Since the Hilbert space we are considering is two dimensional, there are four linearly independent matrices acting on it. Let us include the identity matrix along with the other three Pauli matrices, σ_i's, and denote them collectively as

$$\sigma_\alpha, \quad \alpha = 0, 1, 2, 3, \quad \text{with} \quad \sigma_0 = \mathbb{1}. \tag{14.28}$$

Then, it is clear that

$$\text{Tr}\,(\sigma_\alpha\sigma_\beta) = 2\delta_{\alpha\beta}, \qquad \alpha, \beta = 0, 1, 2, 3. \tag{14.29}$$

Furthermore, we can also show that these four matrices are linearly independent. That is,

$$\sum_{\alpha=0}^{3} c_\alpha \sigma_\alpha = 0, \tag{14.30}$$

where c_α's are constants, implies that

$$c_\alpha = 0, \quad \text{for all } \alpha. \tag{14.31}$$

Proof. We note that (14.30) implies that

$$\sum_{\alpha=0}^{3} c_\alpha \sigma_\alpha = 0,$$

$$\text{or,}\quad \sigma_\beta \sum_{\alpha=0}^{3} c_\alpha \sigma_\alpha = 0,$$

$$\text{or,}\quad \text{Tr} \sum_{\alpha=0}^{3} c_\alpha \left(\sigma_\beta \sigma_\alpha\right) = 0,$$

$$\text{or,}\quad \sum_{\alpha=0}^{3} c_\alpha 2\delta_{\beta\alpha} = 0,$$

$$\text{or,}\quad 2c_\beta = 0,$$

$$\text{or,}\quad c_\beta = 0, \quad \text{for all } \beta, \tag{14.32}$$

where we have used (14.29). This proves that the set of 2×2 matrices consisting of the identity matrix and the three Pauli matrices (see (14.28)), constitutes a set of linearly independent operators in this space. Thus, they define a basis for the operators in this space, namely, any operator M, in this space, can be written as

$$M = \sum_{\alpha=0}^{3} m_\alpha \sigma_\alpha, \tag{14.33}$$

where

$$m_\alpha = \frac{1}{2} \text{Tr}\ \sigma_\alpha M. \tag{14.34}$$

■

14.1 Complete Hilbert space for the electron

The spin operators, in the case of the electron, act on a two dimensional Hilbert space, the two basis vectors of which can be defined as in (14.10)

$$|+\rangle = \begin{pmatrix} 1 \\ 0 \end{pmatrix}, \qquad |-\rangle = \begin{pmatrix} 0 \\ 1 \end{pmatrix}. \tag{14.35}$$

Let us denote this space by $\mathcal{E}^{(\text{spin})}$. This space has no coordinate or momentum dependence. We also know that the electron wave function has spatial dependence. Let us denote by $\mathcal{E}^{(\text{space})}$ the Hilbert space on which the operators $\mathbf{X}$ and $\mathbf{P}$ act. The eigenvectors in this space are denoted by $|\mathbf{r}\rangle$. This is an infinite dimensional Hilbert space. The total electron wave function, of course, depends on both position as well as spin. Since the spin angular momentum commutes with both coordinates and momenta, it is clear that the total Hilbert space for the electron must be a direct product of these two spaces. That is,

$$\mathcal{E}^{(\text{total})} = \mathcal{E}^{(\text{space})} \otimes \mathcal{E}^{(\text{spin})}. \tag{14.36}$$

The basis vectors in this larger space have the form

$$|\mathbf{r}, s_z\rangle = |\mathbf{r}\rangle \otimes |s_z\rangle. \tag{14.37}$$

Explicitly, we can write

$$\begin{aligned}
|\mathbf{r}, +\rangle &= |\mathbf{r}\rangle \otimes \begin{pmatrix} 1 \\ 0 \end{pmatrix} = \begin{pmatrix} |\mathbf{r}\rangle \\ 0 \end{pmatrix}, \\
|\mathbf{r}, -\rangle &= |\mathbf{r}\rangle \otimes \begin{pmatrix} 0 \\ 1 \end{pmatrix} = \begin{pmatrix} 0 \\ |\mathbf{r}\rangle \end{pmatrix}.
\end{aligned} \tag{14.38}$$

The complete space is now two-fold infinite dimensional corresponding to the fact that each element of $|\mathbf{r}, +\rangle$ and $|\mathbf{r}, -\rangle$ is infinite dimensional. But, we can, for simplicity, pretend as if the total space is two dimensional with the understanding that each element in this space has infinite dimensions. We, of course, know the two dimensional structure of the spin operators. They act only on the $|s_z\rangle$ space. Similarly, the coordinate operators act only on the $|\mathbf{r}\rangle$ space and, in the $|\mathbf{r}, s_z\rangle$ basis, have the forms

$$\begin{aligned}
\mathbf{X} &\longrightarrow \begin{pmatrix} \mathbf{X} & 0 \\ 0 & \mathbf{X} \end{pmatrix}, \\
\mathbf{P} &\longrightarrow \begin{pmatrix} -i\hbar\frac{\partial}{\partial \mathbf{X}} & 0 \\ 0 & -i\hbar\frac{\partial}{\partial \mathbf{X}} \end{pmatrix}.
\end{aligned} \tag{14.39}$$

The orthogonality condition, in this basis, is given by

$$\langle \mathbf{r}', s_z' | \mathbf{r}, s_z\rangle = \delta_{s_z', s_z}\delta^3\left(\mathbf{r}' - \mathbf{r}\right). \tag{14.40}$$

Similarly, the completeness relation takes the form

$$\sum_{s_z=-\frac{1}{2}}^{\frac{1}{2}} \int \mathrm{d}^3 r\ |\mathbf{r}, s_z\rangle\langle \mathbf{r}, s_z| = \mathbb{1}, \tag{14.41}$$

so that any vector, in this space, can be written as

$$|\psi\rangle = \sum_{s_z} \int \mathrm{d}^3r\ |\mathbf{r}, s_z\rangle\langle\mathbf{r}, s_z|\psi\rangle. \tag{14.42}$$

Let us next define

$$\begin{aligned}\langle\mathbf{r}, +|\psi\rangle &= \psi_+(\mathbf{r}),\\ \langle\mathbf{r}, -|\psi\rangle &= \psi_-(\mathbf{r}).\end{aligned} \tag{14.43}$$

Thus, we see that, to specify the electron state completely, we must specify both these functions. We combine both these functions into a two component object called the (electron) spinor wave function defined as

$$\psi(\mathbf{r}) = \begin{pmatrix} \psi_+(\mathbf{r}) \\ \psi_-(\mathbf{r}) \end{pmatrix}. \tag{14.44}$$

It is clear that, with this, we can write the wave function in (14.42) as

$$\begin{aligned}|\psi\rangle &= \sum_{s_z} \int \mathrm{d}^3r\ |\mathbf{r}, s_z\rangle\langle\mathbf{r}, s_z|\psi\rangle\\ &= \int \mathrm{d}^3r\,(|\mathbf{r}, +\rangle\psi_+(\mathbf{r}) + |\mathbf{r}, -\rangle\psi_-(\mathbf{r}))\\ &= \int \mathrm{d}^3r \left(|\mathbf{r}\rangle \otimes \begin{pmatrix} \psi_+(\mathbf{r}) \\ 0 \end{pmatrix} + |\mathbf{r}\rangle \otimes \begin{pmatrix} 0 \\ \psi_-(\mathbf{r}) \end{pmatrix}\right)\\ &= \int \mathrm{d}^3r\ |\mathbf{r}\rangle \otimes \begin{pmatrix} \psi_+(\mathbf{r}) \\ \psi_-(\mathbf{r}) \end{pmatrix} = \int \mathrm{d}^3r\ |\mathbf{r}\rangle \otimes \psi(\mathbf{r}),\end{aligned} \tag{14.45}$$

where $\langle\mathbf{r}|\psi\rangle = \psi(\mathbf{r})$.

Furthermore, it is clear that the normalization condition for the wave function now becomes

$$\begin{aligned}\langle\psi|\psi\rangle &= \int \mathrm{d}^3r\mathrm{d}^3r'\ \psi^\dagger(\mathbf{r}') \otimes \langle\mathbf{r}'|\mathbf{r}\rangle \otimes \psi(\mathbf{r})\\ &= \int \mathrm{d}^3r\mathrm{d}^3r'\ \delta^3\left(\mathbf{r}' - \mathbf{r}\right) \psi^\dagger\left(\mathbf{r}'\right) \psi(\mathbf{r})\\ &= \int \mathrm{d}^3r\ \psi^\dagger\left(\mathbf{r}\right) \psi(\mathbf{r})\\ &= \int \mathrm{d}^3r\ \left(\psi_+^*\left(\mathbf{r}\right) \psi_+(\mathbf{r}) + \psi_-^*\left(\mathbf{r}\right) \psi_-\left(\mathbf{r}\right)\right) = 1.\end{aligned} \tag{14.46}$$

We see that the normalization involves both the components of the wave function.

The wave functions

$$\psi_1(\mathbf{r}) = \begin{pmatrix} \psi_+(\mathbf{r}) \\ 0 \end{pmatrix}, \qquad \psi_2(\mathbf{r}) = \begin{pmatrix} 0 \\ \psi_-(\mathbf{r}) \end{pmatrix}, \tag{14.47}$$

can be easily seen to be eigenstates of σ_z with eigenvalues 1 and -1 respectively. Thus, these wave functions correspond to an electron with spin $\frac{\hbar}{2}$ or $-\frac{\hbar}{2}$ along the z-direction (recall that $S_z = \frac{\hbar}{2}\sigma_z$). A general wave function of the electron, however, is a linear combination of these two states with definite spin along the z-direction. It follows that

$$\psi_+^*(\mathbf{r})\,\psi_+(\mathbf{r})\,\mathrm{d}^3r = |\psi_+(\mathbf{r})|^2\mathrm{d}^3r, \tag{14.48}$$

measures the probability for finding the electron in the volume d^3r with spin up (or along the z-direction). Similarly,

$$\psi_-^*(\mathbf{r})\psi_-(\mathbf{r})\mathrm{d}^3r = |\psi_-(\mathbf{r})|^2\mathrm{d}^3r, \tag{14.49}$$

measures the probability that the electron is in the volume d^3r with its spin anti-parallel to the z-direction. It is worth noting here that the introduction of spin has led to a multi-component wave function for the particle. In general, a particle with spin s has a $(2s+1)$ component wave function.

14.2 Identical particles

Two particles are said to be identical if they cannot be distinguished by any physical property. Let us examine a classical phenomenon involving two identical particles.

For example, let us take two billiard balls (see Fig. 14.1) located at 1 and 2. Suppose at $t = 0$, two players hit the balls at 1 and 2 such that at a later time the two balls go into the holes numbered 3 and 4. For someone who is not observing the play, both of the following possibilities are equally likely,

1. Ball 1 goes into the hole 3, while ball 2 goes into the hole 4.
2. Ball 1 goes into the hole 4, while ball 2 goes into the hole 3.

Namely, by definition the two balls are identical and, therefore, we cannot distinguish one process from the other.

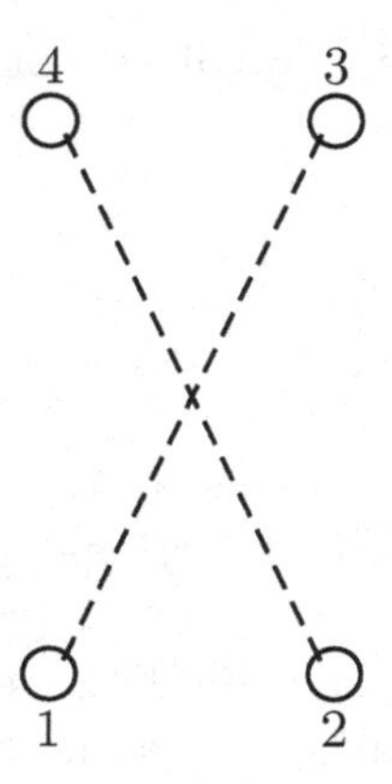

Figure 14.1: Scattering of two identical billiard balls.

In classical mechanics, of course, we can distinguish between the two possibilities. This is simply because we can follow the trajectory of each ball exactly and hence can determine the particular process that takes place. In quantum mechanics, on the other hand, the notion of a trajectory does not exist. In fact, there are uncertainties associated with measurements. Therefore, when we are dealing with identical particles in quantum mechanics, more care is needed to specify the state vectors.

First of all, let us consider two distinguishable particles in quantum mechanics. Let the state

$$|\psi\rangle = |a,b\rangle = |a\rangle \otimes |b\rangle, \tag{14.50}$$

describe the system. That is, this state describes particle 1 at $x = a$ and particle 2 at $x = b$. The state which describes particle 1 at $x = b$ and particle 2 at $x = a$ is then given by

$$|\psi'\rangle = |b,a\rangle = |b\rangle \otimes |a\rangle. \tag{14.51}$$

Since the particles are distinguishable, the states $|\psi\rangle$ and $|\psi'\rangle$ are distinct.

Let us now consider two identical particles. We cannot make very definitive statements anymore. But, we can ask about the state which describes one particle at $x = a$ and the other at $x = b$. Notice that we cannot talk about particle 1 being at $x = a$ and particle 2 being at $x = b$ any more. That is because we cannot distinguish between them. There are, of course, two states $|a,b\rangle$ and $|b,a\rangle$ which describe one particle being at $x = a$ and the other being at $x = b$. Let us assume that the actual state which describes the system is a linear combination of the above two states, i.e.,

$$|\psi(a,b)\rangle = \alpha|a,b\rangle + \beta|b,a\rangle, \tag{14.52}$$

where α and β are constants to be determined.

We note that, since the particles are indistinguishable, exchanging the two of them must still describe the same state. Namely, we must have

$$|\psi(b,a)\rangle = \lambda|\psi(a,b)\rangle = \lambda(\alpha|a,b\rangle + \beta|b,a\rangle), \tag{14.53}$$

where λ is a constant. As a result, we see that under two successive exchanges, we have

$$\begin{aligned} |\psi(a,b)\rangle &\longrightarrow |\psi(b,a)\rangle = \lambda|\psi(a,b)\rangle \\ &\longrightarrow \lambda|\psi(b,a)\rangle = \lambda^2|\psi(a,b)\rangle. \end{aligned} \tag{14.54}$$

Therefore, we have

$$\begin{aligned} &|\psi(a,b)\rangle = \lambda^2|\psi(a,b)\rangle \\ \text{or,}\quad &\lambda^2 = 1, \quad \text{or} \quad \lambda = \pm 1. \end{aligned} \tag{14.55}$$

In turn, this implies, from (14.52) and (14.53), that

$$\beta = \pm\alpha. \tag{14.56}$$

This, therefore, tells us that a quantum mechanical system consisting of two identical particles must be described by a state that is either symmetric or anti-symmetric under the exchange of the two particles. That is, we have two classes of state vectors when dealing with identical particles in quantum mechanics, namely,

$$\begin{aligned} |a,b\rangle_S &= |a,b\rangle + |b,a\rangle, \\ |a,b\rangle_A &= |a,b\rangle - |b,a\rangle. \end{aligned} \tag{14.57}$$

where these states are not normalized. However, this still does not tell us when a symmetric state is to be used and when an anti-symmetric state is preferred.

Postulate. States describing identical bosons are given by symmetric vectors (wave functions) whereas states describing identical fermions are given by anti-symmetric vectors (wave functions).

Let us emphasize here that the symmetry or the anti-symmetry of a quantum mechanical state only refers to the total wave function. Namely, as we have seen, a wave function can be a product of functions which depend on space, spin as well as other quantum numbers that we may not have studied so far. It is only when all these quantum numbers including space and spin are exchanged that the wave function has to be symmetric or anti-symmetric. A wave function does not have to be symmetric or anti-symmetric in each of its quantum numbers. As an example, let us consider a state that depends only on space and spin and is anti-symmetric.

$$|\psi\rangle_A = |\text{space}\rangle \otimes |\text{spin}\rangle. \tag{14.58}$$

This immediately tells us that such a state is possible only if: (a) the space part is anti-symmetric and the spin part is symmetric, (b) the space part is symmetric and the spin part is anti-symmetric under an exchange of particles. But the space and the spin parts do not have to be separately anti-symmetric. In fact, if both the space as well as the spin parts were anti-symmetric under exchange, then, the total wave function would be symmetric and, therefore, would be bosonic. Let us also note here that, in the case of a system containing a large number of identical particles, the wave function has to be symmetric or anti-symmetric under the exchange of any pair of particles, depending on whether they are bosons or fermions.

Let us now consider a state describing two identical fermions. Let us denote by ω_i all the quantum numbers that the particles can have. Thus, the state containing two identical fermions is given by

$$|\omega_1, \omega_2\rangle_A = |\omega_1, \omega_2\rangle - |\omega_2, \omega_1\rangle. \tag{14.59}$$

If we now set $\omega_1 = \omega_2 = \omega$, then we obtain

$$|\omega, \omega\rangle_A = |\omega, \omega\rangle - |\omega, \omega\rangle = 0. \tag{14.60}$$

This is the famous Pauli exclusion principle which says that two identical fermions cannot have the same quantum numbers (Two identical fermions cannot be in the same quantum state).

Normalization of states. Let us next study the question of normalization of such states. First, if we have a symmetric state of two identical particles, we can write

$$\begin{aligned}|a,b\rangle_S &= \alpha\left(|a,b\rangle + |b,a\rangle\right)\\ &= \alpha\left(|a\rangle\otimes|b\rangle + |b\rangle\otimes|a\rangle\right).\end{aligned} \tag{14.61}$$

In this case, normalization would require

$${}_S\langle a,b|a,b\rangle_S = 1 = 2\alpha\alpha^*, \tag{14.62}$$

so that we can choose

$$\alpha = \alpha^* = \frac{1}{\sqrt{2}}. \tag{14.63}$$

Thus, we determine the normalized symmetric state to be

$$|a,b\rangle_S = \frac{1}{\sqrt{2}}\left(|a,b\rangle + |b,a\rangle\right). \tag{14.64}$$

Similarly, for an anti-symmetric state, we have

$$\begin{aligned}|a,b\rangle_A &= \beta(|a,b\rangle - |b,a\rangle)\\ &= \beta(|a\rangle\otimes|b\rangle - |b\rangle\otimes|a\rangle),\end{aligned} \tag{14.65}$$

and normalization gives

$${}_A\langle a,b|a,b\rangle_A = 1 = 2\beta\beta^*, \tag{14.66}$$

so that we can choose

$$\beta = \frac{1}{\sqrt{2}} = \beta^*. \tag{14.67}$$

The normalized anti-symmetric state, therefore, has the form

$$|a,b\rangle_A = \frac{1}{\sqrt{2}}\left(|a,b\rangle - |b,a\rangle\right). \tag{14.68}$$

14.3 General discussion of groups

We had discussed earlier, very briefly, the concept of a group in connection with the properties of translation (see chapter 6). Let us study this a bit more in some detail.

A group G consists of a set of elements with a definite combination (multiplication) rule such that

1. if $g_1, g_2 \in G$, then, $g_1 g_2 = g_3 \in G$. That is, a group is closed under multiplication. (The real numbers 0, 1, 2 ... 10, for example, do not form a group because $5 + 6 = 11$ is not in the set. In this case, combination is simply through addition. On the other hand, the set of all real numbers does form a group.)

2. if $g_1, g_2, g_3 \in G$, then,

$$g_1(g_2 g_3) = (g_1 g_2) g_3.$$

   That is, the multiplication of group elements is associative.

3. There exists an identity element $\mathbb{1} \in G$ such that

$$\mathbb{1} g_i = g_i = g_i \mathbb{1}.$$

   (In case of the example of the real numbers above, 0 is the identity element.)

4. For every element $g \in G$, there exists an inverse element $g^{-1} \in G$ such that

$$g g^{-1} = \mathbb{1} = g^{-1} g.$$

   (In case of the example of the reals above, inverses correspond to negative numbers which are contained in the set.)

The group multiplication (composition) is not necessarily commutative. When it is, the group is called Abelian, otherwise, it is known as a non-Abelian group. If the number of elements in a group is finite, then the group is called a finite group. However, if there are an infinite number of discrete elements, then the group is known as an infinite discrete group. If, however, the elements form a continuum (in a topological sense), then the group is said to be a continuous group (by definition, they are infinite groups). Lie groups are continuous groups where each element possesses derivatives (in the continuous parameters of the group) up to all orders.

It is clear that if we take all the $n \times n$ square matrices with non-vanishing determinants, then they form a group, i.e.,

1. All $n \times n$ matrices close under multiplication.

2. Matrix multiplication is associative.

3. The identity element is the n dimensional identity matrix

$$\mathbb{1} = \begin{pmatrix} 1 & & & & \\ & 1 & & & \\ & & 1 & & \\ & & & \ddots & \\ & & & & 1 \end{pmatrix}.$$

4. Since $\det g \neq 0$ (g's are $n \times n$ matrices), the inverse, g^{-1}, exists such that

$$g^{-1}g = \mathbb{1} = gg^{-1}.$$

A matrix, in addition, can contain functions of parameters as its elements. The functions may be continuous functions of the parameters in which case the group is a continuous group. Furthermore, if the parameters take values over a bounded range, then the group is called compact. It is known as non-compact otherwise.

These groups can be classified into various categories. The most general is the $n \times n$ complex matrices with non-vanishing determinant. This group is known as $GL(n, C)$ or the complex general linear group in n dimensions. It has (n^2 complex) $2n^2$ parameters. If we restrict all the elements to be real, then the group is called $GL(n, R)$ and has n^2 parameters. It is obvious that

$$GL(n, R) \subset GL(n, C). \tag{14.69}$$

Let us note here that $GL(4, R)$ corresponds to the group of coordinate transformations in Einstein's theory.

If one takes all the matrices of $GL(n, C)$ and demands that the determinant of the matrices be unity, then the subset of the matrices form a group called $SL(n, C)$ (special linear). These matrices have $2\left(n^2 - 1\right)$ (real) parameters. If one is only dealing with real matrices with determinant unity, then they form a group known as $SL(n, R)$ which has $\left(n^2 - 1\right)$ parameters. It is clear that

$$SL(n, R) \subset SL(n, C) \subset GL(n, C). \tag{14.70}$$

We note here that $SL(2, C)$ is known as the (homogeneous) Lorentz group and describes spinors naturally. On the other hand, we note that

$$\begin{aligned} &SL(n, R) \subset GL(n, R), \\ &SL(n, C) \not\subset GL(n, R), \end{aligned} \tag{14.71}$$

which implies that Einstein's theory cannot incorporate spinors directly.

All $n \times n$ matrices that are unitary, i.e., those satisfying

$$g^\dagger g = \mathbb{1} = gg^\dagger, \tag{14.72}$$

form the unitary group $U(n)$ and depend on n^2 parameters. Furthermore, the unitary matrices leave the form $\sum_i z_i^* z_i$ unchanged. That is, if

$$z \to z' = gz, \tag{14.73}$$

where g belongs to $U(n)$, then,

$$z'^\dagger z' = z^\dagger g^\dagger g z = z^\dagger \mathbb{1} z = z^\dagger z. \tag{14.74}$$

It is clear that

$$U(n) \subset GL(n, C). \tag{14.75}$$

If we choose, from all the $n \times n$ unitary matrices, only those whose determinant is 1, then they form a group called $SU(n)$, the special unitary group. It has $n^2 - 1$ independent parameters. Clearly,

$$SU(n) = U(n) \cap SL(n, C). \tag{14.76}$$

All $n \times n$ complex orthogonal matrices also form a group called $O(n, C)$. They satisfy the condition

$$g^T g = \mathbb{1} = gg^T, \quad g \in O(n, C), \tag{14.77}$$

and leave invariant the complex quadratic form $\sum_i z_i^2$. It is clear, from (14.77), that

$$\det g = \pm 1. \tag{14.78}$$

Thus, the group $O(n, C)$ decomposes into two disconnected pieces, since we cannot go continuously from one to the other. The matrices with determinant 1 form the special complex orthogonal group denoted by $SO(n, C)$ and it depends on $n(n-1)$ parameters. It is also clear that

$$SO(n, C) = O(n, C) \cap SL(n, C). \tag{14.79}$$

If the set of orthogonal matrices with determinant 1 is real, then they form the special real orthogonal group denoted by $SO(n, R)$. These depend on $\frac{1}{2}n(n-1)$ parameters, leave invariant the real quadratic form $\sum_i x_i^2$ and

$$SO(n, R) = O(n, R) \cap SL(n, R). \tag{14.80}$$

We have already seen, in chapter 7, that rotations in three dimensions are generated by orthogonal matrices whose determinant is equal to 1. By our classification we see, therefore, that $SO(3, R)$ (or, simply $SO(3)$) is the group of proper rotations in 3 dimensional space. Furthermore, we also showed from the form of the infinitesimal rotations that the generators of rotation are the orbital angular momentum operators, which satisfy the commutation relations (Let us set $\hbar = 1$ for simplicity.)

$$[L_i, L_j] = i\epsilon_{ijk}L_k, \qquad i, j, k = 1, 2, 3. \tag{14.81}$$

This is known as the Lie algebra of $SO(3)$ group, where L_i's denote the generator of the group. The algebra is closed in the sense that the commutator of two generators is again a generator. The constants, ϵ_{ijk}, are known as the structure constants of the group. We also know that there exists only one Casimir operator $\mathbf{L}^2$ for this algebra. The number of Casimir operators gives the rank of the group. Thus, we say that $SO(3)$ is a group of rank 1. Let us further note that the Pauli matrices also satisfy the same algebra as the angular momentum operators. They are Hermitian. Thus, we can define a set of 2×2 complex matrices as

$$g = e^{i\boldsymbol{\theta}\cdot\boldsymbol{\sigma}}. \tag{14.82}$$

These would be unitary since $\sigma_i^\dagger = \sigma_i$. Furthermore, since the Pauli matrices are traceless, $\det g = 1$ and the set of matrices $\{g\}$ defines the group $SU(2)$. The Pauli matrices which are three in number are, therefore, the generators of this group and satisfy the same Lie algebra as that of the $SO(3)$ group. When the generators of two different groups satisfy the same algebra, they are said to be locally isomorphic to each other. Thus, we have shown that

$$\mathrm{SO}(3) \simeq \mathrm{SU}(2), \qquad \text{(locally isomorphic)}. \tag{14.83}$$

Local isomorphism simply implies that the two groups are equivalent near the identity element or for infinitesimal transformations. Finite transformations for the two groups, however, may be very different. The other way of saying the same thing is that the global structure of the two groups may be very different.

To note the difference between the two groups, let us consider the following. Let us denote by $\{R\}$ the set of all $SO(3)$ matrices which rotate the 3 dimensional space on to itself. These matrices preserve the length, namely,

$$x' = Rx, \quad \Rightarrow x_1'^2 + x_2'^2 + x_3'^2 = x_1^2 + x_2^2 + x_3^2. \tag{14.84}$$

Furthermore, let us consider the 2 dimensional complex space where length is defined to be

$$|\lambda_1|^2 + |\lambda_2|^2 = \lambda_1^*\lambda_1 + \lambda_2^*\lambda_2 = \lambda_\alpha^*\lambda_\alpha = \lambda^\dagger\lambda. \tag{14.85}$$

Let U denote all the 2×2 unitary matrices with determinant equal to 1 so that they belong to SU(2) and have the form

$$U = \begin{pmatrix} \alpha & \beta \\ -\beta^* & \alpha^* \end{pmatrix}, \qquad |\alpha|^2 + |\beta|^2 = 1. \tag{14.86}$$

These matrices, U, leave the length of a vector invariant in this space. Furthermore, let us consider the Hermitian, traceless matrix (Cayley-Klein parameterization)

$$x = \begin{pmatrix} x_3 & x_1 + ix_2 \\ x_1 - ix_2 & -x_3 \end{pmatrix}. \tag{14.87}$$

Under the action of the matrices U, this changes to x' as

$$x' = \begin{pmatrix} x_3' & x_1' + ix_2' \\ x_1' - ix_2' & -x_3' \end{pmatrix} = x' = UxU^{-1}. \tag{14.88}$$

It follows now that

$$\begin{aligned} \det x' &= \det x, \\ \text{or,} \quad -\left(x_1'^2 + x_2'^2 + x_3'^2\right) &= -\left(x_1^2 + x_2^2 + x_3^2\right). \end{aligned} \tag{14.89}$$

Thus, we can think of the matrices U as rotating the 3 dimensional space such that the length of the vector is invariant. Therefore, we can think of $SU(2)$ rotations as equivalent to $SO(3)$ rotations. There is one difference though. We see, from (14.88), that the matrices U and $-U$ give the same rotation. Therefore, we see that each $SO(3)$ rotation corresponds to two distinct $SU(2)$ rotations. This is also known as the doubly connectedness of the $SO(3)$ group. This can be seen in a different way in the parameter space. An $SO(3)$ rotation is an angle ϕ about some axis with magnitude $0 \leq |\phi| \leq \pi$.

This can be represented as a sphere of radius π. Furthermore, since a rotation about an axis by π is the same as the rotation by $-\pi$, opposite points on the surface of the sphere are identified. Any series of rotations can be represented as a curve in this sphere. If all possible curves representing the same rotation can be reduced to one single curve, then, the space is said to be simply connected. However, in the case of $SO(3)$ rotations, we see that there are two distinct paths for a rotation. Thus, $SO(3)$ is doubly connected.

The parameter space for $SU(2)$, on the other hand, is again a sphere with radius π. However, in this case all points on the surface are identified. This is seen from the fact that

$$U(\boldsymbol{\theta}) = e^{i\boldsymbol{\theta}\cdot\boldsymbol{\sigma}} = \cos\theta + i\hat{\boldsymbol{\theta}}\cdot\boldsymbol{\sigma}\sin\theta, \tag{14.90}$$

where $\theta = |\boldsymbol{\theta}|$. It is clear now that

$$U(\boldsymbol{\theta} = \pi\hat{\boldsymbol{\theta}}) = -1, \tag{14.91}$$

irrespective of the direction of the rotation. As a result, all points on the surface of a sphere can be identified. Hence it is easy to see that each curve in this space can be reduced to just one path. Therefore, $SU(2)$ is simply connected and is known as the covering group of $SO(3)$.

14.4 Addition of angular momentum

Suppose we have a system consisting of two particles. With each particle is associated an angular momentum operator, which we denote by $\mathbf{J}_1$ and $\mathbf{J}_2$ respectively. The commutation relations satisfied by these operators can be written in the compact form

$$\begin{aligned}
&\mathbf{J}_1 \times \mathbf{J}_1 = i\hbar\mathbf{J}_1,\\
&\mathbf{J}_2 \times \mathbf{J}_2 = i\hbar\mathbf{J}_2,\\
&[\mathbf{J}_1, \mathbf{J}_2] = 0.
\end{aligned} \tag{14.92}$$

With each angular momentum is also associated a Casimir operator. Thus, we have two Casimir operators $\mathbf{J}_1^2$ and $\mathbf{J}_2^2$, which commute with all the generators of the algebra. Therefore, we can label the representations of the enlarged algebra by the eigenvalues of these Casimir operators. Let us assume that the values of the angular momenta for the two particles are j_1 and j_2 respectively. Thus,

$$\begin{aligned}
&\mathbf{J}_1^2: \quad \hbar^2 j_1(j_1+1),\\
&\mathbf{J}_2^2: \quad \hbar^2 j_2(j_2+1).
\end{aligned} \tag{14.93}$$

Furthermore, the z-components of the angular momenta can take the range of values

$$\begin{array}{lll} J_{1z}: & \hbar m_1, & j_1 \geq m_1 \geq -j_1, \\ J_{2z}: & \hbar m_2, & j_2 \geq m_2 \geq -j_2. \end{array} \tag{14.94}$$

Let us denote by $\mathcal{E}^{(j_1)}$ the space on which the operators $\mathbf{J}_1$ act. Similarly, $\mathcal{E}^{(j_2)}$ is the space on which the operators $\mathbf{J}_2$ act. Since $\mathbf{J}_1$ and $\mathbf{J}_2$ commute, the total space is a direct product of the two spaces.

$$\mathcal{E} = \mathcal{E}^{(j_1)} \otimes \mathcal{E}^{(j_2)},$$

$$\text{or,} \quad |j_1, m_1; j_2, m_2\rangle = |j_1, m_1\rangle \otimes |j_2, m_2\rangle. \tag{14.95}$$

We can, of course, define the total angular momentum operator for the system as the sum of the individual angular momentum operators, namely,

$$\mathbf{J} = \mathbf{J}_1 + \mathbf{J}_2. \tag{14.96}$$

It is clear that

$$\begin{aligned} &[J_i, J_j] = i\hbar\epsilon_{ijk} J_k, \\ &\left[\mathbf{J}^2, J_i\right] = 0, \\ &\left[\mathbf{J}^2, \mathbf{J}_1^2\right] = 0 = \left[\mathbf{J}^2, \mathbf{J}_2^2\right]. \end{aligned} \tag{14.97}$$

However, it can be checked easily that

$$\begin{aligned} &\left[\mathbf{J}^2, J_{1i}\right] \neq 0, \\ &\left[\mathbf{J}^2, J_{2i}\right] \neq 0. \end{aligned} \tag{14.98}$$

Thus, it is clear that rather than working in the basis in which the operators $\mathbf{J}_1^2, \mathbf{J}_2^2, J_{1z}$ and J_{2z} are diagonal, we can equivalently also work in the basis in which the operators $\mathbf{J}^2$, $J_z, \mathbf{J}_1^2, \mathbf{J}_2^2$ are diagonal. Let us denote the states in this basis by

$$|j, m; j_1, j_2\rangle. \tag{14.99}$$

The question that we would like to address is the possible values of the quantum numbers j and m, given a j_1 and j_2.

First of all, let us note that the space spanned by $|j_1, m_1\rangle$ is $(2j_1 + 1)$ dimensional, since m_1 takes $2j_1 + 1$ values. Similarly, the space spanned by $|j_2, m_2\rangle$ is $(2j_2 + 1)$ dimensional. Thus, the total space, which is a direct product of the two, must have dimension $(2j_1+1)(2j_2+1)$. Furthermore, let us observe that

$$\begin{aligned}
&J_z|j_1,m_1;j_2,m_2\rangle\\
&\quad=(J_{1z}+J_{2z})\,|j_1,m_1;j_2,m_2\rangle\\
&\quad=J_{1z}|j_1,m_1\rangle\otimes|j_2,m_2\rangle+|j_1,m_1\rangle\otimes J_{2z}|j_2,m_2\rangle\\
&\quad=\hbar(m_1+m_2)|j_1,m_1;j_2,m_2\rangle.
\end{aligned}\tag{14.100}$$

Therefore, the eigenvalues of J_z, namely $\hbar m$, are such that

$$m=m_1+m_2. \tag{14.101}$$

But, we know that

$$j_1\geq m_1\geq -j_1,\qquad j_2\geq m_2\geq -j_2. \tag{14.102}$$

It follows, then, that the quantum number m can take values

$$(j_1+j_2)\geq m\geq -(j_1+j_2). \tag{14.103}$$

However, it is also clear that a particular m value may be obtained from different values of m_1 and m_2 so that we expect degeneracy of states in this space.

If we assume that the eigenvalues of $\mathbf{J}^2$ are $\hbar^2 j(j+1)$, then, as we have seen from the algebra of angular momentum, m is expected to take values

$$j\geq m\geq -j, \tag{14.104}$$

for every value of j. Furthermore, we have seen in (14.103) that $m_{\max}=j_1+j_2$. Therefore, it follows from (14.104) that

$$j_{\max}=j_1+j_2. \tag{14.105}$$

We note that $j_{\max}$ cannot be the only value of j because then the dimensionality of the space would be $(2j_{\max}+1)=(2j_1+2j_2+1)$. However, for nontrivial j_1,j_2

$$(2j_1+2j_2+1)\neq(2j_1+1)(2j_2+1). \tag{14.106}$$

Thus, j must take other values also. To determine these, let us note that the state with $m=m_{\max}=j_1+j_2$ is unique and can be identified with

$$|j_1+j_2,j_1+j_2;j_1,j_2\rangle=|j_1,j_1\rangle\otimes|j_2,j_2\rangle. \tag{14.107}$$

However, the state with $m = j_1 + j_2 - 1$ is doubly degenerate in the sense that there are two possible states with this quantum number, namely, both

$$|j_1, j_1\rangle \otimes |j_2, j_2 - 1\rangle, \qquad |j_1, j_1 - 1\rangle \otimes |j_2, j_2\rangle, \tag{14.108}$$

give the same value for m. In the $|j, m; j_1, j_2\rangle$ basis, on the other hand, the state

$$|j_1 + j_2, j_1 + j_2 - 1; j_1, j_2\rangle, \tag{14.109}$$

gives one state with $m = j_1 + j_2 - 1$. The second state, therefore, has to be of the form

$$|j_1 + j_2 - 1, j_1 + j_2 - 1; j_1, j_2\rangle, \tag{14.110}$$

so that $j = j_1 + j_2 - 1$ is another allowed value for the total angular momentum quantum number.

We can carry out this analysis further and show that the quantum number j takes the values

$$j: \quad j_1 + j_2, j_1 + j_2 - 1, j_1 + j_2 - 2, \ldots, j_{\text{min}}. \tag{14.111}$$

The minimum value, j_{min}, should be such that the dimensionality of the total space is equal to $(2j_1 + 1)(2j_2 + 1)$. For each j value, the dimensionality of the space is $(2j + 1)$. Therefore, this constraint leads to

$$\begin{aligned}
&\sum_{j_{\text{min}}}^{j_1+j_2} (2j + 1) = (2j_1 + 1)(2j_2 + 1),\\
\text{or,} \quad &(j_1 + j_2)(j_1 + j_2 + 1) - j_{\text{min}}(j_{\text{min}} - 1) + j_1 + j_2 - j_{\text{min}} + 1\\
&\qquad = (2j_1 + 1)(2j_2 + 1).
\end{aligned} \tag{14.112}$$

Simplifying this, we obtain,

$$\begin{aligned}
j_{\text{min}}^2 &= j_1^2 + j_2^2 + 2j_1 j_2 + j_1 + j_2 + j_1 + j_2 + 1\\
&\quad - (4j_1 j_2 + 2j_1 + 2j_2 + 1)\\
&= j_1^2 + j_2^2 - 2j_1 j_2 = (j_1 - j_2)^2,\\
\text{or,} \quad j_{\text{min}} &= |j_1 - j_2|.
\end{aligned} \tag{14.113}$$

Thus, the total angular momentum operator takes values form j_1+j_2 down to $|j_1-j_2|$, decreasing in steps of unity and we have

$$\mathcal{E} = \mathcal{E}^{(j_1)} \otimes \mathcal{E}^{(j_2)} = \sum_{j=|j_1-j_2|}^{j_1+j_2} \oplus \mathcal{E}^{(j)}. \tag{14.114}$$

This simply means that the $|j,m;j_1,j_2\rangle$ basis defines a reducible space and operators take block diagonal form in this basis.

14.5 Clebsch-Gordan coefficients

We see that a system consisting of two angular momentum operators can be equivalently described in terms of two alternate basis, namely,

$$|j_1,m_1;j_2,m_2\rangle, \quad \text{or}, \quad |j,m;j_1,j_2\rangle. \tag{14.115}$$

We also know that two equivalent basis can be related through a unitary transformation. In fact, we can express one basis completely in terms of the other basis, since each defines a complete basis. Thus, we can write

$$\begin{aligned}
&|j,m;j_1,j_2\rangle\\
&\quad = \sum_{m_1,m_2} |j_1,m_1;j_2,m_2\rangle\langle j_1,m_1;j_2,m_2|j,m;j_1,j_2\rangle\\
&\quad = \sum_{m_1,m_2} C(j,j_1,j_2;m,m_1,m_2)|j_1,m_1;j_2,m_2\rangle.
\end{aligned} \tag{14.116}$$

The coefficients of the expansion are called the Clebsch-Gordan-Wigner coefficients and are nontrivial only if $|j_1-j_2| \leq j \leq j_1+j_2$. We note that, since

$$\begin{aligned}
&J_z|j,m;j_1,j_2\rangle\\
&\quad = \sum_{m_1,m_2} C(j,j_1,j_2;m,m_1,m_2)(J_{1z}+J_{2z})|j_1,m_1;j_2,m_2\rangle,
\end{aligned} \tag{14.117}$$

we have

$$\sum_{m_1,m_2} (m-m_1-m_2)C(j,j_1,j_2;m,m_1,m_2)|j_1,m_1;j_2,m_2\rangle = 0. \tag{14.118}$$

This implies that

$$C(j, j_1, j_2; m, m_1, m_2) = 0, \qquad m \neq m_1 + m_2. \tag{14.119}$$

In other words,

$$C(j, j_1, j_2; m, m_1, m_2) = \langle j_1, m_1; j_2, m_2 | j, m; j_1, j_2\rangle = 0, \tag{14.120}$$

if $m \neq m_1 + m_2$ and $j_1 + j_2 \geq j \geq |j_1 - j_2|$ does not hold. Thus, we can effectively write

$$|j, m; j_1, j_2\rangle = \sum_{m_1} C(j, j_1, j_2; m_1, m - m_1)|j_1, m_1; j_2, m - m_1\rangle. \tag{14.121}$$

The normalization condition for the basis states,

$$\langle j', m'; j_1, j_2 | j, m; j_1, j_2\rangle = \delta_{j'j}\delta_{m'm}, \tag{14.122}$$

further leads to the relation,

$$\sum_{m_1', m_1} C^*\left(j', j_1, j_2; m_1', m' - m_1'\right)$$

$$\times\, C(j, j_1, j_2; m_1, m - m_1)\delta_{m_1' m_1}\delta_{m'm} = \delta_{j'j}\delta_{m'm},$$

$$\text{or,}\quad \sum_{m_1} C^*\left(j', j_1, j_2; m_1, m' - m_1\right)$$

$$\times\, C\left(j, j_1, j_2; m_1, m - m_1\right)\delta_{m'm} = \delta_{j'j}\delta_{m'm}. \tag{14.123}$$

This is the orthogonality relation for the Clebsch-Gordan coefficients. Furthermore, the phases of the coefficients are fixed by demanding that

$$\langle j_1, j_1; j_2, j - j_1 | j, j; j_1, j_2\rangle = C\left(j, j_1, j_2; j_1, j - j_1\right), \tag{14.124}$$

is real and positive, which fixes the phases of all the other coefficients. We can also define recursion relations for the Clebsch-Gordan coefficients by applying the raising and lowering operators and, in this way, can determine all of them. In a particular problem, however, it is much simpler to construct them from first principles rather than look for general results.

Even though we have talked about the addition of angular momentum for two distinct particles, the same analysis holds even if we are considering the addition of, say, **L** and **S** for the same particle. Addition of three angular momenta can be done by first combining any two of them and then adding the third one to the resultant of the first two. Furthermore, just as one can expand $|j, m; j_1, j_2\rangle$ states in the $|j_1, m_1; j_2, m_2\rangle$ basis, one can also invert the expansion and write

$$\begin{aligned} |j_1, m_1; j_2, m_2\rangle &= \sum_{m,j} |j, m; j_1, j_2\rangle\langle j, m; j_1, j_2|j_1, m_1; j_2, m_2\rangle \\ &= \sum_{m,j} C^* (j, j_1, j_2; m, m_1, m_2) \, |j, m; j_1, |j_2\rangle. \end{aligned} \quad (14.125)$$

► **Example.** Let us consider the sum of two angular momenta with eigenvalues $\frac{1}{2}$ each and analyze the resulting eigenvalues and eigenstates. In this case, we have

$$j_1 = \frac{1}{2}, \qquad j_2 = \frac{1}{2}, \quad (14.126)$$

and, therefore,

$$m_1 = -\frac{1}{2}, \frac{1}{2}; \qquad m_2 = -\frac{1}{2}, \frac{1}{2}. \quad (14.127)$$

The basis states for each of the angular momentum operators, in this case, is easily determined to be

$$\begin{aligned} |j_1, m_1\rangle : &\qquad |\frac{1}{2}, \frac{1}{2}\rangle, \qquad |\frac{1}{2}, -\frac{1}{2}\rangle, \\ |j_2, m_2\rangle : &\qquad |\frac{1}{2}, \frac{1}{2}\rangle, \qquad |\frac{1}{2}, -\frac{1}{2}\rangle. \end{aligned} \quad (14.128)$$

Since both $j_1 = j_2 = \frac{1}{2}$, we drop these and denote the states by their m quantum numbers only so that

$$|j_1, m_1; j_2, m_2\rangle = |j_1, m_1\rangle \otimes |j_2, m_2\rangle \Rightarrow \; |m_1, m_2\rangle = |m_1\rangle \otimes |m_2\rangle. \quad (14.129)$$

Clearly, there are four independent basis states in the total space which we can denote by $|+,+\rangle, |+,-\rangle, |-,+\rangle, |-,-\rangle$. Here the $\pm$ signs stand for the signatures of the z-components of the angular momenta. We see that

$$\begin{aligned} J_z|+,+\rangle &= (J_{1z} + J_{2z})\,|+,+\rangle \\ &= (J_{1z}|+\rangle) \otimes |+\rangle + |+\rangle \otimes (J_{2z}|+\rangle) \\ &= \hbar|+,+\rangle. \end{aligned} \quad (14.130)$$

Similarly, we can show that

$$\begin{aligned} J_z|+,-\rangle &= 0, \\ J_z|-,+\rangle &= 0, \\ J_z|-,-\rangle &= -\hbar|-,-\rangle, \end{aligned} \quad (14.131)$$

which implies that, in the product basis, we can write

$$J_z \longrightarrow \begin{array}{c} (+,+) \\ (+,-) \\ (-,+) \\ (-,-) \end{array} \begin{pmatrix} \hbar & 0 & 0 & 0 \\ 0 & 0 & 0 & 0 \\ 0 & 0 & 0 & 0 \\ 0 & 0 & 0 & -\hbar \end{pmatrix}. \tag{14.132}$$

This makes it clear that the allowed values for the azimuthal quantum number m are $+1, 0, -1$. This is consistent with our general result in (14.103).

Furthermore, using the result

$$\begin{aligned} \mathbf{J}^2 &= \mathbf{J}_1^2 + \mathbf{J}_2^2 + 2\mathbf{J}_1 \cdot \mathbf{J}_2 \\ &= \mathbf{J}_1^2 + \mathbf{J}_2^2 + 2J_{1z}J_{2z} + J_{1+}J_{2-} + J_{1-}J_{2+}, \end{aligned} \tag{14.133}$$

we can show that, in the product basis,

$$\mathbf{J}^2 \longrightarrow \hbar^2 \begin{pmatrix} 2 & 0 & 0 & 0 \\ 0 & 1 & 1 & 0 \\ 0 & 1 & 1 & 0 \\ 0 & 0 & 0 & 2 \end{pmatrix}. \tag{14.134}$$

Thus, we see that $\mathbf{J}^2$, the total angular momentum squared, is not diagonal in this basis. In fact, even though the $|+,+\rangle$ and $|-,-\rangle$ states are eigenstates with eigenvalue $j = 1$, the $|+,-\rangle$ and $|-,+\rangle$ states are not. We can diagonalize this matrix and show that the states

$$\begin{aligned} &|+,+\rangle, \\ &\tfrac{1}{\sqrt{2}}\left(|+,-\rangle + \ |-,+\rangle\right), \\ &|-,-\rangle, \end{aligned} \tag{14.135}$$

represent the eigenbasis corresponding to $j = 1$ and

$$\frac{1}{\sqrt{2}}\left(|+,-\rangle - |-,+\rangle\right), \tag{14.136}$$

corresponds to the eigenstate with $j = 0$.

This determines that $j = 0, 1$ which agrees with our general result in (14.111). Furthermore, it is clear that we can identify

$$\begin{aligned} |j=1, m=1\rangle &= |+,+\rangle, \\ |j=1, m=0\rangle &= \frac{1}{\sqrt{2}}\left(|+,-\rangle + ||-,+\rangle\right), \\ |j=1, m=-1\rangle &= |-,-\rangle, \\ |j=0, m=0\rangle &= \frac{1}{\sqrt{2}}\left(|+,-\rangle - ||-,+\rangle\right). \end{aligned} \tag{14.137}$$

The states with $j = 1$ are known as the triplet states, whereas the one with $j = 0$ is called the singlet state, following the conventions of atomic spectra. It is clear that the triplet states are symmetric under exchange whereas the singlet state is anti-symmetric. Furthermore, we can write the relation, (14.137), between the two sets of basis states in the matrix form as

$$\begin{pmatrix} |1,1\rangle \\ |1,0\rangle \\ |1,-1\rangle \\ |0,0\rangle \end{pmatrix} = \begin{pmatrix} 1 & 0 & 0 & 0 \\ 0 & \frac{1}{\sqrt{2}} & \frac{1}{\sqrt{2}} & 0 \\ 0 & 0 & 0 & 1 \\ 0 & \frac{1}{\sqrt{2}} & -\frac{1}{\sqrt{2}} & 0 \end{pmatrix} \begin{pmatrix} |+,+\rangle \\ |+,-\rangle \\ |-,+\rangle \\ |-,-\rangle \end{pmatrix}. \tag{14.138}$$

The elements of the matrix connecting the two sets of basis states are the Clebsch-Gordan coefficients for this problem. Symbolically, we can write the composition of the angular momenta, in this case, as

$$\frac{1}{2}\otimes\frac{1}{2}=1\oplus 0. \tag{14.139}$$

◀

14.6 Selected problems

1. a) Show that any 2×2 matrix that commutes with the three Pauli matrices, $\vec{\sigma}$, is a multiple of the identity matrix.

   b) Show that there is no 2×2 matrix that anti-commutes with the three Pauli matrices.

2. In (14.33)-(14.34), we showed that any 2×2 matrix, M, can be written as

$$M=\sum_{\alpha=0}^{3} m_\alpha\sigma_\alpha. \tag{14.140}$$

   Use this to prove that

$$(\boldsymbol{\sigma}\cdot\mathbf{A})(\boldsymbol{\sigma}\cdot\mathbf{B})=(\mathbf{A}\cdot\mathbf{B})+i(\mathbf{A}\times\mathbf{B})\cdot\boldsymbol{\sigma}, \tag{14.141}$$

   where $\mathbf{A}$ and $\mathbf{B}$ are two vector operators which commute with the Pauli matrices but do not commute among themselves.

3. Find the Clebsch-Gordan coefficients for

$$\frac{1}{2}\otimes 1=\frac{3}{2}\oplus\frac{1}{2}. \tag{14.142}$$

4. Two particles with angular momenta $j_1=j_2=1$ form a state with total angular momentum $j=2$ and the z-component of the total angular momentum $m=1$. Construct such a state as a linear combination of the states $|j_1,m_1;j_2,m_2\rangle$.

5. The ground state electron in a hydrogen atom experiences an interaction because of the magnetic moments of the electron and the proton. This interaction, known as the hyperfine interaction, has the form

$$H'_{\text{hyperfine}} = A\mathbf{S}_1 \cdot \mathbf{S}_2, \qquad A > 0, \tag{14.143}$$

where $\mathbf{S}_1$ and $\mathbf{S}_2$ are the spins of the electron and the proton respectively. Show that this interaction splits the ground state into two levels. Calculate the energy of the two levels to first order in perturbation theory.

6. Let the Hamiltonian describing only the spin of two spin $\frac{1}{2}$ particles be

$$H = H_0 + H', \tag{14.144}$$

where

$$\begin{aligned} H_0 &= \frac{4A}{\hbar^2}\mathbf{S}_1 \cdot \mathbf{S}_2, \\ H' &= \frac{2B}{\hbar}S_{1,z}. \end{aligned} \tag{14.145}$$

Here A, B are constants.

a) Determine the eigenvalues and the eigenstates of H_0.

b) Calculate the lowest order effect of H' on the energy eigenvalues.

CHAPTER 15

Scattering theory

Scattering is a valuable probe to study the structure of particles when we cannot directly see them. For example, it is through scattering experiments that we have learned that the hydrogen atom consists of a nucleus and an electron going around it. We know that electrons are point particles also from the results of scattering experiments. Furthermore, the scattering experiments have told us that protons consist of yet other constituents – the quarks. Therefore, we see that scattering is an essential tool in our understanding of the quantum nature of particles. We will, therefore, spend some time studying this topic. However, let me begin by recapitulating some of the features of scattering theory in classical physics.

Classically, the simplest scattering that we can consider is that of a beam of particles from a fixed center of force as shown in Fig. 15.1. We can think of the fixed center of force to be a charged particle of infinite mass, if the particles that are being scattered are thought of as electrons or protons. Let us assume that a beam of particles is incident on the fixed source of force along the z-axis. The trajectories of the particles change due to the force experienced and the deflection of the trajectory, from the initial direction, is known as the scattering angle.

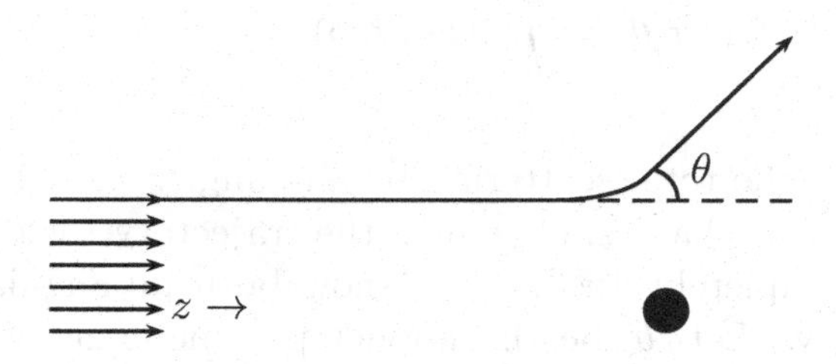

Figure 15.1: Classical scattering of a beam of particles from a scattering source.

Classically, of course, we are not interested in measuring the exact trajectory of each of the particles. In fact, it is impossible if the number of particles involved is very large. We can only measure the initial velocity of the particles (which are assumed to be of the same energy) and their final velocities. Furthermore, this is done statistically. Namely, we measure how many particles scatter into a solid angle $d\Omega$ at θ. Thus, if N is the number of incident particles per unit area per unit time and if n of them scatter by an angle θ into the solid angle $d\Omega$ per unit time, then we define the differential cross section for scattering as

$$\sigma(\theta, \phi) = \frac{n}{N}. \tag{15.1}$$

Note that the cross section has the dimension of an area. The unit in which the scattering cross section is measured in quantum experiments (i.e., nuclear physics, particle physics, etc.) is called a barn, where

$$1 \text{ barn } = 10^{-24} \text{ cm}^2. \tag{15.2}$$

Furthermore, if the interaction is rotationally invariant, then the scattering cross section will be independent of the azimuthal angle and we can define an alternate cross section by integrating out the azimuthal angle, namely,

$$\sigma(\theta) = \int_0^{2\pi} \mathrm{d}\phi\, \sigma(\theta, \phi) = 2\pi\sigma(\theta, \phi). \tag{15.3}$$

We can also define a total cross section by integrating over all scattering angles, namely,

$$\sigma_{\text{tot}} = \int_0^{\pi} \sin\theta \mathrm{d}\theta\, \sigma(\theta) = \int \mathrm{d}\Omega\, \sigma(\theta, \phi), \tag{15.4}$$

which measures the total scattering by any angle.

In classical mechanics, of course, the trajectory (or the orbit) of a particle is completely fixed, if we know both its angular momentum and energy. Let 'b' be the impact parameter associated with a particular trajectory shown in Fig. 15.2. Here v_0 represents the initial velocity of the particle at infinity along the z-axis. Because of rotational invariance and the azimuthal symmetry, we can consider

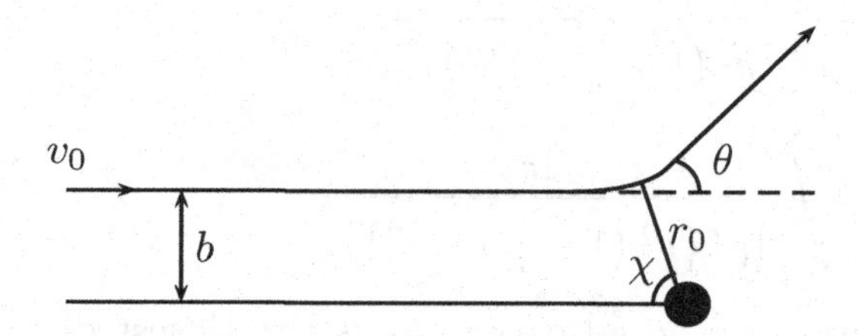

Figure 15.2: Classical trajectory of a single particle in the beam under the influence of a scattering source.

motion to lie on a plane. A point on the trajectory can then be described by the angle it makes with the z-axis. Furthermore, since this angle changes by χ coming from ∞ to r_0 - the point of closest approach to the scattering source - the change in the angle in going from r_0 to ∞ would also be χ so that the scattering angle is given by

$$\theta = \pi - 2\chi. \tag{15.5}$$

Let us note that $0 \leq \theta \leq \pi$, so that $0 \leq \chi \leq \frac{\pi}{2}$.

Let us recall from the definition of the orbital angular momentum that

$$\begin{aligned} \ell &= mv_0 b = b\sqrt{2mE} = mr^2\,\frac{\mathrm{d}\chi}{\mathrm{d}t}, \\ \text{or,}\quad \mathrm{d}\chi &= \frac{\ell \mathrm{d}t}{mr^2}, \end{aligned} \tag{15.6}$$

where we have used $E = \frac{1}{2}mv_0^2$. We also know that

$$\begin{aligned} \frac{\mathrm{d}r}{\mathrm{d}t} &= v(r) = \left[\frac{2}{m}\left(E - V - \frac{\ell^2}{2mr^2}\right)\right]^{\frac{1}{2}}, \\ \text{or,}\quad \mathrm{d}t &= \frac{\mathrm{d}r}{\left[\frac{2}{m}\left(E - V - \frac{\ell^2}{2mr^2}\right)\right]^{\frac{1}{2}}}. \end{aligned} \tag{15.7}$$

As a result, we obtain, from (15.6) and (15.7),

$$\begin{aligned} \mathrm{d}\chi &= \frac{\ell \mathrm{d}t}{mr^2} = \frac{\ell \mathrm{d}r}{mr^2\left[\frac{2E}{mr^2}\left\{r^2(1-\frac{V}{E}) - \frac{\ell^2}{2mE}\right\}\right]^{\frac{1}{2}}} \\ &= \frac{\ell \mathrm{d}r}{\sqrt{2mE}\left[r^2\left\{r^2\left(1-\frac{V}{E}\right) - b^2\right\}\right]^{\frac{1}{2}}} \end{aligned}$$

$$= \frac{b\mathrm{d}r}{\left[r^2\left\{r^2\left(1-\frac{V}{E}\right)-b^2\right\}\right]^{\frac{1}{2}}},$$

$$\text{or,}\quad \chi = b\int_{r_0}^{\infty}\frac{\mathrm{d}r}{\left[r^2\left\{r^2\left(1-\frac{V}{E}\right)-b^2\right\}\right]^{\frac{1}{2}}}. \tag{15.8}$$

Here, as we have noted earlier, r_0 is the closest distance of approach of the trajectory to the scattering source, which is determined from the roots of the equation

$$v(r) = 0. \tag{15.9}$$

(Note that, even if the force is attractive, at short distances, the centrifugal barrier is strong enough to lead to repulsion and, therefore, to a closest distance of approach. For attractive potentials much stronger than the centrifugal barrier, the particle would simply fall into the scattering source.) The scattering angle is, therefore, determined, from (15.5) and (15.8), to be

$$\theta = \pi - 2\chi = \pi - 2b\int_{r_0}^{\infty}\frac{\mathrm{d}r}{\left[r^2\left\{r^2\left(1-\frac{V}{E}\right)-b^2\right\}\right]^{\frac{1}{2}}}. \tag{15.10}$$

From (15.10), it is clear that if we take a beam of mono-energetic particles, particles with a definite impact parameter 'b' would scatter by the same angle, θ. If N is the number of incident particles per unit area per second, then the number of particles with impact parameter between b and $b + \mathrm{d}b$ is given by

$$2\pi N b\mathrm{d}b. \tag{15.11}$$

All these particles would scatter by θ and $\theta - \mathrm{d}\theta$. But from the definition of the cross section in (15.1), we note that the number of such particles is given by

$$N\sigma(\theta,\phi)(-2\pi\sin\theta\mathrm{d}\theta). \tag{15.12}$$

(We note here that particles with a larger impact parameter scatter less, which implies that particles with impact parameters between b and $b + \mathrm{d}b$ would scatter by angles θ and $\theta - \mathrm{d}\theta$ respectively.) Therefore, comparing (15.11) and (15.12), we have

$$2\pi N b\mathrm{d}b = -N\sigma(\theta,\phi)2\pi\sin\theta\mathrm{d}\theta,$$

$$\text{or,}\quad \sigma(\theta,\phi) = \frac{b}{\sin\theta}\left|\frac{\mathrm{d}b}{\mathrm{d}\theta}\right|, \tag{15.13}$$

which determines

$$\sigma(\theta) = 2\pi\sigma(\theta, \phi) = 2\pi \, \frac{b}{\sin\theta} \left|\frac{\mathrm{d}b}{\mathrm{d}\theta}\right| . \tag{15.14}$$

This gives an expression for the differential scattering cross section, once we know the form of θ as a function of the impact parameter (or *vice versa*).

► **Example (3-dimensional hard sphere).** Let us consider scattering from a fixed center of force described by the potential

$$V(r) = \begin{cases} \infty & r < a, \\ 0 & r > a. \end{cases} \tag{15.15}$$

It is clear from (15.15) that if $b > a$, the particle would not experience any force and would pass through without any scattering. Therefore, we conclude that

$$\theta = 0, \qquad \text{if } b > a. \tag{15.16}$$

For $b < a$, on the other hand, the particle cannot approach a distance closer than 'a' classically, since the velocity would then become imaginary. Thus, the closest distance of approach, in this case, is $r_0 = a$ and it follows from (15.10) that

$$\begin{aligned}
\theta &= \pi - 2b \int_a^\infty \frac{\mathrm{d}r}{\left[r^2\left(r^2 - b^2\right)\right]^{\frac{1}{2}}} = \pi - 2b \int_a^\infty \frac{\mathrm{d}r}{r\left(r^2 - b^2\right)^{\frac{1}{2}}} \\
&= \pi - 2b \left[\frac{1}{b}\sec^{-1}\frac{r}{b}\right]_a^\infty = \pi - 2\left[\frac{\pi}{2} - \sec^{-1}\frac{a}{b}\right] \\
&= 2\sec^{-1}\frac{a}{b} = 2\cos^{-1}\frac{b}{a}.
\end{aligned} \tag{15.17}$$

From (15.17), we note that we can write

$$\begin{aligned}
b &= a\cos\frac{\theta}{2}, \\
\frac{\mathrm{d}b}{\mathrm{d}\theta} &= -\frac{a}{2}\sin\frac{\theta}{2}, \\
\frac{b}{\sin\theta} &= \frac{a\cos\frac{\theta}{2}}{\sin\theta} = \frac{a}{2\sin\frac{\theta}{2}}.
\end{aligned} \tag{15.18}$$

Substituting these into (15.14), we obtain

$$\sigma(\theta) = 2\pi \, \frac{b}{\sin\theta}\left|\frac{\mathrm{d}b}{\mathrm{d}\theta}\right| = 2\pi \, \frac{a}{2\sin\frac{\theta}{2}} \, \frac{a}{2}\sin\frac{\theta}{2} = \frac{a^2}{4}\, 2\pi = \frac{\pi a^2}{2}. \tag{15.19}$$

The total scattering cross section is now determined to be

$$\sigma_{\text{tot}} = \int_0^\pi \sin\theta \mathrm{d}\theta \; \sigma(\theta) = \int_0^\pi \sin\theta \mathrm{d}\theta \; \frac{\pi a^2}{2} = \pi a^2. \tag{15.20}$$

Thus, we see that the differential cross section for scattering, in this case, is a constant independent of the angle and the total scattering cross section is equal to the area of a diametrical section of the scattering sphere. ◄

► **Example (Rutherford scattering).** In this case, the force experienced by the scattered particles is given by the Coulomb potential

$$V(r)=\frac{Ze^2}{r}, \tag{15.21}$$

where e denotes the charge of the particles being scattered, while Ze is the charge of the scattering source. The closest distance of approach, in this case, is obtained from the roots of the equation (see (15.7) and (15.9))

$$\begin{aligned}
&E-V-\frac{\ell^2}{2mr^2}=0,\\
\text{or.}\quad &1-\frac{V}{E}-\frac{b^2}{r^2}=0,\\
\text{or,}\quad &r^2-\frac{Ze^2r}{E}-b^2=0,
\end{aligned} \tag{15.22}$$

which determines

$$r=\frac{\frac{Ze^2}{E}\pm\sqrt{\left(\frac{Ze^2}{E}\right)^2+4b^2}}{2}. \tag{15.23}$$

Since the radial vector can only take positive values, we conclude, from (15.23), that

$$r_0=\frac{Ze^2}{2E}\left(1+\sqrt{1+\frac{4b^2E^2}{Z^2e^4}}\right). \tag{15.24}$$

In this case, we obtain, from (15.8), that

$$\chi=b\int_{r_0}^{\infty}\frac{\mathrm{d}r}{\left[r^2\left\{r^2\left(1-\frac{Ze^2}{Er}\right)-b^2\right\}\right]^{\frac{1}{2}}}. \tag{15.25}$$

Let us define $x=\frac{1}{r}$ which implies that $\mathrm{d}x=-\frac{\mathrm{d}r}{r^2}$, or $\mathrm{d}r=-\frac{\mathrm{d}x}{x^2}$ so that we can write (15.25) as

$$\begin{aligned}
\chi&=-b\int_{x_0}^{0}\frac{\mathrm{d}x}{x^2\left[\frac{1}{x^2}\left\{\frac{1}{x^2}\left(1-\frac{Ze^2x}{E}\right)-b^2\right\}\right]^{\frac{1}{2}}}\\
&=-b\int_{x_0}^{0}\frac{\mathrm{d}x}{\left(1-\frac{Ze^2x}{E}-b^2x^2\right)^{\frac{1}{2}}}.
\end{aligned} \tag{15.26}$$

Furthermore, using the standard integral,

$$\int\frac{\mathrm{d}x}{\sqrt{\alpha+\beta x+\gamma x^2}}=\frac{1}{\sqrt{-\gamma}}\cos^{-1}\left(-\frac{\beta+2\gamma x}{\sqrt{\beta^2-4\alpha\gamma}}\right), \tag{15.27}$$

we determine

$$\chi = -b\,\frac{1}{b}\cos^{-1}\left.\frac{\frac{Ze^2}{E}+2b^2x}{\left[\left(\frac{Ze^2}{E}\right)^2+4b^2\right]^{\frac{1}{2}}}\right|_{x_0}^{0} = -\cos^{-1}\left.\frac{1+\frac{2b^2E}{Ze^2}\,x}{\left(1+\frac{4b^2E^2}{Z^2e^4}\right)^{\frac{1}{2}}}\right|_{x_0}^{0}$$

$$= -\cos^{-1}\frac{1}{\left(1+\frac{4b^2E^2}{Z^2e^4}\right)^{\frac{1}{2}}} + \cos^{-1}\frac{1+\frac{2b^2E}{Ze^2}\,\frac{2E}{Ze^2\left[1+\left(1+\frac{4b^2E^2}{Z^2e^4}\right)^{\frac{1}{2}}\right]}}{\left(1+\frac{4b^2E^2}{Z^2e^4}\right)^{\frac{1}{2}}}$$

$$= -\cos^{-1}\frac{1}{\left(1+\frac{4b^2E^2}{Z^2e^4}\right)^{\frac{1}{2}}} + \cos^{-1}\frac{1+\left(1+\frac{4b^2E^2}{Z^2e^4}\right)^{\frac{1}{2}}+\frac{4b^2E^2}{Z^2e^4}}{1+\frac{4b^2E^2}{Z^2e^4}+\left(1+\frac{4b^2E^2}{Z^2e^4}\right)^{\frac{1}{2}}}$$

$$= -\cos^{-1}\frac{1}{\left(1+\frac{4b^2E^2}{Z^2e^4}\right)^{\frac{1}{2}}} + \cos^{-1}(1) = -\cos^{-1}\frac{1}{\left(1+\frac{4b^2E^2}{Z^2e^4}\right)^{\frac{1}{2}}}. \tag{15.28}$$

In this case, therefore, we obtain

$$\theta = \pi - 2\chi = \pi + 2\cos^{-1}\frac{1}{\left(1+\frac{4b^2E^2}{Z^2e^4}\right)^{\frac{1}{2}}},$$

$$\text{or,}\quad \frac{1}{\left(1+\frac{4b^2E^2}{Z^2e^4}\right)^{\frac{1}{2}}} = \cos\left(\frac{\theta}{2}-\frac{\pi}{2}\right) = \cos\left(\frac{\pi}{2}-\frac{\theta}{2}\right) = \sin\frac{\theta}{2}, \tag{15.29}$$

which leads to

$$\frac{2bE}{Ze^2} = \cot\frac{\theta}{2}. \tag{15.30}$$

This, in turn determines

$$b = \frac{Ze^2}{2E}\cot\frac{\theta}{2},$$

$$\frac{\mathrm{d}b}{\mathrm{d}\theta} = -\frac{Ze^2}{4E}\,\mathrm{cosec}^2\,\frac{\theta}{2},$$

$$\frac{b}{\sin\theta} = \frac{Ze^2}{2E}\,\frac{\cot\frac{\theta}{2}}{\sin\theta} = \frac{Ze^2}{4E}\,\mathrm{cosec}^2\,\frac{\theta}{2}. \tag{15.31}$$

Therefore, in this case, the differential cross section is determined to be

$$\sigma(\theta) = 2\pi\,\frac{b}{\sin\theta}\left|\frac{\mathrm{d}b}{\mathrm{d}\theta}\right| = 2\pi\left(\frac{Ze^2}{4E}\right)^2\mathrm{cosec}^4\,\frac{\theta}{2}. \tag{15.32}$$

This is known as the Rutherford scattering formula. If we integrate this to determine the total cross section, we find that the total cross section diverges. This is a consequence of our assumption that the Coulomb force extends to infinite distances. Hence, any particle, no matter how far away it is from the scattering source, suffers a change in its trajectory. In reality, of course, the Coulomb force is screened and hence we have a finite total cross section. ◀

15.1 Laboratory frame and the center of mass frame

We have so far considered the scattering of an incident particle by a fixed source. However, in practice, one shoots beams of particles at fixed targets which, after being hit, are also in motion. Thus, the problem is not quite the same as we have studied so far. However, we can translate the problem to an equivalent one. For example, we have seen, in chapter 9, that the motion of two particles interacting through a potential, which depends on the relative separation of the two particles, can be split into the motion of the center of mass and the motion of a particle with a reduced mass. Furthermore, in the center of mass frame, the center of mass is at rest and, therefore, any scattering can be reduced to that of incident particles with a reduced mass scattering from a fixed source. We can utilize the formulae we have developed in the last section in the center of mass frame and, to obtain quantities in the laboratory frame, we simply have to make a transformation, which we describe below. However, let us first note that scattering can be of two types. If in a scattering the internal energy of the particles change, then it is called an inelastic scattering. However, if there is no change in the internal energy, the scattering is known as an elastic scattering.

Let us consider an elastic scattering of a particle of mass m_1 incident with a velocity v_1 on a stationary target of mass m_2 in the laboratory as shown in Fig. 15.3. This process can be equivalently described in the laboratory frame or in the center of mass frame. Let us assume that there is azimuthal symmetry so that the azimuthal angle is not so important. The center of mass, in this case, moves to the right with a velocity

$$v_{\mathrm{CM}} = \frac{m_1 v_1}{m_1 + m_2}. \tag{15.33}$$

In the center of mass frame, therefore, the particles 1 and 2 move towards each other with speeds (see Fig. 15.4)

$$\begin{aligned} \tilde{v}_1 &= v_1 - v_{\mathrm{CM}} = \frac{m_2 v_1}{m_1 + m_2}, \\ \tilde{v}_2 &= v_{\mathrm{CM}} = \frac{m_1 v_1}{m_1 + m_2}, \end{aligned} \tag{15.34}$$

where we are only talking about the magnitudes of the velocities.

If θ_{lab} and θ_{CM} denote respectively the scattering angles in the laboratory frame and the center of mass frame, then, for the components of the final velocities parallel and perpendicular to the center

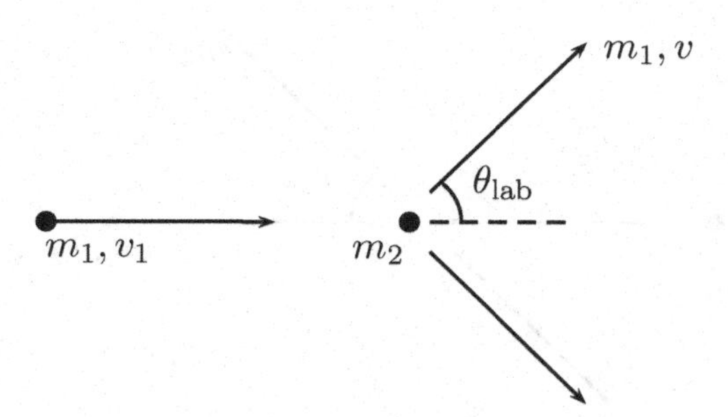

Figure 15.3: Scattering of a particle incident on another particle at rest in the laboratory.

of mass velocity, we must again have

$$\begin{aligned} & v\cos\theta_{\rm lab} - v_{\rm CM} = \tilde{v}_1\cos\theta_{\rm CM}, \\ \text{or,}\quad & v\cos\theta_{\rm lab} = v_{\rm CM} + \tilde{v}_1\cos\theta_{\rm CM}, \end{aligned} \tag{15.35}$$

and

$$v\sin\theta_{\rm lab} = \tilde{v}_1\sin\theta_{\rm CM}. \tag{15.36}$$

It follows from (15.35) and (15.36) that

$$\tan\theta_{\rm lab} = \frac{\sin\theta_{\rm CM}}{\cos\theta_{\rm CM} + \frac{v_{\rm CM}}{\tilde{v}_1}} = \frac{\sin\theta_{\rm CM}}{\gamma + \cos\theta_{\rm CM}}, \tag{15.37}$$

where we have defined

$$\gamma = \frac{v_{\rm CM}}{\tilde{v}_1} = \frac{m_1}{m_2}. \tag{15.38}$$

Equation (15.37), therefore, determines the transformation between the scattering angles in the two frames.

We are discussing so far an elastic scattering. However, if we had an inelastic scattering of the form

$$m_1 + m_2 \to m_3 + m_4,$$

then, a relation of the form (15.37) is still valid except that the quantity γ is now defined to be

$$\gamma = \left(\frac{m_1 m_3}{m_2 m_4}\,\frac{E}{E+Q}\right)^{\frac{1}{2}}, \tag{15.39}$$

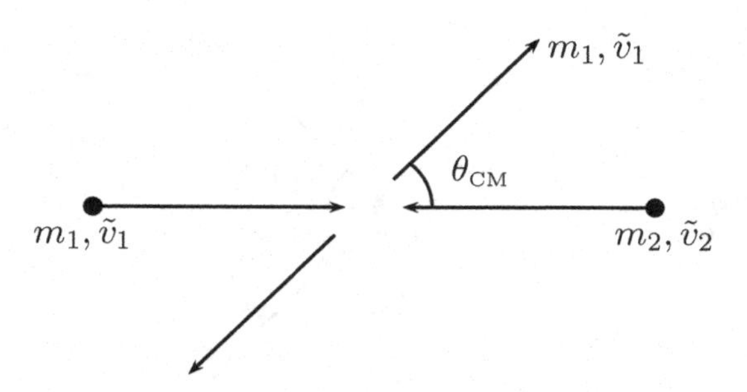

Figure 15.4: The same scatterin as in Fig. 15.3 viewed in the center of mass frame.

where Q denotes the amount of internal energy that is converted into the kinetic energy of the emerging particles and E is the initial energy of both the particles in the center of mass frame. (Let us note that (15.39) reduces to (15.38) – the elastic case – when $Q = 0, m_3 = m_1, m_4 = m_2$.)

The relation between the cross sections in the center of mass frame and the laboratory frame can now be obtained from the fact that the same particles which go into the solid angle $d\Omega_{\rm CM}$ at $\theta_{\rm CM}$ in the center of mass frame go into the solid angle $d\Omega_{\rm lab}$ at $\theta_{\rm lab}$ in the laboratory frame. Therefore, we have

$$\sigma_{\rm lab}(\theta_{\rm lab}) \sin\theta_{\rm lab} d\theta_{\rm lab} = \sigma_{\rm CM}(\theta_{\rm CM}) \sin\theta_{\rm CM} d\theta_{\rm CM}. \tag{15.40}$$

Using (15.37), this can be simplified to give

$$\sigma_{\rm lab}(\theta_{\rm lab}) = \frac{\left(1 + \gamma^2 + 2\gamma\cos\theta_{\rm CM}\right)^{\frac{3}{2}}}{|1 + \gamma\cos\theta_{\rm CM}|}\,\sigma_{\rm CM}(\theta_{\rm CM}). \tag{15.41}$$

This formula holds for both elastic as well as inelastic collisions with the appropriate identification of γ. It should be noted that the total cross section is the same in both the laboratory frame and the center of mass frame, as it should be expected.

Let us note, from (15.37), the following.

1. For $\gamma < 1$, $\theta_{\rm lab}$ increases from 0 to π as $\theta_{\rm CM}$ increases from 0 to π.

2. For $\gamma = 1$, $\theta_{\text{lab}} = \frac{\theta_{\text{CM}}}{2}$, $0 \leq \theta_{\text{lab}} \leq \frac{\pi}{2}$, and $0 \leq \theta_{\text{CM}} \leq \pi$. In this case, (15.41) yields

$$\sigma_{\text{lab}}(\theta_{\text{lab}}) = 4\cos\theta_{\text{lab}}\sigma_{\text{CM}}(2\theta_{\text{lab}}), \tag{15.42}$$

   which shows that there is no backward scattering in this case.

3. For $\gamma > 1$, θ_{lab} increases from 0 to a maximum value as θ_{CM} increases from 0 to $\cos^{-1}\left(-\frac{1}{\gamma}\right)$. Then it decreases to 0 as θ_{CM} increases to π.

15.2 Quantum theory of scattering

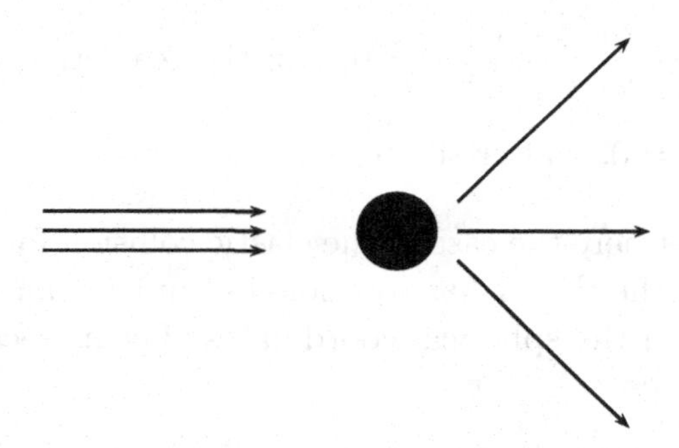

Figure 15.5: The quantum mechanical description of the scattering of a beam of particles from a scattering source.

Unlike the case of classical scattering, where particles with a given initial velocity and angular momentum follow a definite trajectory, in quantum mechanics, the concept of a definite trajectory with a definite velocity does not exist. As we know, for example, if we measure the velocity of a particle to be exactly parallel to the z-axis, then the component of its velocity perpendicular to the z-axis is zero. This would imply that there would be an infinite uncertainty in the value of the impact parameter of the particle. Therefore, we have to pose the question of quantum scattering differently. The scattering set up is still the same as shown in Fig. 15.5. Namely, we still measure the initial velocity of the particles and the number of particles going into the solid angle $d\Omega$ at θ after interaction. The definitions of the differential cross section and the total cross section are still the same as given in (15.1) and (15.4) respectively. However, the solutions are now obtained from the Schrödinger equation.

We will begin with the simplest of scattering problems, namely, scattering from a fixed potential which has rotational symmetry and which is independent of time. Furthermore, the particles being scattered are assumed to be spinless. In addition, we assume that the initial particles are incident along the z-axis. Thus, conceptually, we have a beam of free particles incident along the z-axis. There is interaction and we have outgoing scattered particles at large distance. Thus, we are looking for solutions of the Schrödinger equations (μ denotes the mass of the particle so as to avoid confusion with the azimuthal quantum number)

$$\begin{aligned}
&\left(\boldsymbol{\nabla}^2 + k^2\right)\psi = 0, \quad \text{for the incident particles,}\\
&\left\{\boldsymbol{\nabla}^2 + \left(k^2 - \frac{2\mu}{\hbar^2}V(r)\right)\right\}\psi = 0, \quad \text{in the scattering region,}\\
&\left(\boldsymbol{\nabla}^2 + k^2\right)\psi = 0, \quad \text{after scattering as } r \to \infty. \qquad (15.43)
\end{aligned}$$

Let us also consider here the case of the elastic collision for simplicity.

We note that the theory has rotational symmetry and hence, we look for solutions in the spherical coordinates. Let us recall that, in spherical coordinates,

$$\begin{aligned}
\boldsymbol{\nabla} &= \hat{\mathbf{r}}\,\frac{\partial}{\partial r} + \frac{\hat{\theta}}{r}\,\frac{\partial}{\partial\theta} + \frac{\hat{\phi}}{r\sin\theta}\,\frac{\partial}{\partial\phi},\\
\boldsymbol{\nabla}^2 &= \frac{\partial^2}{\partial r^2} + \frac{2}{r}\,\frac{\partial}{\partial r} - \frac{\mathbf{L}^2}{\hbar^2 r^2} = \frac{1}{r^2}\,\frac{\partial}{\partial r}\left(r^2\,\frac{\partial}{\partial r}\right) - \frac{\mathbf{L}^2}{\hbar^2 r^2}. \qquad (15.44)
\end{aligned}$$

Furthermore, we have already seen that the solutions of the free Schrödinger equation

$$\left(\boldsymbol{\nabla}^2 + k^2\right)\psi = 0, \qquad k^2 = \frac{2\mu E}{\hbar^2},$$

can be written in the spherical coordinates as

$$\psi_{\ell,m}(r,\theta,\phi) = R_\ell(r)Y_{\ell,m}(\theta,\phi), \qquad (15.45)$$

where $R_\ell(r)$ satisfies the radial equation

$$\frac{1}{r^2}\,\frac{\mathrm{d}}{\mathrm{d}r}\left(r^2\frac{\mathrm{d}R_\ell}{\mathrm{d}r}\right) + \left(k^2 - \frac{\ell(\ell+1)}{r^2}\right)R_\ell(r) = 0. \qquad (15.46)$$

The angular part of the solution, $Y_{\ell,m}$, corresponds to the spherical harmonics defined as (see (8.85) and (8.86))

$$Y_{\ell,m}(\theta,\phi) = (-1)^{\frac{m+|m|}{2}} \sqrt{\frac{2\ell+1}{4\pi}\frac{(\ell-|m|)!}{(\ell+|m|)!}} P_{\ell,m}(\cos\theta)e^{im\phi}, \tag{15.47}$$

where the $P_{\ell,m}(\cos\theta)$'s are the associated Legendre functions and are related to the Legendre functions $P_\ell(\cos\theta)$'s through the relation

$$P_{\ell,m}(t) = (1-t^2)^{\frac{|m|}{2}} \frac{\mathrm{d}^{|m|}P_\ell(t)}{\mathrm{d}t^{|m|}}. \tag{15.48}$$

Furthermore, the radial equation, (15.46), is simplified by defining

$$R_\ell(r) = \frac{u_\ell(r)}{r}, \tag{15.49}$$

so that the $u_\ell(r)$'s satisfy the equation

$$\frac{\mathrm{d}^2u_\ell}{\mathrm{d}r^2} + \left(k^2 - \frac{\ell(\ell+1)}{r^2}\right)u_\ell(r) = 0. \tag{15.50}$$

As we have seen, in chapter 8, if the solutions of the Schrödinger equation are to be well behaved (normalizable), $u_\ell(r)$'s must satisfy the boundary condition $u_\ell(0) = 0$ (see (8.107)).

Changing to the new variables $y = kr$, equation (15.50) takes the form

$$\frac{\mathrm{d}^2u_\ell}{\mathrm{d}y^2} + \left(1 - \frac{\ell(\ell+1)}{y^2}\right)u_\ell(y) = 0. \tag{15.51}$$

The two independent solutions of this equation are related to the spherical Bessel functions, namely,

$$u_\ell(y) = yj_\ell(y), \quad \text{or,} \quad u_\ell(y) = yj_{-\ell-1}(y), \tag{15.52}$$

so that the most general solution of (15.51) can be written in the form

$$u_\ell(y) = u_\ell(kr) = kr\,(a_1j_\ell(kr) + a_2j_{-\ell-1}(kr)). \tag{15.53}$$

Furthermore, we know that, near the origin (as $x \to 0$),

$$\begin{aligned} j_\ell(x) &\propto x^\ell & \text{for} \quad \ell > 0,\\ &\propto \tfrac{1}{x^{|\ell|+1}} & \text{for} \quad \ell < 0. \end{aligned} \tag{15.54}$$

Since $u_\ell(kr)$'s have to satisfy the boundary condition $u_\ell(0) = 0$, it follows that

$$a_2 = 0. \tag{15.55}$$

Thus, we have determined that

$$u_\ell(kr) = a_1\ kr j_\ell(kr). \tag{15.56}$$

The radial solution, therefore, becomes

$$R_\ell(y) = \frac{u_\ell(y)}{y} = j_\ell(y), \tag{15.57}$$

so that the complete solution of the Schrödinger equation has the form

$$\begin{aligned}\psi_{\ell,m}(r,\theta,\phi) &= N_\ell\ R_\ell(r)Y_{\ell,m}(\theta,\phi)\\ &= N_\ell\ j_\ell(kr)Y_{\ell,m}(\theta,\phi),\end{aligned} \tag{15.58}$$

where N_ℓ is the normalization constant and can be determined by using the orthonormality relations

$$\begin{aligned}&\int \mathrm{d}\Omega\ Y^*_{\ell,m}(\theta,\phi)Y_{\ell,m}(\theta,\phi) = 1,\\ &\int_0^\infty r^2\mathrm{d}r\ j_\ell(kr)j_\ell(kr) = \frac{\pi}{2k^2}.\end{aligned} \tag{15.59}$$

Namely, using the relations in (15.59), we obtain

$$\begin{aligned}&\int \mathrm{d}^3r\ \psi^*_{\ell,m}(r,\theta,\phi)\psi_{\ell,m}(r,\theta,\phi) = 1,\\ \text{or,}\quad &|N_\ell|^2 \int r^2\mathrm{d}r\mathrm{d}\Omega\ j_\ell(kr)j_\ell(kr)Y^*_{\ell,m}(\theta,\phi)Y_{\ell,m}(\theta,\phi) = 1,\\ \text{or,}\quad &|N_\ell|^2\ \frac{\pi}{2k^2} = 1,\\ \text{or,}\quad &N_\ell = N^*_\ell = \sqrt{\frac{2k^2}{\pi}}.\end{aligned} \tag{15.60}$$

This determines the normalized free particle wave functions to be

$$\psi_{k,\ell,m}(r,\theta,\phi) = \sqrt{\frac{2k^2}{\pi}}\ j_\ell(kr)Y_{\ell,m}(\theta,\phi). \tag{15.61}$$

These free particle wave functions form a complete basis so that any other wave function can be expressed in terms of them. As an example, let us consider the expansion of a plane wave in this basis. That is, we consider a plane wave of wave number k incident along the z-axis,

$$\psi_{(\text{plane wave})} = e^{ikz} = e^{ikr\cos\theta}. \tag{15.62}$$

This does not depend on the azimuthal angle, which simply reflects the fact that the wave has no angular momentum along the z-axis. Therefore, its expansion in the spherical basis would involve only the $m = 0$ components of the spherical harmonics and we can write

$$\begin{aligned} e^{ikz} = e^{ikr\cos\theta} &= \sum_{\ell=0}^{\infty} a_\ell \psi_{k,\ell,0}(r,\theta,\phi) \\ &= \sum_{\ell=0}^{\infty} a_\ell \sqrt{\frac{2k^2}{\pi}}\, j_\ell(kr) Y_{\ell,0}(\theta,\phi) \\ &= \sqrt{\frac{2k^2}{\pi}} \sum_{\ell=0}^{\infty} a_\ell\, j_\ell(kr) \sqrt{\frac{2\ell+1}{4\pi}}\, P_\ell(\cos\theta), \end{aligned} \tag{15.63}$$

where we have used the definition of the spherical harmonics introduced in (15.47).

The expansion coefficients, a_ℓ's, are determined from the orthogonality relations of the Legendre polynomials,

$$\int_0^\pi \sin\theta \mathrm{d}\theta\ P_\ell(\cos\theta) P_{\ell'}(\cos\theta) = \frac{2}{2\ell+1}\, \delta_{\ell\ell'}, \tag{15.64}$$

as well as the integral representation for the spherical Bessel functions,

$$j_\ell(x) = \frac{1}{2i^\ell} \int_0^\pi \sin\theta \mathrm{d}\theta\ e^{ix\cos\theta} P_\ell(\cos\theta). \tag{15.65}$$

For example, using the orthogonality of the Legendre polynomials, (15.64), we obtain, from (15.63),

$$\begin{aligned} \int \mathrm{d}\Omega\ P_\ell(\cos\theta) e^{ikr\cos\theta} &= \sqrt{\frac{2k^2}{\pi}}\, a_\ell j_\ell(kr) \sqrt{\frac{2\ell+1}{4\pi}}\, 2\pi\, \frac{2}{2\ell+1} \\ &= \sqrt{\frac{8k^2}{2\ell+1}}\, a_\ell j_\ell(kr). \end{aligned} \tag{15.66}$$

On the other hand, the use of the integral representation of the spherical Bessel functions, (15.65), leads to

$$\begin{aligned}
\sqrt{\frac{8k^2}{2\ell+1}}\, a_\ell j_\ell(kr) &= \int \mathrm{d}\phi \sin\theta \mathrm{d}\theta\ e^{ikr\cos\theta} P_\ell(\cos\theta) \\
&= 2\pi \int_0^\pi \sin\theta \mathrm{d}\theta\ e^{ikr\cos\theta} P_\ell(\cos\theta) \\
&= 2\pi \times 2i^\ell j_\ell(kr). \qquad (15.67)
\end{aligned}$$

This determines

$$a_\ell = 4\pi i^\ell \sqrt{\frac{2\ell+1}{8k^2}} = 2\pi i^\ell \sqrt{\frac{2\ell+1}{2k^2}}. \qquad (15.68)$$

Therefore, we see that a plane wave incident along the z-axis can be expanded as

$$\begin{aligned}
e^{ikz} &= \sum_{\ell=0}^{\infty} 2\pi i^\ell \sqrt{\frac{2\ell+1}{2k^2}} \sqrt{\frac{2k^2}{\pi}} \sqrt{\frac{2\ell+1}{4\pi}}\, j_\ell(kr) P_\ell(\cos\theta) \\
&= \sum_{\ell=0}^{\infty} (2\ell+1)\ i^\ell\ j_\ell(kr)\ P_\ell(\cos\theta). \qquad (15.69)
\end{aligned}$$

Recalling that, for large r,

$$j_\ell(kr) \longrightarrow \frac{1}{kr} \sin\left(kr - \frac{\ell\pi}{2}\right), \qquad (15.70)$$

we see that a plane wave, incident along the z-axis, is equivalent to a superposition of an infinite number of spherical waves with all angular momentum values. In other words, in a plane wave, components of angular momentum corresponding to all values of ℓ would be present (which is what we had observed earlier, namely, the impact parameter will become infinitely uncertain in this case). Classically, the part of the plane wave corresponding to angular momentum $(\ell(\ell+1))^{\frac{1}{2}}\hbar$ would correspond to an impact parameter $b = \frac{(\ell(\ell+1))^{\frac{1}{2}}}{k}$ and would pass through the scattering region with a definite distance of closest approach $r_{0\ell}$. Quantum mechanically, however, the probability of finding a particle of this angular momentum at a distance $r < r_{0\ell}$ is not zero, but rapidly falls off.

We are now ready to solve the scattering problem. We have an incident plane wave along the z-direction. Thus,

$$\psi_{\rm inc} = e^{ikz}. \tag{15.71}$$

After scattering, we have a spherical scattered wave which is outgoing at large distances. Thus, for large r, we can write

$$\psi_{\rm sc}(r,\theta,\phi) \longrightarrow \frac{e^{ikr}}{r}\, f(\theta,\phi). \tag{15.72}$$

Therefore, the total wave function, at large distances away from the scattering source, can be written as

$$\psi = \psi_{\rm inc} + \psi_{\rm sc} = e^{ikz} + \frac{e^{ikr}}{r}\, f(\theta,\phi). \tag{15.73}$$

The quantity $f(\theta,\phi)$ measures the angular distribution of the outgoing spherical wave.

We can now define the current densities for both the incident as well as the scattered waves. Thus,

$$\mathbf{j}_{\rm inc} = \frac{\hbar}{2i\mu}\left(\psi^*_{\rm inc}\boldsymbol{\nabla}\psi_{\rm inc} - \boldsymbol{\nabla}\psi^*_{\rm inc}\psi_{\rm inc}\right) = \frac{\hbar k}{\mu}\,\hat{\mathbf{z}}, \tag{15.74}$$

where we have used $\psi_{\rm inc} = e^{ikz}$. The incident current measures the probability that an incident particle crosses a unit area per unit time. Similarly, the current density for the scattered beam at large distances is given by

$$\mathbf{j}_{\rm sc} = \frac{\hbar}{2i\mu}\left[\psi^*_{\rm sc}\boldsymbol{\nabla}\psi_{\rm sc} - \boldsymbol{\nabla}\psi^*_{\rm sc}\psi_{\rm sc}\right]. \tag{15.75}$$

Substituting the form of the gradient operator in the spherical basis, (15.44), and taking the form of the scattered wave in (15.72), we have, for large distances,

$$\mathbf{j}_{\rm sc} \longrightarrow \frac{\hbar k}{\mu}\,\frac{|f(\theta,\phi)|^2}{r^2}\,\hat{\mathbf{r}}. \tag{15.76}$$

Thus, we find the flux of scattered particles across an area ds of a sphere of radius r to be

$$\mathbf{j}_{\rm sc}\cdot \mathrm{d}\mathbf{s} = \frac{\hbar k}{\mu}\,\frac{|f(\theta,\phi)|^2}{r^2}\,r^2\mathrm{d}\Omega = \frac{\hbar k}{\mu}|f(\theta,\phi)|^2\mathrm{d}\Omega. \tag{15.77}$$

The other way of saying this is that the flux of particles scattered into the solid angle $\mathrm{d}\Omega$ is

$$\frac{\hbar k}{\mu}\, |f(\theta,\phi)|^2 \mathrm{d}\Omega. \tag{15.78}$$

Thus, we immediately see, from the definition in (15.1), that the differential cross section for scattering is given by

$$\sigma(\theta,\phi) = |f(\theta,\phi)|^2. \tag{15.79}$$

We see that the function $f(\theta,\phi)$, which measures the angular distribution of the outgoing waves, determines the differential scattering cross section. $f(\theta,\phi)$ is correspondingly also known as the scattering amplitude. Our main interest lies in determining the scattering amplitude, which, in turn, would determine the scattering cross section.

We note that the incident plane wave is the solution of the free Schrödinger equation

$$\left(\boldsymbol{\nabla}^2 + k^2\right) \psi_{\mathrm{inc}} = 0,$$

and has the expansion

$$\begin{aligned}\psi_{\mathrm{inc}} = e^{ikz} &= \sum_{\ell=0}^{\infty} (2\ell+1) i^{\ell} j_{\ell}(kr) P_{\ell}(\cos\theta)\\ &\xrightarrow{r\ \mathrm{large}} \sum_{\ell=0}^{\infty} (2\ell+1) i^{\ell}\ \frac{\sin\left(kr - \frac{\ell\pi}{2}\right)}{kr}\ P_{\ell}(\cos\theta).\end{aligned} \tag{15.80}$$

The total wave function, which is the solution of

$$\left[\boldsymbol{\nabla}^2 + \left(k^2 - \frac{2\mu}{\hbar^2}\, V(r)\right)\right] \psi = 0, \tag{15.81}$$

would also have a similar form at large distances, if $V(r) \to 0$ faster than $\frac{1}{r^2}$ for large r. This is because, then the potential term can be neglected (at large distances) compared to the centrifugal barrier and the equation would again have the form of a free equation

$$\left(\boldsymbol{\nabla}^2 + k^2\right) \psi = 0.$$

However, each angular momentum component of the incident wave function would suffer a phase change due to interactions with the scattering source. This can be intuitively seen as follows. If the scattering potential is attractive, then, in the scattering region, the particle will be accelerated and consequently the wavelength would be shorter. Conversely, if the scattering potential is repulsive, then, the particle would be decelerated in the scattering region and, consequently, would suffer a change in its wavelength. In either case, away from the scattering region, the phase of each angular momentum component of the wave function would be different from the one in the absence of scattering as is shown in the Fig. 15.6.

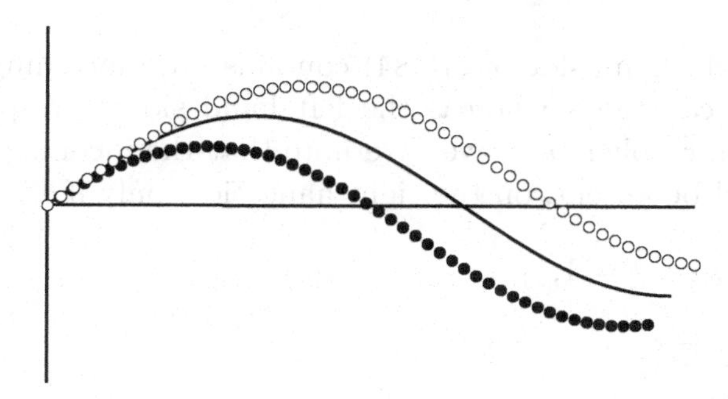

Figure 15.6: Qualitative depiction of phase change in an angular momentum component of the wave due to scattering. The black and white dots show respectively the cases of scattering from an attractive and a repulsive potential.

Thus, the total wave function, at large distances, must have the form

$$\psi \longrightarrow \sum_{\ell=0}^{\infty} A_\ell (2\ell+1) i^\ell \, \frac{\sin\left(kr - \frac{\ell\pi}{2} + \delta_\ell\right)}{kr} \, P_\ell(\cos\theta), \tag{15.82}$$

where A_ℓ's are constants. The quantity δ_ℓ is known as the phase shift in the ℓ th partial wave or the component of the wave function with angular momentum ℓ and δ_ℓ is positive if the potential is attractive, while it is negative for a repulsive potential. We know that, for large r,

$$\psi \longrightarrow e^{ikz} + \frac{e^{ikr}}{r} \, f(\theta, \phi). \tag{15.83}$$

Thus, we expect, for large r,

$$\begin{aligned} \frac{e^{ikr}}{r} \, f(\theta,\phi) &\to \psi - e^{ikz} \\ &= \sum_{\ell=0}^{\infty} A_\ell (2\ell+1) i^\ell \frac{\sin\left(kr - \frac{\ell\pi}{2} + \delta_\ell\right)}{kr} P_\ell(\cos\theta) \end{aligned}$$

$$
\begin{aligned}
&- \sum_{\ell=0}^{\infty}(2\ell+1)i^{\ell}\frac{\sin\left(kr-\frac{\ell\pi}{2}\right)}{kr}\,P_{\ell}(\cos\theta)\\
&= \sum_{\ell=0}^{\infty}(2\ell+1)\frac{i^{\ell}}{kr}\frac{P_{\ell}(\cos\theta)}{2i}\left\{\left[A_{\ell}e^{i\left(kr-\frac{\ell\pi}{2}+\delta_{\ell}\right)}-e^{i\left(kr-\frac{\ell\pi}{2}\right)}\right]\right.\\
&\quad\left.-\left[A_{\ell}e^{-i\left(kr-\frac{\ell\pi}{2}+\delta_{\ell}\right)}-e^{-i\left(kr-\frac{\ell\pi}{2}\right)}\right]\right\}. \qquad (15.84)
\end{aligned}
$$

The right hand side of (15.84) contains both incoming and outgoing spherical waves whereas the left hand side corresponds only to an outgoing spherical wave. We note that the incoming spherical waves would be absent, on the right hand side, only if

$$
\begin{aligned}
&A_{\ell}\,e^{-i\delta_{\ell}} = 1,\\
\text{or,}\quad &A_{\ell} = e^{i\delta_{\ell}}. \qquad (15.85)
\end{aligned}
$$

Substituting this into (15.84), we obtain

$$
\begin{aligned}
&\frac{e^{ikr}}{r}\,f(\theta,\phi)\\
&= \sum_{\ell=0}^{\infty}(2\ell+1)\frac{i^{\ell}}{kr}\frac{1}{2i}\,e^{i\left(kr-\frac{\ell\pi}{2}\right)}\left(e^{2i\delta_{\ell}}-1\right)P_{\ell}(\cos\theta)\\
&= \frac{e^{ikr}}{r}\sum_{\ell=0}^{\infty}\frac{(2\ell+1)}{2ik}(i)^{\ell}(-i)^{\ell}\left(e^{2i\delta_{\ell}}-1\right)P_{\ell}(\cos\theta). \qquad (15.86)
\end{aligned}
$$

Comparing both sides, we determine

$$
f(\theta,\phi) = \sum_{\ell=0}^{\infty}\frac{(2\ell+1)}{2ik}\left(e^{2i\delta_{\ell}}-1\right)P_{\ell}(\cos\theta) = \sum_{\ell=0}^{\infty}f_{\ell}(\theta), \qquad (15.87)
$$

where we can think of $f_{\ell}(\theta)$ as the scattering amplitude for the ℓth partial wave.

The differential scattering cross section can now be determined using (15.79) and the total scattering cross section can be obtained as

$$
\begin{aligned}
\sigma_{\text{tot}} &= \int \mathrm{d}\Omega|f(\theta,\phi)|^{2}\\
&= 2\pi\int_{0}^{\pi}\sin\theta\mathrm{d}\theta\sum_{\ell,\ell'}\frac{1}{4k^{2}}\,(2\ell+1)(2\ell'+1)
\end{aligned}
$$

$$\times \left|e^{2i\delta_\ell}-1\right|\left|e^{2i\delta_{\ell'}}-1\right| P_\ell(\cos\theta)P_{\ell'}(\cos\theta)$$

$$= 2\pi \sum_{\ell,\ell'} \frac{1}{4k^2}\,(2\ell+1)(2\ell'+1)\left|e^{2i\delta_\ell}-1\right|\left|e^{2i\delta_{\ell'}}-1\right|$$

$$\times \int_0^\pi \sin\theta \mathrm{d}\theta\; P_\ell(\cos\theta)P_{\ell'}(\cos\theta)$$

$$= 2\pi \sum_{\ell,\ell'} (2\ell+1)(2\ell'+1)\left|e^{2i\delta_\ell}-1\right|^2 \frac{2\delta_{\ell\ell'}}{2\ell+1}$$

$$= \sum_{\ell=0}^{\infty} \frac{4\pi}{4k^2}\,(2\ell+1)\; 4\sin^2\delta_\ell$$

$$= \frac{4\pi}{k^2}\sum_{\ell=0}^{\infty}(2\ell+1)\sin^2\delta_\ell. \tag{15.88}$$

Thus, we see that if we know the phase shifts for every partial wave, we know everything about the scattering. This is known as the partial wave analysis or the phase shift analysis. In many cases, only the lower angular momentum components of the wave suffer the maximum amount of phase shift. Then, the infinite sum in (15.88) can be approximated by only a few terms.

It is worth noting here that the forward scattering amplitude can be obtained from (15.87) to be

$$f(0) = \sum_{\ell=0}^{\infty} \frac{2\ell+1}{2ik}\left(e^{2i\delta_\ell}-1\right) P_\ell(1)$$

$$= \sum_{\ell=0}^{\infty} \frac{2\ell+1}{k}\; e^{i\delta_\ell}\sin\delta_\ell. \tag{15.89}$$

As a result, we note that

$$\mathrm{Im} f(0) = \sum_{\ell=0}^{\infty} \frac{2\ell+1}{k}\sin^2\delta_\ell. \tag{15.90}$$

Comparing with (15.88), we note that we can identify

$$\sigma_{\mathrm{tot}} = \frac{4\pi}{k}\mathrm{Im} f(0). \tag{15.91}$$

Even though we have derived this relation for elastic scattering, as we will discuss later, this relation is more general and is known as the optical theorem.

In the quantum theory of scattering, we realize that there is no minimum distance of approach. However, it is also true that the probability of finding a particle with angular momentum ℓ is very small at a distance $r < \frac{(\ell(\ell+1))^{\frac{1}{2}}}{k}$. On the other hand, if the potential is such that it becomes negligible for $r > \frac{(\ell(\ell+1))^{\frac{1}{2}}}{k}$, then it would have very little influence on components of wave function with such angular momentum quantum numbers. Thus, the phase shift for such components would be negligible. The number of phase shifts that will influence the scattering in any given case can then be obtained, for a short range potential, by calculating ℓ such that

$$\ell(\ell+1) \leq k^2 r_0^2, \tag{15.92}$$

where r_0 is the distance (or radius) beyond which the potential has become negligible. For short range potentials in nuclear physics, this relation severely restricts the number of phase shifts that contribute to scattering. In fact, in many nuclear scattering phenomena of interest, only the $\ell = 0$ wave suffers any appreciable phase shift. On the other hand, for long range potentials like the Coulomb potential, scattering, even at low energies, contains significant contributions from many ℓ values.

► **Example ($n - p$ scattering at low energies).** The strong force between a neutron and a proton can be represented by a spherical square well of the form

$$V(\mathbf{r}) = \begin{cases} -V_0 & r < a, \\ 0 & r > a, \end{cases} \tag{15.93}$$

which is shown in Fig. 15.7.
In this case, we have

$$r_0 = a \simeq 2 \times 10^{-13} \text{ cm} = 2\text{F}, \tag{15.94}$$

where 'F' stands for a Fermi. This is the range of the nuclear force which is very small. Furthermore, let us consider a neutron of energy 1 MeV incident on a stationary proton so that, in the center of mass, the neutron will have an energy 0.5 MeV. This leads to

$$p = \sqrt{2ME} = \frac{1}{c}\sqrt{2Mc^2E} = \frac{1}{c}\sqrt{2 \times 10^3 \times \frac{1}{2}\text{ MeV}} \simeq \frac{30}{c}\text{ MeV},$$

$$k = \frac{p}{\hbar} = \frac{30\text{ MeV}}{\hbar c} \simeq \frac{30\text{ MeV}}{200\text{ MeV-F}} \simeq \frac{30\text{ MeV}}{2 \times 10^{-11}\text{ MeV-cm}} \simeq \frac{10^{12}}{\text{cm}}, \tag{15.95}$$

so that we obtain

$$k^2 r_0^2 = k^2 a^2 \simeq \left(10^{12} \times 2 \times 10^{-13}\right)^2 = \left(2 \times 10^{-1}\right)^2 \ll 1. \tag{15.96}$$

Comparing with (15.92),

$$\ell(\ell+1) < k^2 a^2, \tag{15.97}$$

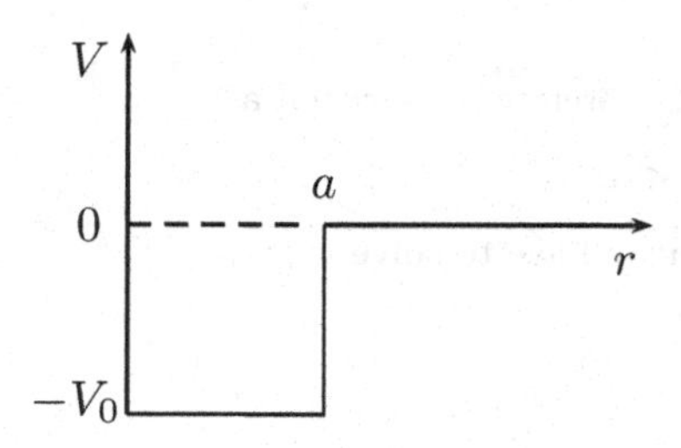

Figure 15.7: The square well potential represents well the strong force between nucleons.

we immediately see, from (15.96), that only the $\ell = 0$ wave or the s-wave would suffer significant phase change. Parenthetically, we note that the relation (15.97) can be seen in a simple manner as follows. Let us consider the radial equation of the form

$$\frac{\mathrm{d}^2 u_\ell}{\mathrm{d}r^2} + \left(k^2 - \frac{\ell(\ell+1)}{r^2}\right) u_\ell = 0. \tag{15.98}$$

It is clear that, at any distance r, the probability of finding a component with angular momentum ℓ is appreciable, only if (otherwise, we will have a damped solution)

$$k^2 r^2 > \ell(\ell+1). \tag{15.99}$$

If the potential has a range r_0, then, the only components that would suffer appreciable scattering are the ones satisfying

$$k^2 r_0^2 > \ell(\ell+1). \tag{15.100}$$

The higher angular momentum components will not feel the effect of the scattering source.

To calculate the phase shift for the s-wave, let us specialize to the case of a shallow potential well where we do not have any discrete levels. The case where there are discrete levels has to be considered separately. Let us write $k_0^2 = \frac{2mV_0}{\hbar^2}$. We want to solve the radial equation, which, for $r < a$, has the form ($\ell = 0$)

$$\frac{\mathrm{d}^2 u}{\mathrm{d}r^2} + \left(k^2 + k_0^2\right) u = 0,$$

$$\text{or,}\quad \frac{\mathrm{d}^2 u}{\mathrm{d}r^2} + k_1^2 u = 0, \tag{15.101}$$

where $k_1^2 = k^2 + k_0^2$ and $k^2 = \frac{2mE}{\hbar^2}$. The general solution is easily seen to be

$$u(r) = A \sin k_1 r + B \cos k_1 r. \tag{15.102}$$

However, the wave function has to satisfy the boundary condition

$$u(r = 0) = 0, \tag{15.103}$$

which implies that

$$B = 0. \tag{15.104}$$

The solution of (15.101) can, therefore, be written as

$$u(r) = A \sin k_1 r, \qquad r < a. \tag{15.105}$$

For $r > a$, the equation we have to solve is

$$\frac{\mathrm{d}^2 v}{\mathrm{d}r^2} + k^2 v = 0, \tag{15.106}$$

which leads to the solution

$$v(r) = C \sin(kr + \delta), \qquad r > a. \tag{15.107}$$

(An alternate way of saying this is that, for large distances, the s-wave component of the wave function for the incoming particle is given by $u_{\text{inc}} \sim \sin kr$, while the total wave function has the form $\sin(kr + \delta)$, namely, it is the same wave with the phase shifted (see (15.82)).)

We have to match the solutions and their first derivatives at the boundary, which leads to

$$\begin{aligned}
&u(a) = v(a),\\
\text{or,}\quad &A \sin k_1 a = C \sin(ka + \delta),\\
&u'(a) = v'(a),\\
\text{or,}\quad &k_1 A \cos k_1 a = kC \cos(ka + \delta).
\end{aligned} \tag{15.108}$$

Taking the ratio of the two relations in (15.108), we have

$$\frac{1}{k_1} \tan k_1 a = \frac{1}{k} \tan(ka + \delta). \tag{15.109}$$

Remembering from (15.96) that $ka \ll 1$, we can write

$$\begin{aligned}
&\frac{1}{k_1} \tan k_1 a = \frac{1}{k} \tan(ka + \delta) = a + \frac{\delta}{k},\\
\text{or,}\quad &\frac{\delta}{k} = \frac{1}{k_1} \tan k_1 a - a = \frac{1}{k_1} (\tan k_1 a - k_1 a),
\end{aligned} \tag{15.110}$$

where we have assumed that the phase shift is small. Thus, we obtain

$$\begin{aligned}
\sigma_{\text{tot}} &= \frac{4\pi}{k^2} \sin^2 \delta \simeq 4\pi \left(\frac{\delta}{k}\right)^2 = 4\pi \left(\frac{\tan k_1 a - k_1 a}{k_1}\right)^2\\
&\simeq 4\pi \left(\frac{\tan k_0 a - k_0 a}{k_0}\right)^2, \qquad \text{for } k_0 \gg k.
\end{aligned} \tag{15.111}$$

This determines the total scattering cross section, which, when $k_0 \gg k$, is seen to be independent of the energy of the incident particles. This formula, of course, becomes inapplicable if the depth and the range of the potential are such that

$$k_1 a \simeq k_0 a \simeq \frac{\pi}{2} (2n + 1). \tag{15.112}$$

We would discuss this special case in the next section. ◀

15.3 Resonance scattering

For low energy scattering by a spherical square well potential, we see, from (15.111), that the scattering cross section enhances considerably if

$$(k_0 a)^2 = \frac{2mV_0 a^2}{\hbar^2} \simeq \left(\frac{2n+1}{2}\right)^2 \pi^2. \tag{15.113}$$

From the solutions of the three dimensional spherical square well in chapter 8, we know that these are precisely the values where bound states can occur for $\ell = 0$ *with zero energy.* Thus, in this case the scattered particle which has almost zero energy would be in resonance with such a level.

Let us assume that there is, among the levels of the well, only one discrete state whose energy, $(-\mathcal{E})$, is close to zero. We also assume that

$$k_0 a \simeq \frac{\pi}{2}, \qquad \mathcal{E} \simeq 0. \tag{15.114}$$

(In fact, one implies the other.) Thus, the Schrödinger equation for the bound state, in this case, is

$$\frac{\mathrm{d}^2 u}{\mathrm{d}r^2} - \frac{2m}{\hbar^2}\left(\mathcal{E} + V(r)\right) u = 0. \tag{15.115}$$

For $r < a$, we note that $V(r)$ is large compared to $\mathcal{E}$ and, therefore, we can write the equation, in this region, as

$$\begin{aligned} &\frac{\mathrm{d}^2 u}{\mathrm{d}r^2} + k_0^2 u = 0,\\ \text{or,}\quad &u(r) = A \sin k_0 r, \end{aligned} \tag{15.116}$$

where $k_0^2 = \frac{2mV_0}{\hbar^2}$. This wave function vanishes at the origin, as it should.

For $r > a$, the equation is

$$\begin{aligned} &\frac{\mathrm{d}^2 u}{\mathrm{d}r^2} - \alpha^2 u = 0,\\ \text{or,}\quad &u(r) = B e^{-\alpha r} + C e^{\alpha r}, \end{aligned} \tag{15.117}$$

where $\alpha^2 = \frac{2m\mathcal{E}}{\hbar^2}$. However, the wave function must vanish at infinity, which determines

$$C = 0, \tag{15.118}$$

so that we have

$$u(r) = Be^{-\alpha r}, \qquad r > a. \tag{15.119}$$

Matching the wave functions at the boundary, we have

$$A \sin k_0 a = Be^{-\alpha a}. \tag{15.120}$$

Furthermore, matching the first derivatives of the wave functions at the boundary, we have

$$k_0 A \cos k_0 a = -\alpha Be^{-\alpha a}. \tag{15.121}$$

Taking the ratio of the two relations in (15.120) and (15.121), we obtain

$$\begin{aligned}
& k_0 \cot k_0 a = -\alpha, \\
\text{or,} \quad & -\cot k_0 a = \frac{\alpha}{k_0}, \\
\text{or,} \quad & k_0 a = \frac{\pi}{2} + \frac{\alpha}{k_0},
\end{aligned} \tag{15.122}$$

where we have used

$$-\cot k_0 a = -\tan\left(\frac{\pi}{2} - k_0 a\right) \simeq -\left(\frac{\pi}{2} - k_0 a\right). \tag{15.123}$$

Such relations between the depth of the potential, the range of the potential and the bound state energy values, as in (15.122), are known as effective range relations.

From the solution of the scattering problem, we know that

$$k \tan k_1 a = k_1 \tan(ka + \delta), \tag{15.124}$$

where

$$k_1 = (k_0^2 + k^2)^{\frac{1}{2}} \simeq k_0 \tag{15.125}$$

Thus, k_0 should be such that both the relations (15.122) and (15.124) are satisfied.

$$k \tan k_1 a = k_1 \tan(ka + \delta) = k_1 \, \frac{\tan ka + \tan\delta}{1 - \tan ka \tan\delta}. \tag{15.126}$$

Since $ka \ll 1$, we approximate $k_1 \simeq k_0$ and use the effective range relation, (15.122), to write

$$k \tan\left(\frac{\pi}{2} + \frac{\alpha}{k_0}\right) = k_0 \, \frac{ka + \tan\delta}{1 - ka\tan\delta},$$

$$\text{or,} \quad -k\cot\frac{\alpha}{k_0} \simeq -\frac{kk_0}{\alpha} = k_0\,\frac{ka + \tan\delta}{1 - ka\tan\delta},$$

$$\text{or,} \quad -k(1 - ka\tan\delta) = \alpha(ka + \tan\delta),$$

$$\text{or,} \quad \tan\delta(-\alpha + k^2 a) = k(\alpha a + 1),$$

$$\text{or,} \quad \cot\delta = \frac{(-\alpha + k^2 a)}{k(\alpha a + 1)} \simeq -\frac{\alpha}{k},$$

$$\text{or,} \quad \operatorname{cosec}^2\delta = 1 + \cot^2\delta = \frac{k^2 + \alpha^2}{k^2}. \tag{15.127}$$

Thus, we can determine the total scattering cross section to be

$$\sigma_{\rm tot} = \frac{4\pi}{k^2}\sin^2\delta = \frac{4\pi}{k^2 + \alpha^2} = \frac{2\pi\hbar^2}{m(E + \mathcal{E})}. \tag{15.128}$$

We note that the scattering cross section has a Breit-Wigner form. We see that, unlike the previous case, now the scattering cross section depends on the energy of the particle being scattered. Furthermore, the scattering cross section is significantly larger in the case of a resonance than in its absence.

The physical way to understand this result is as follows. If the well has a bound state level whose energy is close to zero, then the scattered particles have a tendency to get bound to the well. On the other hand, since the energy is not really negative for the incident particles, they cannot actually form bound states. Rather, they tend to interact much more strongly which leads to the enhanced scattering cross section.

15.4 Examples of scattering

1. Ramsauer-Townsend effect:

   From the formula for the scattering amplitude in (15.87),

$$f(\theta,\phi) = \sum_{\ell=0}^{\infty} \frac{2\ell+1}{k}\, e^{i\delta_\ell}\sin\delta_\ell\, P_\ell(\cos\theta), \tag{15.129}$$

   it is clear that if $\delta_\ell = \pi$ then, the scattering amplitude and, therefore, the scattering cross section vanishes, in all directions.

Such an effect, in fact, occurs in physical systems. When very low energy electrons are scattered against rare gas atoms, there is no scattering at all. The medium behaves completely transparent to the electrons. This effect is known as the Ramsauer-Townsend effect.

This effect can be explained as follows. First of all, we note that if the electrons are of extremely small energy, then only the s wave or $\ell = 0$ wave would suffer any appreciable change in phase. Furthermore, if the attractive potential is strong enough, then it can change the phase of the $\ell = 0$ wave quite a bit without influencing the other wave components at all. In particular, if the strength of the potential and its range are such that the phase of the s-wave changes by $180°$, then clearly, the scattering amplitude as well as the cross section would vanish in all directions. That is, there will be no scattering at all. We can think of this as due to the destructive interference of the waves.

2. Repulsive square well potential:

We have determined, in (15.111), the low energy scattering cross section from an attractive potential to be given by ($k_0 \gg k$)

$$\sigma_{\text{tot}} = 4\pi \left(\frac{\tan k_0 a - k_0 a}{k_0}\right)^2, \qquad k_0^2 = \frac{2mV_0}{\hbar^2}. \tag{15.130}$$

It is then obvious that the scattering cross section, from a repulsive potential of height V_0, is simply obtained by changing

$$k_0 \to i\kappa_0, \tag{15.131}$$

so that, for a repulsive potential, we obtain the total scattering cross section, from (15.130), to be

$$\sigma_{\text{tot}} = 4\pi \left(\frac{\tanh \kappa_0 a - \kappa_0 a}{\kappa_0}\right)^2. \tag{15.132}$$

The total scattering cross section from a hard sphere at low energies is now obtained by letting $\kappa_0 \to \infty$ in (15.132), which leads to

$$\sigma_{\text{tot}}^{(\text{hard sphere})} = 4\pi a^2. \tag{15.133}$$

This shows that, at low energies, the scattering cross section from a hard sphere is independent of the incident energy and is four times the classical value. We would expect that for high energy particles we would get back the classical value for the scattering cross section. If we carry out the calculations, now keeping higher ℓ waves, it turns out that the total scattering cross section, at high energies, is

$$\sigma_{\text{tot}}^{(\text{hard sphere})} = 2\pi a^2. \tag{15.134}$$

This is still twice the classical value. This difference from the classical value arises because of the wave nature of particles. There is diffraction around the edges of the sphere which leads to constructive interference.

Exercise. Solve the Schrödinger equation for the hard sphere where the incident particles have very low energy and verify that the formula, (15.133), is true.

3. Scattering from a delta potential:

   Let us assume that we are scattering low energy particles from a spherically symmetric delta function potential shown in Fig. 15.8,

$$V(\mathbf{r}) = \gamma\delta(r - a), \qquad \gamma > 0. \tag{15.135}$$

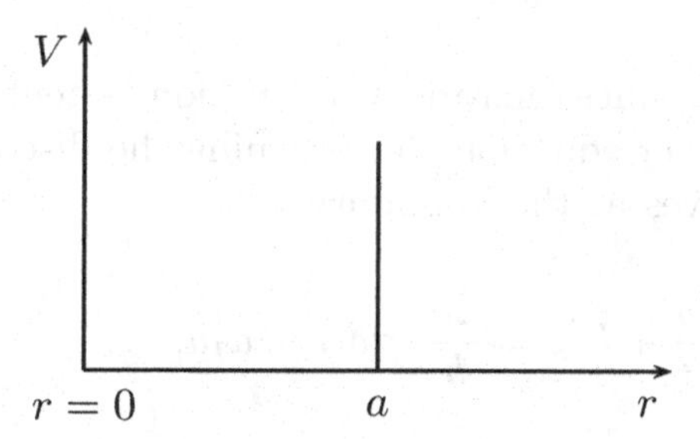

Figure 15.8: A delta potential centered at $r = a$.

Here, γ measures the strength of the potential. Furthermore, since particles have low energy, only the $\ell = 0$ wave or the s wave will be involved in the scattering. Thus, for $r < a$, the radial equation to solve is ($\ell = 0$)

$$\begin{aligned}
&\frac{\mathrm{d}^2 u}{\mathrm{d}r^2} + k^2 u = 0, \\
\text{or,}\quad & u = A \sin kr,
\end{aligned} \tag{15.136}$$

where

$$k^2 = \frac{2mE}{\hbar^2}. \tag{15.137}$$

This solution satisfies the boundary condition that the wave function has to vanish at the origin.

For, $r > a$, the equation is again the same

$$\begin{aligned}
&\frac{\mathrm{d}^2 u}{\mathrm{d}r^2} + k^2 u = 0, \\
\text{or,}\quad & u = B \sin(kr + \delta),
\end{aligned} \tag{15.138}$$

where δ represents the phase shift introduced by scattering.

Matching the solutions in (15.136) and (15.138) at the boundary, we have

$$A \sin ka = B \sin(ka + \delta). \tag{15.139}$$

Furthermore, since there is a δ function potential, we analyze the Schrödinger equation to determine the discontinuity in the first derivatives at the boundary.

$$\begin{aligned}
&\frac{\mathrm{d}^2 u}{\mathrm{d}r^2} + k^2 u = \frac{2m}{\hbar^2}\,\gamma\delta(r-a)u, \\
\text{or,}\quad & \lim_{\epsilon\to 0}\int_{a-\epsilon}^{a+\epsilon} \mathrm{d}r \left(\frac{\mathrm{d}^2 u}{\mathrm{d}r^2} + k^2 u\right) = \int_{a-\epsilon}^{a+\epsilon} \mathrm{d}r \frac{2m\gamma}{\hbar^2}\delta(r-a)u, \\
\text{or,}\quad & \lim_{\epsilon\to 0} \left.\frac{\mathrm{d}u}{\mathrm{d}r}\right|_{a+\epsilon} - \left.\frac{\mathrm{d}u}{\mathrm{d}r}\right|_{a-\epsilon} = \frac{2m\gamma}{\hbar^2}\, u(a), \\
\text{or,}\quad & kB\cos(ka+\delta) - kA\cos ka = \frac{2m\gamma}{\hbar^2}\, B\sin(ka+\delta).
\end{aligned} \tag{15.140}$$

Dividing (15.140) by (15.139), we obtain

$$k\cot(ka+\delta) - k\cot ka = \frac{2m\gamma}{\hbar^2},$$

$$\text{or,}\quad \cot(ka+\delta) = \frac{2m\gamma}{\hbar^2 k} + \cot ka,$$

$$\text{or,}\quad \frac{1-\tan ka\tan\delta}{\tan ka+\tan\delta} = \frac{2m\gamma}{\hbar^2 k} + \cot ka,$$

$$\text{or,}\quad \tan\delta\left(\frac{2m\gamma}{\hbar^2 k} + \cot ka + \tan ka\right) = -\frac{2m\gamma}{\hbar^2 k}\tan ka. \tag{15.141}$$

It follows from (15.141) that

$$\tan\delta = -\frac{\frac{2m\gamma}{\hbar^2 k}\tan ka}{\frac{2m\gamma}{\hbar^2 k} + \frac{1}{\cos ka\sin ka}} = -\frac{\frac{2m\gamma}{\hbar^2 k}\sin^2 ka}{1+\frac{2m\gamma}{\hbar^2 k}\sin ka\cos ka}$$

$$\simeq -\frac{\frac{2m\gamma}{\hbar^2 k}\ k^2a^2}{1+\frac{2m\gamma}{\hbar^2 k}\ ka} \simeq -\frac{\frac{2m\gamma ka^2}{\hbar^2}}{1+\frac{2m\gamma a}{\hbar^2}},$$

$$\cot\delta = \frac{1}{\tan\delta} = -\frac{1+\frac{2m\gamma a}{\hbar^2}}{\frac{2m\gamma ka^2}{\hbar^2}},$$

$$\operatorname{cosec}^2\delta = 1+\cot^2\delta = \frac{\frac{4m^2\gamma^2k^2a^4}{\hbar^4} + \left(1+\frac{2m\gamma a}{\hbar^2}\right)^2}{\left(\frac{2m\gamma ka^2}{\hbar^2}\right)^2}. \tag{15.142}$$

This determines the total scattering cross section, in this case, to be

$$\begin{aligned}
\sigma_{\text{tot}} &= \frac{4\pi}{k^2}\sin^2\delta \\
&= \frac{4\pi}{k^2}\times\frac{\left(\frac{2m\gamma ka^2}{\hbar^2}\right)^2}{1+\frac{4m\gamma a}{\hbar^2}+\frac{4m^2\gamma^2a^2}{\hbar^4}\ (1+k^2a^2)} \\
&\simeq \frac{\frac{16\pi m^2\gamma^2a^4}{\hbar^2}}{1+\frac{4m\gamma a}{\hbar^2}+\frac{4m^2\gamma^2a^2}{\hbar^2}} = 4\pi\left(\frac{\frac{2m\gamma a^2}{\hbar^2}}{1+\frac{2m\gamma a}{\hbar^2}}\right)^2 \\
&= 4\pi a^2\left(\frac{\frac{2m\gamma a}{\hbar^2}}{1+\frac{2m\gamma a}{\hbar^2}}\right)^2.
\end{aligned} \tag{15.143}$$

Once again, we see that the scattering cross section is independent of energy at low energies.

15.5 Inelastic scattering

We have seen, in (15.82) and (15.85), that the total wave function for a scattering process can be written, for large distances, as

$$\begin{aligned}
\psi &= \psi_{\text{inc}} + \psi_{\text{sc}} \\
&\to \sum_{\ell=0}^{\infty} \frac{(2\ell+1)}{2ikr}\, i^{\ell} \left[e^{2i\delta_\ell} e^{i\left(kr-\frac{\ell\pi}{2}\right)} - e^{-i\left(kr-\frac{\ell\pi}{2}\right)}\right] P_\ell(\cos\theta) \\
&= \sum_{\ell=0}^{\infty} \frac{(2\ell+1)}{2ikr} \left[e^{2i\delta_\ell} e^{ikr} - (-1)^{\ell} e^{-ikr}\right] P_\ell(\cos\theta) \\
&= \sum_{\ell=0}^{\infty} \frac{(2\ell+1)}{2ikr} \left[e^{2i\delta_\ell} e^{ikr} + (-1)^{\ell+1} e^{-ikr}\right] P_\ell(\cos\theta).
\end{aligned} \tag{15.144}$$

The phase shifts, δ_ℓ, are all real when the potential is real. Furthermore, let us define

$$S_\ell = e^{2i\delta_\ell}. \tag{15.145}$$

With this, we can write, for large r,

$$\psi \to \sum_{\ell=0}^{\infty} \frac{2\ell+1}{2ikr} \left[S_\ell e^{ikr} + (-1)^{\ell+1} e^{-ikr}\right] P_\ell(\cos\theta),$$

$$\begin{aligned}
\psi^* \frac{\partial}{\partial r}\psi &\to -\sum_{\ell,\ell'} \frac{2\ell'+1}{2ikr}\, \frac{2\ell+1}{2ikr} \\
&\quad \times \left[S^*_{\ell'} e^{-ikr} + (-1)^{\ell'+1} e^{ikr}\right] P_{\ell'}(\cos\theta) \\
&\quad \times (ik) \left[S_\ell e^{ikr} + (-1)^{\ell+2} e^{-ikr}\right] P_\ell(\cos\theta) \\
&= -ik \sum_{\ell,\ell'} \frac{2\ell'+1}{2ikr}\, \frac{2\ell+1}{2ikr} \left[S^*_{\ell'} e^{-ikr} + (-1)^{\ell'+1} e^{ikr}\right] \\
&\quad \times \left[S_\ell e^{ikr} + (-1)^{\ell} e^{-ikr}\right] P_\ell(\cos\theta) P_{\ell'}(\cos\theta) \\
&= -ik \sum_{\ell,\ell'} \frac{2\ell'+1}{2ikr}\, \frac{2\ell+1}{2ikr} \left[S^*_{\ell'} S_\ell + (-1)^{\ell+\ell'+1}\right.
\end{aligned}$$

$$+(-1)^{\ell}S^{*}_{\ell'}e^{-2ikr}-(-1)^{\ell'}S_{\ell}e^{2ikr}\Big]\,P_{\ell}(\cos\theta)P_{\ell'}(\cos\theta). \tag{15.146}$$

Thus, the radial current, at large distances, has the form

$$\begin{aligned}
j_r &= \frac{\hbar}{2mi}\left(\psi^{*}\,\frac{\partial\psi}{\partial r}-\frac{\partial\psi^{*}}{\partial r}\,\psi\right)\\
&\to -\frac{\hbar}{2mi}\,ik\sum_{\ell,\ell'}\frac{2\ell'+1}{2ikr}\,\frac{2\ell+1}{2ikr}\Big[S^{*}_{\ell'}S_{\ell}+(-1)^{\ell+\ell'+1}\\
&\quad+(-1)^{\ell}S^{*}_{\ell'}e^{-2ikr}-(-1)^{\ell'}S_{\ell}e^{2ikr}+S_{\ell'}S^{*}_{\ell}+(-1)^{\ell+\ell'+1}\\
&\quad+(-1)^{\ell}S_{\ell'}e^{2ikr}-(-1)^{\ell'}S^{*}_{\ell}e^{-2ikr}\Big]\,P_{\ell}(\cos\theta)P_{\ell'}(\cos\theta).
\end{aligned} \tag{15.147}$$

Equation (15.147) leads to the flux of probability out of a sphere of large radius R to be

$$\begin{aligned}
\int \mathrm{d}s\; j_r &= \int R^2\sin\theta\mathrm{d}\theta\mathrm{d}\phi\; j_r = 2\pi R^2\int_0^{\pi}\sin\theta\mathrm{d}\theta\; j_r\\
&= 2\pi R^2\frac{(-\hbar)}{2mi}\,ik\sum_{\ell,\ell'}\frac{2\ell'+1}{2ikR}\,\frac{2\ell+1}{2ikR}\left[\;\cdots\;\right]\frac{2}{2\ell+1}\,\delta_{\ell\ell'}\\
&= \frac{4\pi\hbar k}{2m}\sum_{\ell}\frac{2\ell+1}{4k^2}\left(2|S_{\ell}|^2-2\right)\\
&= \frac{\pi\hbar}{mk}\sum_{\ell=0}^{\infty}(2\ell+1)\left(|S_{\ell}|^2-1\right).
\end{aligned} \tag{15.148}$$

We see that if the phase shifts are real,

$$|S_{\ell}|^2 = 1, \tag{15.149}$$

and hence the net flux moving out of a large sphere is zero. This simply is the conservation of probability. It says that the amount of particles that go in is the same as the number of particles that are scattered.

However, there occur, in nature, processes in which the number of particles is not conserved in a scattering process. In fact, when a neutron is scattered off a complex nucleus, two things may happen. The neutron may scatter elastically. It may also scatter inelastically by raising the nucleus to an excited state or may be absorbed by the nucleus. Clearly, this means that the net radial flux out of a large sphere would not vanish in such cases. In fact, it should be negative since we are losing particles. Looking at the expression for the flux in (15.148), it is clear that for this to happen, we must have

$$|S_\ell|^2 < 1. \tag{15.150}$$

This implies that if we parameterize, as in (15.145),

$$S_\ell = e^{2i\delta_\ell}, \tag{15.151}$$

then, δ_ℓ cannot be completely real. In fact, we note that if we write

$$\delta_\ell \to \delta_\ell + i\eta_\ell, \qquad \eta_\ell > 0, \tag{15.152}$$

we would have

$$S_\ell = e^{-2\eta_\ell} e^{2i\delta_\ell}, \tag{15.153}$$

which will lead to $|S_\ell|^2 < 1$, compatible with (15.150).

Thus, we see that a complex phase shift corresponds to non-conservation of probability which is necessary to describe inelastic scattering processes. To understand further what this means, let us go back to the Schrödinger equation (let us assume V is real)

$$\begin{aligned}
i\hbar\,\frac{\partial\psi}{\partial t} &= \left[-\frac{\hbar^2}{2m}\boldsymbol{\nabla}^2 + V\right]\psi,\\
-i\hbar\,\frac{\partial\psi^*}{\partial t} &= \left[-\frac{\hbar^2}{2m}\,\boldsymbol{\nabla}^2 + V\right]\psi^*.
\end{aligned} \tag{15.154}$$

Multiplying the first relation in (15.154) by ψ^* and the second by ψ, and subtracting the two we have

$$\begin{aligned}
&i\hbar\left(\psi^*\frac{\partial\psi}{\partial t} + \psi\frac{\partial\psi^*}{\partial t}\right)\\
&\quad = \psi^*\left[-\frac{\hbar^2}{2m}\boldsymbol{\nabla}^2 + V\right]\psi - \psi\left[-\frac{\hbar^2}{2m}\boldsymbol{\nabla}^2 + V\right]\psi^*,\\
\text{or,}\quad & i\hbar\frac{\partial}{\partial t}(\psi^*\psi) = -\frac{\hbar^2}{2m}\left(\psi^*\boldsymbol{\nabla}^2\psi - \psi\boldsymbol{\nabla}^2\psi^*\right)\\
&\quad = -\frac{\hbar^2}{2m}\,\boldsymbol{\nabla}\cdot\left(\psi^*\boldsymbol{\nabla}\psi - \boldsymbol{\nabla}\psi^*\psi\right),
\end{aligned}$$

$$\text{or,}\quad \frac{\partial}{\partial t}(\psi^*\psi) = -\frac{\hbar}{2mi}\boldsymbol{\nabla}\cdot(\psi^*\boldsymbol{\nabla}\psi - \boldsymbol{\nabla}\psi^*\psi),$$

$$\text{or,}\quad \frac{\partial}{\partial t}\int_{\Omega} \mathrm{d}^3r\ P(\mathbf{r},t) = -\int_{\Omega} \mathrm{d}^3r\ \boldsymbol{\nabla}\cdot\mathbf{j} = -\int_{S} \mathrm{d}\mathbf{s}\cdot\mathbf{j}. \qquad (15.155)$$

This, of course, tells us that if the flux out of a closed area is zero, then the particles are in stationary states and the probability of finding them in a closed volume does not change with time. That is, there are no sources or sinks of particles. This result is, of course, derived by assuming that the potential is real. Suppose we now allow for a complex potential, then, we can follow the earlier derivation and it follows that

$$i\hbar\ \frac{\partial}{\partial t}\ (\psi^*\psi) = -\frac{\hbar^2}{2m}\ \boldsymbol{\nabla}\cdot(\psi^*\boldsymbol{\nabla}\psi - \boldsymbol{\nabla}\psi^*\psi) + (V - V^*)\ \psi^*\psi. \qquad (15.156)$$

Furthermore, if we write

$$V = V_{\mathrm{R}} - iV_{\mathrm{I}}, \qquad (15.157)$$

then, equation (15.156) takes the form

$$\frac{\partial}{\partial t}\ P(\mathbf{r},t) = -\boldsymbol{\nabla}\cdot\mathbf{j} - \frac{2}{\hbar}\ V_{\mathrm{I}}P(\mathbf{r},t),$$

$$\text{or,}\quad \frac{\partial}{\partial t}\ P(\mathbf{r},t) + \boldsymbol{\nabla}\cdot\mathbf{j} = -\frac{2}{\hbar}\ V_{\mathrm{I}}P(\mathbf{r},t). \qquad (15.158)$$

This is the continuity equation in the presence of a complex potential. Furthermore, since $P(\mathbf{r},t)$ is positive it is clear that the potential acts as a source of particles if $V_{\mathrm{I}} < 0$ and as a sink if $V_{\mathrm{I}} > 0$. If we can find a region in which

$$\boldsymbol{\nabla}\cdot\mathbf{j} = 0, \qquad (15.159)$$

then, it follows, from (15.158), that, in that region, we will have

$$P(\mathbf{r},t) \propto e^{-\frac{2}{\hbar}V_{\mathrm{I}}t}. \qquad (15.160)$$

The wave function, therefore, would have a time dependence of the form

$$\psi(\mathbf{r},t) \propto e^{-\frac{i}{\hbar}(E-iV_{\mathrm{I}})t}. \qquad (15.161)$$

This implies that the particle is no longer in a stationary state and the probability of finding the particle in a volume decreases with time.

(This can describe the case where a particle decays. This is one of the ways to look at the non-conservation of probability. An alternate way is through inelastic scattering.)

In scattering theory, however, we assume the wave functions to be stationary states of the form

$$\psi(\mathbf{r},t) \propto e^{-\frac{i}{\hbar}Et}. \tag{15.162}$$

It is clear, therefore, that, in such a case, we cannot find a region where both $V_{\rm I} \neq 0$ and $\boldsymbol{\nabla}\cdot\mathbf{j} = 0$. (This simply means that, to describe processes like absorption in scattering theory, we have to introduce a complex potential, which leads to complex phase shifts which leads to a nonzero flux out of a closed surface.) In fact, integrating the continuity equation, (15.158), over a volume of large dimensions we have

$$\begin{aligned}\frac{\partial}{\partial t}\int_{\Omega} \mathrm{d}^3r\ P(\mathbf{r},t) &= -\int_{\Omega} \mathrm{d}^3r\ \boldsymbol{\nabla}\cdot\mathbf{j} - \frac{2}{\hbar}\int \mathrm{d}^3r\ V_{\rm I}|\psi|^2,\\ \text{or,}\quad 0 &= -\int_S \mathrm{ds}\cdot\mathbf{j} - \frac{2}{\hbar}\int \mathrm{d}^3r\ V_{\rm I}|\psi|^2.\end{aligned} \tag{15.163}$$

The first term on the right hand side simply measures the flux of particles removed from the incident beam. From the definition of the cross section in (15.1) we can, therefore, define the total cross section for absorption as

$$\sigma_{\rm abs} = -\frac{m}{\hbar k}\int_S \mathbf{j}\cdot\mathrm{ds} = -\frac{1}{v}\int_S \mathbf{j}\cdot\mathrm{ds} = \frac{2}{\hbar v}\int_{\Omega} V_{\rm I}|\psi|^2\ \mathrm{d}^3r. \tag{15.164}$$

We have seen, in (15.148), that, for a sphere of large radius, we can write

$$\int_S \mathbf{j}\cdot\mathrm{ds} = \frac{\pi\hbar}{mk}\sum_{\ell=0}^{\infty}(2\ell+1)\left[|S_\ell|^2 - 1\right]. \tag{15.165}$$

The total cross section for absorption can, therefore, be written as

$$\begin{aligned}\sigma_{\rm abs} &= -\frac{m}{\hbar k}\int_S \mathrm{ds}\cdot\mathbf{j} = -\frac{m}{\hbar k}\frac{\pi\hbar}{mk}\sum_{\ell=0}^{\infty}(2\ell+1)\left[|S_\ell|^2 - 1\right]\\ &= \frac{\pi}{k^2}\sum_{\ell=0}^{\infty}(2\ell+1)\left[1 - |S_\ell|^2\right].\end{aligned} \tag{15.166}$$

We also know, from (15.87), that the scattering amplitude is obtained from the phase shift analysis to be

$$f(\theta,\phi) = \sum_{\ell=0}^{\infty} \frac{2\ell+1}{2ik} \left(e^{2i\delta_\ell} - 1\right) P_\ell(\cos\theta)$$
$$= \sum_{\ell=0}^{\infty} \frac{2\ell+1}{2ik} \left(S_\ell - 1\right) P_\ell(\cos\theta), \tag{15.167}$$

so that the total cross section, for elastic scattering, is

$$\sigma_{\rm el} = \int \sin\theta {\rm d}\theta {\rm d}\phi\, |f(\theta,\phi)|^2$$
$$= \frac{\pi}{k^2} \sum_{\ell=0}^{\infty} (2\ell+1)|S_\ell - 1|^2. \tag{15.168}$$

The total cross section, which is the sum of the elastic as well as the absorptive cross sections, is then obtained to be

$$\sigma_{\rm tot} = \sigma_{\rm el} + \sigma_{\rm abs}$$
$$= \frac{\pi}{k^2} \sum_{\ell=0}^{\infty} (2\ell+1) \left[1 - |S_\ell|^2 + |S_\ell - 1|^2\right]$$
$$= \frac{\pi}{k^2} \sum_{\ell=0}^{\infty} (2\ell+1) \left[2 - S_\ell - S_\ell^*\right]$$
$$= \frac{2\pi}{k^2} \sum_{\ell=0}^{\infty} (2\ell+1) \left[1 - {\rm Re}\ S_\ell\right]. \tag{15.169}$$

It is clear that this expression reduces to the familiar expression for elastic scattering in (15.88) when the phase shifts are real.

15.6 Generalized optical theorem

Rather than considering the incident particles to be moving along the z-axis, let us consider a more general situation. Let the incident particle have a momentum $\mathbf{k}$ along some arbitrary direction. The beam is scattered with a momentum $\mathbf{k}'$ along some other arbitrary direction as shown in Fig. 15.9. Thus, for large distances, we can write the total wave function as

$$\psi_k(\mathbf{r}) \rightarrow e^{i\mathbf{k}\cdot\mathbf{r}} + f(\mathbf{k}',\mathbf{k})\, \frac{e^{ikr}}{r}. \tag{15.170}$$

Here $f(\mathbf{k}', \mathbf{k})$ is the scattering amplitude which measures scattering from $\mathbf{k}$ to $\mathbf{k}'$. Furthermore, k is the magnitude of the wave number.

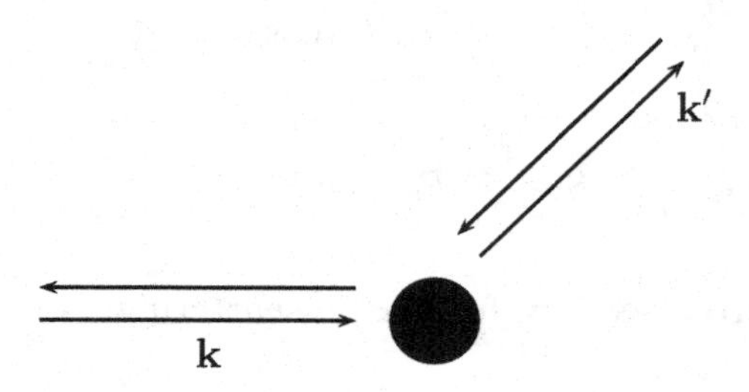

Figure 15.9: An incident particle with momentum $\mathbf{k}$ scattering with momentum $\mathbf{k}'$.

Let us emphasize here that $f(\mathbf{k}', \mathbf{k})$ is general, in the sense that it can account for both elastic as well as inelastic scattering (absorption).

Next, let us consider the time reversed process. Namely, the particle is incident with momentum $-\mathbf{k}'$ and scatters with momentum $-\mathbf{k}$. The wave function for such a process can again be written, for large distances, as

$$\psi_{-k'}(\mathbf{r}) \rightarrow e^{-i\mathbf{k}'\cdot\mathbf{r}} + f(-\mathbf{k}, -\mathbf{k}')\, \frac{e^{ikr}}{r}. \tag{15.171}$$

Since time reversal is a good symmetry of the system, it is clear that the scattering amplitude for the original scattering process must be the same as that for the time reversed process. In other words,

$$f(\mathbf{k}', \mathbf{k}) = f(-\mathbf{k}, -\mathbf{k}'). \tag{15.172}$$

This is known as the reciprocity relation for the scattering amplitude.

The Hamiltonian for scattering is, in addition, invariant under space reflections or the parity operation. Thus, we also have

$$f(\mathbf{k}, \mathbf{k}') = f(-\mathbf{k}, -\mathbf{k}'). \tag{15.173}$$

Comparing (15.172) and (15.173), we obtain

$$f(\mathbf{k}', \mathbf{k}) = f(\mathbf{k}, \mathbf{k}'). \tag{15.174}$$

We know from the continuity equation that, when integrated

over a large sphere of radius r,

$$-\int_S \mathrm{d}\mathbf{s}\cdot\mathbf{j} - \frac{2}{\hbar}\int_\Omega \mathrm{d}^3r\ V_{\mathrm{I}}|\psi_k|^2 = 0,$$

$$\text{or,}\quad r^2\int_0^{2\pi}\mathrm{d}\phi\int_0^{\pi}\sin\theta\mathrm{d}\theta\left(\psi_k^*\ \frac{\partial\psi_k}{\partial r} - \frac{\partial\psi_k^*}{\partial r}\ \psi_k\right) + \frac{4mi}{\hbar^2}\int |\psi_k|^2 V_{\mathrm{I}}\mathrm{d}^3r = 0. \tag{15.175}$$

Putting in the asymptotic form for the wave functions in (15.170), we obtain

$$\begin{aligned} r^2\int_0^{2\pi}\mathrm{d}\phi\int_0^{\pi}\sin\theta\mathrm{d}\theta\Bigg[2ik\cos\theta \\ &+ \left[\frac{ik}{r}\ (1+\cos\theta) - \frac{1}{r^2}\right] f(\mathbf{k}',\mathbf{k})e^{ikr(1-\cos\theta)} \\ &+ \left[\frac{ik}{r}\ (1+\cos\theta) + \frac{1}{r^2}\right] f^*(\mathbf{k}',\mathbf{k})e^{-ikr(1-\cos\theta)} \\ &+ \frac{2ik}{r^2} f^*(\mathbf{k}',\mathbf{k})f(\mathbf{k}',\mathbf{k})\Bigg] \\ &+ \frac{4mi}{\hbar^2}\int \mathrm{d}^3r\ |\psi_k|^2 V_{\mathrm{I}} = 0. \end{aligned} \tag{15.176}$$

The first term in the integral vanishes. The $\frac{1}{r^2}$ terms are negligible for large r compared to the $\frac{1}{r}$ terms in the second and the third parenthesis. Furthermore,

$$\begin{aligned} \int_0^{\pi}\sin\theta\mathrm{d}\theta\ (1+\cos\theta)\ e^{ikr(1-\cos\theta)} &= \int_{-1}^{+1}\mathrm{d}x\ (1+x)\ e^{ikr(1-x)} \\ &= \frac{(1+x)}{-ikr}\ e^{ikr(1-x)}\Big|_{-1}^{+1} + \frac{1}{ikr}\int_{-1}^{+1}\mathrm{d}x\ e^{ikr(1-x)} \\ &= -\frac{2}{ikr} + \frac{1}{(kr)^2}\left(1-e^{2ikr}\right) \\ &\simeq -\frac{2}{ikr}, \qquad \text{for large } r. \end{aligned} \tag{15.177}$$

Thus, we have, from (15.176),

$$2\pi r^2 \left[\frac{ik}{r}\left(-\frac{2}{ikr}\right) f(\mathbf{k},\mathbf{k}) + \frac{ik}{r}\left(\frac{2}{ikr}\right) f^*(\mathbf{k},\mathbf{k})\right] + \frac{2ikr^2}{r^2}\int \mathrm{d}\Omega\ f^*(\mathbf{k}',\mathbf{k}) f(\mathbf{k}',\mathbf{k}) + \frac{4mi}{\hbar^2}\int \mathrm{d}^3r\ |\psi_k|^2 V_{\mathrm{I}} = 0, \tag{15.178}$$

which leads to

$$2ik\int \mathrm{d}\Omega |f|^2 + \frac{4mi}{\hbar^2}\int \mathrm{d}^3r\ |\psi_k|^2 V_{\mathrm{I}} = -4\pi\left[f^*(\mathbf{k},\mathbf{k}) - f(\mathbf{k},\mathbf{k})\right],$$

$$\text{or,}\quad \int \mathrm{d}\Omega |f|^2 + \frac{2m}{\hbar^2 k}\int \mathrm{d}^3r\ |\psi_k|^2 V_{\mathrm{I}} = -\frac{2\pi}{ik}\left(f^* - f\right),$$

$$\text{or,}\quad \sigma_{\mathrm{el}} + \sigma_{\mathrm{abs}} = \frac{4\pi}{k}\ \mathrm{Im}\ f(\mathbf{k},\mathbf{k}),$$

$$\text{or,}\quad \sigma_{\mathrm{tot}} = \frac{4\pi}{k}\ \mathrm{Im}\ f(\mathbf{k},\mathbf{k}) = \frac{4\pi}{k}\ \mathrm{Im}\ f(0). \tag{15.179}$$

This proves the generalized optical theorem which says that, even in the presence of inelastic scattering, the total cross section is related to the imaginary part of the forward scattering amplitude.

15.7 Integral equations for scattering

The central idea behind this method is that rather than analyzing each angular momentum component separately, we can try to obtain the scattering amplitude as a whole by solving an integral equation. It is clear that such an approach is quite useful if a significant number of angular momentum components suffer appreciable scattering. Thus, we try to solve the Schrödinger equation

$$\boldsymbol{\nabla}^2\psi(\mathbf{r}) + k^2\psi(\mathbf{r}) = \frac{2m}{\hbar^2}\ V(\mathbf{r})\psi(\mathbf{r}). \tag{15.180}$$

where $k^2 = \frac{2mE}{\hbar^2}$.

We know how to solve such an equation. This is, in fact, similar to the Poisson equation in electrostatics,

$$\boldsymbol{\nabla}^2\phi = -4\pi\rho. \tag{15.181}$$

We can solve this equation simply by defining the Green's function for the Laplacian,

$$\boldsymbol{\nabla}^2 G(\mathbf{r}) = -4\pi\delta(\mathbf{r}). \tag{15.182}$$

Furthermore, we know the solution of equation (15.182) to be

$$G(\mathbf{r}) = \frac{1}{|\mathbf{r}|}, \tag{15.183}$$

from which it follows that we can write the solution of the Poisson equation, (15.181), as

$$\phi(\mathbf{r}) = \int \mathrm{d}^3 r' \; G\left(\mathbf{r} - \mathbf{r}'\right) \rho\left(\mathbf{r}'\right) = \int \mathrm{d}^3 r' \; \frac{\rho\left(\mathbf{r}'\right)}{|\mathbf{r} - \mathbf{r}'|}. \tag{15.184}$$

Thus, we see that the solution is given by an integral of the sources over all space.

We can extend the same method to our present problem also. We simply have to define a Green's function for the problem. Furthermore, our solution for large spatial distances must have the form

$$\psi(\mathbf{r}) \rightarrow \; e^{i\mathbf{k}\cdot\mathbf{r}} + f(\theta, \phi) \; \frac{e^{ikr}}{r}. \tag{15.185}$$

Let us assume that $G(\mathbf{r}, \mathbf{r}')$ is the Green's function for the operator $(\boldsymbol{\nabla}^2 + k^2)$. That is,

$$\left(\boldsymbol{\nabla}^2 + k^2\right) G\left(\mathbf{r}, \mathbf{r}'\right) = \delta\left(\mathbf{r} - \mathbf{r}'\right). \tag{15.186}$$

We, of course, have to determine the form of $G(\mathbf{r}, \mathbf{r}')$. But, let us note that if such an object can be determined, the solution of the Schrödinger equation can be written as

$$\psi(\mathbf{r}) = \frac{2m}{\hbar^2} \int \mathrm{d}^3 r' \; G\left(\mathbf{r}, \mathbf{r}'\right) V\left(\mathbf{r}'\right) \psi\left(\mathbf{r}'\right). \tag{15.187}$$

This is easily checked as

$$\begin{aligned} \left(\boldsymbol{\nabla}^2 + k^2\right) \psi(\mathbf{r}) &= \frac{2m}{\hbar^2} \int \mathrm{d}^3 r' \left(\boldsymbol{\nabla}^2 + k^2\right) G\left(\mathbf{r}, \mathbf{r}'\right) V\left(\mathbf{r}'\right) \psi\left(\mathbf{r}'\right) \\ &= \frac{2m}{\hbar^2} \int \mathrm{d}^3 r' \; \delta\left(\mathbf{r} - \mathbf{r}'\right) V\left(\mathbf{r}'\right) \psi\left(\mathbf{r}'\right) \\ &= \frac{2m}{\hbar^2} \; V(\mathbf{r}) \psi(\mathbf{r}). \end{aligned} \tag{15.188}$$

Thus, this is like the solution for the Poisson equation except that, in the present case, the solution $\psi(\mathbf{r})$ depends on the source which is a function of $\psi(\mathbf{r})$ itself. Such a solution is known as an integral equation. Furthermore, the above solution is not unique in the sense that

$$\psi(\mathbf{r}) = \psi^{(0)}(\mathbf{r}) + \frac{2m}{\hbar^2} \int \mathrm{d}^3r' \, G\left(\mathbf{r}, \mathbf{r}'\right) V\left(\mathbf{r}'\right) \psi\left(\mathbf{r}'\right), \tag{15.189}$$

is also a solution of the Schrödinger equation if

$$\left(\boldsymbol{\nabla}^2 + k^2\right) \psi^{(0)}(\mathbf{r}) = 0. \tag{15.190}$$

Namely, we can always add any homogeneous solution of a differential equation to a given solution. This non-uniqueness can however, be fixed from the physical requirement that when $V = 0$ (i.e., when there is no scattering), the solution has the form

$$\psi(\mathbf{r}) \rightarrow e^{i\mathbf{k}\cdot\mathbf{r}}. \tag{15.191}$$

Therefore, we can write

$$\begin{aligned}
\psi(\mathbf{r}) &= e^{i\mathbf{k}\cdot\mathbf{r}} + \frac{2m}{\hbar^2} \int \mathrm{d}^3r' \, G(\mathbf{r}, \mathbf{r}') V\left(\mathbf{r}'\right) \psi\left(\mathbf{r}'\right) \\
&= e^{i\mathbf{k}\cdot\mathbf{r}} + \psi_{\mathrm{sc}} \rightarrow e^{i\mathbf{k}\cdot\mathbf{r}} + f(\theta, \phi) \frac{e^{ikr}}{r},
\end{aligned} \tag{15.192}$$

for large distances. By comparison, it is clear that the particular solution of the differential equation represents the scattered wave, namely,

$$\psi_{\mathrm{sc}} = \frac{2m}{\hbar^2} \int \mathrm{d}^3r' \, G(\mathbf{r}, \mathbf{r}') V(\mathbf{r}') \psi(\mathbf{r}'). \tag{15.193}$$

The integral solution of the Schrödinger equation is known as the Lippman-Schwinger solution. However, it is not always easy to solve an integral equation exactly. But, we can solve it iteratively. That is, let us introduce the notation

$$\begin{aligned}
\psi(\mathbf{r}) &= \psi^{(0)}(\mathbf{r}) + \frac{2m}{\hbar^2} \int \mathrm{d}^3r' \, G(\mathbf{r}, \mathbf{r}') V(\mathbf{r}') \psi(\mathbf{r}') \\
&= \psi^{(0)}(\mathbf{r}) + \frac{2m}{\hbar^2} \, G(\mathbf{r}, \mathbf{r}') V(\mathbf{r}') \psi(\mathbf{r}'),
\end{aligned} \tag{15.194}$$

where $\psi^{(0)}(\mathbf{r}) = e^{i\mathbf{k}\cdot\mathbf{r}}$ and integration over the intermediate variables is understood. If we substitute the lowest order form of the solution on the right hand side, we obtain to first order

$$\psi^{(1)}(\mathbf{r}) = \psi^{(0)}(\mathbf{r}) + \frac{2m}{\hbar^2} \, G(\mathbf{r}, \mathbf{r}') V(\mathbf{r}') \psi^{(0)}(\mathbf{r}'). \tag{15.195}$$

Putting this in the right hand side of the integral equation we obtain, to second order,

$$\psi^{(2)}(\mathbf{r}) = \psi^{(0)}(\mathbf{r}) + \frac{2m}{\hbar^2}\, G(\mathbf{r},\mathbf{r}')V(\mathbf{r}')\psi^{(0)}(\mathbf{r}')$$
$$+ \left(\frac{2m}{\hbar^2}\right)^2 G(\mathbf{r},\mathbf{r}')V(\mathbf{r}')G(\mathbf{r}',\mathbf{r}'')V(\mathbf{r}'')\psi^{(0)}(\mathbf{r}''). \tag{15.196}$$

Thus, we can iterate to as many orders as is desirable. This kind of an approximation is known as the Born approximation and is extremely useful if the strength of the potential is not large enough so that only a few orders can be kept in the expansion. (This is, in fact, a perturbative solution for the wave function where the perturbation is the potential.)

15.8 Green's functions

Let us now determine the Green's function for the differential operator governing scattering, which we need for any calculation. The Green's function satisfies the equation

$$\left(\boldsymbol{\nabla}^2 + k^2\right) G\left(\mathbf{r},\mathbf{r}'\right) = \delta\left(\mathbf{r}-\mathbf{r}'\right). \tag{15.197}$$

From the translational invariance of the equation, it is clear that $G(\mathbf{r},\mathbf{r}')$ has to have the form

$$G\left(\mathbf{r},\mathbf{r}'\right) = G\left(\mathbf{r}-\mathbf{r}'\right). \tag{15.198}$$

Furthermore, let us define the Fourier transforms

$$G\left(\mathbf{r}-\mathbf{r}'\right) = \int \mathrm{d}^3q\; e^{i\mathbf{q}\cdot(\mathbf{r}-\mathbf{r}')}\tilde{G}(\mathbf{q}),$$
$$\delta\left(\mathbf{r}-\mathbf{r}'\right) = \frac{1}{(2\pi)^3}\int \mathrm{d}^3q\; e^{i\mathbf{q}\cdot(\mathbf{r}-\mathbf{r}')}. \tag{15.199}$$

Substituting these relations into the differential equation, we obtain an algebraic equation for $\tilde{G}(\mathbf{q})$, namely,

$$\left(-\mathbf{q}^2 + k^2\right)\tilde{G}(\mathbf{q}) = \frac{1}{(2\pi)^3},$$
$$\text{or,}\quad \tilde{G}(q) = -\frac{1}{(2\pi)^3}\frac{1}{q^2-k^2}. \tag{15.200}$$

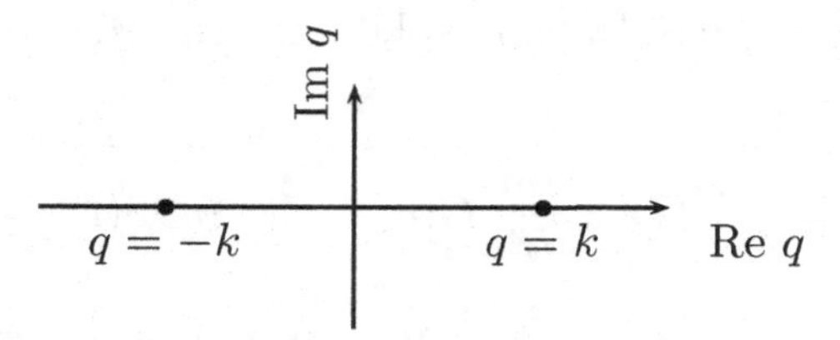

Figure 15.10: The poles of the integrand of the Green's function on the real axis.

Here $q = |\mathbf{q}|$ and we note that the Fourier transform of the Green's function depends only on the magnitude of the momentum vector.

First of all, we note that the function in (15.200) has simple poles at $q = \pm k$ as shown in Fig. 15.10. Thus, to obtain a unique $G(\mathbf{r} - \mathbf{r}')$, we must specify the contour of integration. Each choice of the contour would lead to a distinct Green's function and some of the possible forms of the contour are shown in Fig. 15.11-Fig. 15.13. In the case of a space-time dependent Green's function, the contours in Fig. 15.11 lead to the retarded and advanced Green's functions and are quite useful in the study of classical mechanics. Furthermore, we can also choose the principal value in evaluating the integral which would lead to another Green's function (known as the stationary Green's function). In quantum mechanics, on the other hand, a different Green's function – known as the causal Green's function (also called the Feynman Green's function in the space-time dependent case) – plays an important role and we will study this particular Green's function in the following.

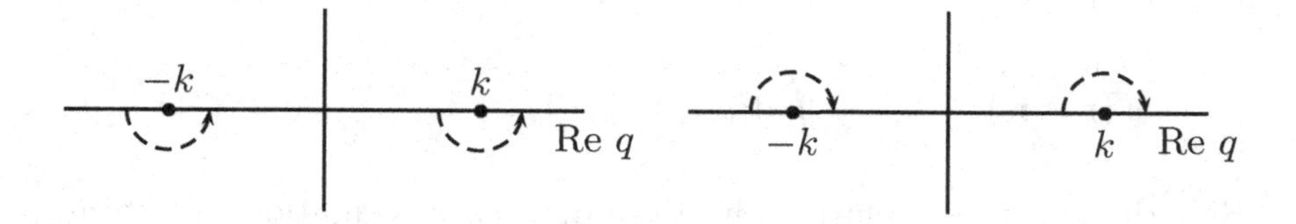

Figure 15.11: Contours yielding advanced and retarded Green's functions.

First, let us note that we can simplify the Green's function by evaluating the angular integrals as follows.

$$G(\mathbf{r}-\mathbf{r}') = \int \mathrm{d}^3q\; e^{i\mathbf{q}\cdot(\mathbf{r}-\mathbf{r}')}\tilde{G}(q)$$
$$= -\frac{1}{(2\pi)^3}\int \mathrm{d}^3q\; \frac{e^{i\mathbf{q}\cdot(\mathbf{r}-\mathbf{r}')}}{q^2-k^2}. \tag{15.201}$$

Let us define

$$\mathbf{r}-\mathbf{r}' = \mathbf{R}, \tag{15.202}$$

so that we have

$$\mathbf{q}\cdot(\mathbf{r}-\mathbf{r}') = qR\cos\theta, \tag{15.203}$$

and this leads to (see (15.201)),

$$G(\mathbf{r}-\mathbf{r}') = -\frac{1}{(2\pi)^3}\int \mathrm{d}\phi\sin\theta\mathrm{d}\theta q^2\mathrm{d}q\;\frac{e^{iqR\cos\theta}}{q^2-k^2}$$
$$= -\frac{1}{(2\pi)^3}\;2\pi\int q^2\mathrm{d}q\sin\theta\mathrm{d}\theta\;\frac{e^{iqR\cos\theta}}{q^2-k^2}$$
$$= -\frac{1}{(2\pi)^2}\int_0^\infty q^2\mathrm{d}q\;\frac{1}{iqR}\,\frac{1}{q^2-k^2}\left(e^{iqR}-e^{-iqR}\right)$$
$$= -\frac{1}{(2\pi)^2 iR}\int_0^\infty \mathrm{d}q\;\frac{q}{q^2-k^2}\left(e^{iqR}-e^{-iqR}\right)$$
$$= -\frac{1}{(2\pi)^2 iR}\int_{-\infty}^{\infty} \mathrm{d}q\;\frac{qe^{iqR}}{q^2-k^2}. \tag{15.204}$$

This integral, of course can be evaluated using Cauchy's residue theorem. The contour has to be closed in the upper half plane because only then will the exponential be damped. If we choose the contour in the way shown in Fig. 15.12, that is, if we enclose only the pole at $q=-k$ then the value of the integral in (15.204) becomes

$$G(\mathbf{r}-\mathbf{r}') = \lim_{q\to -k} -\frac{1}{(2\pi)^2 iR}2\pi i\frac{(q+k)qe^{iqR}}{q^2-k^2}$$
$$= -\frac{1}{(2\pi)^2 iR}2\pi i\frac{(-k)e^{-ikR}}{(-2k)}$$
$$= -\frac{e^{-ikR}}{4\pi R} = -\frac{e^{-ik|\mathbf{r}-\mathbf{r}'|}}{4\pi|\mathbf{r}-\mathbf{r}'|}. \tag{15.205}$$

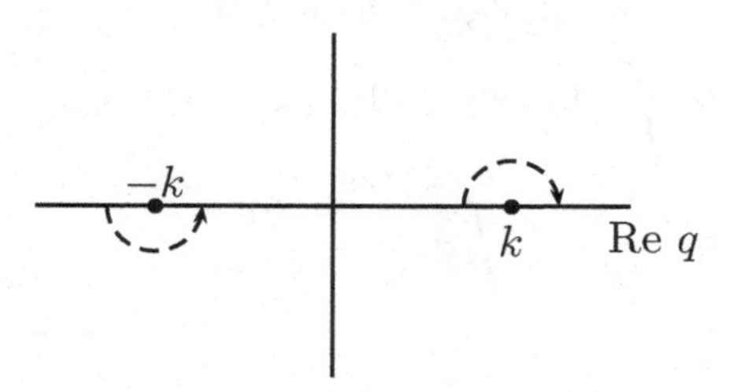

Figure 15.12: An alternate contour in the complex plane for the Green's function.

On the other hand, we can choose to enclose only the pole at $q = k$ by choosing the contour shown in Fig. 15.13 and then the integral in (15.204) would become

$$\begin{aligned} G(\mathbf{r}-\mathbf{r}') &= \lim_{q\to k} -\frac{1}{(2\pi)^2 iR} 2\pi i \frac{(q-k)qe^{iqR}}{q^2-k^2} \\ &= -\frac{1}{(2\pi)^2 iR} 2\pi i \frac{ke^{ikR}}{(2k)} \\ &= -\frac{e^{ikR}}{4\pi R} = -\frac{e^{ik|\mathbf{r}-\mathbf{r}'|}}{4\pi|\mathbf{r}-\mathbf{r}'|}. \end{aligned} \tag{15.206}$$

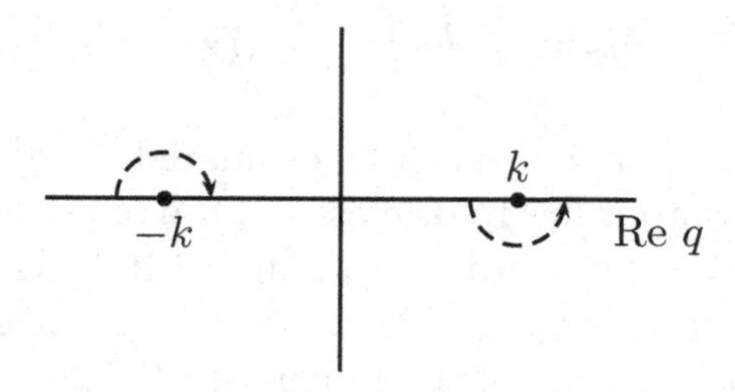

Figure 15.13: Yet another contour for the Green's function.

For the present case, we note that the Green's function is the wave function at $\mathbf{r}$ due to a delta source at $\mathbf{r}'$. Furthermore, since we want the scattered wave function to be outgoing, the contour for the Green's function must be chosen such that it has an outgoing form. Thus, we see that the proper boundary condition is imposed by choosing the second contour, in which case the appropriate Green's function has the form, (15.206),

$$G(\mathbf{r}-\mathbf{r}') = -\frac{e^{ik|\mathbf{r}-\mathbf{r}'|}}{4\pi|\mathbf{r}-\mathbf{r}'|}. \tag{15.207}$$

There is another way we can obtain the correct Green's function. This method is used extensively in quantum field theory and is due to Feynman. The idea is to change the denominator of the momentum space Green's function by an infinitesimal imaginary amount as in Fig. 15.14. That is, let us define

$$\begin{aligned}
\tilde{G}(q) &\propto \lim_{\epsilon\to 0^+} \frac{1}{q^2-k^2-i\epsilon} \\
&\simeq \lim_{\epsilon\to 0^+} \frac{1}{q-\left(k+\frac{i\epsilon}{2k}\right)^2} = \lim_{\eta\to 0^+} \frac{1}{q^2-(k+i\eta)^2} \\
&= \lim_{\eta\to 0^+} \frac{1}{(q+k+i\eta)(q-k-i\eta)}.
\end{aligned} \tag{15.208}$$

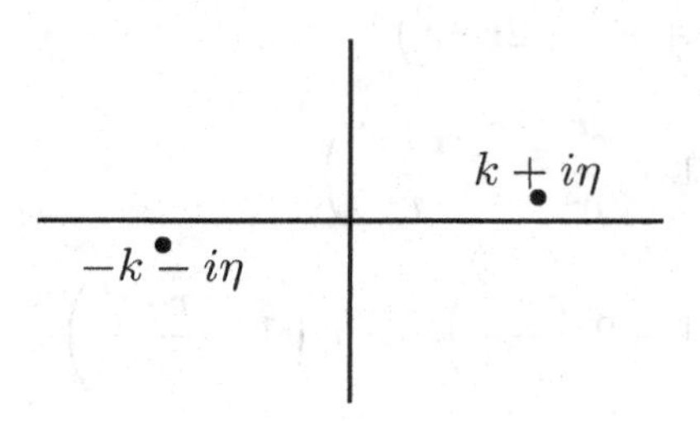

Figure 15.14: An equivalent way of denoting the contour in Fig. 15.13 by shifting the poles into the upper and the lower half planes by giving an infinitesimal imaginary part to the momentum.

Thus, now the poles of the Green's function move away from the real axis and there is no ambiguity in choosing the contour.

$$G(\mathbf{r}-\mathbf{r}') = \lim_{\eta\to 0^+} \lim_{q\to k+i\eta} -\frac{1}{(2\pi)^2 iR} 2\pi i \frac{(q-k-i\eta)qe^{iqR}}{q^2-(k+i\eta)^2}$$

$$= \lim_{\eta\to 0^+} -\frac{1}{(2\pi)^2 iR}\, 2\pi i(k+i\eta)\, \frac{e^{i(k+i\eta)R}}{2(k+i\eta)}$$

$$= -\frac{e^{ikR}}{4\pi R} = -\frac{e^{ik|\mathbf{r}-\mathbf{r}'|}}{4\pi|\mathbf{r}-\mathbf{r}'|}. \tag{15.209}$$

This prescription gives the correct Green's function and is known as Feynman's $i\epsilon$ prescription.

Now that we know the form of the Green's function, we can write down the solution of the Schrödinger equation, (15.192) as

$$\psi(\mathbf{r}) = e^{i\mathbf{k}\cdot\mathbf{r}} + \frac{2m}{\hbar^2}\int \mathrm{d}^3r' G(\mathbf{r}-\mathbf{r}')V(\mathbf{r}')\psi(\mathbf{r}')$$

$$= e^{i\mathbf{k}\cdot\mathbf{r}} + \frac{2m}{\hbar^2}\left(-\frac{1}{4\pi}\right)\int \mathrm{d}^3r'\, \frac{e^{ik|\mathbf{r}-\mathbf{r}'|}}{|\mathbf{r}-\mathbf{r}'|}\, V(\mathbf{r}')\psi(\mathbf{r}'). \tag{15.210}$$

We see that the first Born approximation gives

$$\psi(\mathbf{r}) = e^{i\mathbf{k}\cdot\mathbf{r}} - \frac{m}{2\pi\hbar^2}\int \mathrm{d}^3r'\, \frac{e^{ik|\mathbf{r}-\mathbf{r}'|}}{|\mathbf{r}-\mathbf{r}'|}\, V(\mathbf{r}')e^{i\mathbf{k}\cdot\mathbf{r}'}. \tag{15.211}$$

We are interested in the form of the wave function as $r\to\infty$. Since the range over which the potential is appreciable is finite, $r' \ll r$. Thus, we can expand

$$|\mathbf{r}-\mathbf{r}'| = \left(r^2 + r'^2 - 2\mathbf{r}\cdot\mathbf{r}'\right)^{\frac{1}{2}}$$

$$= r\left(1 + \frac{r'^2}{r^2} - 2\,\frac{\mathbf{r}\cdot\mathbf{r}'}{r^2}\right)^{\frac{1}{2}}$$

$$\simeq r\left(1 - 2\,\frac{\mathbf{r}\cdot\mathbf{r}'}{r^2}\right)^{\frac{1}{2}} \simeq r\left(1 - \frac{\mathbf{r}\cdot\mathbf{r}'}{r^2}\right). \tag{15.212}$$

This leads to

$$\frac{1}{|\mathbf{r}-\mathbf{r}'|} \simeq \frac{1}{r\left(1-\frac{\mathbf{r}\cdot\mathbf{r}'}{r^2}\right)} \simeq \frac{1}{r}\left(1 + \frac{\mathbf{r}\cdot\mathbf{r}'}{r^2}\right). \tag{15.213}$$

Thus we see that, for large r, we can replace

$$\frac{1}{|\mathbf{r}-\mathbf{r}'|} \simeq \frac{1}{r}. \tag{15.214}$$

On the other hand,

$$e^{ik|\mathbf{r}-\mathbf{r}'|} \simeq e^{ikr\left(1-\frac{\mathbf{r}\cdot\mathbf{r}'}{r^2}\right)} \simeq e^{ikr - i\mathbf{k}_f\cdot\mathbf{r}'}, \tag{15.215}$$

where $\mathbf{k}_f = k\hat{\mathbf{r}}$ is the momentum of the outgoing particle. It has the same magnitude as the initial momentum but is along the radial direction.

Substituting (15.214) and (15.215) into (15.211), we get, for large r,

$$\begin{aligned}\psi(\mathbf{r}) &= e^{i\mathbf{k}\cdot\mathbf{r}} - \frac{m}{2\pi\hbar^2}\int \mathrm{d}^3r' \ \frac{e^{ikr}}{r} \ e^{-i\mathbf{k}_f\cdot\mathbf{r}'}V(\mathbf{r}')e^{i\mathbf{k}\cdot\mathbf{r}'} \\ &= e^{i\mathbf{k}\cdot\mathbf{r}} - \frac{e^{ikr}}{r}\ \frac{m}{2\pi\hbar^2}\int \mathrm{d}^3r' \ e^{-i\mathbf{k}_f\cdot\mathbf{r}'}V(\mathbf{r}')e^{i\mathbf{k}\cdot\mathbf{r}'} \\ &= e^{i\mathbf{k}\cdot\mathbf{r}} + f(\theta,\phi)\ \frac{e^{ikr}}{r}. \end{aligned} \tag{15.216}$$

Thus, comparing, we determine the form of the scattering amplitude in the first Born approximation to be

$$\begin{aligned} f\left(\mathbf{k}_f,\mathbf{k}_i\right) &= -\frac{m}{2\pi\hbar^2}\int \mathrm{d}^3r' \ e^{-i\mathbf{k}_f\cdot\mathbf{r}'}V(\mathbf{r}')e^{i\mathbf{k}_i\cdot\mathbf{r}'} \\ &= -\frac{m}{2\pi\hbar^2}\int \mathrm{d}^3r' \, e^{i(\mathbf{k}_i-\mathbf{k}_f)\cdot\mathbf{r}'}\ V(\mathbf{r}'). \end{aligned} \tag{15.217}$$

This shows that the scattering amplitude, in the first Born approximation, is proportional to the Fourier transform of the potential with respect to the momentum transfer $\mathbf{q} = \mathbf{k}_i - \mathbf{k}_f$. The angular dependence of the scattering amplitude is contained in the factor $e^{i\mathbf{q}\cdot\mathbf{r}'}$.

15.9 Validity of the Born approximation

As we have seen, the total wave function is given by

$$\begin{aligned}\psi(\mathbf{r}) &= e^{i\mathbf{k}\cdot\mathbf{r}} + \psi_{\text{sc}} \\ &= e^{i\mathbf{k}\cdot\mathbf{r}} - \frac{m}{2\pi\hbar^2}\int \mathrm{d}^3r' \ \frac{e^{ik|\mathbf{r}-\mathbf{r}'|}}{|\mathbf{r}-\mathbf{r}'|}\ V(\mathbf{r}')\psi(\mathbf{r}'). \end{aligned} \tag{15.218}$$

In the first Born approximation, we replace the wave function on the right hand side of (15.218), under the integral, by the incident wave. Thus,

$$\psi_B(\mathbf{r}) = e^{i\mathbf{k}\cdot\mathbf{r}} - \frac{m}{2\pi\hbar^2}\int \mathrm{d}^3r' \ \frac{e^{ik|\mathbf{r}-\mathbf{r}'|}}{|\mathbf{r}-\mathbf{r}'|}\ V(\mathbf{r}')e^{i\mathbf{k}\cdot\mathbf{r}'}. \tag{15.219}$$

We expect this approximation to be a good approximation if in the range in which the potential influences strongly, we have

$$|\psi_{\text{sc}}| \ll |e^{i\mathbf{k}\cdot\mathbf{r}}| = 1. \tag{15.220}$$

Since the influence of the potential is the maximum at the origin, if $|\psi_{\rm sc}(0)| \ll 1$, then this approximation would be a good approximation.

Let us assume, for simplicity, that $V(\mathbf{r}') = V(r')$, namely, it is rotationally invariant. Then, the Born approximation will be valid if

$$
\begin{aligned}
& |\psi_{\rm sc}(0)| \ll 1, \\
\text{or,}\quad & \left| \frac{m}{2\pi\hbar^2} \int \mathrm{d}^3r' \ \frac{e^{ikr'}}{r'} \ V(r')e^{i\mathbf{k}\cdot\mathbf{r}'} \right| \ll 1, \\
\text{or,}\quad & \left| \frac{m}{2\pi\hbar^2} 2\pi \int r'^2 \mathrm{d}r' \ \frac{e^{ikr'}}{r'} V(r') \frac{1}{ikr'} \left(e^{ikr'} - e^{-ikr'}\right) \right| \ll 1, \\
\text{or,}\quad & \frac{2m}{\hbar^2 k} \left| \int_0^\infty \mathrm{d}r' \ e^{ikr'} V(r') \sin kr' \right| \ll 1.
\end{aligned} \tag{15.221}
$$

This is the condition for validity of the Born approximation.

At low energies, $kr' \to 0$ which implies that $\sin kr' \simeq kr'$ and $e^{ikr'} \simeq 1$. Thus, the condition for validity, (15.221), becomes

$$
\begin{aligned}
& \frac{2m}{\hbar^2 k} \left| \int_0^\infty \mathrm{d}r' \ V(r') \ kr' \right| \ll 1, \\
\text{or,}\quad & \frac{2m}{\hbar^2} \left| \int_0^\infty \mathrm{d}r' \ r'V(r') \right| \ll 1.
\end{aligned} \tag{15.222}
$$

Therefore, we see that if the potential has a height V_0 and range r_0, then the condition for validity at low energies, (15.222), becomes

$$
\frac{m|V_0|r_0^2}{\hbar^2} \ll 1. \tag{15.223}
$$

At high energies, $kr' \to \infty$, on the other hand, the exponential in (15.221) oscillates rapidly and picks up contribution only for $r' \sim \frac{1}{k}$. Thus, the condition for the validity of the Born approximation , in this case, becomes (for a square well as before)

$$
\frac{2m}{\hbar^2 k} \left| \int_0^{\frac{1}{k}} \mathrm{d}r' \, e^{ikr'} \sin kr' V(r') \right| \ll 1,
$$

$$\text{or,}\quad \frac{2m}{\hbar^2 k}\left|\int_0^{\frac{1}{k}} \mathrm{d}r'\, e^{ikr'}\left(\frac{e^{ikr'}-e^{-ikr'}}{2i}\right)V(r')\right| \ll 1,$$

$$\text{or,}\quad \frac{2m}{\hbar^2 k}\,\frac{|V_0|}{2}\,\frac{1}{k} \ll 1,$$

$$\text{or,}\quad \frac{m|V_0|r_0^2}{\hbar^2} \ll (kr_0)^2. \tag{15.224}$$

We see, from (15.223) and (15.224), that if the Born approximation is valid at low energies, it is valid at all energies.

► **Example (Square well potential).** Let us consider low energy scattering from a square well of the form

$$V(\mathbf{r}) = \begin{cases} V_0 & r < a, \\ 0 & r > a. \end{cases} \tag{15.225}$$

In this case, the Born approximation, (15.217), gives

$$f_B(\mathbf{k}_f, \mathbf{k}_i) = -\frac{m}{2\pi\hbar^2}\int \mathrm{d}^3r'\, e^{i\mathbf{q}\cdot\mathbf{r}'}V(\mathbf{r}'), \tag{15.226}$$

where we have defined

$$\mathbf{q} = \mathbf{k}_i - \mathbf{k}_f, \tag{15.227}$$

so that we have

$$q^2 = \left(k_i^2 + k_f^2 - 2\mathbf{k}_i\cdot\mathbf{k}_f\right). \tag{15.228}$$

On the other hand, we have

$$k_i^2 = k_f^2 = k^2, \qquad \mathbf{k}_i\cdot\mathbf{k}_f = k^2\cos\theta. \tag{15.229}$$

Using (15.229), we obtain

$$\begin{aligned} q^2 &= \left(k^2 + k^2 - 2k^2\cos\theta\right) \\ &= 2k^2(1-\cos\theta) = 4k^2\sin^2\frac{\theta}{2}, \quad \Rightarrow \quad q = 2k\sin\frac{\theta}{2}. \end{aligned} \tag{15.230}$$

Consequently, we now obtain from (15.226)

$$\begin{aligned} f_B(\mathbf{k}_f, \mathbf{k}_i) &= -\frac{m}{2\pi\hbar^2}\int_0^{2\pi}\mathrm{d}\phi'\int_0^{\pi}\sin\theta'\mathrm{d}\theta'\int_0^a r'^2\mathrm{d}r' e^{iqr'\cos\theta'}V_0 \\ &= -\frac{mV_0}{2\pi\hbar^2}\,2\pi\int_0^a r'^2\mathrm{d}r'\,\frac{1}{iqr'}\left(e^{iqr'} - e^{-iqr'}\right) \\ &= -\frac{mV_0}{iq\hbar^2}\int_0^a \mathrm{d}r' r'\left(e^{iqr'} - e^{-iqr'}\right) \end{aligned}$$

$$= -\frac{mV_0}{iq\hbar^2} \int_{-a}^{a} \mathrm{d}r' r' e^{iqr'}$$

$$= -\frac{mV_0}{iq\hbar^2} \left[\frac{1}{iq} \, r' e^{iqr'} \Big|_{-a}^{a} - \frac{1}{iq} \int_{-a}^{a} \mathrm{d}r' e^{iqr'} \right]$$

$$= -\frac{mV_0}{iq\hbar^2} \left[\frac{1}{iq} \, a \left(e^{iqa} + e^{-iqa} \right) - \frac{1}{(iq)^2} \left(e^{iqa} - e^{-iqa} \right) \right]$$

$$= \frac{2mV_0}{q^2\hbar^2} \left[a \cos qa - \frac{1}{q} \sin qa \right]. \tag{15.231}$$

Since k is small at low energies, it follows from (15.230) that $qa \ll 1$. Thus, we can write

$$f_B \simeq \frac{2mV_0}{q^2\hbar^2} \left[a \left(1 - \frac{q^2 a^2}{2!} \right) - \frac{1}{q} \left(qa - \frac{(qa)^3}{3!} \right) \right]$$

$$= \frac{2mV_0}{q^2\hbar^2} \left[-\frac{q^2 a^3}{2} + \frac{q^2 a^3}{6} \right]$$

$$= \frac{2mV_0}{q^2\hbar^2} \left(-\frac{2q^2 a^3}{6} \right) = -\frac{2mV_0 a^3}{3\hbar^2}. \tag{15.232}$$

Therefore, in this approximation, the total scattering cross section becomes

$$\sigma_{\text{tot}} = \int \mathrm{d}\Omega |f_B|^2 = 4\pi |f_B|^2 = 4\pi \left(-\frac{\kappa^2 a^3}{3} \right)^2, \tag{15.233}$$

where

$$\kappa^2 = \frac{2mV_0}{\hbar^2}. \tag{15.234}$$

The condition for validity of the Born approximation at low energies, in this case, is

$$\kappa a \ll 1. \tag{15.235}$$

◀

Exercise. Show that under this assumption the phase shift analysis also gives the same scattering cross section.

► **Example (Gaussian potential).** As a second example, let us consider scattering from the potential

$$V(\mathbf{r}) = V_0 e^{-\alpha^2 r^2}. \tag{15.236}$$

The Born amplitude, in this case, is obtained to be

$$
\begin{aligned}
f_B &= -\frac{m}{2\pi\hbar^2}\int \mathrm{d}^3r'\ e^{i\mathbf{q}\cdot\mathbf{r}'}\ V(\mathbf{r}') \\
&= -\frac{m}{2\pi\hbar^2}\int_0^{2\pi}\mathrm{d}\phi'\int_0^{\pi}\sin\theta'\mathrm{d}\theta'\int_0^{\infty} r'^2\mathrm{d}r'\ e^{iqr'\cos\theta'}\ V_0e^{-\alpha^2r'^2} \\
&= -\frac{mV_0}{2\pi\hbar^2}\ 2\pi\int_0^{\infty}\mathrm{d}r' r'^2 e^{-\alpha^2r'^2}\frac{1}{iqr'}\left(e^{iqr'}-e^{-iqr'}\right) \\
&= -\frac{mV_0}{i\hbar^2 q}\int_{-\infty}^{\infty}\mathrm{d}r'\ r' e^{-\alpha^2r'^2+iqr'} \\
&= -\frac{mV_0}{i\hbar^2 q}\int_{-\infty}^{\infty}\mathrm{d}r'\ r' e^{-\alpha^2(r'-\frac{iq}{2\alpha^2})^2-\frac{q^2}{4\alpha^2}} \\
&= -\frac{mV_0}{i\hbar^2 q}e^{-\frac{q^2}{4\alpha^2}}\int_{-\infty}^{\infty}\mathrm{d}r'\ (r'+\frac{iq}{2\alpha^2})e^{-\alpha^2r'^2} \\
&= -\frac{mV_0}{i\hbar^2 q}e^{-\frac{q^2}{4\alpha^2}}\frac{iq}{2\alpha^2}\frac{\sqrt{\pi}}{\alpha} \\
&= -\frac{\sqrt{\pi}mV_0}{2\hbar^2\alpha^3}e^{-\frac{q^2}{4\alpha^2}}.
\end{aligned}
\tag{15.237}
$$

Thus, the differential cross section, in this approximation, is given by

$$
\begin{aligned}
\sigma(\theta,\phi) = |f_B|^2 &= \frac{\pi m^2V_0^2}{4\hbar^4\alpha^6}\exp\left(-\frac{q^2}{2\alpha^2}\right) \\
&= \frac{\pi m^2V_0^2}{4\hbar^4\alpha^6}\exp\left(-\frac{2k^2\sin^2\frac{\theta}{2}}{\alpha^2}\right).
\end{aligned}
\tag{15.238}
$$

We can show that the Born approximation is valid for all energies in this case. Therefore, this is a good formula for the differential scattering cross section. Furthermore, the total cross section which is obtained by integrating this quantity over all angles is also finite. ◀

15.10 Coulomb scattering

The Coulomb potential is probably the most familiar of all potentials. It has a long range. In fact, the range of the potential is infinite. Therefore, the conventional phase shift analysis does not apply in the case of scattering from a Coulomb potential. The reason is not very hard to understand. In the usual phase shift analysis, we assume that the incident wave is a plane wave. However, because the Coulomb potential has an infinite range, it is clear that even at infinite separation the incident particles feel the force and, therefore, cannot be represented by a plane wave. Furthermore, the long range also has the consequence that even at low energies, quite a large number of partial waves suffer appreciable scattering. Thus, we see that the partial wave analysis is not the proper way to handle Coulomb scattering.

Let us note here that because of the special form of the Coulomb potential, Coulomb scattering can be exactly solved in the parabolic coordinates. It is also clear from (15.222) that, even at low energies, the Born approximation for Coulomb scattering is not valid. The condition for validity of Born approximation at low energies is

$$\frac{2m}{\hbar^2}\left|\int_0^\infty \mathrm{d}r\ rV(r)\right| \ll 1. \tag{15.239}$$

However, since $V(r) = \frac{Ze^2}{r}$,

$$\int_0^\infty \mathrm{d}r\ rV(r) \to \infty. \tag{15.240}$$

As a result, the Born approximation breaks down. (This simply means that we have to go to higher orders in the iteration.) However, we get around this difficulty in the following way. Consider the scattering potential to be

$$V(r) = \frac{\alpha e^{-\mu r}}{r}, \tag{15.241}$$

where α and μ are constants. It is clear now that $\frac{1}{\mu}$ defines the range of the potential. This potential is known as the Yukawa potential because of Yukawa's postulate that the form of the potential due to exchange of mesons between nucleons is the one given above.

If we are considering low energy scattering from the Yukawa potential, the condition for validity of the Born approximation is

$$\begin{aligned}\frac{2m}{\hbar^2}\left|\int_0^\infty \mathrm{d}r\ rV(r)\right| &= \frac{2m}{\hbar^2}\left|\int_0^\infty \mathrm{d}r\ r\ \frac{\alpha e^{-\mu r}}{r}\right| \\ &= \frac{2m}{\hbar^2}\left|\frac{\alpha e^{-\mu r}}{-\mu}\right|_0^\infty = \left|\frac{2m\alpha}{\hbar^2\mu}\right| \ll 1. \end{aligned} \tag{15.242}$$

Thus, we see that the Born approximation can be valid in this case. We can show that, in this case, the Born approximation is also valid at high energies.

Let us now calculate the Born amplitude for scattering from the Yukawa potential.

$$\begin{aligned}
f_B &= -\frac{m}{2\pi\hbar^2}\int \mathrm{d}^3r'\ e^{i\mathbf{q}\cdot\mathbf{r}'}V(r') \\
&= -\frac{m}{2\pi\hbar^2}\ 2\pi\int_0^\infty \mathrm{d}r' r'^2\ \frac{\alpha e^{-\mu r'}}{r'}\ \frac{1}{iqr'}\left(e^{iqr'}-e^{-iqr'}\right) \\
&= -\frac{m\alpha}{i\hbar^2 q}\int_0^\infty \mathrm{d}r'\left(e^{(-\mu+iq)r'}-e^{(-\mu-iq)r'}\right) \\
&= -\frac{m\alpha}{i\hbar^2 q}\left[\frac{e^{(-\mu+iq)r'}}{-\mu+iq}-\frac{e^{(-\mu-iq)r'}}{-\mu-iq}\right]_0^\infty \\
&= -\frac{m\alpha}{i\hbar^2 q}\left[\frac{1}{\mu-iq}-\frac{1}{\mu+iq}\right] \\
&= -\frac{2m\alpha}{\hbar^2\left(\mu^2+q^2\right)} = -\frac{2m\alpha}{\hbar^2\left(\mu^2+4k^2\sin^2\frac{\theta}{2}\right)},
\end{aligned} \tag{15.243}$$

where we have used

$$q^2 = 4k^2\sin^2\frac{\theta}{2}. \tag{15.244}$$

Thus, we obtain the differential scattering cross section, in this approximation, to be

$$\sigma(\theta,\phi) = |f_B(\theta,\phi)|^2 = \frac{4m^2\alpha^2}{\hbar^4\left(\mu^2+4k^2\sin^2\frac{\theta}{2}\right)^2}. \tag{15.245}$$

Let us note, from (15.241), that the Coulomb potential can be obtained as a limit of the Yukawa potential, namely, it is a potential with an infinite range. In other words, when

$$\begin{aligned}
&\mu = 0, \quad \alpha = Ze^2, \\
&V_{\text{Yukawa}} \to \frac{Ze^2}{r} = V_{\text{Coulomb}}.
\end{aligned} \tag{15.246}$$

Thus, in this limit, we obtain the differential cross section for the Coulomb scattering, from (15.245), to be

$$\begin{aligned}\sigma_{\text{Coulomb}}(\theta,\phi) &= \frac{4m^2(Ze^2)^2}{\hbar^4\left(4k^2\sin^2\frac{\theta}{2}\right)^2}\\ &= \left(\frac{mZe^2}{2\hbar^2k^2}\right)^2 \operatorname{cosec}^4\frac{\theta}{2}\\ &= \left(\frac{Ze^2}{4E}\right)^2 \operatorname{cosec}^4\frac{\theta}{2}.\end{aligned} \tag{15.247}$$

We recognize this to be the exact classical formula for the differential cross section for a Coulomb potential (see (15.32)). This also happens to be the exact quantum mechanical result. This is again one of the accidental results associated with the Coulomb potential. Let us note here that although this trick of obtaining Coulomb scattering gives the correct differential cross section, the Born amplitude is not the correct Coulomb amplitude. Rather, the exact scattering amplitude is given by

$$f_C(\theta,\phi) = f_B(\theta,\phi)\exp\left(-i\,\frac{mZe^2}{\hbar^2k}\,\ln\sin^2\frac{\theta}{2} + i\eta\right), \tag{15.248}$$

where η is a constant. Namely, the exact scattering amplitude is the Born amplitude multiplied by a phase factor whose effect is irrelevant in the differential cross section.

15.11 Scattering of identical particles

So far, we have assumed that the scattering process involves distinguishable particles. However, if the particles being scattered are indistinguishable, i.e., suppose that we are scattering electrons off electrons, then one has to take into account the symmetry properties that the wave function has to satisfy. Thus, for example, suppose we are scattering identical bosons, then the total wave function has to be symmetric under the interchange of coordinates. For simplicity, let us assume that the bosons are spinless so that the wave function only depends on the spatial coordinates.

$$\psi_{\text{tot}}\left(\mathbf{r}_1,\mathbf{r}_2\right) = \psi_{\text{CM}}(\mathbf{r}_{\text{CM}})\psi(\mathbf{r}), \tag{15.249}$$

where

$$\mathbf{r}_{\text{CM}} = \frac{\mathbf{r}_1+\mathbf{r}_2}{2}, \qquad \mathbf{r} = \mathbf{r}_1 - \mathbf{r}_2. \tag{15.250}$$

The total wave function has to be symmetric. We see that, under an exchange of coordinates,

$$\mathbf{r}_{\rm CM} \to \mathbf{r}_{\rm CM}, \quad \psi_{\rm CM}\left(\mathbf{r}_{\rm CM}\right) \to \psi_{\rm CM}\left(\mathbf{r}_{\rm CM}\right). \tag{15.251}$$

Under an exchange of the coordinates,

$$\mathbf{r} \to -\mathbf{r}, \tag{15.252}$$

we must, therefore, have

$$\psi(\mathbf{r}) \to \psi(-\mathbf{r}). \tag{15.253}$$

We note that

$$\mathbf{r} \to -\mathbf{r} \quad \Rightarrow \quad r \to r,\ \theta \to \pi - \theta,\ \phi \to \pi + \phi. \tag{15.254}$$

Since we have to symmetrize the wave function, for large distances, we can write

$$\psi(\mathbf{r}) \to \ e^{ikz} + e^{-ikz} + f(\theta,\phi)\frac{e^{ikr}}{r} + f(\pi-\theta,\pi+\phi)\frac{e^{ikr}}{r}. \tag{15.255}$$

In this case, we define a symmetrized scattering amplitude as

$$f_{\rm sym}(\theta,\phi) = [f(\theta,\phi) + f(\pi-\theta,\pi+\phi)]. \tag{15.256}$$

The differential scattering cross section, in this case, becomes

$$\begin{aligned} \sigma(\theta,\phi) &= |f_{\rm sym}(\theta,\phi)|^2 \\ &= \Big[|f(\theta,\phi)|^2 + |f(\pi-\theta,\pi+\phi)|^2 \\ &\quad + 2{\rm Re} f(\theta,\phi) f^*(\pi-\theta,\pi+\phi)\Big]. \end{aligned} \tag{15.257}$$

The first two terms are, of course, what we would obtain if we had two distinguishable particles. The cross terms represent the quantum interference that accompanies whenever identical particles scatter.

Let us next consider the scattering of two identical fermions. Suppose that the fermions are two electrons. Then, $s = \frac{1}{2}$ for each of them and they can be in a state with total angular momentum equal to one or zero. Correspondingly, we say that they are in the triplet state or in the singlet state. The triplet state is symmetric in the spin space. Since the total wave function has to be anti-symmetric, this implies that the wave function has to be anti-symmetric in its space coordinates. But the CM wave function is symmetric under exchange of coordinates and, therefore, we must have

$$\psi_{\text{triplet}}(\mathbf{r}) = -\psi_{\text{triplet}}(-\mathbf{r}). \tag{15.258}$$

This leads to the form of the scattering amplitude in the triplet state to be

$$f_{\text{triplet}}(\theta,\phi) = [f(\theta,\phi) - f(\pi-\theta,\pi+\phi)]. \tag{15.259}$$

As a result, the differential cross section takes the form

$$\begin{aligned}\sigma_{\text{triplet}} &= |f_{\text{triplet}}|^2 \\ &= \left[|f(\theta,\phi)|^2 + |f(\pi-\theta,\pi+\phi)|^2\right. \\ &\quad \left. -2\text{Re}f(\theta,\phi)f^*(\pi-\theta,\pi+\phi)\right].\end{aligned} \tag{15.260}$$

On the other hand, if the electrons are in the singlet state, then the wave function is anti-symmetric in the spin space. This means that the wave function has to be symmetric in its spatial coordinates. Since $\psi_{CM}(\mathbf{r}_{CM})$ is already symmetric, this implies

$$\psi_{\text{singlet}}(\mathbf{r}) = \psi_{\text{singlet}}(-\mathbf{r}). \tag{15.261}$$

As a result, the form of the singlet state scattering amplitude is determined to be

$$f_{\text{singlet}}(\theta,\phi) = [f(\theta,\phi) + f(\pi-\theta,\pi+\phi)], \tag{15.262}$$

so that

$$\begin{aligned}\sigma_{\text{singlet}}(\theta,\phi) &= |f_{\text{singlet}}(\theta,\phi)|^2 \\ &= \left[|f(\theta,\phi)|^2 + |f(\pi-\theta,\pi+\phi)|^2\right. \\ &\quad \left. +2\text{Re}f(\theta,\phi)f^*(\pi-\theta,\pi+\phi)\right].\end{aligned} \tag{15.263}$$

In most scattering experiments, however, the particles can form either the singlet or the triplet state. Consequently, we define, in such cases, a spin averaged cross section. Thus, for example, in the case of two electrons, there are four final states available out of which three belong to the triplet state and one to the singlet. Therefore, the triplet state is three times as likely as the singlet state and a spin averaged differential cross section is given by

$$\begin{aligned}
\sigma_{\rm av}(\theta,\phi) &= \frac{1}{4}\left[3\sigma_{\rm triplet}+\sigma_{\rm singlet}\right] \\
&= \frac{1}{4}\left[4|f(\theta,\phi)|^2+4|f(\pi-\theta,\pi+\phi)|^2\right. \\
&\qquad \left. -4{\rm Re}f(\theta,\phi)f^*(\pi-\theta,\pi+\phi)\right] \\
&= \left[|f(\theta,\phi)|^2+|f(\pi-\theta,\pi+\phi)|^2\right. \\
&\qquad \left. -{\rm Re}f(\theta,\phi)f^*(\pi-\theta,\pi+\phi)\right].
\end{aligned} \tag{15.264}$$

For example, if we take the exact form of the Coulomb scattering amplitude in (15.248),

$$\begin{aligned}
f_C(\theta,\phi) &= f_B(\theta,\phi)\exp\left(-\frac{ime^2}{\hbar^2 k}\ln\sin^2\frac{\theta}{2}+i\eta\right) \\
&= -\frac{me^2}{2\hbar^2k^2\sin^2\frac{\theta}{2}}\exp\left(-\frac{ime^2}{\hbar^2 k}\ln\sin^2\frac{\theta}{2}+i\eta\right) \\
&= -\frac{e^2}{4E\sin^2\frac{\theta}{2}}\exp\left(-\frac{ime^2}{\hbar^2 k}\ln\sin^2\frac{\theta}{2}+i\eta\right),
\end{aligned} \tag{15.265}$$

we obtain

$$\begin{aligned}
\sigma_{\rm av}^{\rm Coulomb}(\theta,\phi) &= \left(\frac{e^2}{4E}\right)^2 \\
&\times\left[\frac{1}{\sin^4\frac{\theta}{2}}+\frac{1}{\cos^4\frac{\theta}{2}}-\frac{\cos\left(\frac{me^2}{\hbar^2 k}\ln\tan^2\frac{\theta}{2}\right)}{\sin^2\frac{\theta}{2}\cos^2\frac{\theta}{2}}\right].
\end{aligned} \tag{15.266}$$

Note that even though the Born approximation gives the exact cross section for the Coulomb scattering in the simple case, since the amplitude is not exact, it would lead to a wrong cross section in this case, simply because the interference terms would be different.

15.12 Selected problems

1. A particle is scattered by a potential at a sufficiently low energy that $\delta_\ell = 0$ for $\ell > 1$. Assume that the potential is invariant under rotations.

   a) Show that the differential scattering cross section has the form

$$\sigma(\theta,\phi) = A + B\cos\theta + C\cos^2\theta, \tag{15.267}$$

and determine A, B, C in terms of the phase shifts.

b) Determine the total scattering cross section in terms of A, B, C.

c) Assume that the differential cross section is known for $\theta = 90°$ ($\sigma = \alpha^2$), $\theta = 180°$ ($\sigma = \beta^2$) and $\theta = 45°$ ($\sigma = \gamma^2$). Determine $\sigma(\theta,\phi)$ for $\theta = 0°$ in terms of α, β, γ.

d) Obtain the imaginary part of the forward scattering amplitude in terms of α, β, γ.

2. What must V_0a^2 be for a three dimensional square well potential (attractive) so that the scattering cross section is zero at zero bombarding energy (Ramsauer-Townsend effect)?

3. Determine the total scattering cross section for particles of low energy (namely, keep only $\ell = 0$) in a potential

$$V(\mathbf{r}) = \frac{\alpha}{r^4}, \qquad \alpha > 0, \tag{15.268}$$

where $r = |\mathbf{r}|$.

4. Consider a spherically symmetric repulsive potential

$$V(\mathbf{r}) = \frac{\alpha}{r^2}, \qquad \alpha > 0, \tag{15.269}$$

where $r = |\mathbf{r}|$. Use the first Born approximation to calculate the angular dependence as well as the energy dependence of the differential cross section.

5. Derive the Greens functions for one dimensional scattering,

$$\frac{\mathrm{d}^2G(x,x')}{\mathrm{d}x^2} + k^2G(x,x') = \delta(x-x'), \tag{15.270}$$

satisfying the property that

$$G(x,x') \sim \begin{cases} f_+ e^{ik(x-x')}, & x-x' \to \infty, \\ f_- e^{-ik(x-x')}, & x-x' \to -\infty. \end{cases} \tag{15.271}$$

6. Let $V(x,y,z) = 0$ everywhere except inside a cube of length a defined by $x^2 \leq (\frac{a}{2})^2, y^2 \leq (\frac{a}{2})^2, z^2 \leq (\frac{a}{2})^2$. Inside the cube, the potential is a constant V_0. A plane wave e^{ikz} is incident on the cube. What is the differential cross section for scattering as a function of the angles (θ, ϕ) in the Born approximation?

7. Consider a situation where, after scattering, particles are "ingoing". Find the wave function for such a scattering problem as a series in $P_\ell(\cos\theta)$ for spherically symmetric potentials. The "ingoing" solutions have the form

$$\psi^{(-)} = e^{ikz} + \psi_{\rm sc}^{(-)} \to e^{ikz} + f^{(-)}(\theta)\,\frac{e^{-ikr}}{r}, \tag{15.272}$$

whereas the outgoing solutions discussed in this chapter have the form

$$\psi^{(+)} = e^{ikz} + \psi_{\rm sc}^{(+)} \to e^{ikz} + f^{(+)}(\theta)\,\frac{e^{ikr}}{r}. \tag{15.273}$$

What relation exists between $f^{(+)}(\theta)$ and $f^{(-)}(\theta)$?

CHAPTER 16

Relativistic one particle equations

We have so far studied only non-relativistic quantum mechanical systems. In the following lectures, we would like to extend the discussion to a relativistic quantum mechanical system describing a single particle.

16.1 Klein-Gordon equation

In non-relativistic quantum mechanics, we start from a classical Hamiltonian system and promote each observable to an operator. As we have seen, the Schrödinger equation, for a free particle, for example, has the form

$$i\hbar\frac{\partial}{\partial t} = H\psi = \frac{\mathbf{p}^2}{2m}\psi. \tag{16.1}$$

In fact, this method is not relativistic in the sense that we have started from a Hamiltonian formalism which is not relativistically invariant. This is reflected in the fact that the time variable is singled out in the Schrödinger equation.

A relativistic formulation of a dynamical system, on the other hand, must treat both space and time symmetrically. A wave function which would give a relativistic description of the system must, therefore, satisfy an equation which is the same in different Lorentz frames. Before, proceeding any further, however, let us introduce some standard notation. The length interval that is invariant under a Lorentz transformation is given by

$$c^2t^2 - \mathbf{x}^2. \tag{16.2}$$

This is different from rotations in three dimensions which leave the length $\mathbf{x}^2$ unchanged.

The three dimensional space, that we are used to, is called a Euclidean space where the metric for the space is the Kronecker delta and hence we do not distinguish between covariant and contravariant vectors. In the case of Lorentz transformations, on the other hand, we note that if we combine time and space into a four vector $(ct, \mathbf{x})$, in the definition of the length, there is a relative negative sign between the time and the space components. Such a space is known as a Minkowski space and the metric is defined to be

$$\eta_{\mu\nu} = \text{diagonal}\,(1,-1,-1,-1) = \begin{pmatrix} 1 & 0 & 0 & 0 \\ 0 & -1 & 0 & 0 \\ 0 & 0 & -1 & 0 \\ 0 & 0 & 0 & -1 \end{pmatrix}. \tag{16.3}$$

The inverse metric also has the same form, namely,

$$\eta^{\mu\nu} = \text{diagonal}\,(1,-1,-1,-1), \tag{16.4}$$

so that

$$\eta^{\mu\lambda}\eta_{\lambda\nu} = \delta^{\mu}_{\nu}. \tag{16.5}$$

Since the metric is not positive definite, in this case, we have to distinguish between a covariant and a contravariant vector. The contravariant coordinate four vector, x^{μ}, is defined to be

$$x^{\mu} = (ct, \mathbf{x}), \qquad \mu = 0, 1, 2, 3. \tag{16.6}$$

The covariant four vector x_{μ} is then obtained from (16.6) to be

$$x_{\mu} = \eta_{\mu\nu}x^{\nu} = (ct, -\mathbf{x}), \tag{16.7}$$

since the metric raises and lowers the indices. (The covariant and the contravariant vectors transform differently under a Lorentz transformation – something we will not go into.) The contravariant derivative (contragradient) is defined to be

$$\partial^{\mu} = \frac{\partial}{\partial x_{\mu}} = \left(\frac{1}{c}\frac{\partial}{\partial t}, -\boldsymbol{\nabla}\right). \tag{16.8}$$

Similarly, the covariant derivative (cogradient) is defined as

$$\partial_{\mu} = \frac{\partial}{\partial x^{\mu}} = \left(\frac{1}{c}\frac{\partial}{\partial t}, \boldsymbol{\nabla}\right). \tag{16.9}$$

The scalar product, in this space, is defined to be

$$\begin{aligned} a \cdot b &= a^{\mu}b^{\nu}\eta_{\mu\nu} = a_{\mu}b_{\nu}\eta^{\mu\nu} = a_{\mu}b^{\mu} = a^{\mu}b_{\mu} \\ &= a^{0}b^{0} - \mathbf{a}\cdot\mathbf{b}. \end{aligned} \tag{16.10}$$

Here summation over repeated indices is understood. In this space, the invariant length of an arbitrary vector, a^{μ}, is, then, obtained to be

$$a^2 = a \cdot a = (a^0)^2 - \mathbf{a} \cdot \mathbf{a}. \tag{16.11}$$

An expression is Lorentz invariant if and only if all the Lorentz indices in it have been contracted. That is, there is no free Lorentz index available.

We can define an invariant length interval in this space as

$$\mathrm{d}s^2 = \eta_{\mu\nu}\mathrm{d}x^{\mu}\mathrm{d}x^{\nu} = (\mathrm{d}x^0)^2 - (\mathrm{d}x^1)^2 - (\mathrm{d}x^2)^2 - (\mathrm{d}x^3)^2. \tag{16.12}$$

A vector, a^{μ}, is said to be time like, if

$$a^2 = a_{\mu}a^{\mu} = (a^0)^2 - \mathbf{a} \cdot \mathbf{a} > 0. \tag{16.13}$$

If

$$a^2 < 0, \tag{16.14}$$

then it is said to be space like. An intermediate case is when

$$a^2 = 0. \tag{16.15}$$

In this case, we say that the vector is light like.

To develop the relativistic equation for a free scalar particle, let us recall that the non-relativistic formulation of quantum mechanics corresponds to starting with the classical relation $E = \frac{\mathbf{p}^2}{2m}$ in the absence of any interaction. We then promote both E and $\mathbf{p}$ to be operators and denote them as

$$E \to i\hbar\frac{\partial}{\partial t}, \qquad \mathbf{p} \to -i\hbar\boldsymbol{\nabla}. \tag{16.16}$$

Relativistically, however, the classical relation between energy and momentum is given by

$$E^2 = \mathbf{p}^2c^2 + m^2c^4. \tag{16.17}$$

Here, m is the rest mass of the particle. Thus, we see that the relativistically invariant relation in (16.17) can be written as

$$\begin{aligned} & E^2 - \mathbf{p}^2c^2 = m^2c^4, \\ \text{or,} \quad & \frac{E^2}{c^2} - \mathbf{p}^2 = m^2c^2. \end{aligned} \tag{16.18}$$

This relation is true in all Lorentz frames and is known as Einstein's relation. Furthermore, we can define a momentum four vector (consisting of energy and momentum) as

$$p^{\mu} = \left(\frac{E}{c}, \mathbf{p}\right),$$

$$p_{\mu} = \eta_{\mu\nu} p^{\nu} = \left(\frac{E}{c}, -\mathbf{p}\right), \tag{16.19}$$

so that the relation, (16.18), can be written as

$$p_{\mu} p^{\mu} = \frac{E^2}{c^2} - \mathbf{p}^2 = m^2 c^2. \tag{16.20}$$

The Lorentz invariant nature of this relation is now manifest. Furthermore, we note that, since $E \to i\hbar\frac{\partial}{\partial t}$ and $\mathbf{p} \to -i\hbar\boldsymbol{\nabla}$, we can write (see (16.19))

$$p^{\mu} \to i\hbar\partial^{\mu} = i\hbar\frac{\partial}{\partial x_{\mu}} = \left(\frac{i\hbar}{c}\frac{\partial}{\partial t}, -i\hbar\boldsymbol{\nabla}\right),$$

$$p_{\mu} \to i\hbar\partial_{\mu} = i\hbar\frac{\partial}{\partial x^{\mu}} = \left(\frac{i\hbar}{c}\frac{\partial}{\partial t}, i\hbar\boldsymbol{\nabla}\right). \tag{16.21}$$

Thus, the simplest relativistic wave equation that we can write down is

$$p_{\mu} p^{\mu} \psi = m^2 c^2 \psi. \tag{16.22}$$

Since ψ is assumed to be a scalar wave function and since $p_{\mu}p^{\mu}$ is a Lorentz invariant quantity, this equation holds true in all Lorentz frames. Putting in the differential forms for the four momentum operators in (16.21), we have

$$i\hbar\partial_{\mu}\left(i\hbar\partial^{\mu}\right)\psi = m^2 c^2 \psi,$$

$$\text{or,}\quad -\hbar^2\left(\frac{1}{c^2}\frac{\partial^2}{\partial t^2} - \boldsymbol{\nabla}^2\right)\psi = m^2 c^2 \psi,$$

$$\text{or,}\quad \left(\frac{1}{c^2}\frac{\partial^2}{\partial t^2} - \boldsymbol{\nabla}^2\right)\psi = -\frac{m^2 c^2}{\hbar^2}\psi,$$

$$\text{or,}\quad \Box\psi = -\frac{m^2 c^2}{\hbar^2}\psi,$$

$$\text{or,}\quad \left(\Box + \frac{m^2 c^2}{\hbar^2}\right)\psi = 0, \tag{16.23}$$

where we have defined

$$\square = \partial_\mu \partial^\mu = \frac{1}{c^2}\frac{\partial^2}{\partial t^2} - \boldsymbol{\nabla}^2. \tag{16.24}$$

This is known as the D'Alembertian operator (It is the analog of the Laplacian in the four dimensional Minkowski space.) and the above equation is known as the Klein-Gordon equation. If $m = 0$, namely, if the particle is massless, then we know that this equation represents the wave equation describing traveling waves.

In fact, the Klein-Gordon equation, (16.24), also has plane wave solutions (much like the wave equation) of the form

$$\psi(\mathbf{r}, t) \sim e^{-i(\omega t - \mathbf{k}\cdot\mathbf{r})}. \tag{16.25}$$

It is clear that these plane waves are eigenfunctions of $i\hbar\frac{\partial}{\partial t}$ and $-i\hbar\boldsymbol{\nabla}$ with eigenvalues

$$E = \hbar\omega, \qquad \mathbf{p} = \hbar\mathbf{k}. \tag{16.26}$$

Furthermore, substitution of the plane wave solutions in (16.25) into the Klein-Gordon equation, (16.24), gives the relation

$$\begin{aligned} (\hbar\omega)^2 &= (\hbar c\mathbf{k})^2 + m^2c^4, \\ \text{or,} \qquad \hbar\omega &= \pm\sqrt{(\hbar c\mathbf{k})^2 + m^2c^4}. \end{aligned} \tag{16.27}$$

Thus, we see that the Klein-Gordon equation has plane wave solutions with positive as well as negative energy. This is a consequence of the fact that the Klein-Gordon equation, unlike the Schrödinger equation, is a second order equation in the time derivative.

Writing out explicitly, the Klein-Gordon equation, (16.24), has the form

$$\left(\boldsymbol{\nabla}^2 - \frac{1}{c^2}\frac{\partial^2}{\partial t^2}\right)\psi = \frac{m^2c^2}{\hbar^2}\psi. \tag{16.28}$$

Similarly, the complex conjugate of this equation gives

$$\left(\boldsymbol{\nabla}^2 - \frac{1}{c^2}\frac{\partial^2}{\partial t^2}\right)\psi^* = \frac{m^2c^2}{\hbar^2}\psi^*. \tag{16.29}$$

Multiplying equation (16.28) with ψ^* and equation (16.29) with ψ, and subtracting the two, we obtain

$$\psi^*\left(\boldsymbol{\nabla}^2-\frac{1}{c^2}\frac{\partial^2}{\partial t^2}\right)\psi-\psi\left(\boldsymbol{\nabla}^2-\frac{1}{c^2}\frac{\partial^2}{\partial t^2}\right)\psi^*=0,$$

$$\text{or,}\quad \frac{1}{c^2}\left(\psi^*\frac{\partial^2\psi}{\partial t^2}-\psi\frac{\partial^2\psi^*}{\partial t^2}\right)=\left(\psi^*\boldsymbol{\nabla}^2\psi-\psi\boldsymbol{\nabla}^2\psi^*\right),$$

$$\text{or,}\quad \frac{1}{c^2}\frac{\partial}{\partial t}\left(\psi^*\frac{\partial}{\partial t}-\psi\frac{\partial\psi^*}{\partial t}\right)=\boldsymbol{\nabla}\cdot\left(\psi^*\boldsymbol{\nabla}\psi-\psi\boldsymbol{\nabla}\psi^*\right),$$

$$\text{or,}\quad \frac{\partial}{\partial t}\left(-\frac{\hbar}{2mic^2}\left(\psi^*\frac{\partial}{\partial t}-\psi\frac{\partial\psi^*}{\partial t}\right)\right)$$

$$=-\frac{\hbar}{2mi}\boldsymbol{\nabla}\cdot\left(\psi^*\boldsymbol{\nabla}\psi-\psi\boldsymbol{\nabla}\psi^*\right),$$

$$\text{or,}\quad \frac{\partial}{\partial t}\rho(\mathbf{r},t)=-\boldsymbol{\nabla}\cdot\mathbf{J}(\mathbf{r},t), \tag{16.30}$$

where

$$\rho(\mathbf{r},t)=\frac{i\hbar}{2mc^2}\left(\psi^*\frac{\partial}{\partial t}-\psi\frac{\partial\psi^*}{\partial t}\right), \tag{16.31}$$

and $\mathbf{J}(\mathbf{r},t)$ is the usual probability current density in (3.103).

Putting in the form of the plane wave solution in (16.25), it is clear that

$$\rho(\mathbf{r},t)=\frac{\hbar\omega}{mc^2}. \tag{16.32}$$

This quantity is not positive definite since $\hbar\omega$ can be positive as well as negative (see (16.27)). Thus, unlike the case of the Schrödinger equation, $\rho(\mathbf{r},t)$, in the present case, cannot be thought of as a probability density, which has to be strictly non-negative. For this reason, the Klein-Gordon equation was abandoned for a long time as being inadequate. Pauli and Weisskopf resurrected it long after the Dirac equation by reinterpreting the quantities $\rho(\mathbf{r},t)$ and $\mathbf{J}(\mathbf{r},t)$. The interpretation is roughly as follows. If we do not consider the Klein-Gordon equation as a single particle equation, but consider it as a field equation, then after quantizing the fields properly, the associated energy comes out to be positive even though the parameter $\hbar\omega$ can have both positive and negative values. Furthermore, the quantities $\rho(\mathbf{r},t)$ and $\mathbf{J}(\mathbf{r},t)$, when multiplied by the electric charge associated with the field, simply represent respectively the electric charge density and the electric current density associated with the theory. These quantities can, of course, become negative without leading to any inconsistency.

16.2 Dirac equation

It is clear that the difficulty with the Klein-Gordon equation arises because it is second order in the time derivative. Thus, to avoid such difficulties, we look for a first order equation of the form

$$i\hbar\frac{\partial\psi}{\partial t} = H\psi. \tag{16.33}$$

Relativistic invariance would require that the equation be symmetrical in space and time coordinates. Thus, H is required to be linear in the momentum operators. If we take the usual energy momentum relation in classical physics (see (16.17)), we obtain

$$E = \sqrt{\mathbf{p}^2c^2 + m^2c^4}, \tag{16.34}$$

where we are keeping only the positive root, it is clear that this is not a linear function of the momenta. Thus, Dirac introduced the following form of the Hamiltonian in (16.33)

$$H = c\boldsymbol{\alpha}\cdot\mathbf{p} + \beta mc^2, \tag{16.35}$$

where the quantities $\boldsymbol{\alpha}$ and β are assumed to be independent of coordinates and momenta. With this, equation (16.33) becomes

$$i\hbar\frac{\partial\psi}{\partial t} = c\left(\boldsymbol{\alpha}\cdot\mathbf{p} + \beta mc\right)\psi, \tag{16.36}$$

is symmetrical in space and time. Furthermore, since $\boldsymbol{\alpha}$ and β are independent of coordinates and momenta, they commute with all such operators. However, they need not commute among themselves, i.e.,

$$[\alpha_i, \alpha_j] \neq 0, \qquad [\alpha_i, \beta] \neq 0. \tag{16.37}$$

Furthermore, if H is the correct Hamiltonian, then by squaring it, we should get back the relation from relativity (16.17) (the Einstein relation). That is,

$$\begin{aligned}
c^2\mathbf{p}^2 + m^2c^4 = H^2 &= c^2\left(\boldsymbol{\alpha}\cdot\mathbf{p} + \beta mc\right)^2 \\
&= c^2\left(\alpha_i p_i \alpha_j p_j + \left(\alpha_i p_i \beta mc + \beta mc\alpha_i p_i\right) + \beta^2 m^2 c^2\right) \\
&= c^2\left(p_i p_j \frac{1}{2}\left\{\alpha_i, \alpha_j\right\} + mcp_i\left\{\alpha_i, \beta\right\} + \beta^2 m^2 c^2\right).
\end{aligned} \tag{16.38}$$

Thus, we see that the left hand side of (16.38) will be equal to the right hand side, only if

$$\{\alpha_i, \alpha_j\} = 2\delta_{ij}\mathbb{1}, \quad \{\alpha_i, \beta\} = 0, \quad \beta^2 = \mathbb{1}, \tag{16.39}$$

where the curly bracket denotes an anti-commutator. The four quantities, α_i and β, must, therefore, anti-commute in pairs and their squares must be equal to unity. Since these quantities anti-commute rather than commute, they cannot be scalars. In fact, they are matrices.

Let us determine the dimensionality of these matrices. We know, from (16.39), that

$$\alpha_i\beta + \beta\alpha_i = 0, \qquad \text{for any} \quad i. \tag{16.40}$$

Let us further assume that these matrices are non-singular, i.e., their inverses exist. Then, we can write (16.40) also as

$$\alpha_i\beta\alpha_i^{-1} = -\beta, \quad \text{for any } i. \tag{16.41}$$

If we assume the matrix β to be diagonal, that is, of the form

$$\beta = \begin{pmatrix} b_1 & & & \\ & b_2 & & \\ & & \ddots & \\ & & & b_n \end{pmatrix}, \tag{16.42}$$

then, since $\beta^2 = 1$, this implies that all the diagonal elements can only be ± 1.

If we now take the trace of the relation in (16.41), we obtain

$$\text{Tr}\left(\alpha_i\beta\alpha_i^{-1}\right) = -\text{Tr}\ \beta, \tag{16.43}$$

which, upon using the cyclicity of the trace

$$\text{Tr}\ (ABC) = \text{Tr}\ (CAB), \tag{16.44}$$

leads to

$$\text{Tr}\ \beta = -\text{Tr}\ \beta, \quad \text{or} \quad \text{Tr}\ \beta = 0. \tag{16.45}$$

That is, all the diagonal elements in β must add up to zero. Since each element is ± 1, this is possible only if the matrices are even dimensional. Thus, we conclude that $\boldsymbol{\alpha}$ and β are $2n \times 2n$ matrices.

If $n = 1$, the matrices are 2-dimensional. As we already know, there are four linearly independent 2×2 matrices which we can represent by $(\mathbb{1}, \boldsymbol{\sigma})$. Although σ_i's anti-commute among themselves, the identity matrix commutes with every matrix. In other words, we cannot find four anti-commuting matrices in two dimensions.

The next possibility is $n = 2$. In this case, we can, in fact, find four anti-commuting 4×4 matrices. They lead to the simplest form of the Dirac equation. It is also clear that since the Hamiltonian is now a 4×4 matrix operator, the wave functions on which it acts must be four component column matrices. We can contrast this with the non-relativistic electron wave function which has only two components. We will solve the Dirac equation and study the physical significance of the four components shortly.

Thus, we see that, for Dirac equation to be compatible with relativity, the minimum dimensionality of the $\boldsymbol{\alpha}$ and β matrices must be 4×4. Furthermore, since

$$H = c\boldsymbol{\alpha} \cdot \mathbf{p} + \beta mc^2, \tag{16.46}$$

it is clear that $\boldsymbol{\alpha}$ and β must be Hermitian so that the Hamiltonian is Hermitian. It is not necessary to know the exact forms of the matrices $\boldsymbol{\alpha}$ and β for most physical calculations (In fact, the forms are not unique.). All we really need to know is that

$$\begin{aligned}
&\alpha_i^\dagger = \alpha_i, \qquad \beta^\dagger = \beta,\\
&\alpha_i^2 = \mathbb{1} = \beta^2, \quad \text{for each } i,\\
&\{\alpha_i, \beta\} = 0, \quad \text{for all } i,\\
&\{\alpha_i, \alpha_j\} = 0 \quad i \neq j.
\end{aligned} \tag{16.47}$$

However, to have an idea of the forms of these matrices, let us define

$$\begin{aligned}
\alpha_i &= \begin{pmatrix} 0 & \sigma_i \\ \sigma_i & 0 \end{pmatrix},\\
\beta &= \begin{pmatrix} \mathbb{1} & 0 \\ 0 & \mathbb{1} \end{pmatrix},
\end{aligned} \tag{16.48}$$

where the identity submatrices in (16.48) correspond to 2×2 matrices. It is clear from the properties of the Pauli matrices that

$$\alpha_i^2 = \beta^2 = \mathbb{1}, \quad \text{for every} \quad i.$$

Furthermore,

$$\{\alpha_i, \alpha_j\} = \begin{pmatrix} \{\sigma_i, \sigma_j\} & 0 \\ 0 & \{\sigma_i, \sigma_j\} \end{pmatrix} = 0, \quad \text{for } i \neq j,$$

$$\{\alpha_i, \beta\} = \begin{pmatrix} 0 & -\sigma_i \\ \sigma_i & 0 \end{pmatrix} + \begin{pmatrix} 0 & \sigma_i \\ -\sigma_i & 0 \end{pmatrix} = 0, \quad \text{for all } i.$$

Let us note here that the forms of the matrices in (16.48) are not unique. If we make a similarity transformation

$$\alpha_i' = S^{-1}\alpha_i S,$$
$$\beta' = S^{-1}\beta S. \tag{16.49}$$

clearly, α_i' and β' would also satisfy all the relations in (16.47) which α_i and β satisfy.

It is clear that since the Hamiltonian is now a 4×4 matrix, the wave function will be a four component column matrix. Thus, let us write

$$\psi(\mathbf{r}, t) = \begin{pmatrix} \psi_1(\mathbf{r}, t) \\ \psi_2(\mathbf{r}, t) \\ \psi_3(\mathbf{r}, t) \\ \psi_4(\mathbf{r}, t) \end{pmatrix}. \tag{16.50}$$

The Dirac equation is given by

$$\left(c\boldsymbol{\alpha} \cdot \mathbf{p} + \beta mc^2\right) \psi(\mathbf{r}, t) = i\hbar \frac{\partial \psi(\mathbf{r}, t)}{\partial t}. \tag{16.51}$$

This is a matrix equation or equivalently four simultaneous partial differential equations. First of all, we note that since the wave function has more than one component, it must be connected with a particle with a nontrivial spin and we recall from our discussions in chapter 14 (see (14.36)) that

$$\mathcal{E}^{(\text{total})} = \mathcal{E}^{(\text{space})} \otimes \mathcal{E}^{(\text{spin})}. \tag{16.52}$$

We can, therefore, write the wave function in (16.50) as a product of a matrix and a plane wave.

$$\psi(\mathbf{r}, t) = u\, e^{-\frac{i}{\hbar}(Et - \mathbf{p}\cdot\mathbf{r})}, \tag{16.53}$$

where u is a coordinate independent four component matrix. This will represent a solution of the free Dirac equation if

$$\left(c\boldsymbol{\alpha} \cdot \mathbf{p} + \beta mc^2\right) u = Eu. \tag{16.54}$$

Remembering the non-relativistic case of the electron, we think of u as a four component spinor. To solve for u, we simplify the problem by assuming that the particle is moving along the z-axis. In this case, equation (16.54) becomes

$$\begin{aligned}&\left(c\alpha_z p_z + \beta mc^2\right) u = Eu,\\ \text{or,}\quad &\begin{pmatrix} mc^2\mathbb{1} & cp_z\sigma_3 \\ cp_z\sigma_3 & -mc^2\mathbb{1}\end{pmatrix} u = Eu.\end{aligned} \tag{16.55}$$

This can be written explicitly as

$$\begin{pmatrix} mc^2 - E & 0 & cp_z & 0 \\ 0 & mc^2 - E & 0 & -cp_z \\ cp_z & 0 & -mc^2 - E & 0 \\ 0 & -cp_z & 0 & -mc^2 - E \end{pmatrix} \begin{pmatrix} u_1 \\ u_2 \\ u_3 \\ u_4 \end{pmatrix} = 0. \tag{16.56}$$

There would exist a nontrivial solution of the coupled homogeneous set of equations in (16.56) only if the determinant of the coefficient matrix vanishes. That is, for a nontrivial solution, we must have

$$\begin{aligned}&\det\begin{pmatrix} \left(mc^2 - E\right)\mathbb{1} & cp_z\sigma_3 \\ cp_z\sigma_3 & -\left(mc^2 + E\right)\mathbb{1}\end{pmatrix} = 0,\\ \text{or,}\quad &\det\left(-\left(mc^2 - E\right)\left(mc^2 + E\right)\mathbb{1} - c^2p_z^2\mathbb{1}\right) = 0,\\ \text{or,}\quad &\det\left(\left(E^2 - c^2p_z^2 - m^2c^4\right)\mathbb{1}\right) = 0,\\ \text{or,}\quad &\left(E^2 - c^2p_z^2 - m^2c^4\right)^2 = 0,\end{aligned} \tag{16.57}$$

which leads to

$$E = E_\pm = \pm\sqrt{c^2p_z^2 + m^2c^4}. \tag{16.58}$$

Thus, we see that there are four eigenvalues which are degenerate in pairs, namely,

$$E = E_+, E_+, E_-, E_-, \tag{16.59}$$

and a nontrivial solution exists only if the energy coincides with one of these eigenvalues (namely, only if the Einstein relation is satisfied).

If $E = E_+$, the set of equations to solve follows from (16.56) to be

$$\begin{aligned}
&\left(mc^2 - E_+\right) u_1 + cp_z u_3 = 0, \\
&\left(mc^2 - E_+\right) u_2 - cp_z u_4 = 0, \\
&cp_z u_1 - \left(mc^2 + E_+\right) u_3 = 0, \\
&- cp_z u_2 - \left(mc^2 + E_+\right) u_4 = 0.
\end{aligned} \tag{16.60}$$

Any solution of these equations must extrapolate smoothly to the zero momentum limit. The other way of saying this is that one can solve this set of equations in the rest frame and then Lorentz boost the solutions to obtain the momentum dependence. It is clear from looking at the equations that at the zero momentum limit the first two equations do not give us any information on the unknowns. Thus we have to solve the second two equations. The two independent solutions of (16.60), corresponding to the eigenvalue E_+, are

$$u_1 = 1, \quad u_2 = 0, \quad u_3 = \frac{cp_z}{E_+ + mc^2}, \quad u_4 = 0, \tag{16.61}$$

and

$$u_1 = 0, \quad u_2 = 1, \quad u_3 = 0, \quad u_4 = -\frac{cp_z}{E_+ + mc^2}. \tag{16.62}$$

Similarly, for $E = E_-$, the two independent solutions of (16.60) are

$$u_1 = -\frac{cp_z}{mc^2 - E_-}, \quad u_2 = 0, \quad u_3 = 1, \quad u_4 = 0, \tag{16.63}$$

and

$$u_1 = 0, \quad u_2 = \frac{cp_z}{mc^2 - E_-}, \quad u_3 = 0, \quad u_4 = 1, \tag{16.64}$$

so that we can write the four normalized solutions of the free Dirac equation as

$$u_\uparrow^{(+)} = \sqrt{\frac{E_+ + mc^2}{2E_+}} \begin{pmatrix} 1 \\ 0 \\ \frac{cp_z}{E_+ + mc^2} \\ 0 \end{pmatrix},$$

$$u_\downarrow^{(+)} = \sqrt{\frac{E_+ + mc^2}{2E_+}} \begin{pmatrix} 0 \\ 1 \\ 0 \\ -\frac{cp_z}{E_+ + mc^2} \end{pmatrix},$$

$$u_{\uparrow}^{(-)} = \sqrt{\frac{E_- - mc^2}{2E_-}} \begin{pmatrix} -\frac{cp_z}{mc^2-E_-} \\ 0 \\ 1 \\ 0 \end{pmatrix},$$

$$u_{\downarrow}^{(-)} = \sqrt{\frac{E_- - mc^2}{2E_-}} \begin{pmatrix} 0 \\ \frac{cp_z}{mc^2-E_-} \\ 0 \\ 1 \end{pmatrix}. \tag{16.65}$$

The notation is suggestive. Namely, if the momentum is zero, then the first two solutions look like the spin states of the non-relativistic theory (see (14.35)). They are degenerate and have energy eigenvalue E_+ which is positive. In the same limit, the last two solutions also look like the non-relativistic spin states, but they belong to the energy eigenvalue E_- which is negative. Let us note here that the normalization of the wave functions, in this case, is carried out by requiring

$$\int \mathrm{d}^3x\, \psi_k^\dagger(x)\psi_{k'}(x) = \delta^3(k-k'). \tag{16.66}$$

16.3 Continuity equation

Since the Dirac equation has both positive as well as negative energy solutions, it is worth investigating whether we can define a meaningful probability density in this theory. We note that, in the coordinate space, the Dirac equation, (16.36), has the form

$$i\hbar\frac{\partial}{\partial t} = H\psi = \big(-i\hbar c\boldsymbol{\alpha}\cdot\boldsymbol{\nabla} + \beta mc^2\big)\psi. \tag{16.67}$$

Taking the Hermitian conjugate of (16.67), we obtain

$$-i\hbar\frac{\partial\psi^\dagger}{\partial t} = \psi^\dagger\big(i\hbar c\boldsymbol{\alpha}\cdot\overleftarrow{\boldsymbol{\nabla}} + \beta mc^2\big), \tag{16.68}$$

where the gradient, on the right hand side of (16.68), is assumed to act on $\psi^\dagger$. Multiplying (16.67) by $\psi^\dagger$ on the left and (16.68) by ψ on the right and subtracting the second from the first, we obtain (we drop the overall factor of $\hbar$ on both sides)

$$
\begin{aligned}
& i\psi^\dagger \frac{\partial}{\partial t} + i\frac{\partial \psi^\dagger}{\partial t}\psi = -ic(\psi^\dagger \boldsymbol{\alpha}\cdot\boldsymbol{\nabla}\psi + (\boldsymbol{\nabla}\psi^\dagger)\cdot\boldsymbol{\alpha}\psi),\\
\text{or,}\quad & i\frac{\partial}{\partial t}(\psi^\dagger\psi) = -i\boldsymbol{\nabla}\cdot(\psi^\dagger c\boldsymbol{\alpha}\psi),\\
\text{or,}\quad & \frac{\partial}{\partial t}(\psi^\dagger\psi) = -\boldsymbol{\nabla}\cdot(\psi^\dagger c\boldsymbol{\alpha}\psi).
\end{aligned}
\tag{16.69}
$$

This is the continuity equation associated with the Dirac equation and we note that we can identify

$$
\begin{aligned}
& \rho = \psi^\dagger\psi = \text{ probability density},\\
& \jmath = \psi^\dagger c\boldsymbol{\alpha}\psi = \text{ probability current density},
\end{aligned}
\tag{16.70}
$$

to write the continuity equation in (16.69) as

$$
\frac{\partial\rho}{\partial t} = -\boldsymbol{\nabla}\cdot\jmath. \tag{16.71}
$$

This suggests that we can define a current four vector as

$$
\jmath^\mu = (\rho, \jmath) = (\psi^\dagger\psi, \psi^\dagger c\boldsymbol{\alpha}\psi), \tag{16.72}
$$

so that the continuity equation can be written in the manifestly covariant form

$$
\partial_\mu \jmath^\mu = 0. \tag{16.73}
$$

This, in fact, shows that the probability density, ρ, is the time component of $\jmath^\mu$ and, therefore, must transform like the time coordinate under a Lorentz transformation. (We are, of course, yet to show that $\jmath^\mu$ transforms like a four vector.) On the other hand, the total probability

$$
P = \int \mathrm{d}^3x\,\rho = \int \mathrm{d}^3x\,\psi^\dagger\psi, \tag{16.74}
$$

is a constant independent of any particular Lorentz frame. We have, of course, already used this Lorentz property of ρ in defining the normalization of the wave function in (16.66).

Let us conclude this discussion by noting that since the Dirac equation is first order in the time derivative, the probability density is independent of time derivatives. Consequently, the probability density, as we have checked explicitly, can be defined to be positive semi-definite even in the presence of negative energy solutions.

16.4 Dirac's hole theory

It is clear that, as in the case of the Klein-Gordon equation, the Dirac equation also leads to negative energy solutions. It is because of this that the number of components in the relativistic theory doubles compared with the non-relativistic counterpart, since each solution is possible for both positive as well as negative energy eigenvalues. The negative energy is a difficult concept to accept for various reasons. We note that the positive energy solutions and the negative energy solutions are separated by a gap as shown in Fig. 16.1. Classically, of course, we do not expect a system to make transitions through the gap. Thus, we can restrict the energy to be positive classically.

Figure 16.1: Spectrum of a free Dirac particle.

Quantum mechanically, on the other hand, transitions between discrete states can occur. In fact, since negative energy solutions are lower in energy eigenvalue and since a system always prefers to go into the lowest energy state, it is obvious that any kind of interaction – even when it is attractive – when applied to a system would repel it. Namely, the system would prefer to jump down into the negative energy levels. This is completely counter intuitive and would lead to the collapse of all our known models like the hydrogen atom.

At this point, we may raise the question as to why we don't simply rule out the negative energy solutions as being unphysical. It is not such an easy thing. Quantum mechanically our system lives in a Hilbert space. The Hilbert space contains, as a complete set of basis states, the states corresponding to both positive as well as negative energy values. If we restrict ourselves only to the positive energy states, then we are restricting ourselves to only a subspace of the complete Hilbert space and this causes major difficulty.

The conclusion so far is that, for a relativistic theory, we need both positive as well as negative energy states. There is no way to escape this and when we have negative energy states, issues such as the stability of matter will naturally arise. Thus, Dirac put forward the following hypothesis to get around the instability of matter.

The hypothesis is that, unlike what our naive intuition tells us, the vacuum state of the theory is a state where all the negative energy states are filled with electrons. Furthermore, these are all passive electrons in the sense that they do not produce any observable electromagnetic field etc. Once this new definition of the vacuum is accepted, it is clear that the problem of instability does not arise anymore. This is because of the fact that since all the negative energy states are filled with electrons and since electrons are fermions, Pauli principle forbids any positive energy electron to cascade down. We should contrast this with the situation in the case of the Klein-Gordon equation which describes bosons.

Dirac's hypothesis, furthermore, has far reaching consequences. For example, if we give sufficient energy to the system, then we can excite one of the negative energy electrons into a positive energy state. Furthermore, since the negative energy states are all filled, the absence of the electron would appear as a hole with opposite charge and positive energy. This hole has the same mass as the electron but opposite charge. This is what we know as the positron and this process is called pair production. Thus Dirac's equation predicts that for every particle there must exist an anti-particle of identical mass and opposite charge.

The above hypothesis of Dirac is known as the hole theory. Let us note here that Dirac's theory is not really a one particle theory. In fact, it is an infinitely many particle theory simply because the vacuum has been redefined to contain infinitely many particles. Thus any physical wave function has to contain this information. This is a general feature of combining relativity with quantum mechanics. That is, we cannot avoid dealing with many particle states. This leads us to the study of quantum field theory in a natural manner. Furthermore, unlike the case of the Dirac equation where the vacuum is so unsymmetrical in the charges, the Dirac field leads to a charge symmetrical description of the vacuum.

16.5 Spin of the electron

The Dirac equation automatically incorporates in it the spin angular momentum of the electron. To see this, let us define the generalized

Pauli matrix (spin operator)

$$\tilde{\alpha}_i = \begin{pmatrix} \sigma_i & 0 \\ 0 & \sigma_i \end{pmatrix}. \tag{16.75}$$

Let ρ be the matrix which connects the two matrices α_i and $\tilde{\alpha}_i$ (it should not be confused with the probability density defined in (16.70)), namely,

$$\alpha_i = \rho\tilde{\alpha}_i. \tag{16.76}$$

We can easily determine the form of ρ to be

$$\rho = \begin{pmatrix} 0 & \mathbb{1} \\ \mathbb{1} & 0 \end{pmatrix}. \tag{16.77}$$

In fact, we see that

$$\begin{aligned} \rho\tilde{\alpha}_i &= \begin{pmatrix} 0 & \mathbb{1} \\ \mathbb{1} & 0 \end{pmatrix} \begin{pmatrix} \sigma_i & 0 \\ 0 & \sigma_i \end{pmatrix} = \begin{pmatrix} 0 & \sigma_i \\ \sigma_i & 0 \end{pmatrix} = \alpha_i, \\ \tilde{\alpha}_i\rho &= \begin{pmatrix} \sigma_i & 0 \\ 0 & \sigma_i \end{pmatrix} \begin{pmatrix} 0 & \mathbb{1} \\ \mathbb{1} & 0 \end{pmatrix} = \begin{pmatrix} 0 & \sigma_i \\ \sigma_i & 0 \end{pmatrix} = \alpha_i. \end{aligned} \tag{16.78}$$

Thus, we conclude that ρ commutes with $\tilde{\alpha}_i$. Let us also note the following useful relations

$$\begin{aligned} &[\rho, \tilde{\alpha}_i] = 0, \\ &\rho^2 = \mathbb{1}, \\ &\alpha_i = \rho\tilde{\alpha}_i, \\ &[\tilde{\alpha}_i, \tilde{\alpha}_j] = 2i\epsilon_{ijk}\tilde{\alpha}_k, \end{aligned} \tag{16.79}$$

which can be easily checked. Furthermore,

$$\begin{aligned} [\tilde{\alpha}_i, \beta] &= \begin{pmatrix} \sigma_i & 0 \\ 0 & \sigma_i \end{pmatrix} \begin{pmatrix} \mathbb{1} & 0 \\ 0 & -\mathbb{1} \end{pmatrix} - \begin{pmatrix} \mathbb{1} & 0 \\ 0 & -\mathbb{1} \end{pmatrix} \begin{pmatrix} \sigma_i & 0 \\ 0 & \sigma_i \end{pmatrix} \\ &= 0. \end{aligned} \tag{16.80}$$

In terms of these matrices, we can write down the Dirac Hamiltonian as

$$\begin{aligned} H &= c\boldsymbol{\alpha} \cdot \mathbf{p} + \beta mc^2 + V(r) \\ &= c\rho\tilde{\boldsymbol{\alpha}} \cdot \mathbf{p} + \beta mc^2 + V(r) \\ &= c\rho\tilde{\alpha}_i p_i + \beta mc^2 + V(r). \end{aligned} \tag{16.81}$$

Here we have added a spherically symmetric potential to allow for a rotationally invariant interaction. We know that since the Hamiltonian is invariant under rotations, angular momentum must be conserved. The z-component of the orbital angular momentum is given by

$$L_3 = x_1 p_2 - x_2 p_1, \tag{16.82}$$

so that we can calculate

$$\begin{aligned}
[L_3, H] &= \left[L_3, c\rho\tilde{\alpha}_i p_i + \beta mc^2 + V(r)\right] \\
&= [x_1 p_2 - x_2 p_1, c\rho\tilde{\alpha}_i p_i] \\
&= c\rho\tilde{\alpha}_i\,[x_1, p_i]\,p_2 - c\rho\tilde{\alpha}_i\,[x_2, p_i]\,p_1 \\
&= c\rho\tilde{\alpha}_i i\hbar\delta_{i1} p_2 - c\rho\tilde{\alpha}_i i\hbar\delta_{i2} p_1 \\
&= i\hbar c\rho\left(\tilde{\alpha}_1 p_2 - \tilde{\alpha}_2 p_1\right),
\end{aligned} \tag{16.83}$$

where we have used the fact that the potential commutes with L_3 since it is spherically symmetric.

Thus, we see that L_3 does not commute with the Hamiltonian and, therefore, cannot be conserved. That is, the orbital angular momentum is no longer a constant of motion. Let us also note that

$$\begin{aligned}
[\tilde{\alpha}_3, H] &= \left[\tilde{\alpha}_3, c\rho\tilde{\alpha}_i p_i + \beta mc^2 + V(r)\right] \\
&= c\rho\,[\tilde{\alpha}_3, \tilde{\alpha}_i]\,p_i + mc^2\,[\tilde{\alpha}_3, \beta] \\
&= c\rho 2i\epsilon_{3ij}\tilde{\alpha}_j p_i \\
&= 2ic\rho\left(\tilde{\alpha}_2 p_1 - \tilde{\alpha}_1 p_2\right).
\end{aligned} \tag{16.84}$$

Thus, from (16.83) and (16.84), we see that the operator

$$L_3 + \frac{1}{2}\hbar\tilde{\alpha}_3, \tag{16.85}$$

commutes with the Hamiltonian and hence must be conserved. As a result, the total angular momentum which is conserved must have the form

$$\mathbf{J} = \mathbf{L} + \frac{1}{2}\hbar\tilde{\boldsymbol{\alpha}}. \tag{16.86}$$

Thus, it follows that the spin of the electron can be identified with

$$\mathbf{S} = \frac{1}{2}\hbar\tilde{\boldsymbol{\alpha}}. \tag{16.87}$$

Furthermore, since the eigenvalues of the matrix $\tilde{\alpha}_i$ are ± 1, this tells us that the spin of the electron is $\frac{1}{2}$.

16.6 Selected problems

1. Consider a simple Lorentz transformation along the x-axis defined by

$$\begin{aligned}
t' &= \gamma\left(t - \frac{\beta}{c}x\right),\\
x' &= \gamma\left(-\beta ct + x\right),\\
y' &= y,\\
z' &= z,
\end{aligned}\tag{16.88}$$

where $\beta = \frac{v}{c}$ and $\gamma = \frac{1}{\sqrt{1-\beta^2}}$. The transformations can be written in the proper tensor language as

$$x'^{\mu} = \Lambda^{\mu}{}_{\nu}x^{\nu}, \qquad x'_{\mu} = \tilde{\Lambda}_{\mu}{}^{\nu}x_{\nu}. \tag{16.89}$$

a) Determine the 4×4 matrices $\Lambda^{\mu}{}_{\nu}$ and $\tilde{\Lambda}_{\mu}{}^{\nu}$. Show that

$$\tilde{\Lambda}_{\mu}{}^{\nu} = \Lambda_{\mu}{}^{\nu}. \tag{16.90}$$

b) Calculate $\det\Lambda^{\mu}{}_{\nu}$ and $\det\tilde{\Lambda}_{\mu}{}^{\nu}$.

c) Determine the product $\Lambda^{\mu}{}_{\nu}\Lambda_{\mu}{}^{\lambda}$ (μ is summed) and, from this, show that $x^2 = x^{\mu}x_{\mu}$ is the invariant length.

2. Consider the Klein-Gordon equation in the presence of a static Coulomb potential of the form (obtained through minimal coupling)

$$(\Box + m^2 + 2ie\phi\frac{\partial}{\partial t} - e^2\phi^2)\psi(t,\mathbf{x}) = 0, \tag{16.91}$$

where, for simplicity, we will assume that the scalar potential, ϕ, depends only on z and has the form

$$\phi = \begin{cases} 0 & \text{for } z \leq 0, \\ \phi_0 & \text{for } z \geq 0. \end{cases} \tag{16.92}$$

Here ϕ_0 is a constant. (Because of the presence of the potential, the form of ρ will be different from the one derived in class.)

Consider a plane wave incident along the z-axis with a positive energy ω. Assuming that $e\phi_0 > \omega + m$, and that there is an incident as well as a reflected wave in the left region while only a transmitted wave is in the region to the right, show that ρ is positive in the left region while it is negative in the region to the right. Namely, even if we start from a state with a positive ρ, the interaction can take us to a state with a negative ρ so that we cannot avoid such states.

3. a) Show that the solutions of the Dirac equation, for arbitrary $\mathbf{k}$, have the forms

$$u_+(k) = \begin{pmatrix} \tilde{u}(k) \\ \frac{\boldsymbol{\sigma}\cdot\mathbf{k}}{\omega+m}\,\tilde{u}(k) \end{pmatrix}, \quad u_-(k) = \begin{pmatrix} -\frac{\boldsymbol{\sigma}\cdot\mathbf{k}}{\omega+m}\,\tilde{v}(k) \\ \tilde{v}(k) \end{pmatrix}, \tag{16.93}$$

where $\omega = \sqrt{\mathbf{k}^2 + m^2}$.

b) Show explicitly that these are eigenstates of the Hamiltonian operator. What are the corresponding eigenvalues?

CHAPTER 17

Path integral quantum mechanics

Let us recapitulate very briefly what we have learnt so far in quantum mechanics. We have seen that, given a set of dynamical variables in classical mechanics, we go over to quantum mechanics by promoting these variables to operators. In particular the Hamiltonian or the total energy of a classical system also becomes an operator. These operators operate on an infinite dimensional Hilbert space labeled by time.

The operators do not commute in general. Their commutation relations can be obtained from the classical Poisson brackets (when classical Poisson brackets are defined). We can represent the operators as matrices in certain basis. The knowledge of the matrix elements, then, gives all the information about an operator. We can also determine the eigenstates and the eigenvalues of an operator. Thus, for example, for the coordinate operator, we have

$$X|x\rangle = x|x\rangle. \tag{17.1}$$

The eigenvalues of the operators are the results of measurement corresponding to that operator. They are the diagonal elements in their eigenbasis. The off-diagonal elements are known as transition amplitudes and are responsible for transition between different quantum states.

The time evolution of quantum states, describing a physical system, is given by the Schrödinger equation, namely, they satisfy

$$i\hbar\frac{\mathrm{d}}{\mathrm{d}t}\,|\psi(t)\rangle = H|\psi(t)\rangle. \tag{17.2}$$

We can write (17.2), in the x-basis, as

$$i\hbar\frac{\partial\psi(x,t)}{\partial t} = H\psi(x,t) = \left(-\frac{\hbar^2}{2m}\frac{\partial^2}{\partial x^2} + V\right)\psi(x,t), \tag{17.3}$$

where the wave function $\psi(x,t)$ is defined as (we are assuming, for simplicity, that the system is one dimensional)

$$\psi(x,t) = \langle x|\psi(t)\rangle, \tag{17.4}$$

and gives the probability amplitude for the system to be at the coordinate x at time t. When the Hamiltonian is independent of time, we have seen, in chapter 3, that the wave function separates, in a simple manner, to factors depending on time and space separately (see (3.94)),

$$\psi(x,t) \sim u(x)\; e^{-\frac{i}{\hbar}Et}. \tag{17.5}$$

Such states are known as stationary states, since the probability for a system to be in such a state is independent of time. In general, of course, this need not be true. As in classical mechanics, we look for solutions of the Schrödinger equation with a given initial state. In other words, the purpose behind solving the Schrödinger equation is to obtain the time evolution operator, which takes the state at $t = 0$ to a state at an arbitrary time t so that

$$|\psi(t)\rangle = U(t)|\psi(0)\rangle. \tag{17.6}$$

In general the time evolution operator, U, depends on an initial and a final time, namely,

$$U = U(t_2, t_1), \tag{17.7}$$

such that

$$|\psi(t_2)\rangle = U(t_2,t_1)|\psi(t_1)\rangle. \tag{17.8}$$

Furthermore, as we have noted earlier in (3.88), the time evolution operator satisfies the following relations:

$$\begin{aligned}
&U(t_2,t_1)U^{\dagger}(t_2,t_1) = \mathbb{1},\\
&U^{-1}(t_2,t_1) = U(t_1,t_2),\\
&U(t_2,t_3)U(t_3,t_1) = U(t_2,t_1).
\end{aligned} \tag{17.9}$$

The knowledge of the time evolution operator is the ultimate goal in solving the Schrödinger equation because once this is known the wave functions at any time can be obtained. The wave function, of course, contains all the information about the system. In the simple case when the Hamiltonian is time independent, we have seen in (3.81) that

$$U(t_2, t_1) = e^{-\frac{i}{\hbar}H(t_2-t_1)}. \tag{17.10}$$

The knowledge of this operator implies knowing all the matrix elements in a given basis. For example, in the coordinate basis, we have

$$\langle x_2|U(t_2,t_1)|x_1\rangle = \langle x_2|e^{-\frac{i}{\hbar}H(t_2-t_1)}|x_1\rangle = U(x_2,t_2;x_1,t_1). \tag{17.11}$$

All of what we have said so far is in the Schrödinger picture where the state of the system evolves with time but the operators are time independent. The Heisenberg picture, on the other hand, assumes that the state of the system is fixed in time and that the operators evolve with time. Thus, we recall from (3.110) and (3.112) that

$$\begin{aligned} &|\psi\rangle_H = |\psi(0)\rangle_S \neq |\psi(t)\rangle, \\ &\Omega_H = \Omega_H(t) = e^{\frac{i}{\hbar}Ht}\Omega_S\, e^{-\frac{i}{\hbar}Ht}, \end{aligned} \tag{17.12}$$

where we are assuming that

$$\Omega_S \neq \Omega_S(t). \tag{17.13}$$

It is clear that the eigenstates of the operators (in particular, the coordinate operator) would now depend on time, namely,

$$X_H(t)|x,t\rangle_H = x|x,t\rangle_H, \tag{17.14}$$

and that the wave function for the system would be given by

$$\psi(x,t) = {}_H\langle x,t|\psi\rangle_H. \tag{17.15}$$

It is clear, therefore, that the eigenstates for the Heisenberg operators are related to those of the Schrödinger operators as

$$|x,t\rangle_H = e^{\frac{i}{\hbar}Ht}|x\rangle_S. \tag{17.16}$$

If we now calculate in the Heisenberg picture, the probability amplitude for a system at the coordinate x_1 at time t_1 to go to x_2 at time t_2 is obtained to be

$$\begin{aligned} {}_H\langle x_2,t_2|x_1,t_1\rangle_H &= {}_S\langle x_2|e^{-\frac{i}{\hbar}Ht_2}e^{\frac{i}{\hbar}Ht_1}|x_1\rangle_S \\ &= {}_S\langle x_2|e^{-\frac{i}{\hbar}H(t_2-t_1)}|x_1\rangle_S \\ &= U(x_2,t_2;x_1,t_1), \end{aligned} \tag{17.17}$$

where we have used the identification in (17.11).

We note that the operator formalism for quantum mechanics, which we have studied so far and which is also known as the canonical formalism, is completely non-relativistic in the sense that it singles out the time coordinate in a special way. Thus, it is not suited for relativistic problems. Furthermore, in this formalism we solve the Schrödinger equation and iteratively obtain the dynamics of a system. In that sense, it is analogous to Newton's or Euler's method in classical mechanics. On the other hand, the action principle in classical mechanics (see chapter **1**), which leads to the Euler-Lagrange equations, determines the entire classical trajectory as the one with the least action. This is like a global method. In quantum mechanics, there is also a corresponding method and is known as the Path Integral method.

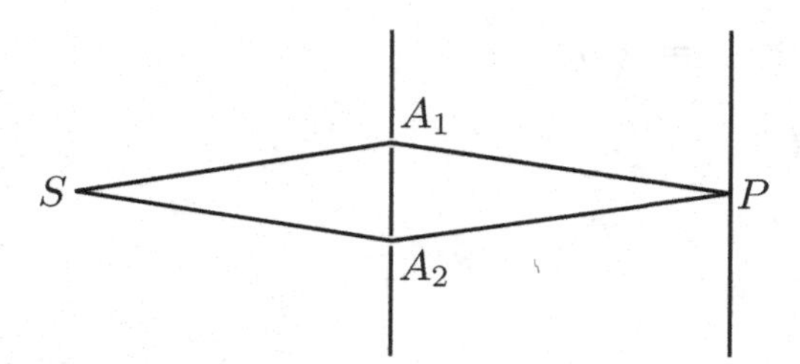

Figure 17.1: Double slit experiment in quantum mechanics where the process can take place through two different paths.

To understand path integral method let us go back again to the double slit experiment. As we have noted earlier, in the microscopic domain, we obtain an interference pattern independent of whether we use a light source or a particle source. This can, of course, be explained by saying that there is a probability amplitude associated with each path. Thus for example, suppose ϕ_1 is the probability amplitude associated with the path SA_1P and ϕ_2 is the probability amplitude associated with the path SA_2P shown in Fig. 17.1. Thus, the total probability amplitude for the particle to go from S to I, given by ϕ has the form

$$\phi = \phi_1 + \phi_2. \tag{17.18}$$

Consequently, the intensity, which is proportional to the probability, takes the form

$$I \propto |\phi|^2 = |\phi_1 + \phi_2|^2 \neq |\phi_1|^2 + |\phi_2|^2. \tag{17.19}$$

This, therefore, leads to interference patterns rather than particle behavior. Thus we know one thing that the total probability amplitude associated with a particle moving from one point to another is the sum of probability amplitudes associated with each possible path between those two points. In general, we can write

$$\phi = \sum_i \phi_i, \tag{17.20}$$

where ϕ_i denotes the probability amplitude corresponding to the ith possible path (see Fig. 17.2).

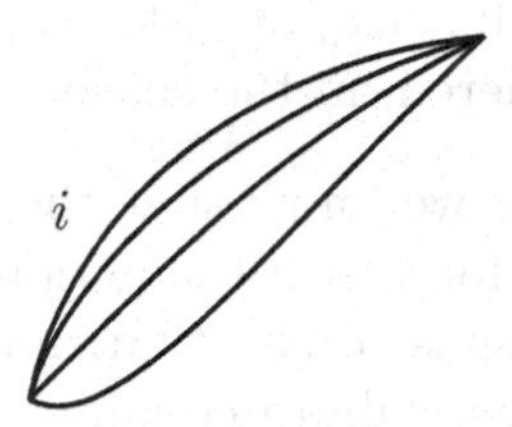

Figure 17.2: A quantum mechanical process that can take place along many different paths.

Feynman's postulate. Feynman's postulate says that the probability amplitude associated with the transition from the point (x_a, t_a) to (x_b, t_b) is the sum over all paths with the action as a phase angle, namely,

$$\text{Amplitude} = \sum_{\text{all paths}} e^{\frac{i}{\hbar}S}, \tag{17.21}$$

where S is the classical action associated with each path. Comparing with (17.17), we can, therefore, write

$$\begin{aligned} {}_H\langle x_b, t_b | x_a, t_a\rangle_H &= U(x_b, t_b; x_a, t_a) \\ &= \sum_{\text{all paths}} e^{\frac{i}{\hbar}\int_{t_a}^{t_b} dt\ L}. \end{aligned} \tag{17.22}$$

We would show that this is indeed equivalent to Schrödinger equation and conversely we will also show that the operator formalism also leads to this. But assuming that this is true let us discuss some of its features.

1. First of all we note that the action is a scalar quantity. Thus, this can also be applied to a relativistic description of systems. In fact, the path integral is more suited for description of relativistic systems as well as very complicated physical theories like the gauge theories.

2. The quantities that one deals with in this formalism are classical quantities – for example the classical action associated with a particular path between the initial and the final points.

3. This is quite a physical picture of the propagation of the system. The system moves from x_a at t_a to x_b at t_b. Thus the postulate is already time ordered, in the sense that $t_b > t_a$.

4. Whereas classically we know that the actual trajectory of the particle is the one for which the action is the minimum, Feynman's postulate seems to say that, for a quantum mechanical system, all paths contribute equally. A natural question that immediately arises is how does one obtain the classical trajectory as the unique trajectory in the classical limit when $\hbar \to 0$, or $S \gg \hbar$.

   This question can be answered in the following way. Let us suppose that $S \gg \hbar$ and that we are considering a path that is far away from the classical path as shown in Fig. 17.3. Let us denote this path by x_1. If we change the path slightly,

   then, we have

$$x_1 \to x_2, \qquad S(x_1) \to S(x_2). \tag{17.23}$$

   Clearly if Δx is very small on a classical scale, then, the change in the action $\Delta S = S(x_2) - S(x_1)$ will also be small on a classical scale. But on a quantum scale or in units of $\hbar$ this would be large. Thus $\frac{\Delta S}{\hbar}$ would be very large and, for every path, there would be a path which contributes negatively and so all such paths far away from the classical trajectory would average out to zero.

   On the other hand, if we are considering the paths near the classical trajectory, an infinitesimal change would lead to no

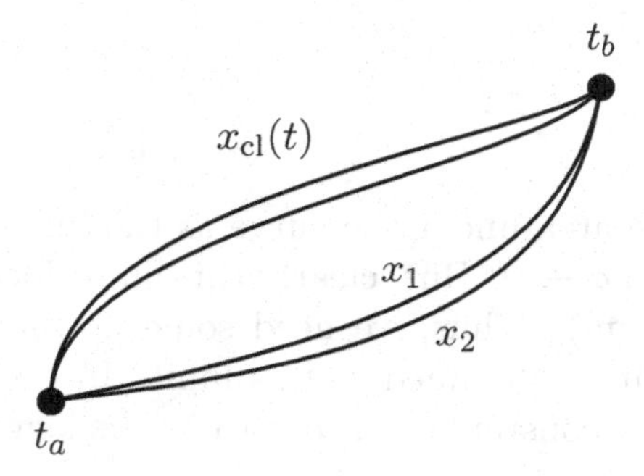

Figure 17.3: Several distinct paths connecting two points among which $x_{\rm cl}(t)$ denotes the classical path.

change in the action, which follows from the principle of least action. Thus, all paths near the classical trajectory would contribute coherently. In fact, all trajectories which differ from $x_{\rm cl}$ contribute as long as their contribution is within $\pi\hbar$ of $S_{\rm cl}$. It is clear, therefore, that for all practical purposes only the classical trajectory is picked up in the classical limit.

Constructing the sum over paths. The number of paths between any two points is obviously infinitely many. Therefore, we have to first define what we mean by constructing the sum over all paths.

Let us begin with the Riemann definition of an integral. Let us consider a curve given by a function $f(x)$. To find out the area under this curve (see Fig. 17.4), we divide the interval between x_0 and x_N to N equal intervals of length a and define

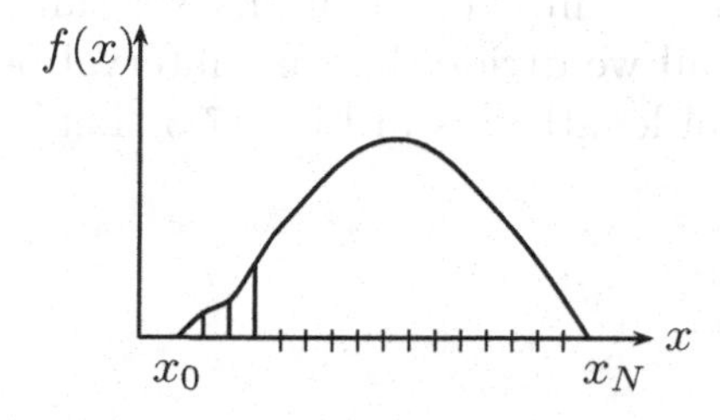

Figure 17.4: Riemann's definition of an integral as the area under a curve.

$$A \sim \sum_i f(x_i). \tag{17.24}$$

We can define the area under the curve as the limit of (17.24) when the interval length $a \to 0$. But, clearly this limit does not exist (since it is an infinite sum). Thus, we need some normalization constant which will make it well defined in this limit. In the case of the Riemann integral, this constant turns out to be the length of the interval so that

$$A = \lim_{a \to 0} \left(a \sum_i f(x_i) \right), \tag{17.25}$$

represents the area under the curve. We recognize that, in this continuum limit, the number of interval increases but the length of each interval decreases so that the product is finite.

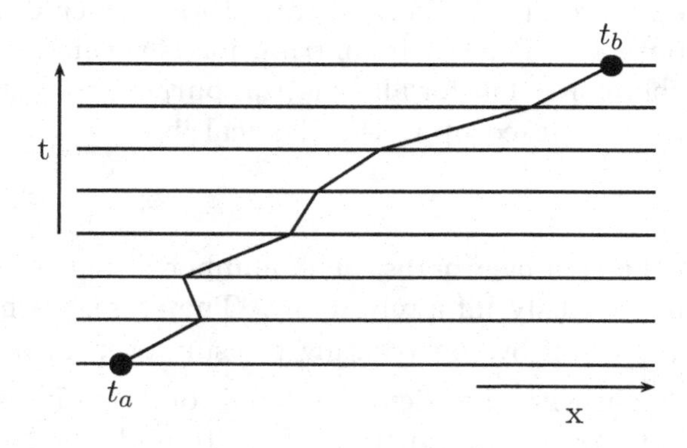

Figure 17.5: Discretized representation of a given path between an initial and a final point.

To construct the sum over all paths we can follow the same procedure. First of all we divide the time interval between t_b and t_a into N equal parts of length ϵ as in Fig. 17.5. Let us identify

$$t_a = t_0, \qquad t_b = t_N, \tag{17.26}$$

so that

$$\begin{aligned}
\epsilon &= t_{i+1} - t_i, \\
N\epsilon &= t_b - t_a, \\
x(t_i) &= x_i.
\end{aligned} \tag{17.27}$$

In particular $x_a = x_0$ and $x_b = x_N$. With this notation, therefore, the action associated with a particular path (which is the integral of the Lagrangian) can be written, following (17.25) as

$$S = \lim_{\epsilon \to 0} \epsilon \sum_i L[x_i]. \tag{17.28}$$

All possible paths between the initial and the final points are, of course, automatically generated if we let x_i take all possible values (keeping x_0 and x_N fixed). Therefore, we can define the sum over all paths in (17.22) as

$$U(x_b, t_b; x_a, t_a) = \lim_{\substack{\epsilon \to 0 \\ N \to \infty \\ N\epsilon = t_b - t_a}} A \int \cdots \int \mathrm{d}x_1 \cdots \mathrm{d}x_{N-1}\, e^{\frac{i\epsilon}{\hbar} \sum_i L[x_i]}, \tag{17.29}$$

where the integration $\mathrm{d}x_i$ is over all possible values of the coordinates, x_i. We note that we do not integrate over the end points since they are held fixed. Furthermore, we have also put a normalization constant A in front so that the limit indeed exists.

17.1 Free particle

To understand the path integral formalism better, let us study the simplest example of a free particle in one dimension. In this case, we know that the Lagrangian has the simple form

$$L = \frac{1}{2} m \dot{x}^2. \tag{17.30}$$

As a result, we obtain, from (17.28),

$$\begin{aligned} S &= \lim_{\epsilon \to 0} \epsilon \sum_i \frac{1}{2} m \dot{x}_i^2 = \lim_{\epsilon \to 0} \frac{m\epsilon}{2} \sum_i \left(\frac{x_{i+1} - x_i}{\epsilon} \right)^2 \\ &= \lim_{\epsilon \to 0} \frac{m}{2\epsilon} \sum_i (x_{i+1} - x_i)^2. \end{aligned} \tag{17.31}$$

It now follows from (17.29) that, in this case,

$$\begin{aligned} &U(x_b, t_b; x_a, t_a) \\ &= \lim_{\substack{\epsilon \to 0 \\ N \to \infty \\ N\epsilon = t_b - t_a}} A \int_{-\infty}^{\infty} \cdots \int_{-\infty}^{\infty} \mathrm{d}x_1 \cdots \mathrm{d}x_{N-1}\, e^{\frac{im}{2\hbar\epsilon} \sum_i (x_{i+1} - x_i)^2}. \end{aligned} \tag{17.32}$$

Let us scale the variable of integration as

$$y_i = \left(\frac{m}{2\hbar\epsilon}\right)^{\frac{1}{2}} x_i. \tag{17.33}$$

With this, the expression, (17.32), becomes

$$\begin{aligned} U(x_b, t_b; x_a, t_a) = & \lim_{\substack{\epsilon\to 0\\ N\to\infty\\ N\epsilon = t_b - t_a}} A \left(\frac{2\hbar\epsilon}{m}\right)^{\frac{(N-1)}{2}} \\ & \times \int_{-\infty}^{\infty} \cdots \int_{-\infty}^{\infty} \mathrm{d}y_1 \cdots \mathrm{d}y_{N-1}\, e^{\sum_i i(y_{i+1}-y_i)^2}. \end{aligned} \tag{17.34}$$

Let us look at the term, inside the integrand, of the form

$$\begin{aligned} & \int_{-\infty}^{\infty} \mathrm{d}y_1\, e^{i[(y_1-y_0)^2+(y_2-y_1)^2]} \\ & = \int_{-\infty}^{\infty} \mathrm{d}y_1\, e^{i\left[2\left(y_1 - \frac{(y_2+y_0)}{2}\right)^2 + \frac{1}{2}(y_2-y_0)^2\right]} \\ & = e^{\frac{i}{2}(y_2-y_0)^2} \left(\frac{i\pi}{2}\right)^{\frac{1}{2}}. \end{aligned} \tag{17.35}$$

Let us next consider the term

$$\begin{aligned} & \iint_{-\infty}^{\infty} \mathrm{d}y_1 \mathrm{d}y_2\, e^{i[(y_1-y_0)^2+(y_2-y_1)^2+(y_3-y_2)^2]} \\ & = \left(\frac{i\pi}{2}\right)^{\frac{1}{2}} \int_{-\infty}^{\infty} \mathrm{d}y_2\, e^{\frac{i}{2}(y_2-y_0)^2 + i(y_3-y_2)^2} \\ & = \left(\frac{i\pi}{2}\right)^{\frac{1}{2}} \int_{-\infty}^{\infty} \mathrm{d}y_2\, e^{\frac{3i}{2}\left(y_2 - \frac{(y_0+2y_3)}{3}\right)^2 + \frac{i}{3}(y_3-y_0)^2} \\ & = \left(\frac{i\pi}{2}\right)^{\frac{1}{2}} \left(\frac{2i\pi}{3}\right)^{\frac{1}{2}} e^{\frac{i}{3}(y_3-y_0)^2} \\ & = \left(\frac{(i\pi)^2}{3}\right)^{\frac{1}{2}} e^{\frac{i}{3}(y_3-y_0)^2}. \end{aligned} \tag{17.36}$$

Thus, a pattern now follows and we see that we can write (17.34) as

$$
\begin{aligned}
&U(x_b,t_b;x_a,t_a)\\
&= \lim_{\substack{\epsilon\to 0\\ N\to\infty\\ N\epsilon=t_b-t_a}} A\left(\frac{2\hbar\epsilon}{m}\right)^{\frac{N-1}{2}}\left[\frac{(i\pi)^{N-1}}{N}\right]^{\frac{1}{2}} e^{\frac{i}{N}(y_N-y_0)^2}\\
&= \lim_{\substack{\epsilon\to 0\\ N\to\infty\\ N\epsilon=t_b-t_a}} A\left(\frac{2\hbar\epsilon}{m}\right)^{\frac{N}{2}}(i\pi)^{\frac{(N-1)}{2}}\left(\frac{m}{2\hbar N\epsilon}\right)^{\frac{1}{2}} e^{\frac{im}{2\hbar N\epsilon}(x_N-x_0)^2}\\
&= \lim_{\substack{\epsilon\to 0\\ N\to\infty\\ N\epsilon=t_b-t_a}} A\left(\frac{2i\pi\hbar\epsilon}{m}\right)^{\frac{N}{2}}\left(\frac{m}{2i\pi\hbar N\epsilon}\right)^{\frac{1}{2}} e^{\frac{im}{2\hbar N\epsilon}(x_N-x_0)^2}\\
&= \lim_{\substack{\epsilon\to 0\\ N\to\infty\\ N\epsilon=t_b-t_a}} A\left(\frac{2i\pi\hbar\epsilon}{m}\right)^{\frac{N}{2}}\left[\frac{m}{2i\pi\hbar(t_b-t_a)}\right]^{\frac{1}{2}} e^{\frac{im}{2\hbar}\frac{(x_b-x_a)^2}{t_b-t_a}}.
\end{aligned}
\tag{17.37}
$$

The normalization factor A is determined by noticing that as $t_b\to t_a$ (see (17.11) or (17.17))

$$U(x_b,t_b;x_a,t_a)\to\delta(x_b-x_a). \tag{17.38}$$

which leads to (see (2.145))

$$A=\left(\frac{m}{2i\pi\hbar\epsilon}\right)^{\frac{N}{2}}=\left(\frac{2i\pi\hbar\epsilon}{m}\right)^{-\frac{N}{2}}=B^{-N}. \tag{17.39}$$

Conventionally, one associates a factor of $\frac{1}{B}$ with each integration variable to write

$$
\begin{aligned}
&U(x_b,t_b;x_a,t_a)\\
&= \lim_{\substack{\epsilon\to 0\\ N\to\infty\\ N\epsilon=t_b-t_a}} \frac{1}{B}\int\cdots\int_{-\infty}^{\infty}\frac{\mathrm{d}x_1}{B}\cdots\frac{\mathrm{d}x_{N-1}}{B}\, e^{\frac{i\epsilon}{\hbar}\sum\limits_i L[x_i]}\\
&= \int_a^b \mathcal{D}x(t)\, e^{\frac{i}{\hbar}S[x(t)]}.
\end{aligned}
\tag{17.40}
$$

In the simple case of the free particle, therefore, we have

$$U(x_b, t_b; x_a, t_a) = \left[\frac{m}{2i\pi\hbar(t_b - t_a)}\right]^{\frac{1}{2}} e^{\frac{im}{2\hbar}\frac{(x_b - x_a)^2}{(t_b - t_a)}}. \tag{17.41}$$

From the definition of the U in (17.17), it is obvious that

$$\begin{aligned} {}_H\langle x_b, t_b|\psi\rangle_H = \psi(x_b, t_b) &= \int \mathrm{d}x_a \; {}_H\langle x_b, t_b|x_a, t_a\rangle_H \; {}_H\langle x_a, t_a|\psi\rangle_H \\ &= \int \mathrm{d}x_a \, U(x_b, t_b; x_a, t_a)\psi(x_a, t_a). \end{aligned} \tag{17.42}$$

Thus if at $t_a = 0$ we choose $\psi(x_a, 0) = \delta(x_a)$, then, we obtain from (17.41) and (17.42)

$$\psi(x_b, t_b) = U(x_b, t_b; 0, 0) = \left[\frac{m}{2i\pi\hbar t_b}\right]^{\frac{1}{2}} e^{\frac{im}{2\hbar}\frac{x_b^2}{t_b}}. \tag{17.43}$$

This can be compared with (4.4) (with $a = 0$), where we had obtained the wave function by solving the Schrödinger equation.

It is clear from the form of U in (17.41) that

$$\begin{aligned} \frac{\partial U}{\partial t_b} &= -\frac{1}{2(t_b - t_a)}\,U - \frac{m}{2i\hbar}\,(x_b - x_a)^2\left(-\frac{1}{2}\,\frac{1}{(t_b - t_a)^2}\right)U \\ &= \left[-\frac{1}{2(t_b - t_a)} + \frac{m}{2i\hbar}\,\frac{(x_b - x_a)^2}{(t_b - t_a)^2}\right]U, \\ \frac{\partial U}{\partial x_b} &= -\frac{m}{2i\hbar}\,\frac{2(x_b - x_a)}{t_b - t_a}\,U, \\ \frac{\partial^2 U}{\partial x_b^2} &= \left[-\frac{m}{i\hbar}\,\frac{1}{t_b - t_a} + \left(\frac{m}{i\hbar}\right)^2\frac{(x_b - x_a)^2}{(t_b - t_a)^2}\right]U. \end{aligned} \tag{17.44}$$

It follows now, from (17.44), that

$$i\hbar\,\frac{\partial U}{\partial t_b} = -\frac{\hbar^2}{2m}\,\frac{\partial^2 U}{\partial x_b^2}. \tag{17.45}$$

Thus the probability amplitude satisfies the Schrödinger equation. Furthermore, since we know (see (17.42))

$$\psi(x_b, t_b) = \int \mathrm{d}x_a \; U(x_b, t_b; x_a, t_a)\psi(x_a, t_a), \tag{17.46}$$

we obtain

$$i\hbar\,\frac{\partial\psi(x_b, t_b)}{\partial t_b} = -\frac{\hbar^2}{2m}\,\frac{\partial^2\psi(x_b, t_b)}{\partial x_b^2}. \tag{17.47}$$

Namely, the wave function obtained in the path integral formalism obeys the Schrödinger equation.

Gaussian path integrals. The simplest path integral corresponds to the case where the dynamical variables appear at the most up to quadratic order in the Lagrangian (the free particle is an example of such a system). Such systems are, therefore, described by path integrals where the integrands are Gaussians. We can write the most general (one dimensional) Lagrangian which is quadratic in the dynamical variables as

$$L = a_1\dot{x}^2 + a_2\dot{x}x + a_3x^2 + a_4\dot{x} + a_5x + a_6(t). \tag{17.48}$$

Here we have allowed for an explicit time dependent term. This is of course a highly generalized Lagrangian and, in physical situations, the quadratic Lagrangians have much simpler forms.

For a quantum mechanical system described by a Lagrangian of the form (17.48), the amplitude, (17.17), can be written as (see also (17.40))

$$U(x_b, t_b; x_a, t_a) = \int_a^b \mathcal{D}x(t)\ e^{\frac{i}{\hbar}S[x]}. \tag{17.49}$$

This concise notation, of course, implicitly implies the limiting procedure we talked about earlier. For a Gaussian action the integrations can be performed. However, rather than doing it tediously let us evaluate it in a different manner.

First of all, let $x_{\text{cl}}(t)$ be the classical trajectory so that $\delta S = 0$ around the classical trajectory. Furthermore, rather than labelling the paths by coordinates from an arbitrary origin, let us measure them from $x_{\text{cl}}(t)$. Thus, for any path, we define

$$\begin{aligned} x(t) &= x_{\text{cl}}(t) + y, \\ S[x] &= S[x_{\text{cl}} + y] \\ &= S[x_{\text{cl}}] + \text{linear in } y + \text{quadratic in } y. \end{aligned} \tag{17.50}$$

We can show that the coefficients of the linear term vanishes since $\delta S|_{x_{\text{cl}}} = 0$ (principle of least action). Furthermore, the form of the quadratic terms follows from (17.48) to be

$$a_1\dot{y}^2 + a_2\dot{y}y + a_3y^2. \tag{17.51}$$

As a result, we can write

$$S[x] = S[x_{\text{cl}}] + \int_{t_a}^{t_b} \text{d}t \left(a_1\dot{y}^2 + a_2\dot{y}y + a_3y^2\right). \tag{17.52}$$

We can now integrate over all possible values of y which will generate all possible paths between the initial and the final points. Correspondingly, we have

$$\mathcal{D}x(t) \to \mathcal{D}y(t). \tag{17.53}$$

Using this as well as (17.52), the amplitude, (17.49), takes the form

$$\begin{aligned} U(x_b, t_b; x_a, t_a) &= \int_0^0 \mathcal{D}y(t)\, e^{\frac{i}{\hbar}\left[S[x_{\rm cl}] + \int_{t_a}^{t_b} {\rm d}t\ (a_1\dot{y}^2 + a_2\dot{y}y + a_3y^2)\right]} \\ &= e^{\frac{i}{\hbar}S[x_{\rm cl}]} \int_0^0 \mathcal{D}y(t)\, e^{\frac{i}{\hbar}\int_{t_a}^{t_b} {\rm d}t\ (a_1\dot{y}^2 + a_2\dot{y}y + a_3y^2)}. \end{aligned} \tag{17.54}$$

The integration limits on $\mathcal{D}y(t)$ imply that since the end points are held fixed y starts from and returns to zero. Clearly the integral can only depend on the end points t_a and t_b since there is no dependence on x_a and x_b any more. Thus, we can write

$$U(x_b, t_b; x_a, t_a) = e^{\frac{i}{\hbar}S[x_{\rm cl}]}\ F(t_b, t_a). \tag{17.55}$$

Furthermore, if the Lagrangian has no explicit time dependence, then time translation must be a symmetry of the system and, consequently, we must have

$$F(t_b, t_a) = F(t_b - t_a). \tag{17.56}$$

Thus, for a time independent quadratic Lagrangian (for which $a_6(t) = 0$), we have

$$U(x_b, t_b; x_a, t_a) = e^{\frac{i}{\hbar}S[x_{\rm cl}]} F(t_b - t_a). \tag{17.57}$$

Comparing (17.57) with the result obtained for the free particle in (17.41), we obtain

$$F^{(\text{free particle})}(t_b - t_a) = \left[\frac{m}{2i\pi\hbar(t_b - t_a)}\right]^{\frac{1}{2}}. \tag{17.58}$$

Gaussian path integrals can, in fact, be exactly evaluated and we will work out the harmonic oscillator later in this chapter from the path integral point of view.

17.2 Equivalence with Schrödinger equation

The Schrödinger equation, which has the form

$$i\hbar \frac{\mathrm{d}}{\mathrm{d}t}|\psi(t)\rangle = H|\psi(t)\rangle, \tag{17.59}$$

implies that, for an infinitesimal time interval ϵ, we can write

$$|\psi(\epsilon)\rangle - |\psi(0)\rangle = -\frac{i\epsilon}{\hbar} H|\psi(0)\rangle. \tag{17.60}$$

In the coordinate basis, (17.60) leads to

$$\psi(x,\epsilon) - \psi(x,0) = -\frac{i\epsilon}{\hbar}\left(-\frac{\hbar^2}{2m}\frac{\partial^2}{\partial x^2} + V(x,0)\right)\psi(x,0). \tag{17.61}$$

We would now show that the path integral also predicts this behavior for the wave function. First of all, we note, from (17.42), that the path integral predicts

$$\psi(x,\epsilon) = \int_{-\infty}^{\infty} \mathrm{d}x'\ U(x,\epsilon;x',0)\psi(x',0), \tag{17.62}$$

where for an infinitesimal time interval ϵ, the transition amplitude is given by (see (17.29))

$$\begin{aligned} U(x,\epsilon;x',0) &= \left(\frac{m}{2i\pi\hbar\epsilon}\right)^{\frac{1}{2}} e^{\frac{i\epsilon}{\hbar}L\left(\frac{x-x'}{\epsilon},\frac{x+x'}{2},\epsilon\right)} \\ &= \left(\frac{m}{2i\pi\hbar\epsilon}\right)^{\frac{1}{2}} e^{\frac{i\epsilon}{\hbar}\left[\frac{1}{2}m\frac{(x-x')^2}{\epsilon^2} - V\left(\frac{x+x'}{2},\epsilon\right)\right]}. \end{aligned} \tag{17.63}$$

It follows from (17.62) and (17.63) that

$$\psi(x,\epsilon) = \left(\frac{m}{2i\pi\hbar\epsilon}\right)^{\frac{1}{2}} \int_{-\infty}^{\infty} \mathrm{d}x'\ e^{\left[\frac{im}{2\hbar\epsilon}(x-x')^2 - \frac{i\epsilon}{\hbar}V\left(\frac{x+x'}{2},\epsilon\right)\right]}\ \psi(x',0). \tag{17.64}$$

We note, from (17.64), that if x' is appreciably different from x, then because the infinitesimal interval ϵ is in the denominator of the first term (in the exponential), the integrand oscillates rapidly. The potential has an ϵ multiplying it because of which it is a smooth function and so also is $\psi(x',0)$. Thus, all such contributions where x' is very different from x would average to zero. The only substantial contribution would, therefore, come from paths around $x' = x$. Hence let us define

$$x' - x = \eta, \tag{17.65}$$

in terms of which, the expression, (17.64), becomes

$$\psi(x,\epsilon) = \left(\frac{m}{2i\pi\hbar\epsilon}\right)^{\frac{1}{2}} \int_{-\infty}^{\infty} \mathrm{d}\eta\; e^{\frac{im\eta^2}{2\hbar\epsilon} - \frac{i\epsilon}{\hbar}V(x+\frac{\eta}{2},\epsilon)}\; \psi(x+\eta,0). \tag{17.66}$$

Let us note that the dominant contribution, in (17.66), comes from the region

$$\frac{m\eta^2}{2\epsilon\hbar} \leq \pi, \quad \text{or,} \quad |\eta| \leq \left(\frac{2\epsilon\hbar\pi}{m}\right)^{\frac{1}{2}}. \tag{17.67}$$

We are interested in terms up to order ϵ in the expansion of the right hand side of (17.66) (see (17.61)). Furthermore, from (17.67), we see that the terms that contribute are those for which $\eta \sim \epsilon^{\frac{1}{2}}$. Thus, we can expand the integrand keeping terms up to order η^2 in the expansion. As a result, we obtain,

$$\begin{aligned}
&\epsilon V(x+\frac{\eta}{2},\epsilon) = \epsilon V(x,0) + O(\epsilon^{\frac{3}{2}}),\\
&\psi(x+\eta,0) = \psi(x,0) + \eta\frac{\partial\psi(x,0)}{\partial x} + \frac{\eta^2}{2!}\,\frac{\partial^2\psi(x,0)}{\partial x^2}.
\end{aligned} \tag{17.68}$$

Putting these back into (17.66), we have

$$\begin{aligned}
\psi(x,\epsilon) &= \left(\frac{m}{2i\pi\hbar\epsilon}\right)^{\frac{1}{2}} \int_{-\infty}^{\infty} \mathrm{d}\eta\; e^{\frac{im\eta^2}{2\epsilon\hbar}} \left(1 - \frac{i\epsilon}{\hbar}V(x,0)\right)\\
&\quad \times \left(\psi(x,0) + \eta\frac{\partial\psi(x,0)}{\partial x} + \frac{\eta^2}{2!}\,\frac{\partial^2\psi(x,0)}{\partial x^2}\right)\\
&= \left(\frac{m}{2i\pi\hbar\epsilon}\right)^{\frac{1}{2}} \int_{-\infty}^{\infty} \mathrm{d}\eta\; e^{\frac{im\eta^2}{2\hbar\epsilon}} \left[\psi(x,0) + \eta\frac{\partial\psi(x,0)}{\partial x}\right.\\
&\quad \left. + \frac{\eta^2}{2!}\,\frac{\partial^2\psi(x,0)}{\partial x^2} - \frac{i\epsilon}{\hbar}V(x,0)\psi(x,0)\right].
\end{aligned} \tag{17.69}$$

Using the standard results

$$\int_{-\infty}^{\infty} d\eta\; e^{\frac{im\eta^2}{2\hbar\epsilon}} = \left(\frac{2i\pi\hbar\epsilon}{m}\right)^{\frac{1}{2}},$$

$$\int_{-\infty}^{\infty} d\eta\; \eta\; e^{\frac{im\eta^2}{2\hbar\epsilon}} = 0,$$

$$\int_{-\infty}^{\infty} d\eta\; \eta^2\; e^{\frac{im\eta^2}{2\hbar\epsilon}} = \frac{i\hbar\epsilon}{m}\left(\frac{2i\pi\hbar\epsilon}{m}\right)^{\frac{1}{2}}, \tag{17.70}$$

we obtain, from (17.69),

$$\begin{aligned}\psi(x,\epsilon) &= \left(\frac{m}{2i\pi\hbar\epsilon}\right)^{\frac{1}{2}}\left[\left(\frac{2i\pi\hbar\epsilon}{m}\right)^{\frac{1}{2}}\left(1-\frac{i\epsilon}{\hbar}\,V(x,0)\right)\psi(x,0)\right.\\ &\qquad\left.+\frac{1}{2!}\,\frac{i\hbar\epsilon}{m}\left(\frac{2i\pi\hbar\epsilon}{m}\right)^{\frac{1}{2}}\,\frac{\partial^2\psi(x,0)}{\partial x^2}\right]\\ &= \left(1-\frac{i\epsilon}{\hbar}\,V(x,0)+\frac{i\epsilon\hbar}{2m}\,\frac{\partial^2}{\partial x^2}\right)\psi(x,0).\end{aligned} \tag{17.71}$$

It follows from (17.71) that

$$\psi(x,\epsilon)-\psi(x,0) = -\frac{i\epsilon}{\hbar}\left(-\frac{\hbar^2}{2m}\,\frac{\partial^2}{\partial x^2}+V(x,0)\right)\psi(x,0), \tag{17.72}$$

which is the same as (17.61). Thus we have shown that the path integral formalism leads to the Schrödinger equation for infinitesimal time intervals. Since any finite interval can be thought of as a series of successive infinitesimal intervals the equivalence would still be true.

We would next show how starting from the operator formalism one can recover Feynman's formula. We start with the definition in (17.17)

$$U(x_b,t_b;x_a,t_a) = {}_H\langle x_b,t_b|x_a,t_a\rangle_H = \langle x_b|e^{-\frac{i}{\hbar}H(t_b-t_a)}|x_a\rangle. \tag{17.73}$$

Let us now divide the time interval into N equal parts of length ϵ so that

$$t_b-t_a = N\epsilon. \tag{17.74}$$

Therefore, we can write

$$
\begin{aligned}
U(x_b, t_b; x_a, t_a) &= \langle x_b | e^{-\frac{i}{\hbar} H N\epsilon} | x_a \rangle \\
&= \lim_{\substack{\epsilon \to 0 \\ N \to \infty \\ N\epsilon = t_b - t_a}} A \langle x_b | \left(1 - \frac{i\epsilon}{\hbar} H\right)^N | x_a \rangle \\
&= \lim_{\substack{\epsilon \to 0 \\ N \to \infty \\ N\epsilon = t_b - t_a}} A \int \mathrm{d}x_1 \cdots \mathrm{d}x_{N-1} \, \langle x_b | \left(1 - \frac{i\epsilon}{\hbar} H\right) | x_{N-1} \rangle \\
&\quad \times \langle x_{N-1} | \left(1 - \frac{i\epsilon}{\hbar} H\right) | x_{N-2} \rangle \cdots \langle x_1 | \left(1 - \frac{i\epsilon}{\hbar} H\right) | x_a \rangle .
\end{aligned} \tag{17.75}
$$

There is a time ordering involved in (17.75), namely, we are assuming that

$$
t_b > t_{N-1} > t_{N-2} \cdots t_1 > t_a . \tag{17.76}
$$

Furthermore, let us note that

$$
\langle p | H | x \rangle = \langle p | x \rangle H(x, p), \tag{17.77}
$$

where $H(x,p)$ is the classical Hamiltonian and we also know that (see (2.177))

$$
\langle p | x \rangle = \frac{1}{\sqrt{2\pi\hbar}} \, e^{-\frac{i}{\hbar} p x} . \tag{17.78}
$$

Thus each element of (17.75) can be rewritten as

$$
\begin{aligned}
\langle x_2 | \left(1 - \frac{i\epsilon}{\hbar} H\right) | x_1 \rangle &= \int \mathrm{d}p_1 \, \langle x_2 | p_1 \rangle \langle p_1 | \left(1 - \frac{i\epsilon}{\hbar} H\right) | x_1 \rangle \\
&= \int \frac{\mathrm{d}p_1}{2\pi\hbar} \left(1 - \frac{i\epsilon}{\hbar} H(x_1, p_1)\right) e^{\frac{i}{\hbar} p_1 (x_2 - x_1)} .
\end{aligned} \tag{17.79}
$$

Using these formulae, therefore, we can write

$$
\begin{aligned}
&U(x_b, t_b; x_a, t_a) \\
&= \lim_{\substack{\epsilon \to 0 \\ N \to \infty \\ N\epsilon = t_b - t_a}} A \int \frac{\mathrm{d}x_1 \mathrm{d}p_1}{2\pi\hbar} \cdots \int \frac{\mathrm{d}x_{N-1} \mathrm{d}p_{N-1}}{2\pi\hbar} \\
&\quad \times e^{\frac{i}{\hbar} \sum\limits_{n=0}^{N-1} p_n (x_{n+1} - x_n)} \prod_{n=0}^{N-1} \left(1 - \frac{i\epsilon}{\hbar} H(x_n, p_n)\right) .
\end{aligned} \tag{17.80}
$$

Let us next use the definition of the exponential

$$\lim_{N\to\infty}\prod_{n=0}^{N}\left(1+\frac{Z_n}{N}\right)=\lim_{N\to\infty}\prod_{n=0}^{N}e^{\frac{Z_n}{N}}. \tag{17.81}$$

With this, the amplitude in (17.80) becomes

$$\begin{aligned}
&U(x_b,t_b;x_a,t_a)\\
&= \lim_{\substack{\epsilon\to 0\\ N\to\infty\\ N\epsilon=t_b-t_a}} A\int\frac{\mathrm{d}x_1\mathrm{d}p_1}{2\pi\hbar}\cdots\int\frac{\mathrm{d}x_{N-1}\mathrm{d}p_{N-1}}{2\pi\hbar}\\
&\quad\times e^{\frac{i}{\hbar}\sum\limits_{n=0}^{N-1}[p_n(x_{n+1}-x_n)-\epsilon H(x_n,p_n)]}\\
&= \lim_{\substack{\epsilon\to 0\\ N\to\infty\\ N\epsilon=t_b-t_a}} A\int\frac{\mathrm{d}x_1\mathrm{d}p_1}{2\pi\hbar}\cdots\int\frac{\mathrm{d}x_{N-1}\mathrm{d}p_{N-1}}{2\pi\hbar}\\
&\quad\times e^{\frac{i\epsilon}{\hbar}\sum\limits_{n=0}^{N-1}[p_n(\frac{x_{n+1}-x_n}{\epsilon})-H(x_n,p_n)]}.
\end{aligned} \tag{17.82}$$

In the limit $\epsilon\to 0, N\to\infty$, we can write the exponent of the exponential as

$$\frac{i}{\hbar}\int_{t_a}^{t_b}\mathrm{d}t\ (p\dot{x}-H(x,p)), \tag{17.83}$$

so that, in this limit, we have

$$U(x_b,t_b;x_a,t_a)=A\int\mathcal{D}x\mathcal{D}p\ \exp\left\{\frac{i}{\hbar}\int_{t_a}^{t_b}\mathrm{d}t\ \ (p\dot{x}-H(x,p))\right\}, \tag{17.84}$$

where we have used the notation

$$\mathcal{D}x=\prod_{n=1}^{N-1}\frac{\mathrm{d}x_n}{\sqrt{2\pi\hbar}},\qquad \mathcal{D}p=\prod_{n=1}^{N-1}\frac{\mathrm{d}p_n}{\sqrt{2\pi\hbar}}. \tag{17.85}$$

Let us now specialize to the case where the Hamiltonian has the standard form

$$H=\frac{p^2}{2m}+V(x), \tag{17.86}$$

with $V(x)$ representing an arbitrary potential. We can, in this case, perform the p-integration in (17.84).

$$\begin{aligned}
&\int \mathcal{D}p\, e^{\frac{i}{\hbar}\int\limits_{t_a}^{t_b} \mathrm{d}t\left(p\dot{x}-\frac{p^2}{2m}-V(x)\right)} \\
&= \int \mathcal{D}p\, e^{-\frac{i}{2m\hbar}\int\limits_{t_a}^{t_b}\mathrm{d}t(p-m\dot{x})^2+\frac{i}{\hbar}\int\limits_{t_a}^{t_b}\mathrm{d}t\left(\frac{1}{2}m\dot{x}^2-V(x)\right)} \\
&= \left(\frac{2m\hbar\pi}{i}\right)^{\frac{N}{2}}\left(\frac{1}{2\pi\hbar}\right)^{\frac{N}{2}} e^{\frac{i}{\hbar}\int\limits_{t_a}^{t_b}\mathrm{d}t\ L(x,\dot{x})} \\
&= \left(\frac{m}{i}\right)^{\frac{N}{2}} e^{\frac{i}{\hbar}\int\limits_{t_a}^{t_b}\mathrm{d}t\ L(x,\dot{x})},
\end{aligned} \tag{17.87}$$

where we have followed a derivation as in (17.37) as well as the definition in (17.85). Putting this result back into (17.84), we have

$$U(x_b,t_b;x_a,t_a) = A'\int \mathcal{D}x\, e^{\frac{i}{\hbar}\int\limits_{t_a}^{t_b}\mathrm{d}t\ L(x,\dot{x})} = A'\int \mathcal{D}x\, e^{\frac{i}{\hbar}S}. \tag{17.88}$$

This is Feynman's formula (see (17.40)) and the constant A' is an infinite normalization constant which drops out in the calculation of relative probability amplitudes of the form $\frac{\langle x_b,t_b|O|x_a,t_a\rangle}{\langle x_b,t_b|x_a,t_a\rangle}$.

The knowledge of U helps in calculating the matrix elements of various operators. Thus, for example,

$$\begin{aligned}
&\langle x_b,t_b|T(X(t_1)X(t_2)\cdots X(t_n))|x_a,t_a\rangle \\
&= A'\int \mathcal{D}x\ x(t_1)x(t_2)\cdots x(t_n)\ e^{\frac{i}{\hbar}\int \mathrm{d}t\ L(x,\dot{x})}.
\end{aligned} \tag{17.89}$$

Exercise. Prove formula (17.89) for $n=2$. Use complete intermediate states to derive the formula.

Even the form, (17.89), in which we have derived the matrix elements, is not very useful because it is extremely difficult to evaluate the integrals. So, let us develop some simplifications. Let us assume that the Lagrangian does not depend on time explicitly. Let $\phi_n(x) = \langle x|n\rangle$ be the wave function associated with the nth energy state of the system. Thus, $\phi_0(x)$ would represent the ground state wave function of the system. Let us assume that the system is in the ground state at a time T in the distant past. We want to calculate the amplitude for the system to be in the ground state at some time T' in the distant future, when an arbitrary external source $J(t)x(t)$ is added to the Lagrangian in some intermediate time interval. Thus, we take $J(t)$ to be an arbitrary function with the constraint that it is non-vanishing only between t' and t such that

$$T' > t' > t > T. \tag{17.90}$$

We denote this amplitude as

$$\langle x_b, T'|x_a, T\rangle^J = \int \mathcal{D}x\mathcal{D}p\, e^{\frac{i}{\hbar}\int_T^{T'} \mathrm{d}t\, [p\dot{x}-H(x,p)+Jx]}. \tag{17.91}$$

On the other hand, using the completeness relation of the coordinate basis states, we can write the left hand side of (17.91) as

$$\langle x_b, T'|x_a, T\rangle^J = \int \mathrm{d}x'\mathrm{d}x \langle x_b, T'|x', t'\rangle \langle x', t'|x, t\rangle^J \langle x, t|x_a, T\rangle. \tag{17.92}$$

Here, we have used the fact that since J is non-vanishing only between t' and t, only the second factor, on the right hand side of (17.92), would have a J dependence. Furthermore, using a complete basis of energy eigenstates, we obtain

$$\begin{aligned}
\langle x, t|x_a, T\rangle &= \langle x|e^{-\frac{i}{\hbar}H(t-T)}|x_a\rangle \\
&= \sum_n \langle x|e^{-\frac{i}{\hbar}H(t-T)}|n\rangle\langle n|x_a\rangle \\
&= \sum_n e^{-\frac{i}{\hbar}E_n(t-T)}\phi_n(x)\phi_n^*(x_a).
\end{aligned} \tag{17.93}$$

If we now consider the limit $T \to i\infty$ (namely, we are considering the amplitude in the Euclidean space, a technicality that we will not get into), then, clearly all terms with $n > 0$ drop out. Thus, we have

$$\lim_{T\to i\infty} \langle x, t|x_a, T\rangle = e^{\frac{i}{\hbar}E_0 T}\phi_0(x, t)\phi_0^*(x_a). \tag{17.94}$$

Similarly, we can show that

$$\lim_{T'\to -i\infty} \langle x_b, T'|x', t'\rangle = e^{\frac{i}{\hbar}E_0 T'}\phi_0(x_b)\phi_0^*(x', t'). \tag{17.95}$$

Substituting (17.94) and (17.95) into (17.92), we obtain

$$\begin{aligned}
&\lim_{\substack{T\to i\infty\\ T'\to -i\infty}} \frac{\langle x_b,T'|x_a,T\rangle^J}{e^{-\frac{i}{\hbar}E_0(T'-T)}\phi_0^*(x_a)\phi_0(x_b)}\\
&= \int \mathrm{d}x\mathrm{d}x'\ \phi_0^*(x',t')\langle x',t'|x,t\rangle^J\phi_0(x,t)\\
&= Z[J] = e^{iW[J]}. \qquad (17.96)
\end{aligned}$$

The right hand side of (17.96) simply represents the ground state to ground state transition amplitude. Since

$$\langle x',t'|x,t\rangle^J = \int \mathcal{D}x\mathcal{D}p\ e^{\frac{i}{\hbar}\int_t^{t'}\mathrm{d}\tau\ [p\dot{x}-H(x,p)+Jx]}, \qquad (17.97)$$

we see that the effect of varying J in Z is to bring down a factor of x. Thus, let us consider

$$\begin{aligned}
\left.\frac{\delta^n Z[J]}{\delta J(t_1)\cdots\delta J(t_n)}\right|_{J=0} &= \frac{(i)^n}{\hbar^n}\int \mathrm{d}x\mathrm{d}x'\ \phi_0^*(x',t')\phi_0(x,t)\\
&\times \int \mathcal{D}x\mathcal{D}p\ x(t_1)\cdots x(t_n)\ e^{\frac{i}{\hbar}\int_t^{t'}\mathrm{d}\tau\ (p\dot{x}-H(x,p))}. \qquad (17.98)
\end{aligned}$$

We recognize this as the matrix element of the time ordered product $T(X(t_1)\cdots X(t_n))$ between the ground state at time t and the ground state at t'. Thus we can obtain the ground state expectation value of operators quite easily. Such expectation values are called correlation functions in statistical mechanics. In quantum field theory they are known as vacuum Green's functions and are directly related to physically observable quantities. In field theory, one is cavalier about infinite normalization constants, mainly because they drop out in normalized amplitudes and writes

$$\begin{aligned}
Z(J) &\sim \lim_{\substack{T\to i\infty\\ T'\to -i\infty}} \langle x_b,T'|x_a,T\rangle^J\\
&= \lim_{\substack{T\to i\infty\\ T'\to -i\infty}} \int \mathcal{D}x\ e^{\frac{i}{\hbar}\int_T^{T'}\mathrm{d}t\ (L[x]+J(t)x(t))}. \qquad (17.99)
\end{aligned}$$

We note here that $Z[J]$ is known as the generating functional for Green's functions while $W[J]$ is called the generating functional for connected Green's functions.

17.3 Harmonic oscillator

Let us now consider a harmonic oscillator with a source term. In this case, the Lagrangian has the form

$$L = \frac{1}{2}m\dot{x}^2 - \frac{1}{2}m\omega^2 x^2 + Jx. \tag{17.100}$$

We recognize that this is a system described by a quadratic Lagrangian. Let us again define

$$x = x_{\text{cl}} + \eta, \tag{17.101}$$

so that we can write

$$\begin{aligned} L[x] &\equiv L[x_{\text{cl}} + \eta] \\ &= \frac{1}{2}m\left(\dot{x}_{\text{cl}} + \dot{\eta}\right) - \frac{1}{2}m\omega^2\left(x_{\text{cl}} + \eta\right)^2 + J\left(x_{\text{cl}} + \eta\right) \\ &= \frac{1}{2}m\dot{x}_{\text{cl}}^2 - \frac{1}{2}m\omega^2 x_{\text{cl}}^2 + Jx_{\text{cl}} + m\dot{x}_{\text{cl}}\dot{\eta} - m\omega^2 x_{\text{cl}}\eta + J\eta \\ &\quad + \frac{1}{2}m\dot{\eta}^2 - \frac{1}{2}m\omega^2\eta^2. \end{aligned} \tag{17.102}$$

Remembering the fact that the classical trajectory satisfies the equation

$$m\ddot{x}_{\text{cl}} + m\omega^2 x_{\text{cl}} - J = 0, \tag{17.103}$$

we note that we can write the action, in this case, to be

$$S = \int \mathrm{d}t\ L[x_{\text{cl}} + \eta] = \int \mathrm{d}t \left[L[x_{\text{cl}}] + \frac{m}{2}\dot{\eta}^2 - \frac{m}{2}\omega^2\eta^2\right], \tag{17.104}$$

where we have used integration by parts. In other words, the linear terms in η drop out and we have identified

$$L[x_{\text{cl}}] = \frac{1}{2}m\dot{x}_{\text{cl}}^2 - \frac{1}{2}m\omega^2 x_{\text{cl}}^2 + Jx_{\text{cl}}. \tag{17.105}$$

In this case, therefore, we can write the transition amplitude as

$$\begin{aligned} \langle x', t'|x, t\rangle^J &\sim \int \mathcal{D}\eta\ e^{\frac{i}{\hbar}S[x_{\text{cl}}]}\ e^{\frac{i}{\hbar}\int_t^{t'}\mathrm{d}\tau\ \left(\frac{1}{2}m\dot{\eta}^2 - \frac{1}{2}m\omega^2\eta^2\right)} \\ &= e^{\frac{i}{\hbar}S[x_{\text{cl}}]} \int \mathcal{D}\eta\ e^{\frac{i}{\hbar}\int_t^{t'}\mathrm{d}\tau\ \left(\frac{1}{2}m\dot{\eta}^2 - \frac{1}{2}m\omega^2\eta^2\right)}. \end{aligned} \tag{17.106}$$

The integrand, in (17.106), is a Gaussian and, therefore, the path integral can be explicitly evaluated. Let us do it in the following way. First of all, let us note that the Lagrangian does not depend on time explicitly. Therefore, we can shift the limits of integration from 0 to $T = t' - t$. Furthermore, η's satisfy the condition $\eta = 0$ at $t = 0$ as well as at $t = T$ (see (17.101)). Thus, we can expand η as

$$\eta(t) = \sum_n a_n \sin \frac{n\pi t}{T}, \qquad n \text{ integers}. \tag{17.107}$$

It follows now that

$$\begin{aligned}
\int_0^T \mathrm{d}t\ \dot{\eta}^2 &= \sum_{n,m} \int_0^T \mathrm{d}t\ \frac{n\pi}{T}\frac{m\pi}{T}\ a_n a_m \cos\frac{n\pi t}{T} \cos\frac{m\pi t}{T} \\
&= \frac{T}{2}\sum_n \left(\frac{n\pi}{T}\right)^2 a_n^2, \\
\int_0^T \mathrm{d}t\ \eta^2 &= \sum_{n,m} \int_0^T \mathrm{d}t\ a_n a_m \sin\frac{n\pi t}{T} \sin\frac{m\pi t}{T} \\
&= \frac{T}{2}\sum_n a_n^2.
\end{aligned} \tag{17.108}$$

If we assume that the time interval is divided into $N+1$ subintervals, then, there will only be a finite number (N) of coefficients, a_n. Thus, we can write

$$\begin{aligned}
&\int \mathcal{D}\eta\ e^{\frac{i}{\hbar}\int_0^T \mathrm{d}t\ \left(\frac{m}{2}\dot{\eta}^2 - \frac{m}{2}\omega^2\eta^2\right)} \\
&= A \int \mathrm{d}a_1 \mathrm{d}a_2 \cdots \mathrm{d}a_N\ e^{\frac{i}{\hbar}\frac{mT}{4}\sum_{n=1}^{N}\left(\left(\frac{n\pi}{T}\right)^2 - \omega^2\right)a_n^2}.
\end{aligned} \tag{17.109}$$

Each term in the integrand of (17.109) is a Gaussian and gives

$$\begin{aligned}
&\int \mathrm{d}a_n\ e^{\frac{imT}{4\hbar}\left(\frac{n^2\pi^2}{T^2} - \omega^2\right)a_n^2} \\
&= \left(\frac{4i\pi\hbar}{mT}\right)^{\frac{1}{2}} \left(\frac{n^2\pi^2}{T^2} - \omega^2\right)^{-\frac{1}{2}} \\
&= \left(\frac{4i\pi\hbar}{mT}\right)^{\frac{1}{2}} \left(\frac{n^2\pi^2}{T^2}\right)^{-\frac{1}{2}} \left(1 - \frac{\omega^2 T^2}{n^2\pi^2}\right)^{-\frac{1}{2}}.
\end{aligned} \tag{17.110}$$

Therefore, we have

$$A\int \mathrm{d}a_1\mathrm{d}a_2\cdots\mathrm{d}a_N\ e^{\frac{i}{\hbar}\frac{mT}{4}\sum\limits_{n=1}^{N}((\frac{n\pi}{T})^2-\omega^2)a_n^2}$$

$$= A'\prod_{n=1}^{N}\left(1-\frac{\omega^2T^2}{n^2\pi^2}\right)^{-\frac{1}{2}}, \tag{17.111}$$

where we have lumped the other factors of the integral into the constant A'. We note that if the time interval ϵ is made vanishingly small so that $N\to\infty$ (keeping $N\epsilon$ fixed), then, in that limit

$$\lim_{\substack{\epsilon\to 0\\ N\to\infty\\ N\epsilon=T}}\prod_{n=1}^{N}\left(1-\frac{\omega^2T^2}{n^2\pi^2}\right) = \frac{\sin\omega T}{\omega T}. \tag{17.112}$$

With this, our expression, (17.111) becomes

$$A\int \mathrm{d}a_1\cdots\mathrm{d}a_N\ e^{\frac{i}{\hbar}\frac{mT}{4}\sum\limits_{n=1}^{N}((\frac{n\pi}{T})^2-\omega^2)a_n^2} = A'\left(\frac{\sin\omega T}{\omega T}\right)^{-\frac{1}{2}}. \tag{17.113}$$

Furthermore, we note that in the limit $\omega=0$, our path integral reduces to that of a free particle, which we have evaluated earlier (see (17.58)) and which has the value $\sqrt{\frac{m}{2\pi i\hbar T}}$. The constant A' does not depend on the frequency and, consequently, we determine

$$\lim_{\substack{\epsilon\to 0\\ N\to\infty\\ N\epsilon=T}} A' = \sqrt{\frac{m}{2\pi i\hbar T}}. \tag{17.114}$$

As a result, we obtain,

$$\int\mathcal{D}\eta\ e^{\frac{i}{\hbar}\int\limits_t^{t'}\mathrm{d}\tau\ (\frac{m}{2}\dot\eta^2-\frac{m}{2}\omega^2\eta^2)}$$

$$= \left[\frac{m}{2\pi i\hbar T}\right]^{\frac{1}{2}}\left[\frac{\omega T}{\sin\omega T}\right]^{\frac{1}{2}} = \left[\frac{m\omega}{2\pi i\hbar\sin\omega T}\right]^{\frac{1}{2}}, \tag{17.115}$$

so that we can write the transition amplitude, (17.106), as

$$\langle x,t'|x,t\rangle^J = e^{\frac{i}{\hbar}S[x_{\rm cl}]}\int\mathcal{D}\eta\ e^{\frac{i}{\hbar}\int\mathrm{d}t\ (\frac{m}{2}\dot\eta^2-\frac{m}{2}\omega^2\eta^2)}$$

$$= \left[\frac{m\omega}{2\pi i\hbar\sin\omega T}\right]^{\frac{1}{2}} e^{\frac{i}{\hbar}S[x_{\rm cl}]}. \tag{17.116}$$

To determine $S[x_{\rm cl}]$, we note that the equation of motion satisfied by $x_{\rm cl}$ is (see (17.103))

$$
\begin{aligned}
& m\ddot{x}_{\rm cl} + m\omega^2 x_{\rm cl} = J(t),\\
\text{or,}\quad & \left(\frac{\mathrm{d}^2}{\mathrm{d}t^2} + \omega^2\right) x_{\rm cl} = \frac{J(t)}{m}.
\end{aligned}
\tag{17.117}
$$

The solution of this equation consists of two parts

$$x_{\rm cl} = x_H + x_I, \tag{17.118}$$

where the homogeneous part of the solution can be written as

$$x_H = Ae^{i\omega t} + Be^{-i\omega t}. \tag{17.119}$$

The inhomogeneous part (or the particular solution) of (17.117) can be obtained from the Green's function of the differential equation (see chapter 15). For example, if

$$\left(\frac{\mathrm{d}^2}{\mathrm{d}t^2} + \omega^2\right) G(t-t') = -\delta(t-t'), \tag{17.120}$$

then, we can write the particular solution as

$$x_I(t) = -\int_{t_a}^{t_b} \mathrm{d}t'\ G(t-t')\ \frac{J(t')}{m}. \tag{17.121}$$

As we have discussed earlier, $G(t-t')$ is known as the Green's function for the operator $\left(\frac{\mathrm{d}^2}{\mathrm{d}t^2} + \omega^2\right)$ and is determined in the following way. Let us look at the differential equation for the Green's function in the Fourier transformed space. Thus, defining,

$$G(t-t') = \int \mathrm{d}k\ e^{ik(t-t')}\ G(k), \tag{17.122}$$

and recalling the Fourier transform of the delta function,

$$\delta(t-t') = \frac{1}{2\pi}\int \mathrm{d}k\ e^{ik(t-t')}, \tag{17.123}$$

we obtain, from (17.120),

$$
\begin{aligned}
& (-k^2 + \omega^2)\, G(k) = -\frac{1}{2\pi},\\
\text{or,}\quad & G(k) = \frac{1}{2\pi}\,\frac{1}{k^2 - \omega^2}.
\end{aligned}
\tag{17.124}
$$

This leads to

$$G(t-t') = \frac{1}{2\pi}\int \mathrm{d}k\, \frac{e^{ik(t-t')}}{k^2-\omega^2}. \tag{17.125}$$

The integral in (17.125), however, is not well defined because of the singularities at $k = \pm\omega$. Therefore, a prescription for the contour of integration has to be given for a unique expression. Depending on whether $t-t' > 0$ or $t-t' < 0$ one defines two Green's functions known as retarded and advanced Green's functions as discussed in chapter **15**. However, in quantum mechanics, as we have noted earlier, it is the Feynman Green's function (or the causal Green's function) that plays an important role and the prescription for the contour, in this case, is as shown in Fig. 17.6.

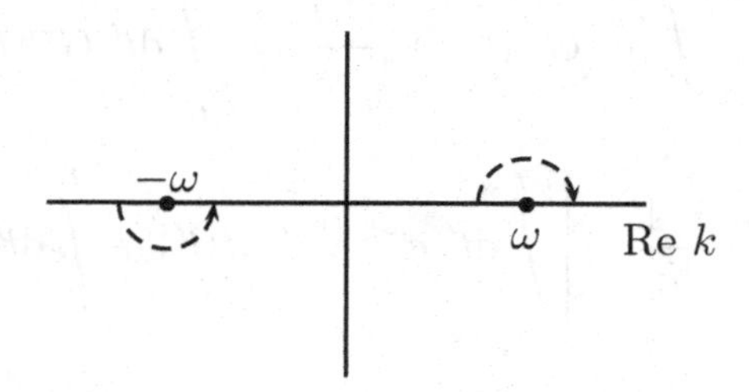

Figure 17.6: Contour in the complex plane for the Feynman Green's function.

For $t - t' > 0$ the contour is closed in the upper half plane so that the Green's function becomes

$$\begin{aligned} G^{+}(t-t') &= 2\pi i\, \frac{1}{2\pi} \lim_{k\to-\omega} (k+\omega)\frac{e^{ik(t-t')}}{(k+\omega)(k-\omega)} \\ &= i\, \frac{e^{i\omega(t-t')}}{-2\omega} = \frac{1}{2i\omega}\, e^{i\omega(t-t')}, \quad t-t' > 0. \end{aligned} \tag{17.126}$$

For $t - t' < 0$, on the other hand, the contour is closed in the lower half plane and the Green's function becomes

$$\begin{aligned} G^{-}(t-t') &= -2\pi i\, \frac{1}{2\pi} \lim_{k\to\omega} (k-\omega)\, \frac{e^{ik(t-t')}}{(k+\omega)(k-\omega)} \\ &= -i\, \frac{e^{i\omega(t-t')}}{2\omega} = \frac{1}{2i\omega}\, e^{i\omega(t-t')}, \quad t-t' < 0. \end{aligned} \tag{17.127}$$

The Feynman Green's function can be written in terms of G^+ and G^-, defined in (17.126) and (17.127) respectively, as

$$G_{\mathrm{F}}(t-t') = \theta(t-t')G^+(t-t') + \theta(t'-t)G^-(t-t'). \quad (17.128)$$

It is an even function and satisfies

$$\left(\frac{\mathrm{d}^2}{\mathrm{d}t^2} + \omega^2\right) G_{\mathrm{F}}(t-t') = -\delta(t-t'). \quad (17.129)$$

We can, therefore, write down the inhomogeneous part (particular solution) of the solution now as (see (17.121))

$$\begin{aligned} x_I(t) &= -\int_{t_a}^{t_b} \mathrm{d}t'\ G_{\mathrm{F}}(t-t')\frac{J(t')}{m} \\ &= -\int_{t_a}^{t} \mathrm{d}t'\ G^+(t-t')\frac{J(t')}{m} - \int_{t}^{t_b} \mathrm{d}t'\ G^-(t-t')\frac{J(t')}{m} \\ &= -\frac{1}{2im\omega}\left[\int_{t_a}^{t} \mathrm{d}t'\ e^{-i\omega(t-t')}J(t') + \int_{t}^{t_b} \mathrm{d}t'\ e^{i\omega(t-t')}J(t')\right]. \end{aligned} \quad (17.130)$$

Using this, we can write the complete solution, (17.118), as

$$\begin{aligned} x_{\mathrm{cl}}(t) &= x_H(t) + x_I(t) \\ &= Ae^{i\omega t} + Be^{-i\omega t} \\ &\quad - \frac{1}{2im\omega}\left[\int_{t_a}^{t} \mathrm{d}t'\ e^{-i\omega(t-t')}J(t') + \int_{t}^{t_b} \mathrm{d}t'\ e^{i\omega(t-t')}J(t')\right]. \end{aligned} \quad (17.131)$$

Furthermore, the boundary conditions lead to

$$\begin{aligned} x_{\mathrm{cl}}(t=t_a) = x_a &= Ae^{i\omega t_a} + Be^{-i\omega t_a} \\ &\quad - \frac{1}{2im\omega}\int_{t_a}^{t_b} \mathrm{d}t'\ e^{i\omega(t_a-t')}J(t'), \\ x_{\mathrm{cl}}(t=t_b) = x_b &= Ae^{i\omega t_b} + Be^{-i\omega t_b} \\ &\quad - \frac{1}{2im\omega}\int_{t_a}^{t_b} \mathrm{d}t' e^{-i\omega(t_b-t')}J(t'). \end{aligned} \quad (17.132)$$

From the relations in (17.132), we obtain

$$
\begin{aligned}
& x_a e^{i\omega t_b} - x_b e^{i\omega t_a} \\
& \quad = B\left[e^{i\omega(t_b-t_a)}\ e^{-i\omega(t_b-t_a)}\right] \\
& \quad\quad - \frac{1}{2im\omega}\int_{t_a}^{t_b} \mathrm{d}t'\ J(t')\left[e^{i\omega(t_b-t_a-t')} - e^{-i\omega(t_b-t_a-t')}\right] \\
& \quad = 2iB\sin\omega T - \frac{e^{i\omega t_a}}{m\omega}\int_{t_a}^{t_b} \mathrm{d}t' J(t')\sin\omega(t_b-t'),
\end{aligned}
\tag{17.133}
$$

where we have identified $T = t_b - t_a$. Equation (17.133) determines

$$
\begin{aligned}
B = \frac{1}{2i\sin\omega T}\Bigg[&\left(x_a e^{i\omega t_b} - x_b e^{i\omega t_a}\right) \\
&+ \frac{e^{i\omega t_a}}{m\omega}\int_{t_a}^{t_b} \mathrm{d}t'\ J(t')\sin\omega(t_b-t')\Bigg].
\end{aligned}
\tag{17.134}
$$

Similarly, from the relations in (17.132), we also obtain

$$
\begin{aligned}
& x_b e^{i\omega t_a} - x_a e^{-i\omega t_b} \\
& \quad = A\left[e^{i\omega(t_b-t_a)} - e^{-i\omega(t_b-t_a)}\right] \\
& \quad\quad - \frac{1}{2im\omega}\int_{t_a}^{t_b} \mathrm{d}t'\ J(t')\left[e^{i\omega(t_b+t_a-t')} - e^{-i\omega(t_a-t_b-t')}\right] \\
& \quad = 2iA\sin\omega T - \frac{e^{-i\omega t_b}}{m\omega}\int_{t_a}^{t_b} \mathrm{d}t'\ J(t')\sin\omega(t_a-t'),
\end{aligned}
\tag{17.135}
$$

which determines

$$
\begin{aligned}
A = \frac{1}{2i\sin\omega T}\Bigg[&\left(x_b e^{-i\omega t_a} - x_a e^{-i\omega t_b}\right) \\
&+ \frac{e^{-i\omega t_b}}{m\omega}\int_{t_a}^{t_b} \mathrm{d}t'\ J(t')\sin\omega(t'-t_a)\Bigg].
\end{aligned}
\tag{17.136}
$$

Using (17.133) and (17.135), we obtain

$$\begin{aligned}
&Ae^{i\omega t} + Be^{-i\omega t}\\
&\quad = \frac{1}{2i\omega T}\Big[x_b\left(e^{i\omega(t-t_a)} - e^{-i\omega(t-t_a)}\right)\\
&\qquad + x_a\left(e^{i\omega(t_b-t)} - e^{-i\omega(t_b-t)}\right)\Big]\\
&\qquad + \frac{1}{m\omega}\int_{t_a}^{t_b} \mathrm{d}t'\ J(t')\Big[e^{-i\omega(t_b-t)}\sin\omega(t'-t_a)\\
&\qquad + e^{-i\omega(t-t_a)}\sin\omega(t_b-t')\Big]\\
&\quad = \frac{1}{\sin\omega T}\Big[x_b\sin\omega(t-t_a) + x_a\sin\omega(t_b-t)\\
&\qquad + \frac{1}{2m\omega}\int_{t_a}^{t_b} \mathrm{d}t'\ J(t')\Big[e^{-i\omega T}\cos\omega(t-t')\\
&\qquad - \cos\omega(t_b+t_a-t-t')\Big]\Big].
\end{aligned} \tag{17.137}$$

As a result, we can write the complete solution in (17.131) as

$$\begin{aligned}
x_{\text{cl}} = {}& \frac{1}{\sin\omega T}\Big[x_b\sin\omega(t-t_a) + x_a\sin\omega(t_b-t)\\
& + \frac{1}{2m\omega}\int_{t_a}^{t_b} \mathrm{d}t'\ J(t')\Big[e^{-i\omega T}\cos\omega(t-t')\\
& - \cos\omega(t_b+t_a-t-t')\Big]\Big] \qquad\qquad (17.138)\\
& - \frac{1}{2im\omega}\left[\int_{t_a}^{t} \mathrm{d}t'\ J(t')e^{-i\omega(t-t')} + \int_{t}^{t_b} \mathrm{d}t'\ J(t')e^{i\omega(t-t')}\right].
\end{aligned}$$

We can now determine

$$S[x_{\text{cl}}] = \frac{m\omega}{2\sin\omega T}\left[\left(x_a^2 + x_b^2\right)\cos\omega T - 2x_a x_b\right]$$

$$
\begin{aligned}
&+ \frac{x_b}{\sin \omega T} \int_{t_a}^{t_b} \mathrm{d}t' \; J(t') \sin \omega (t' - t_a) \\
&+ \frac{x_a}{\sin \omega T} \int_{t_a}^{t_b} \mathrm{d}t' \; J(t') \sin \omega (t_b - t') \\
&- \frac{1}{m\omega \sin \omega T} \int_{t_a}^{t_b} \mathrm{d}t \int_{t_a}^{t} \mathrm{d}\tau \; J(t) J(\tau) \sin \omega (t_b - t) \\
&\qquad\qquad\qquad\qquad \times \sin \omega (\tau - t_a),
\end{aligned} \tag{17.139}
$$

so that, with (17.139), the transition amplitude, (17.116), is completely determined to be

$$
\langle x', t' | x, t \rangle^J = \left[\frac{m\omega}{2\pi i \hbar \sin \omega T} \right]^{\frac{1}{2}} e^{\frac{i}{\hbar} S[x_{\rm cl}]}. \tag{17.140}
$$

Furthermore, from the definition in (17.96),

$$
Z[J] = \int \mathrm{d}x \mathrm{d}x' \; \phi_0^*(x', t') \langle x', t' | x, t \rangle^J \phi_0(x, t), \tag{17.141}
$$

as well as the form for the ground state wave function for the oscillator (see (5.56))

$$
\phi_0(x, \tau) = \left(\frac{m\omega}{\pi \hbar} \right)^{\frac{1}{4}} e^{-\frac{m}{2\hbar} \omega x^2} e^{-\frac{i}{2} \omega \tau}, \tag{17.142}
$$

we can show that (we are interested only in the dependence on J)

$$
\begin{aligned}
Z(J) \sim \exp &\left[\frac{i}{\hbar} \int_t^{t'} \mathrm{d}\sigma \int_t^{\sigma} \mathrm{d}\tau \; J(\sigma) \right. \\
&\left. \times \left\{ \frac{i}{2\omega m} \exp[-i\omega(\sigma - \tau)] \right\} J(\tau) \right].
\end{aligned} \tag{17.143}
$$

Any correlation function for the system can now be obtained by differentiating the generating functional in (17.143) with respect to an appropriate number of J's.

17.4 Selected problems

1. Find the transition amplitude $U(\mathbf{x}_b, t_b; \mathbf{x}_a, t_a)$ for the quantum mechanical systems described by the Lagrangians

i) $L = \frac{1}{2}m\dot{x}^2 + fx$, where f is a constant independent of x.

ii) $L = \frac{1}{2}m\left(\dot{x}^2 + \dot{y}^2 + \dot{z}^2\right) + \frac{eB}{2C}\left(x\dot{y} - y\dot{x}\right)$, wheree, B and C are constants.

2. Prove the relation (17.112), namely,

$$\lim_{N\to\infty} \prod_{n=1}^{N-1} \left(1 - \left(\frac{x}{n\pi}\right)^2\right) = \frac{\sin x}{x}. \tag{17.144}$$

3. From the transition amplitude for the harmonic oscillator discussed in this chapter, obtain the energy levels (spectrum) of the system.

4. If the wave function for a harmonic oscillator at $t = 0$ is given by

$$\psi(x, 0) = \exp\left[-\frac{m\omega}{2\hbar}(x - a)^2\right], \tag{17.145}$$

then show that at a subsequent time T

$$\psi(x, T) = \exp\left[-\frac{i\omega T}{2} - \frac{m\omega}{2\hbar}\left\{x^2 - 2axe^{-i\omega T} + \frac{1}{2}a^2\left(1 + e^{-2i\omega T}\right)\right\}\right]. \tag{17.146}$$

Index